SOLID STATE IMAGING

NATO ADVANCED STUDY INSTITUTES SERIES

Proceedings of the Advanced Study Institute Programme, which aims at the dissemination of advanced knowledge and the formation of contacts among scientists from different countries.

The series is published by an international board of publishers in conjunction with NATO Scientific Affairs Division

A	Life Sciences	Plenum Publishing Corporation
B	Physics	London and New York
C	Mathematical and Physical Sciences	D. Reidel Publishing Company Dordrecht and Boston
D	Behavioural and Social Sciences	Sijthoff International Publishing Company Leyden, The Neth. and Reading, Mass., USA
E	Applied Sciences	Noordhoff International Publishing Leyden, The Neth. and Reading, Mass., USA

Series E: Applied Science - No. 16

SOLID STATE IMAGING

edited by

PAUL G. JESPERS
professor of electrical engineering
Université Catholique de Louvain, Belgium

FERNAND VAN DE WIELE
professor of electrical engineering
Université Catholique de Louvain, Belgium

MARVIN H. WHITE
advisory engineer
Westinghouse Corporation
Systems Development Division
Baltimore, USA

NOORDHOFF - LEYDEN - 1976

Proceedings of the NATO Advanced Study Institute
on Solid State Imaging, held in Louvain-la-Neuve,
Belgium, September 3-12, 1975.

ISBN-13: 978-94-010-1536-3 e-ISBN-13: 978-94-010-1534-9
DOI: 10.1007/978-94-010-1534-9

Copyright © 1976 by Noordhoff International Publishing, division of A. W. Sijthoff International
Publishing Company bv

All rights reserved. No part of this publication may be reproduced, stored in a retrieval system,
or transmitted, in any form or by any means, electronic, mechanical, photocopying, recording,
or otherwise, without the prior permission of the copyright owner.

Preface

An Advanced Study Institute on solid-state imaging was held in Louvain-la-Neuve, Belgium on September 3-12, 1975 under the auspices of the Scientific Affairs Division of NATO. The Institute was organized by a scientific organizing committee consisting of Professor Paul Jespers and Professor Fernand Van de Wiele of the Universite' Catholique de Louvain and Dr. Marvin H. White of the Westinghouse Electric Corporation. This book represents the contributions of the lecturers at the Institute and the chapters present, for the first time, a concise treatment of a very timely subject, namely, solid-state imaging. The organization of the book parallels the program at the Institute with an introduction comprised of historical development and applications. This is followed by the physics of photosensors which leads quite naturally into the various solid-state photosensor arrays. The subject of signal extraction, which is often an overlooked area, follows and the last part of the book is devoted to the various system's considerations.

The subject matter of this book is suitable for a wide range of interests from the advanced student, through the practicing physcist and engineer, to the research worker. Although a novice may find some difficulty with the mathematical development, he can acquire a perspective into the field of solid-state imaging with this book. Likewise, portions of this book may be used as a textbook since the chapters are instructional and self-contained.

The editors would like to express their appreciation to Monsignore Massaux, Rector of the Unversite' Catholique de Louvain for the facilities and accomodations which made the Institute a success, and to Dr. Kester of the Scientific Affairs Division of NATO for his encouragement and support. The personnel of the Microelectronics Laboratory of the Universite' Catholique de Louvain are to be commended for their help during the Institute. In particular, the work of Dr. Claire Stivenaat with the local arrangements made the Institute an enormous success for all the participants. Finally, we would like to thank M.V.M. Galvez for the technical preparation of the U.S.A. manuscripts and to Ruthie Herrick for the excellent typing.

TABLE OF CONTENTS

Preface	V
Section I: Introduction	1
M.H. White	
History of Solid-State Imaging	3
J.A. Hall	
Solid-State Imaging Applications	9
Section II: Physics of Photosensors	27
F. van de Wiele	
Anti-reflection Films and Multilayer Structures	29
F. van de Wiele	
Photodiode Quantum Efficiency	47
A. Goetzberger	
Bulk Trapping	91
A. Goetzberger	
Surface Trapping	107
Section III: Diode and Transistor Arrays	129
G.P. Weckler	
Charge Storage Operation of Silicon Photo-detectors	131
P. Jespers	
XY Addressing	143
M.H. White	
Photodiode Sensor Arrays	165
P. Jespers	
Phototransistor Arrays	195

VIII

Section IV: CTD Arrays   217

D.F. Barbe
  Charge Integration and Storage in MOS Photo-
  sensors   219
C.H. Séquin
  Introduction to Charge-Coupled Devices   233
C.H. Séquin
  Organization of Charge-Coupled Image Sensors   261
M.H. White
  Charge Transport without Traps   275
G.F. Amelio and R.H. Dyck
  Charge Transport with Traps   295
C.H. Séquin
  Image Sensors Using Surface Channel Charge-
  Coupled Devices   305
G.F. Amelio and R.H. Dyck
  Buried Channel CCD's   331
L.J.M. Esser
  Peristaltic Charge-Coupled Devices   343
C.H. Séquin
  Electrical Charge Injection into CCD's   427

Section V: CID Arrays   445

G.J. Michon and H.K. Burke
  Charge-Injection Devices for Solid State
  Imaging   447
P. Jespers and J.M. Millet
  Three-Terminal Charge-Injection Device   463

Section VI: Signal Extraction   483

M.H. White
  Design of Solid-State Imaging Arrays   485
C.H. Séquin
  Interlacing in Solid-State Image Sensors   523
J.A. Hall
  Amplifier and Amplifier Noise Considerations   535
R.R. Buss, S.C. Tanaka, G.P. Weckler
  Principles of Low-Noise Signal Extraction from
  Photodiode Arrays   561
G.F. Amelio and R.H. Dyck
  Distributed Floating Gate Amplifier   605
G.F. Amelio and R.H. Dyck
  Fixed Pattern Noise and Cooled Photosensor
  Arrays   615

Section VII: Systems 621

D.F. Barbe and S.B. Campana
   Aliasing and MTF Effects in Photosensor Arrays 623

J.A. Hall
   Signal and Noise in the Display of Images 637

D.F. Barbe
   Time Delay and Integration Image Sensors 659

D.F. Barbe
   Solid State Infrared Imaging 673

J.A. Hall
   Low Light Level Performance of Charge-Coupled
   Area Imaging Devices 689

J.A. Hall
   Comparison of Solid-State Imagers and Electron
   Beam Scanning Imagers 705

List of Participants 733

Section I

INTRODUCTION

HISTORY OF SOLID-STATE IMAGING

Marvin H. White

Westinghouse Electric Corporation,
Advanced Technology Laboratory,
Baltimore, Maryland U.S.A.

ABSTRACT. In the past few years there has been renewed interest in the maturing technology of totally solid-state imaging. Semiconductor technology has advanced rapidly in the decade of the 60's, and in the early 70's it stands at the point where high resolution, low-noise, integrated imaging systems are a practical reality. In this chapter we will present a discussion of photodiode, phototransistor, and charge-coupled device (CCD) imaging from a historical perspective.

Prior to the rapid advancement of semiconductor technology in the 1960's, the primary method of interrogating a large number of photosensors was the electron beam. In general, when accelerated electrons strike a target, we may have two cases: (1) in the case of high energy electrons, where the secondary electrons liberated outnumber the primary electrons, we have high-velocity or "collector-potential" stabilization. In television pick-up tubes like the iconoscope[1] and the image iconoscope[2], high-velocity stabilization is employed to clamp the free surface of the target to the collector of the electron gun; (2) in the case of low energy electrons, where the secondary emission coefficient is small, we have low-velocity or "cathode-potential" stabilization. In television pick-up tubes like the orthicon[3], image orthicon[4], and vidicon[5], low-velocity stabilization clamps the free surface of the target to the cathode of the electron gun. Thus, we notice an electron beam performs the function of a fast, low-noise "switch" to connect and disconnect each picture element (pixel) successively to either the collector or cathode of the electron gun.

The vidicon camera tube target consists of a thin, high resistive, photoconductor (typically 10 μm) deposited on the window of the tube on which a thin, transparent, conductive layer has been deposited. The latter is connected to a positive d-c potential and a-c coupled to a preamplifier for the signal path. The free surface of the photoconductor is clamped to cathode potential by the scanning electron beam and an electric field is established across the photoconductor. A picture element may be considered as a capacitor shunted by a light-sensitive resistor and if the RC time constant of the pixel is considerably larger than the time which elapses between two successive scans (e.g., a frame time τ_f = 1/30 sec), then the signal current is related linearly to the photocurrent. Under these conditions the incident photon flux density creates charge carriers which discharge the pixel capacitance over a frame time. In essence, the photon flux is integrated or "stored" over a frame time which greatly enhances the signal-to-noise in readout. This "storage principle" is used to some extent in all pick-up tubes. Only a limited number of photoconductors have been used successfully in the vidicon; for example, Sb_2S_3, PbO, CdSe, and amorphous selenium.

With the vidicon background to provide experience, a number of investigators in the late 1950's and early 1960's began to examine the "marriage" of a new, emerging, semiconductor technology to the experienced vacuum tube technology. In particular, the military requirements in the infrared placed the emphasis on doped photoconductors in silicon and germanium. The homogeneous doped, single crystal, photoconductor had similar characteristics to the amorphous photoconductor films, except the former had to be cooled to achieve low dark current and storage mode operation. Amorphous "smoke" films were also examined to provide room temperature infrared imaging with a so-called thermicon camera tube. Other workers pursued the direction of mosaics of P-N junctions in single crystal silicon, germanium, and indium antinonide to provide signal detection and storage. Much of this early work on the combination of semiconductor and vacuum tube technologies was classified and published in conferences such as IRIS; however, over the years through declassification and other workers' findings, the basic results have been published in the open literature.

In the early 1960's two different approaches to replace the electron-beam scanning of image sensors with electronic solid-state switches was examined in considerable detail: (1) thin-film technology[6] and (2) silicon monolithic integrated circuit technology.[7] The latter, due to the continuous growth of the integrated circuit industry and government investment, has emerged the winner, although the concept of self-scanning with X-Y address was developed through the former technology. In parallel with these approaches the combination of vacuum tube and semiconductor technologies proceeded with the culmination of the silicon diode array target[8] for the vidicon. The video signal is taken from the N^+/N contact of the silicon substrate which provides the same function as the transparent conductive layer in the photoconductive vidicon. The electron beam reverse-biases the mosaic of P-N junctions and provides the switch for interrogation. The advanced silicon technology required for these silicon diode targets (i.e., bulk gettering with phosphorous and surface state annealing with hydrogen) was a stepping stone to the subsequent realization of self-scanned image sensor arrays.

The initial work with silicon concentrated on bipolar transistor sensors in X-Y addressable matrix arrays with external scan generators.[7] The formation of X-Y arrays required column to column isolation and diffused epitaxial isolation was employed with back-to-back P-N junction isolation between collectors. The row and column address strips were connected to external scan circuits and large arrays of 400 x 500 elements were incorporated into operational cameras.[9] Other types of arrays were built with a common collector where a series MOS transistor at each element[10] provided isolation. These phototransistor arrays have been useful in small sizes (e.g. 50 x 50) for such applications as reading aids for the blind;[11] however, the nonuniformity in transistor current gain and low level threshold caused by emitter off-set have restricted the development. The streaking in the image caused by β variations limits the achievable noise equivalent signal (NES) and dynamic range of phototransistor sensor arrays. Recent developments in ion-implantation techniques may be used to control the β of phototransistors and thus reduce the fixed pattern noise caused by gain streaking. Other limitations result from the low effective quantum efficiency in the short wavelength response.[12] All of the early phototransistor arrays required the use of external scan registers and a considerable number of bonds were required to bridge the interface between sensor and scan electronics. This resulted in a major packaging problem, degraded performance, reduced reliability, and increased cost.

After considerable experience with phototransistor arrays and the problems of self-scanning with external scan electronics, many workers began to favor the simple photodiode sensor. Early work was carried out with a self-scanned photodiode array called the scanistor[13]; however, the large breakdown voltage requirements and the fact the photodiodes did not operate in the charge storage mode limited the development. The first practical discussion of the storage mode operation in monolithic integrated structures was directed towards photodiode sensors.[14] This structure was the forerunner of today's self-scanned photodiode arrays; however, initial structures used off-chip scan registers.

The scanning function was incorporated with a 10 x 10 element photodiode array[15] and present-day arrays of 64 x 64 photodiode elements are available in solid-state cameras.[16] The major advantages of the photodiode sensor are (1) no variations in gain to provide low gain streaking, (2) low leakage currents for reduced temperature streaking, (3) high uniform quantum efficiency for low spectral streaking, (4) linear light level response, and (5) high packing density in line arrays for high resolution. The development of the MOS-FET electronic switch not only replaced the electron beam switch but also served as the basic element in self-scanned shift registers. The simplicity of MOS-FET fabrication and high packing density are ideal factors to construct on-chip scan electronics for solid-state photosensors.

In the late 1960's two discoveries were introduced which provided a new technique for self-scanned solid-state imaging. The principle of charge transfer in the form of bucket brigade devices[17] (BBD's) and charge-coupled devices[18,19] (CCD's) provided analog shift registers to transfer the optically-created charge pattern over long distances across the silicon surface to a readout amplifier located at the edge of the sensor. With the charge transfer principle significant improvements were realized in signal-to-noise due to (1) the elimination of switching spikes within the array, and (2) the signal detection at a common, low-capacitance node. Further improvements were realized in uniformity because of reduced fixed pattern noise caused by "point defects." Localized bright spots caused by point defects are smeared over several pixels with a loss in resolution but not the objectionable gain streaking. The CCD may be operated with charge transfer along the silicon surface[20] (surface channel CCD) or with charge transfer within the silicon[21] (bulk channel CCD).

The resetting of the CCD photosensor may be accomplished through charge transport from the photosite or with charge injection. The introduction of the charge injection device[22] (CID) provided an alternative method of image sensor array construction. This alternative approach uses the simplicity of MOS-FET scan registers at the periphery of the sensor array, while the sensors may operate with CCD/CID principles. The CID may be used to provide readout of the photosensor or it may sever to simply reset the photosensor for the next integration period. Advanced signal processing methods[23] are used to remove the switching off-sets which appear in X-Y addressed arrays and provide a low noise, uniform, video output signal. Recently, a new type of bulk channel CCD called the "Peristaltic CCD (PCCD)" has been introduced [24] which offers the advantage of high speed video signal processing.

REFERENCES

1. V.K. Zworykin, Proc. IRE, 22, 16, 1934.

2. H. Iams, G.A. Morton, V.K. Zworykin, Proc. IRE, 27, 541, 1939.

3. A. Rose and H. Iams, Proc. IRE, 27, 547, 1939.

4. A. Rose, P.K. Weimer, and H.B. Law, Proc. IRE, 34, 424, 1946.

5. P.K. Weimer, S.V. Forque, and R.R. Goodrich, Electronics, 23, 70, 1950.

6. P.K. Weimer, H. Borkan, G. Sadasiv, L. Meray-Horvath, and F.V. Shallcross, Proc. IEEE, 52, 1479, 1964.

7. M.A. Schuster and G. Strull, IEEE Trans. Electron Devices, 13, 906, 1966.

8. M.H. Crowell, T.M. Buck, E.F. Labuda, J.V. Dalton, and E.J. Walsh, Bell Syst. Tech. J., 46, 491, 1967.

9. D.L. Farnsworth, E.L. Irwin, and C.T. Huggins, Government Microcircuit Application Conference (GOMAC), 1972, San Diego, Calif.

10. R.H. Dyck and G.P. Weckler, IEEE Trans. on Electron Devices, 15, 196, 1968.

11. J.S. Brugler, J.D. Meindl, J.D. Plummer, P.J. Salsbury, and W.T. Young, IEEE J. of Solid-State Circuits, 4, 304, 1969.

12. M.H. White and D.R. Lampe, Intercon "74", New York City, N.Y., 1974.

13. J.W. Horton, R.V. Mazza, I.L. Dym, Proc. IEEE, 52, 1513, 1961.

14. G.P. Weckler, Int'l Electron Device Meeting (IEDM), Washington, D.C., 1965.

15. P.J.W. Noble, IEEE Trans. Electron Devices, 15, 202, 1968.

16. Reticon Corporation, Sunnyvale, Calif. (U.S.A.)

17. F.L.J. Sagnster and K. Teev, IEEE J. of Solid-State Circuits, 4, 131, 1969.

18. W.S. Boyle and G.E. Smith, Bell Syst. Tech. J., 49, 587, 1970.

19. G.F. Amelio, M.F. Tompsett, and G.E. Smith, Bell Syst. Tech. J., 49, 593, 1970.

20. C.H. Séquin, D.A. Sealer, W.J. Betram, M.F. Tompsett, R.R. Buckley, T.A. Shankoff, and W.J. McNamara, IEEE Trans. Electron Devices, 29, 244, 1971.

21. G. Amelio, IEEE Intercon (1971), New York City, N.Y.

22. G.J. Michon and H.K. Burke, IEEE Solid-State Circuits Conf. (1973), 138, Philadelphia, Pa.

23. M.H. White, D.R. Lampe, F.C. Blaha, and I.A. Mack, IEEE J. of Solid-State Circuits, 9, 1, 1974.

24. L.J.M. Esser, IEEE International Solid-State Circuits Conf. (1974), 28, Philadelphia, Pa.

SOLID-STATE IMAGING APPLICATIONS

James A. Hall

Westinghouse Electric Corporation,
Advanced Technology Laboratories,
Baltimore, Maryland U.S.A.

ABSTRACT. Three applications which conventional television has not satisfied are imaging from moving platforms like weather satellites, infra-red imaging, and quantitative astronomical imaging, especially from space platforms. Solid state detectors or arrays of detectors in mechanical scanners provide basically better performance, and time delay and integration techniques are increasing their advantage. Faint object astronomy today has turned to "photon counting", an operating mode better satisfied by the ICCD (Intensified Charge Coupled Device) than by tube type television systems.

1. INTRODUCTION

Solid-state image sensors are used for such diverse applications as single detector mechanically scanned imagers for either the visible or infrared, line arrays of sensors with mechanical scanning, and mechanically scanned x-y arrays used with time delay and integration to effectively increase detector exposure without reducing the data rate. The application at which many charge-coupled area imaging devices are directed, that of replacing camera tubes in semi-conventional television, will be treated only briefly here both because it will be treated in later chapters and because that application is already well served by existing systems. In this chapter we shall concentrate instead on applications where conventional television sensors are not entirely satisfactory. These include imaging from a moving platform or viewing a scene whose aspect ratio is very far from square, or where contrast is very small.

2. METEOROLOGICAL SATELLITES

For example, in applications like weather observation from satellites, conventional framing television techniques are not the best approach. The earliest weather satellites of the Tiros family used vidicon-type television cameras to take and transmit a series of "snapshots" of the earth's surface and its cloud cover. For best resolution, image motion had to be compensated mechanically during each exposure to avoid blurring. Each frame was printed separately, and the observer had to cut and paste to provide area coverage beyond the 300 to 500 element square raster of each frame. In contrast, current state of the weather satellite art is represented by the U.S. Defense Meteorological Satellite, which uses a single silicon detector in each of two mechanical scanners to provide both 2 mile resolution and 1/3 mile resolution imaging at the same time. Cross track scan is provided by a pair of rotating mirrors geared together, as shown in figure 1, while vehicle motion provides along track scan. The sensor package also includes a pair of mercury cadmium telluride detectors scanning identical images in the 8-13μm band in the infra-red.

As shown in figure 2, the satellites operate in nearly polar orbits, inclined at 8.75 degrees to maintain orientation with respect to the sun-earth radius vector. One vehicle crosses the equator from south to north at local noon, and a second at about 7:30 in the morning.

The output data is the image of a continuous strip 1655 nautical miles wide and is printed on long strips of 9 1/2" film, each containing the record of three orbits.

The ground irradiance for the noon-midnight sensor varies from a maximum at the subsolar position, approximately the south-to-north crossing of the equator, to a minimum at the midnight north to south crossing, over at least 10^8:1. The silicon sensor operating as a photoconductor is linear over this range, and output signal level is maintained relatively constant up to 84 degrees from the sun-earth line by use of an upward looking comparison sensor actuating an amplifier gain control. For sun elevation less than 6 degrees, the sun elevation angle is measured and the gain control adjusted in 1.3-dB steps from computed irradiance values. The latter scheme is used exclusively on the early morning satellite, which as shown in figure 2 must compensate for scene irradiance variations along a single scanning line of up to 3×10^5:1 when the scan line spans the terminator. Scene reflectance variations which constitute the information add one to two orders of magnitude to the range of sensor irradiances which must be accommodated.

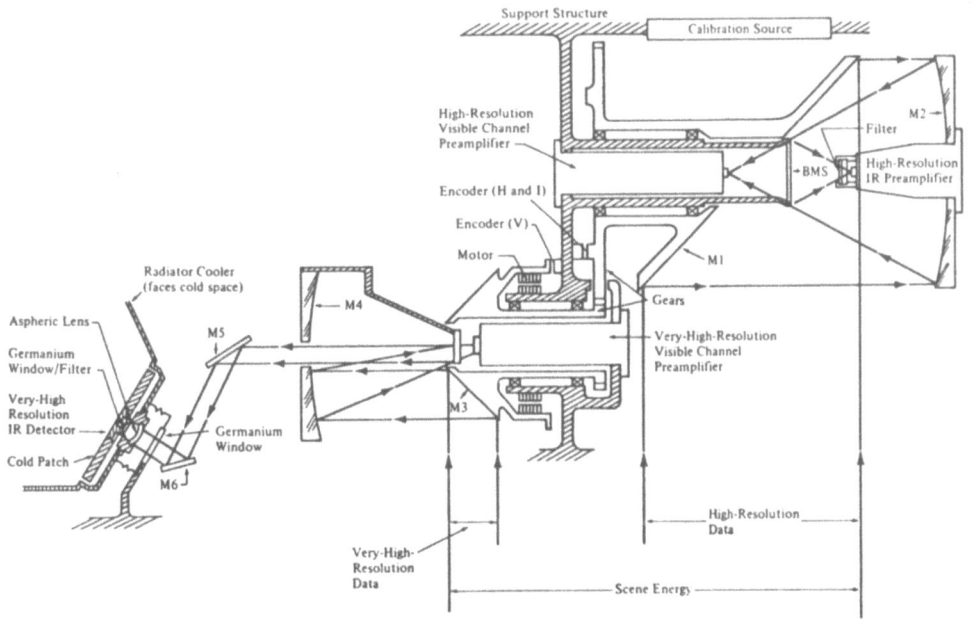

Figure 1. Simplified Optical Schematic of Sensors in the Defense Meteorological Satellite Shows High and Very High Resolution Channels with 6:1 Ratio Mirror Drive for Coincidence of Scanned Areas.

a

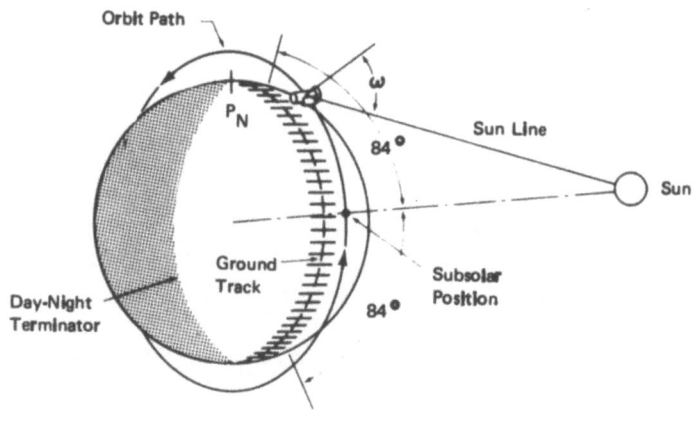

b

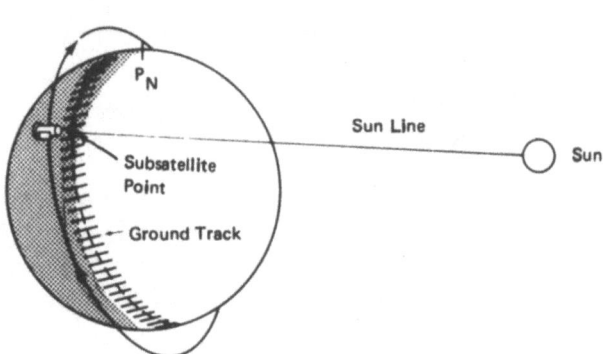

75-0440-VA-92

Figure 2. Meteorological Satellites are Placed in Sun-Synchronous Near Polar Orbits. For a Noon-Midnight Orbit (a), Illumination Variation is Primarily Along Track. For an Early Morning Orbit (b), Illumination can Vary by 10^5:1 Along a Single Scanning Line Spanning the Terminator.

To use a conventional T.V. sensor for the early morning satellite would thus require accommodating a dynamic range of $10^7:1$ in a single scene, a manifestly impossible task.

Some samples of imagery obtained are given in figures 3 through 7. Figures 3 and 4 are taken over Italy by the noon and early morning satellites respectively. Figure 3 shows a few clouds mostly over Corsica but few striations due to gain changes. Land and water are clearly differentiated by the silicon spectral response, but features other than lakes are not readily visible in this 2 nmi resolution picture. Figure 4 is an early morning image of Italy at 2 nmi resolution. The crossed grid lines occur as the amplifier gain is changed along each scanning line and along track. Nevertheless, the image is more than adequate to record the bright cloud cover to the east over Yugoslavia and Albania where the sun is higher and the absence of clouds over Sardinia, Corsica, and Western Italy near Torino where the illumination is at least 5 orders of magnitude dimmer. This picture also shows the Apennines and the Alps in pseudo-relief because the early morning sun illuminates only the eastern side of each mountain.

Figure 5 shows a 1/3 nmi image of cloud cover over Western Europe, and is included primarily to show detail of the low countries where this conference was held. Figure 6 is a picture of the Eastern United States taken near midnight three days before the full moon. The clouds and terrain are illuminated by moonlight, giving useful picture quality, while the cities are recorded by their own lights. Thus sensitivity of this nonintegrating single cell scanner is adequate in the 2 mile mode for some night operation. Anyone from a city in the Eastern United States can probably find his home town with a little patience. Finally, Figure 7 shows an unusual view of a total eclipse. The moon's shadow is plainly visible on the clouds over central Africa. But these pictures, while interesting to the laymen, are shown here to illustrate one application very well served by a solid-state sensor in a scanner where conventional frame integrating television could not have performed nearly as well. One reason for success here is the comparatively modest data rates of 10^4 to 6×10^4 samples per second.

3. INFRARED SCANNERS

The second application is infrared imaging. Historically, successful infrared imaging systems have used single solid-state detectors in mechanical scanners which perform a full raster scan in X and Y, and attempts to use more sensitive infrared television camera tubes with full frame integration

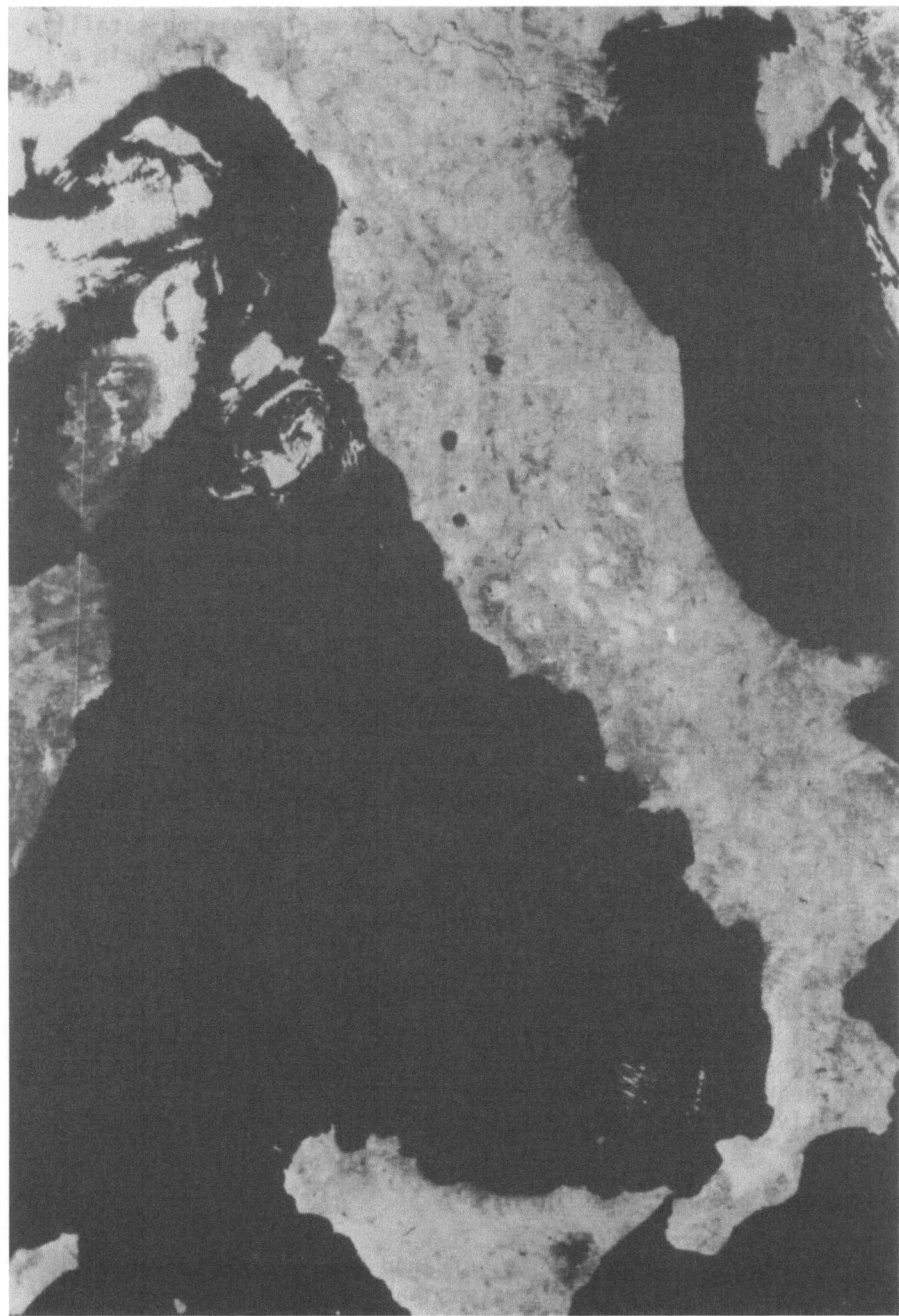

Figure 3. A Near Noon Satellite Image of Italy Shows a Few Gain Changes as the Satellite Moves North.

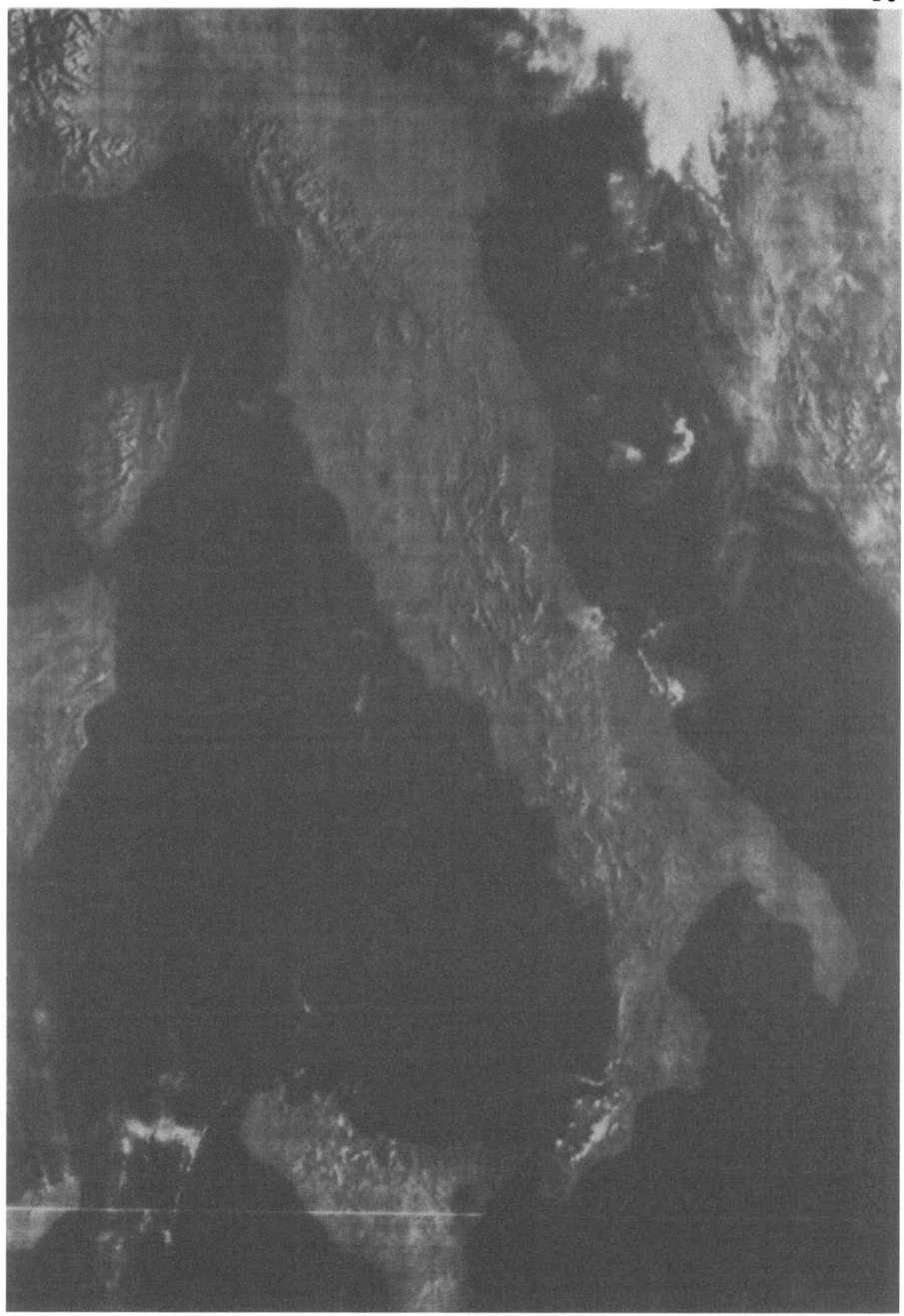

Figure 4. An Early Morning Satellite Image of Italy Shows Crossed Grid Striations as Amplifier Gain is Switched to Compensate for Scene Irradiance Changes Along Scan and Along Track. The Low Lying Sun Illuminates Only the Eastern Side of Each Mountain, Forming an Apparent Relief Map of the Alps and Apennines.

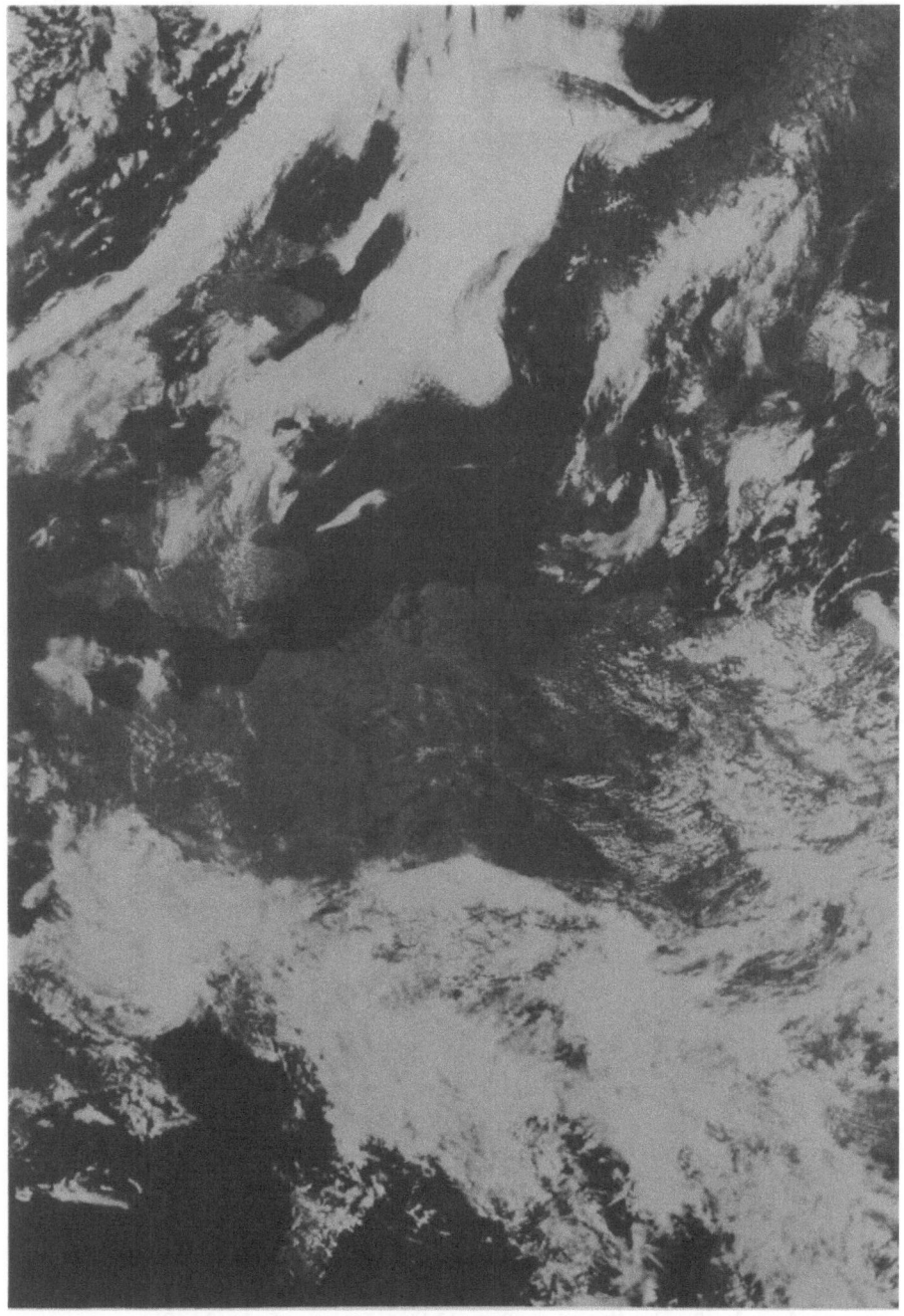

Figure 5. A 1/3 Nautical Mile Near Noon Image of Western Europe Shows a Frontal System Over Great Britain, but Clear Skies Over the Low Countries.

Figure 6. The Eastern United States at Midnight Three Days Before Full Moon. The High Population Density Along the Washington-Boston Corridor and on the Florida Gold Coast is Emphasized by the Image of City Lights.

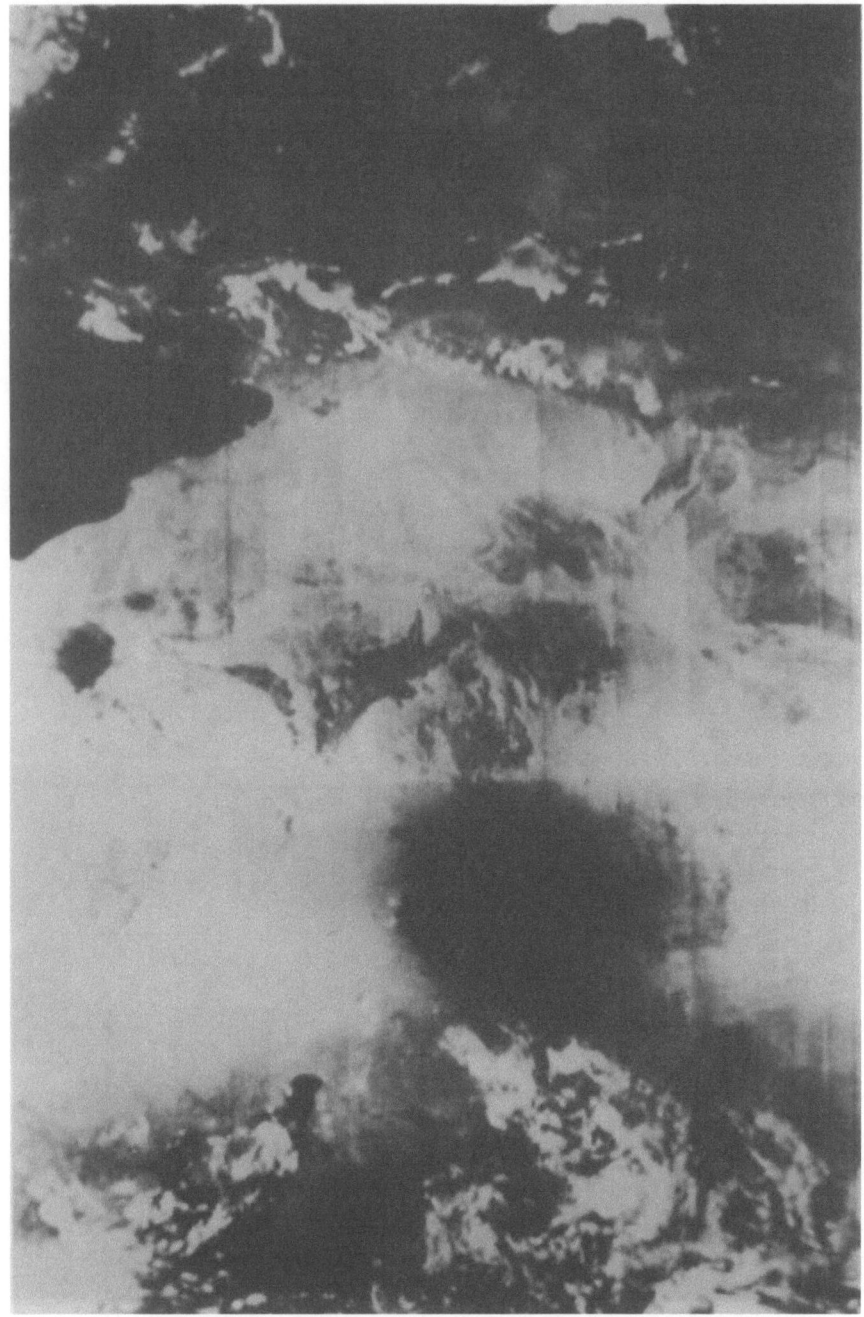

Figure 7. A Total Eclipse of the Sun is Indicated by the Shadow of the Moon on the Clouds Over Central Africa.

have been relatively unsatisfactory. The reasons were explored in "The Problem of Infra-red Television-Camera Tubes vs Infrared Scanners", Applied Optics, 10, p. 838 (April 1971) by this author. Basically, the problem is that the thermal photon flux from room temperature objects is large, but the contrast between scene objects is very small, on the order of fractions of a percent to a few percent for any realistic target.

Thus the infrared imaging sensor does not need sensitivity, but rather the ability to generate useful signals from very small scene contrasts while handling a very large total signal. This requirement implies either extreme uniformity of response across the field of view, or "ac coupling", the ability to ignore the large stready state average irradiance from the scene while generating a signal from the small point-to-point variations. Normal full frame integrating camera tubes have response non-uniformities as high as a prohibitive 10 or 20 percent, and would be saturated by the flux from the scene through a reasonably fast cold shielded optical system even though long wavelength response was limited to 5μm. The 8-13μm window would be hopeless.

In contrast, a solid-state single detector sensor with some optical scanning systems can have essentially complete uniformity of response over the image area, and the simple and normal use of an ac coupled amplifier does ignore the average current from the detector and amplifies only the signal from point-to-point irradiance variations in the image. Finally, the solid-state detector has a linear transfer characteristic over many orders of signal magnitude and can faithfully reproduce small signal variations on top of a large average signal, although of course the detector also reproduces the signal shot noise associated with the background.

To maximize the signal-to-signal shot noise ratio one employs a fast optical system to maximize both time varying and average signal, since the shot noise varies as the square root of the latter, cools the lens and detector housing and the detector itself to limit the background photon flux to that which comes through the lens, and if possible uses a detector whose long wavelength response limit falls just beyond the atmospheric spectral window for the target radiation being sensed. This much is done in all IR scanners.

To further improve performance, one can first use a linear array scanner in which the scene image is scanned in one direction only by moving a linear array of 100, 200, or more solid-state detectors across the image. This arrangement, shown in figure 8, can be thought of either as reducing the

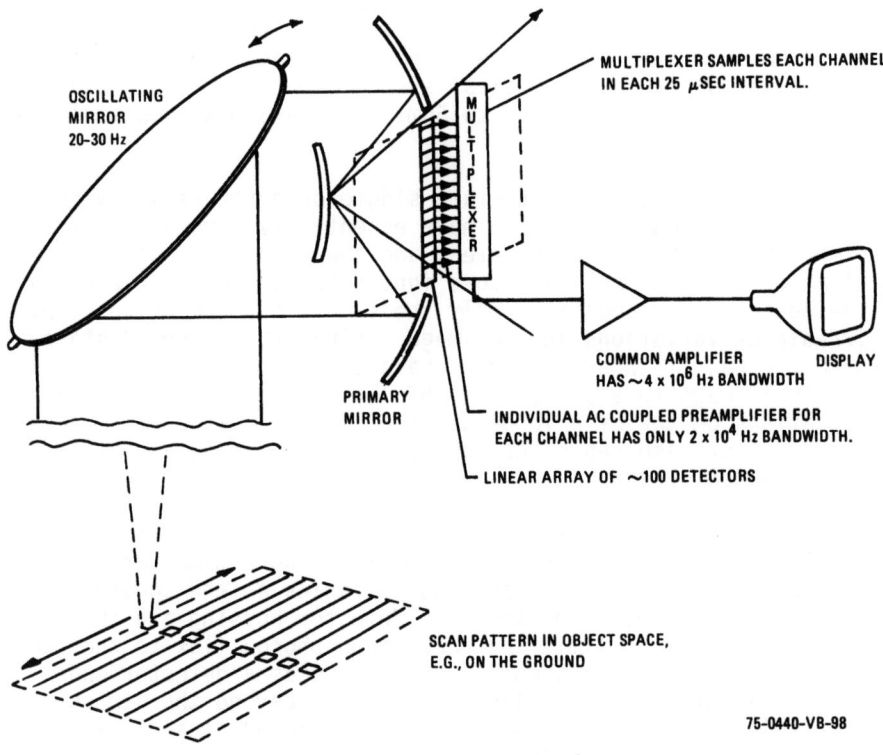

Figure 8. Linear Array Scan Reduces Data Rates to Improve Signal to Noise Ratio, but Can Give Nonuniform Raster Line Interference Pattern if the Channels are not Perfectly Matched.

signal bandwidth required of each detector, hence band limiting
the shot noise at a lower value, or as increasing the effective
sensor area at the image. In either case, the signal-to-background shot noise ratio improves as the square root of the
number of detectors. Since each detector is ac coupled through
its own preamplifier, it senses irradiance variations and ignores
the average background value along its scanning path. The
tradeoff here is that each line of the image is scanned by a
different preamplifier detector combination so channel-to-channel
variations in either average output signal or in gain can produce
line-to-line brightness variations in the reproduced image which
can significantly mask the visibility of low contrast detail in
the reproduced image. Three schemes are used or are under consideration to reduce image disturbance. First, one can manually
adjust the gain and average signal level to match each channel
to its neighbors as closely as patience and stability allow.
Second, one can devise compensation and calibration circuitry
to equalize all channels automatically to some tolerance.
Thirdly, one can alter the scanning pattern to include a small
excursion along the sensor array perpendicular to the scan lines
at both the sensor and image display devices. If this low
frequency excursion completes one or more cycles in less than
the 0.1 to 0.2 second integration time of the observer's eye,
the eye will average out the remaining response and level differences. In fact, one may lose the impression of a raster structure
in the display, and see small signal contrast differences clearly.
In this scheme, the image of the scene is stationary, it is only
the raster structure which is moving.

As an alternate approach, a larger effective sensor area
may be provided by placing detectors in series rather than in
parallel. As shown in figure 9, 8, 12, 16, or any other convenient number of detectors may be placed one behind the other
along a scanning line, using an analog delay line or a CCD to
combine the 8 or more individual output currents in the time
delay and integration mode to form a single coherent output.
Provided that the scanning speed accurately matches the electrical
time delay, this scheme should give resolving power equal that
from a single detector scanner but signal-to-noise ratio is
improved as the square root of the number of elements. Further,
since each element scans the entire scene, except for the
extreme edges, the signal response and background level fed to
the display does not vary instrumentally from scan line to scan
line, but is a function only of scene changes, as it is with a
single detector scanner, so uniformity is excellent. The tradeoff here is the demand for much detail in a flicker free presentation which can require video band widths of 10^7 Hz or higher,
difficult to obtain with some solid-state sensor devices. In
addition, the mechanical design of a high data rate x-y raster

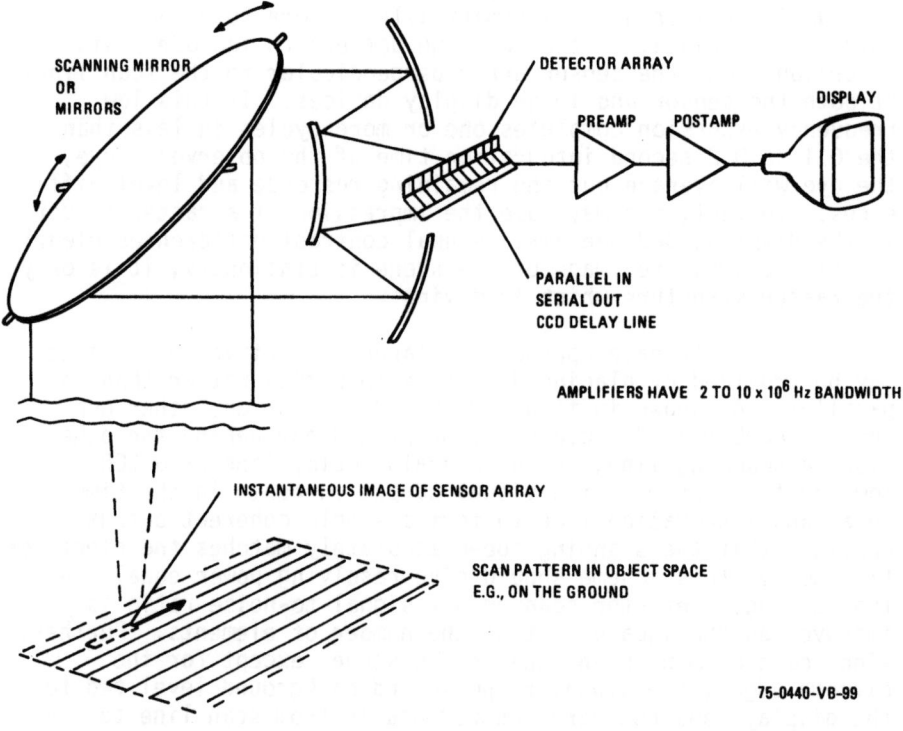

Figure 9. The Signal from Several Detectors may be Combined in Phase by Time Delay and Integration to Provide a Larger Effective Sensor Area in a Serial Scanned Mode.

scanner is a difficult one, although scanners of this type have been made at least experimentally by the Hughes Aircraft Company under the trade name "Discoid".

The latest concept, which is still experimental, is to combine time delay and integration with a linear array scanner to reduce data rates by two orders of magnitude, to improve signal-to-noise ratio by 50 or more compared to a single detector scanner, and hopefully to achieve acceptable uniformity since each channel presents the average signal from 8 or more detectors, and channel to channel variation should be considerably reduced.

These applications are discussed in a later chapter. They are described briefly here since these are again applications to which solid-state image detecting and signal handling technology is well adapted, while traditional television techniques are not, because of the characteristics of the infrared image.

4. SPACE ASTRONOMY

In both the previous applications the small size, low power, and long mean time before failure of solid-state imaging detectors were important advantages. In each case, however, these advantages were not stressed, because there was another dominant requirement which could not be satisfied with conventional framing television. The last application given here is that of astronomical imaging, especially in satellite-borne observatories like the large space telescope now under study by the U.S. National Aeronautics and Space Administration. The astronomer only occasionally looks at star fields or at planetary surfaces. Much of the time he wishes to measure spectra, and often these spectra or those star fields of interest require measuring photon fluxes from very faint objects which are only slightly above background, the low contrast problem which seems to occur in most scientific and technical applications. Here television camera tubes can and are being considered, but solid-state line and x-y array sensors with full frame integration offer some important advantages and should displace camera tubes as the solid-state sensor art advances. The advantages include unvarying well defined geometry, stability of characteristics, and a linear transfer characteristic, as well as wide dynamic range which permits measuring small contrast differences reliably by integrating the signal to high levels by use of longer exposures, as well as the obvious advantages of small size, weight, and power, and long operating and shelf life. The disadvantages are small array size limited to a few hundred small sensor elements in both x and y for an overall array size of a centimeter or less, rather than the several centimeter chip dimensions and 25 to 50 µm element sizes which would better match optical system blur circles, and of course, the sensor element dark current which for silicon arrays requires significant cooling for time

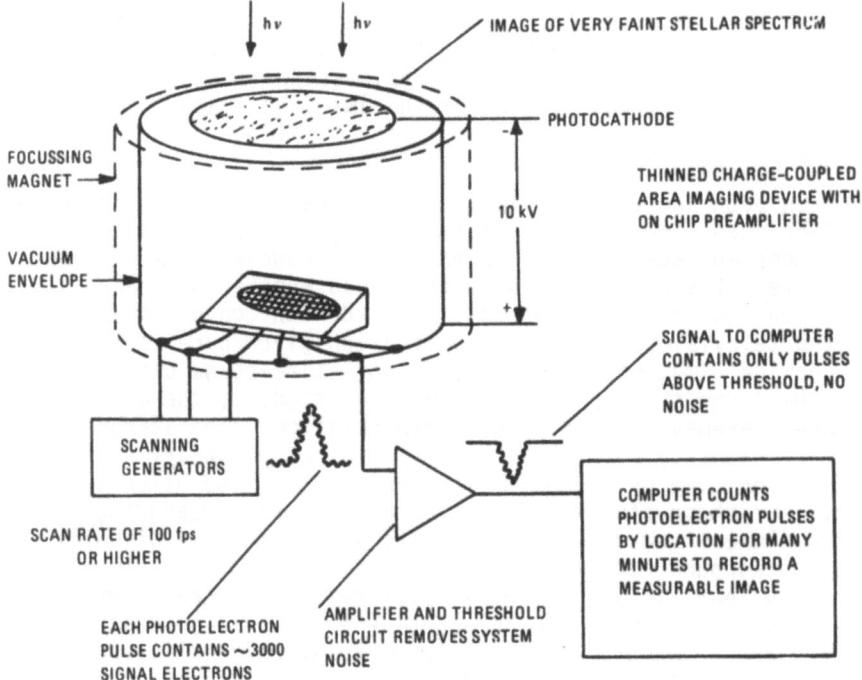

Figure 10. In Photon Counting Imaging of Very Faint Objects, Photoelectrons Accelerated Through 10 kV or More Bombard an Electron Sensitive Area Imaging Device Like a Thinned CCD. The Resulting 3000 Electron Pulses can be Clearly Distinguished from System Noise, and are Counted by Location in an External Register. Scanning is Rapid to Avoid Pulse Pileup at a Single Site.

exposures of even several seconds or tens of seconds. The trade-offs between tube and solid-state array sensors are discussed in a later chapter, but space astronomy is again an application where conventional television techniques with camera tubes have significant problems and where present or prospective solid-state array sensors should have a technical advantage. This is especially true in the photon-counting mode described in figure 10 where a silicon array sensor is bombarded with photo electrons to produce pulses amplified to several thousand electrons, and scanned at high rates so that individual photo electron events can be counted by location in external data handling circuitry rather than integrated in the device itself. Here a modest CCD in a vacuum tube with a photocathode and electron optical imaging system promises the ability to count photoelectrons, while even advanced television camera tubes require one or more auxiliary image intensifiers and high voltages of 20 to 45 kV to accomplish the same task because of the higher system noise of the tube type imager.

Section II

PHYSICS OF PHOTOSENSORS

ANTI-REFLECTION FILMS AND MULTILAYER STRUCTURES

by

F. Van de Wiele

Laboratoire de Microélectronique, Louvain-la-Neuve,
Belgium.

Abstract

The transmittance of multilayer structures is calculated for light incident normally on the active area of a photosensing device. The case of a single anti-reflection silicon dioxide film on a silicon substrate and the case of an Air-SiO_2-Polysilicon-SiO_2-Si structure are discussed in detail.

Notation

d_i thickness of layer i.
$d_{1,m}$ particular value of d_1 for which T_2 becomes maximum or minimum.
g_i see equ. between (16) and (17).
h_i see equ. between (16) and (17).
k_i extinction coefficient of layer i.
n_i real part of the index of refraction of layer i.
r_{m+1} Fresnel coefficient of reflection, see equ. (6).
t_{m+1} Fresnel coefficient of transmission, see equ. (7).
E_m^+ electric field vector incident, at the interface (m+1), upon the layer (m+1).
E_m^- reflected component of the electric field vector.
E_{s+1}^+ electric field vector of light transmitted at interface (s+1).
M_{m+1} see equ. (11).

N_i complex index of refraction of layer i.
R_{m+1}^i ratio of reflected power to incident power at interface (m+1).
T_{m+1} ratio of transmitted power to incident power at interface (m+1).
α_i absorption coefficient of layer i.
δ_{m+1} see equ. (4).
λ free space wavelength of light.

Anti-reflection films and multilayer structures

The active area of a photosensing device is usually covered with a single or multilayer structure. In this case, the light being sensed by the device has to pass through optically different thin layers before reaching the semiconductor. The light sensitivity of the device is to a great extent limited by surface reflection and absorption of the incident radiation.

Reflection losses can be reduced, in a given wavelength range, by coating the reflecting surface with thin films of appropriate antireflection material and thickness. Other structures require, for normal operation, a multilayer structure. For example, the structure on the gate region of Charge Coupled Device (CCD) and Charge Injection Device (CID) image sensors is composed of three layers : a gate oxide on the silicon substrate, a doped polysilicon film and finally a silicon dioxide film.

Only a few papers in the literature [1, 2, 3,. 4] are dealing with the transmittance of multilayer structures including a polysilicon film. The purpose of this section is to develop a scheme relating the light intensity effectively reaching the semiconductor surface to the incident light intensity; light absorption in each thin layer and multiple light reflections at the interface between two adjacent layers are taken into account. In order to simplify the mathematical treatment, only the case of light incident normally on the active area of the photosensing device will be considered.

Assume a semiconductor substrate is covered by s thin layers (Fig.1). The various media from air to the substrate are numbered respectively 0 to s + 1. Each absorbing layer m of thickness d_m has a complex index of refraction N_m :

$$N_m = n_m - ik_m \qquad (1)$$

n_m is the real part of the index of refraction. The imaginary part k_m is called the extinction coefficient and is related to the absorption coefficient $\alpha_m(\lambda)$ at the wavelength λ through the relation

$$\alpha_m(\lambda) = 4\pi k_m/\lambda \qquad (2)$$

λ being the free space wavelength of light.

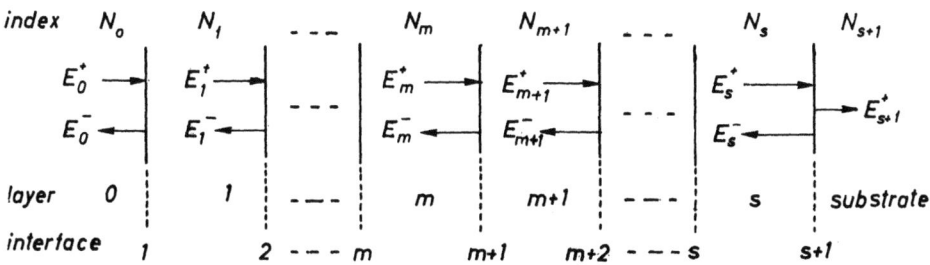

Fig. 1 Multilayer structure.

The matrix method of Abeles is used for the calculation of the reflection and transmission coefficients of the multilayer structure. Consider the interface (m+1) between the media (m) and (m+1). The multiple reflections yield an electric field light vector denoted in the following way. Within the layer (m) the electric field vector incident, at the interface (m+1), upon the layer (m+1) is represented by the complex quantity E_m^+, while the reflected component is denoted by E_m^-. The same terminology is used for all media, except for the substrate. The transmitted light at the interface (s+1) is denoted by E_{s+1}^+; since the substrate is assumed to be semi-infinite the amplitude of the reflected component is taken to be zero.

The continuity conditions for the electric and magnetic field at the interface (m+1) impose the following relations :

$$E_m^+ + E_m^- = E_{m+1}^+ \exp(i\delta_{m+1}) + E_{m+1}^- \exp(-i\delta_{m+1})$$

$$N_m(E_m^+ - E_m^-) = N_{m+1}\left[E_{m+1}^+ \exp(i\delta_{m+1}) - E_{m+1}^- \exp(-i\delta_{m+1})\right]$$

(3)

where

$$\delta_{m+1} = (2\pi/\lambda) N_{m+1} d_{m+1} = (2\pi/\lambda)(n_{m+1} - ik_m) d_{m+1}$$

(4)

These equations are valid for all interfaces, except the last one for which

$$E_s^+ + E_s^- = E_{s+1}^+$$

$$N_s(E_s^+ - E_s^-) = N_{s+1} E_{s+1}^+$$

(5)

At an interface (m+1) between materials with refractive indices N_m and N_{m+1}, the ratio of the reflected (transmitted) electric field amplitude to the incident electric field amplitude is given by the Fresnel coefficient of reflection r_{m+1} (transmission t_{m+1}) :

$$r_{m+1} = (N_m - N_{m+1})/(N_m + N_{m+1})$$

(6)

$$t_{m+1} = 2N_m/(N_m + N_{m+1})$$

(7)

At the same interface the ratio of the reflected (transmitted) power $R_{m+1}(T_{m+1})$ to the incident power is given by,

$$R_{m+1} = |r_{m+1}|^2$$

(8)

$$T_{m+1} = |N_{m+1}/N_m| \cdot |t_{m+1}|^2$$

(9)

When solving (3) for E_m^+ and E_m^- and taking the Fresnel relations into account, one obtains :

$$\begin{vmatrix} E_m^+ \\ \\ E_m^- \end{vmatrix} = \frac{1}{t_{m+1}} M_{m+1} \begin{vmatrix} E_{m+1}^+ \\ \\ E_{m+1}^- \end{vmatrix}$$

(10)

with

$$M_{m+1} = \begin{vmatrix} \exp(i\delta_{m+1}) & r_{m+1}\exp(-i\delta_{m+1}) \\ r_{m+1}\exp(i\delta_{m+1}) & \exp(-i\delta_{m+1}) \end{vmatrix} \quad (11)$$

Similarly one obtains from (5)

$$\begin{vmatrix} E_s^+ \\ E_s^- \end{vmatrix} = \frac{1}{t_{s+1}} \begin{vmatrix} 1 & r_{s+1} \\ r_{s+1} & 1 \end{vmatrix} \cdot \begin{vmatrix} E_{s+1}^+ \\ 0 \end{vmatrix} \quad (12)$$

For the complete structure composed of s layers on a substrate we have the final result

$$\begin{vmatrix} E_o^+ \\ E_o^- \end{vmatrix} = \left[\prod_{j=1}^{s+1} t_j \right]^{-1} \cdot \left[\prod_{j=1}^{s} M_j \right] \cdot \begin{vmatrix} 1 & r_{s+1} \\ r_{s+1} & 1 \end{vmatrix} \cdot \begin{vmatrix} E_{s+1}^+ \\ 0 \end{vmatrix} \quad (13)$$

From (13) one can determine the ratio of the reflected power R_1 on the first interface to the incident power

$$R_1 = |E_o^-/E_o^+|^2 \quad (14)$$

and the ratio of the transmitted power T_{s+1} in the substrate to the incident power on the first interface

$$T_{s+1} = (n_{s+1}/n_o) \, |E_{s+1}^+/E_o^+|^2 \quad (15)$$

Anti-reflection silicon dioxide film on a silicon substrate

As a first example let us consider the case of an Air - SiO_2 - Si structure. The single oxide film on top of the semi-infinite silicon substrate has a thickness d_1; its index of refraction is assumed real and equal to 1.46 [5, 6, 7] :

$$N_o = n_o = 1 \quad , \quad N_1 = n_1 = 1.46 \quad , \quad k_o = k_1 = 0$$

According to the general relation (13) we obtain for s = 1

$$\begin{vmatrix} E_o^+ \\ E_o^- \end{vmatrix} = (t_1 t_2)^{-1} \begin{vmatrix} \exp(i\delta_1)+r_1 r_2 \exp(-i\delta_1) & r_2 \exp(i\delta_1)+r_1 \exp(-i\delta_1) \\ r_1 \exp(i\delta_1)+r_2 \exp(-i\delta_1) & r_1 r_2 \exp(i\delta_1)+\exp(-i\delta_1) \end{vmatrix} \cdot \begin{vmatrix} E_2^+ \\ 0 \end{vmatrix} \quad (16)$$

In the case of a non absorbing medium of thickness d_1, we have

$$\delta_1 = (2\pi/\lambda) n_1 d_1$$
$$r_1 = (n_o - n_1)/(n_o + n_1)$$
$$r_2 = (n_1 - n_2 + i k_2)/(n_1 + n_2 - i k_2) = g_2 + i h_2$$
$$g_2 = (n_1^2 - n_2^2 - k_2^2)/[(n_1+n_2)^2 + k_2^2]$$
$$h_2 = 2 n_1 k_2 / [(n_1+n_2)^2 + k_2^2]$$
$$t_1 = 2 n_o/(n_o + n_1) = 1 + r_1$$
$$t_2 = 2 n_1/(n_1+n_2-ik_2) = 1 + g_2 + i h_2$$

From (16) one calculates the ratio of the reflected power R_1 on the first interface to the incident power

$$R_1 = |E_o^- / E_o^+|^2$$

Taking the previous definitions into account one obtains

$$R_1 = \frac{r_1^2 + g_2^2 + h_2^2 + 2 r_1 (h_2 \sin 2\delta_1 + g_2 \cos 2\delta_1)}{1 + r_1^2 (g_2^2 + h_2^2) + 2 r_1 (g_2 \cos 2\delta_1 + h_2 \sin 2\delta_1)} \quad (17)$$

In a similar way one obtains the ratio of the transmitted power T_2 in the substrate to the incident power on the fist interface

$$T_2 = (n_2/n_0) \cdot |E_2^+/E_0^+|^2$$

Its value is

$$T_2 = \frac{n_2}{n_0} \cdot \frac{(1+r_1)^2 \left[(1+g_2)^2 + h_2^2\right]}{1 + r_1^2 (g_2^2 + h_2^2) + 2r_1(g_2\cos 2\delta_1 + h_2\sin 2\delta_1)}$$

(18)

Clearly the transmittance T_2 is a periodic function of the thickness d_1 of the non absorbing thin film; the period $\lambda/2n_1$ depends on the wavelength λ of the incident light and on the corresponding value n_1 of the real part of the index of refraction of the film. This implies that an antireflecting coating of thickness d_1 will only be active within a finite wavelength range.

The extreme values of T_2 considered as function of d_1 are obtained when

$$g_2 \sin 2\delta_1 = h_2 \cos 2\delta_1$$

Solving for the thickness of the thin film one gets the particular values $d_{1,m}$ for which T_2 becomes maximum or minimum at a given wavelength λ :

$$d_{1,m} = (\lambda/4\pi n_1) \tan^{-1}\left[2n_1 k_2/(n_1^2 - n_2^2 - k_2^2)\right]$$

Figure 2 shows, for the case of a silicon substrate covered with a silicon dioxide film of thickness d_1, the transmittance T_2 calcultad from (18) as a function of the thickness d_1 for different wavelengths. The real and imaginary part of the index of refraction of silicon used for computation [8, 9, 10] are shown in Figure 3. Clearly the commonly used oxide thickness of $d_1 = 0.1$ μ maximizes the transmittance over a wide spectral range centred around $\lambda = 0.6$ μ . By taking larger values of d_1, one can optimize the transmittance within a narrow wavelength band. The last fact is clearly illustrated by Figure 4 in which the values of $d_{1,m}$ given by (19) are plotted as a function of the wavelength λ . For example the transmittance T_2 at $\lambda = 0.6$ μ attains a maximum for $d_1 = 0.308$ μ ; at $\lambda = 0.9$ μ and $\lambda = 0.45$ μ T_2 becomes minimum for the same value of d_1.

From these figures we conclude that a judicious choice of the thickness of the single thin film can enhance the transmittance considerably; the particular choice depends on the application under consideration. Fig. 5 shows the transmittance as a function of the wavelength λ for different values of d_1.

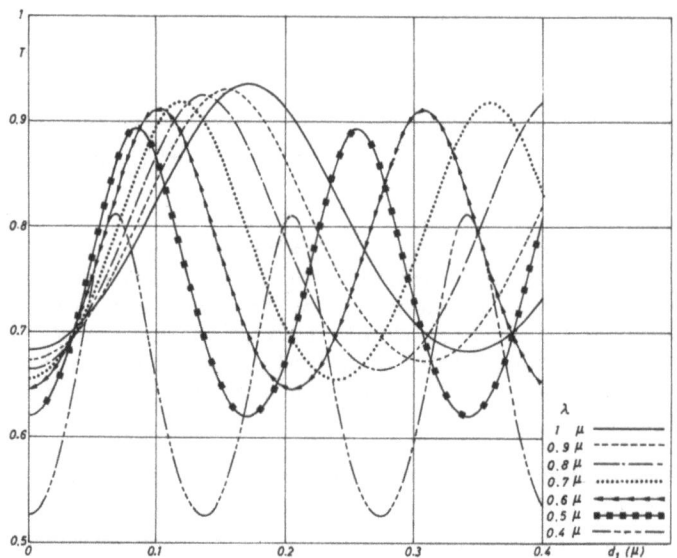

Fig. 2   Transmittance as a function of the thickness d_1 of the SiO_2 layer on a Si substrate for different wavelengths λ.

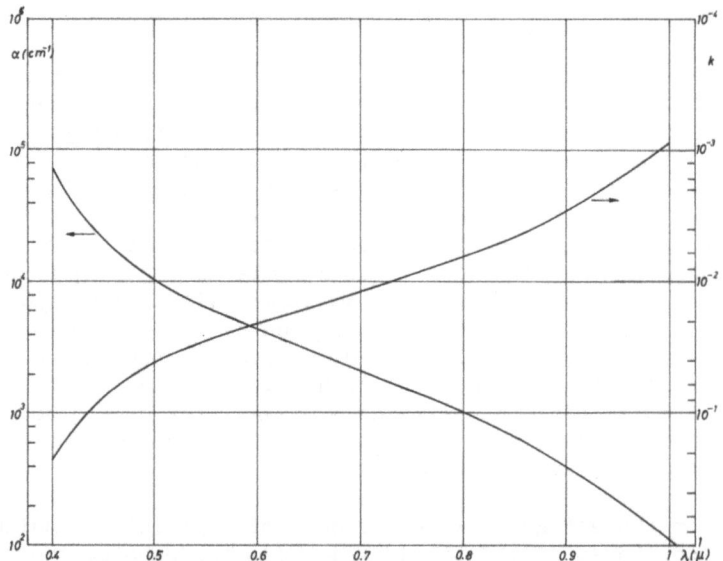

Fig. 3.A.   Absorption coefficient α and extinction coefficient k of Si vs. wavelength λ.

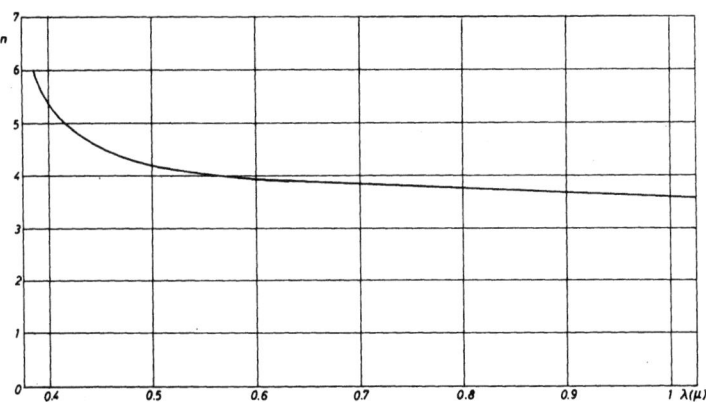

Fig. 3.B. Real part n of the index of refraction of Si vs. wavelength λ.

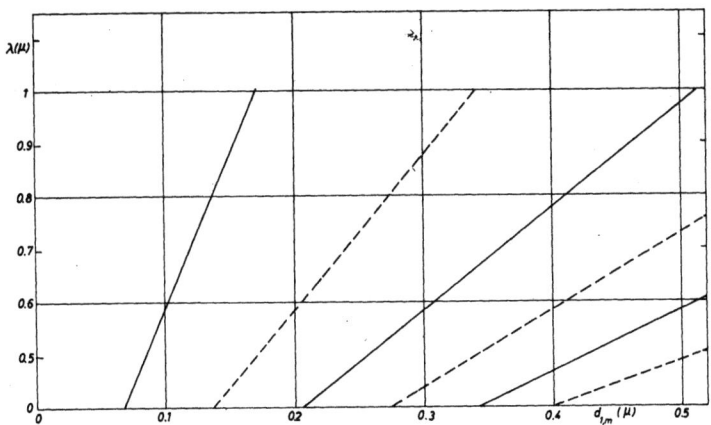

Fig. 4. Dependence of $d_{1,m}$ on wavelength λ for Si; —— maximum transmittance; --- minimum transmittance.

The uncoated Si substrate corresponds to $d_1=0$. Narrow wavelength bands of transmission may be selected by increasing the value of d_1.

Finally we pay attention to the influence of the real part of the index of refraction of the thin film. Consider the ideal case of low light absorption in the substrate, i.e. assume the influence of h_2 on T_2 is negligible. In this case a simple formula can be deduced from (18) for $h_2=0$:

$$T_2 = (n_2/n_0) \cdot (1+r_1)^2 \cdot (1+r_2)^2 / \left[1+r_1^2 r_2^2 + 2r_1 r_2 \cos 2\delta_1 \right]$$

with

$$r_2 = (n_1-n_2)/(n_1+n_2)$$

The reflection R_1 on the first interface can be totally cancelled and the transmittance T_2 made equal to unity by a judicious choice of the index of refraction n_1 and the thickness d_1 of the thin film:

$$n_1 = \sqrt{n_0 n_2} \qquad (20)$$

$$d_{1,m} = (2l+1)\,\lambda/4n_1 \qquad (l = 1, 3, 5, \ldots) \qquad (21)$$

It is worthwhile to note that for the case of an Air-SiO_2-Si structure the condition (20) is not satisfied; however the expression (21) is an excellent approximation for the general expression (19). This is due to the fact, as noticed by C. Anagnostopoulos and G. Sadasiv [4], that the extinction coefficient k of silicon is much smaller than the difference of the refractive indices of silicon and silicon dioxide; since no noticeable phase shift occurs due to k, the positions of the maxima and minima of T_2 vs. λ are not affected by the values of k for the case of silicon.

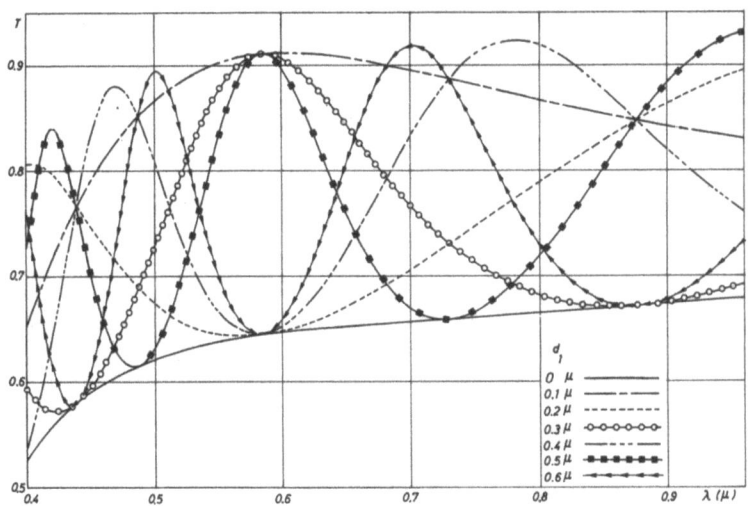

Fig. 5. Transmittance of a SiO_2 layer of thickness d_1, on a Si substrate, as a function of the wavelength λ.

Multilayer structure on silicon substrate

The transmittance of Silicon gate CCD or CID has been treated by R.W. Brown and S.G. Chamberlain [3] ; a detailed study was made by C. Anagostopoulos and G. Sadasiv [4]. The last authors have derived general expressions, particularly suited for implementation in a digital computer, for the reflectance and transmittance of multilayer structures. In this section we first briefly present some of their results.

According to the previous authors the structure Air-SiO_2-Polysilicon-SiO_2-Si shows an optimum transmittance in the visible region of the spectrum when an 0.1 µ SiO_2 film on top of the polysilicon film is used. However, for a polysilicon film of 0.15 µ, the thickness of the gate oxide must be taken equal to 0.18 µ to produce an overall enhancement of the transmittance. Furthermore the thickness of the polysilicon film should be minimized, since a considerable amount of incident visible radiation is absorbed within it.

We have used the scheme presented by Anagnostopoulos and Sadasiv [4] to compute from the general expressions (15) and (13) the transmittance of the structure Air-SiO_2-Polysilicon-SiO_2-Si for different thicknesses of the polysilicon film, and of the oxide films.

We found for example that an optimum transmittance in the region of λ = 0.6 µ can be obtained with the following figures :

thickness of top oxide d_1 = 0.1 + (n x 0.206)µ (n=0,1,2,...)
thickness of polysilicon d_2 = 0.06 + (n x 0.0765)µ (n=0,1,2,...)
thickness of gate oxide d_3 = 0.19 + (n x 0.206)µ (n=0,1,2,...)

Figures 6 and 7 show examples of the resulting transmittance; clearly the number of maxima and minima within the wavelength interval from 0.4 to 0.9 µ increases for increasing values of the thickness of d_2 of the polysilicon film while the transmittance at λ = 0.6 µ slowly decreases. Furthermore the optimization of the transmittance at a given wavelength may introduce an unwanted maximum transmission at an other wavelength.

Figure 8 shows for an optimum set of parameters (d_1, d_2, d_3) the variation of the transmittance at λ = 0.6 µ as a function of one of the parameters. Clearly if an optimum choice is made for d_1 and d_2, a deviation of the thickness d_3 of the gate oxide from its optimum value may drastically alter the transmission at λ = 0.6 µ. This implies technological limitations to maintain a good transmission in the wavelength range under consideration.

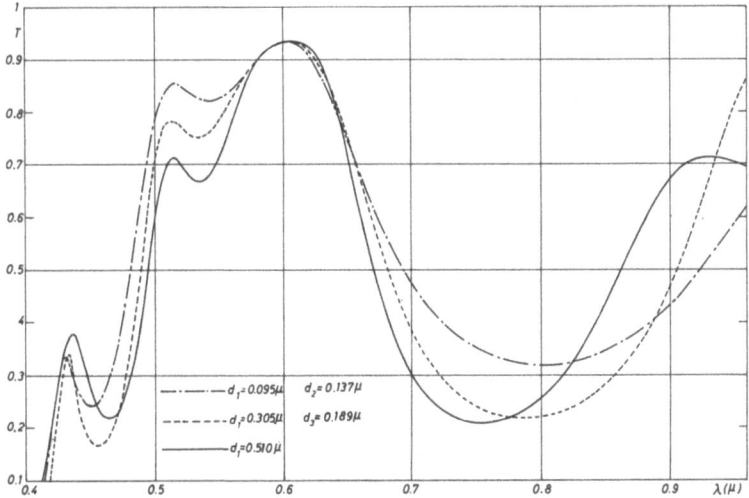

Fig. 6. Transmittance vs. wavelength λ of the structure Air-SiO_2 (d_1)-Polysilicon (d_2) - SiO_2 (d_3) on a Si substrate for three values of d_1 and constant values of d_2 and d_3.

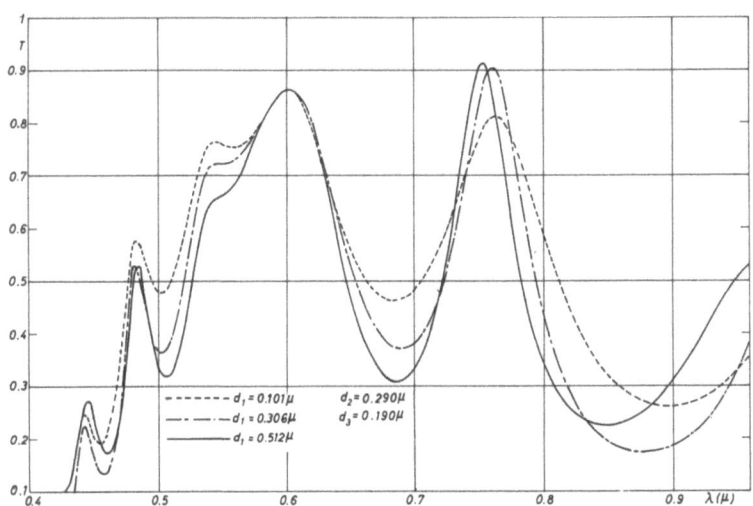

Fig. 7. Transmittance vs. wavelength λ of the structure Air-$SiO_2(d_1)$-Polysilicon (d_2) - $SiO_2(d_3)$ on a Si substrate for three values of d_1 and constant values of d_2 and d_3.

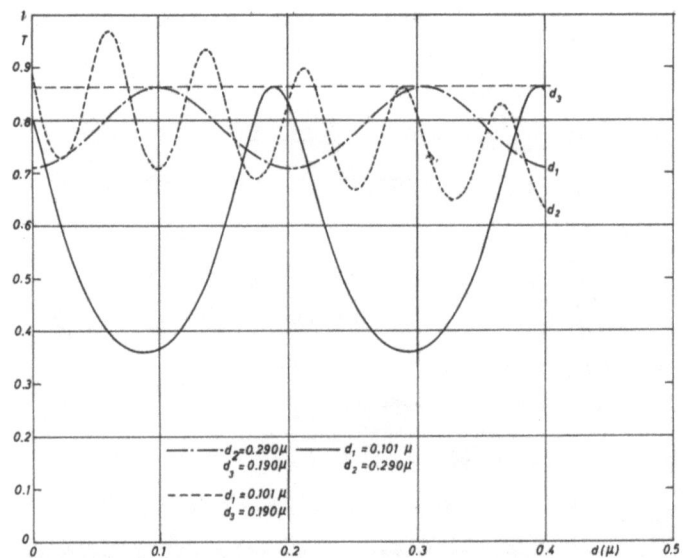

Fig. 8. Transmittance at $\lambda = 0.6$ μ of the structure Air-$SiO_2(d_1)$-Polysilicon (d_2)-$SiO_2(d_3)$ on a Si substrate vs. one of the parameters (d_1, d_2, d_3).

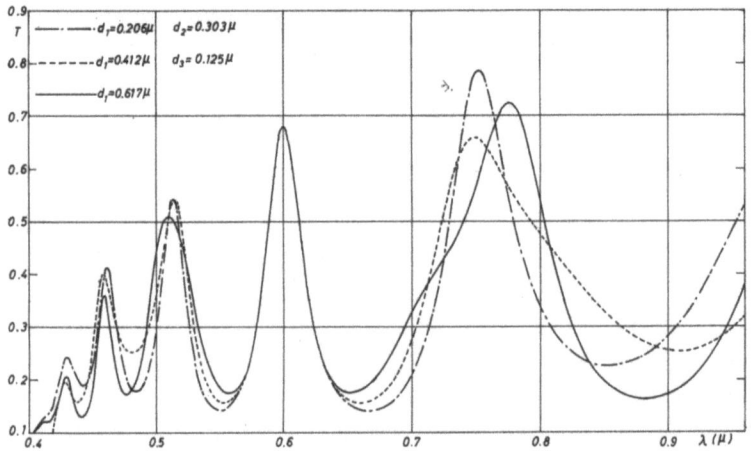

Fig. 9. Transmittance vs. wavelength λ of the structure Air-$SiO_2(d_1)$-Polysilicon(d_2)-$SiO_2(d_3)$ on a Si substrate for three values of d_1 and constant values of d_2 and d_3.

The thickness of 0.19 µ for the gate oxide used in the previous examples may be unpractible for technological reasons. Therefore we have calculated as an example the optimum values of d_1 and d_2 for a fixed oxide thickness of 0.125 µ . We found for $\lambda = 0.6$

$$d_1 = 0.206 + (n \times 0.206)\mu \qquad (n = 0, 1, 2, ...)$$

$$d_2 = 0,073 + (n \times 0.0765)\mu \qquad (n = 0, 1, 2, ...)$$

Figure 9 shows the resulting transmittance for a few values; compared to the ideal case of figure 7 the transmission band around $\lambda = 0.6$ µ is much sharper while its maximum is lower. Finally figure 10 shows the variation of the transmittance at $\lambda = 0.6$ µ as a function of one of the parameters (d_1, d_2, d_3), the two other ones remaining fixed. In this case, for fixed values of $d_1 = 0.206$ µ and $d_3 = 0.125$ µ , any deviation of the thickness d_2 of the polysilicon film from its optimum value drastically modifies the transmittance.

For the particular case of multilayer structures with a polysilicon film the multiple reflections at the interfaces cause rapid variations of the transmission in the range $\lambda = 0.4$ to 1 µ , especially for polysilicon films larger than 0.3 µ [2] . The mismatch between the index of refraction of polysilicon and the index of silicon dioxide enhances the multiple interference characteristics for film thicknesses comparable to the wavelengths of visible light [11] . Furthermore the transmission for wavelenghts below $\lambda = 0.55$ µ decreases rapidly. These factors imply a serious limitation for the use of polysilicon multilayer structures in the visible range of the light spectrum [1] .

Recently D.M. Brown, H. Ghezzo and M. Garfinkel reported the use of transparent metal oxide (tin, antimony, indium oxide) electrodes in CID imaging arrays [11] . According to these authors improvement of the transmission of the multilayer structure results from the following physical properties.
The index of refraction of the metal oxide ($\simeq 2$) is a close match to silicon dioxide ($\simeq 1.46$) and also to silicon nitride ($\simeq 2$). Metal oxides have a high transmittivity ($\simeq 80\%$) in the range 0.4 to 0.9 µ .
The electrical conductivity of the metal oxides is as good as that of doped polysilicon.
The figures reported by the previous authors show a nearly uniform optical transmission for the multilayer structure in the range of $\lambda = 0.4$ to 0.8 µ.

We conclude that the general expressions (15) and (13) for the transmission of multilayer structures allow to specify the conditions of an optimum transmission in a given wavelength interval, taking the physical properties of the layers into account and also the technological limitations.

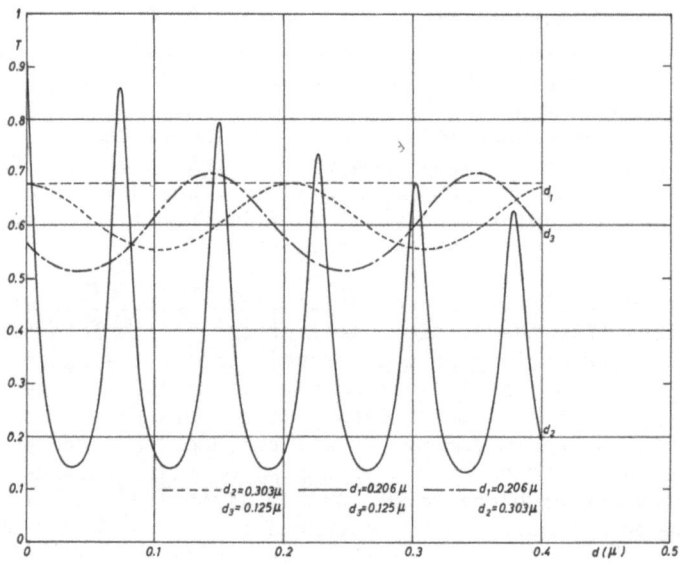

Fig. 10 Transmittance at $\lambda = 0.6$ µ of the structure Air-$SiO_2(d_1)$-Polysilicon(d_2)-$SiO_2(d_3)$ on a Si substrate vs. one of the parameters (d_1, d_2, d_3).

1   C.H. Sequin, F.J. Morris, T.A. Shankoff, M.F. Tompsett and E.J. Zimany, IEEE Trans. Electron Devices, ED-21, pp. 712-720, (1974).

2   R.D. Melen and J.C. Meindl, IEEE J. Solid-State Circuits, SC-9, pp. 41-49, (1974).

3   R.W. Brown and S.G. Chamberlain, Physica Status Solidi (a)-20, pp. 675-685,(1973).

4   C. Anagnostopoulos and G. Sadasiv, IEEE J. Solid-State Circuits, pp. 177-179, June (1975).

5   O.S. Heavens, Optical properties of thin solid films, Butterworth, London (1955).

6   F. Reizman and W. Van Gelder, Solid-State Electron., 10, pp. 625-632, (1967).

7   E.A. Corl and H. Wimpfheimer, Solid-State Electron., 7, pp. 755-761, (1964).

8   W.C. Dash and R. Newman, Phys. Rev. 99, pp. 1151-1155, (1955).

9   H.R. Philipp and E.A. Taft, Phys. Rev. 120, pp. 37-38, (1960).

10 C.D. Salzberg, J.J. Villa, J. Opt. Soc. Am., 47, pp. 244-246, (1957).

11 D.M. Brown, M. Ghezzo and M. Garfinkel, ISSCC Digest of Technical Papers, p. 34-35, (1975).

Acknowledgment

The author thanks C. Anagnostopoulos and G. Sadasiv for making a preprint of their article available, and A.M. Trullemans for writing the computer program.

PHOTODIODE QUANTUM EFFICIENCY

by

F. Van de Wiele

Laboratoire de Microélectronique, Louvain-la-Neuve, Belgium.

Abstract

The concept of quantum efficiency η for a photodiode is introduced and a scheme is presented allowing the determination of η as a function of the wavelength of the incident light and the structural parameters of the photodiode.
The particular cases of field induced junctions and abrupt junctions are discussed in detail, using appropriate one-dimensional approximations.

Notation

- c   velocity of light.
- c_n probability for capture of an electron by an unfilled center.
- c_p probability for capture of a hole by a filled center.
- $g(x)$ optical generation rate at x.
- h   Planck's constant.
- k   Boltzmann's constant.
- n_1 electron concentration when the Fermi level falls to the recombination level.
- n_i intrinsic carrier concentration.
- n_o equilibrium electron concentration.
- n_d dark current concentration.
- n_{ph} excess concentration.
- n_r concentration of electron-occupied recombination centers.
- p_1 hole concentration when the Fermi level falls to the recombination level.
- p_o equilibrium hole concentration.

q   charge of an electron.
p_d dark current concentration.
p_{ph} excess concentration.
s   surface recombination velocity.
x_j depth of the metallurgical junction.
x_n boundary of the depletion region in the n region.
x_p boundary of the depletion region in the p region.
y_n lateral depletion width.
ξ electric field.
E_r recombination level.
D_n electron diffusion constant.
D_p hole diffusion constant.
G_n external generation rate of electrons.
G_p external generation rate of holes.
I(x)   radiant flux density at x.
$I_o(\lambda)$ radiant flux density at wavelength λ at x = o.
J   total current density.
J_d dark current density.
J_{ph} photocurrent density.
L   thickness of the substrate or of the epitaxial layer.
L'   lateral thickness of the substrate.
L_n diffusion length of electrons.
L_p diffusion length of holes.
N_A acceptor concentration.
N_D donor concentration.
N_r concentration of recombination centers.
T   temperature.
U_n electron recombination rate.
U_p hole recombination rate.
V_a applied voltage.
α absorption coefficient.
ε permittivity of the semiconductor.
λ free space wavelength of light.
μ_n electron mobility.

μ_p hole mobility.
η total quantum efficiency.
η_{DR} partial quantum efficiency of the depletion region.
η_n partial quantum efficiency of the n region.
η_p partial quantum efficiency of the p region.
τ_{no} electron lifetime.
τ_{po} hole lifetime.
ψ electrostatic potential.
ψ_b built-in potential.

Photodiode quantum efficiency

The purpose of this section is to introduce the concept of quantum efficiency η for a photodiode and to present a scheme allowing the determination of η as a function of the wavelength λ of the incident light and the structural parameters of the photodiode.
The particular cases of field induced junctions and abrupt junctions will be studied in detail using appropriate approximations.

Consider the case of a one-dimensional p-n photodiode, as shown in Figure 1. The metallurgical junction is located at a depth x_j from the surface on which light is incident. Assume that the surface region is p type and that the n type substrate is bounded by an ohmic contact at $x = L$. The length L may correspond to the thickness of the substrate or of an epitaxial layer. The intensity of the incident light at $x = 0$ is supposed to be constant over the entire area of the photodiode; denote by $I_o(\lambda)$ the radiant flux density at wavelength λ within the semiconductor at $x = 0$. The light absorption mechanisms of the semiconductor, characterized by the wavelength dependent absorption coefficient $\alpha(\lambda)$, yield an exponentially decreasing radiant flux density $I(x)$ as a function of the depth x within the semiconductor :

$$I(x) = I_o \cdot e^{-\alpha x}$$

The decrease of $I(x)$ within an elementary interval dx around the point x is

$$dI(x) = - \alpha I_o e^{-\alpha x} dx$$

The corresponding number of photons of wavelength λ disappearing by absorption in the same interval is

$$\alpha I_0 e^{-\alpha x} \lambda dx/hc$$

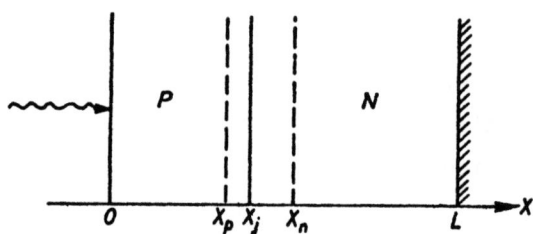

Fig. 1 One-dimensional photodiode.

h being Planck's constant and c the velocity of light. We assume, for simplicity, that each disappearing photon produces a single electron-hole pair, i.e. we assume a quantum yield equal to unity. In this case, the optical generation rate g(x)dx of electrons and holes within the interval dx of the substrate at a distance x beneath the surface is given by

$$g(x) \, dx = \alpha I_0 \lambda \, e^{-\alpha x} \, dx/hc$$

We now define the quantum efficiency η of the photodiode. Assume a dark current density $J_d(V_a)$ is flowing through the junction when a voltage difference V_a is applied in the absence of incident light. Denote by $J(V_a)$ the current density when the constant external generation source is switched on; one may write :

$$J(V_a) = J_d(V_a) + J_{ph}(V_a)$$

where $J_{ph}(V_a)$ is called the photocurrent due to the incident radiant flux density $I_0(\lambda)$ at x = 0 within the semiconductor. By definition the quantum efficiency of the photodiode is the ratio :

$$\eta = - h c J_{ph}(V_a)/qI_0 \lambda$$

In order to better understand the meaning of η , consider the following case. Assume the incident light is totally absorbed within the semiconductor and that each generated electron-hole pair is perfectly collected by the junction. In this ideal case the resulting photocurrent is $J_{ph} = - qI_0 \lambda/hc$ and, according to its definition, the quantum efficiency η is unity.

In practice η is smaller than unity because

- not every absorbed photon produces an electron-hole pair, i.e. the quantum yield is smaller than unity;
- not all photons of wavelength λ are absorbed within the semiconductor having a finite lenght L;
- not all generated carriers are collected by the junction, since they have a finite lifetime and consequently may suffer a recombination process before reaching the active collection area of the junction.

In order to calculate the quantum efficiency of a photodiode, we have to solve the following continuity equations for the electron n and hole p density.

$$\partial n/\partial t = G_n - U_n + (1/q) \text{ div } J_n$$

$$\partial p/\partial t = G_p - U_p - (1/q) \text{ div } J_p$$

where U_n (U_p) is the electron (hole) recombination term, J_n (J_p) the electron (hole) current density and G_n (G_p) the external generation rate of electrons (holes). We assume that the recombination process occurs through recombination centers located at a level E_r within the energy gap of the semiconductor. The continuity equation for the density n_r of electron-occupied recombination centers is

$$\partial n_r/\partial t = G_p - G_n - U_p + U_n$$

According to the Shockley-Read single recombination level model [1], U_n and U_p are given by,

$$U_n = c_n \left[n (N_r - n_r) - n_1 n_r \right]$$

$$U_p = c_p \left[p n_r - p_1 (N_r - n_r) \right]$$

for nondegenerate semiconductors, where c_n (c_p) is the probability for capture of an electron (hole) by an unfilled (filled) center, N_r is the density of recombination centers, and n_1 (p_1) is the density of free electrons (holes) when the Fermi level falls to the recombination level E_r. Under steady conditions, div $(J_n + J_p) = 0$, and

$$G_n - G_p = U_n - U_p$$

From this relation we may solve for n_r as follows [2]:

$$n_r = N_r \frac{n\tau_{po} + p_1 \tau_{no} + (G_p - G_n)\tau_{no}\tau_{po}}{\tau_{no}(p+p_1) + \tau_{po}(n+n_1)}$$

where $\tau_{no} = (c_n N_r)^{-1}$ and $\tau_{po} = (c_p N_r)^{-1}$. If we substitute this value back into U_n and U_p, we obtain

$$U_n = \frac{pn - n_i^2 - (n+n_1)(G_p - G_n)\tau_{po}}{\tau_{no}(p+p_1) + \tau_{po}(n+n_1)}$$

$$U_p = \frac{pn - n_i^2 + (p+p_1)(G_p - G_n)\tau_{no}}{\tau_{no}(p+p_1) + \tau_{po}(n+n_1)}$$

where n_i is the intrinsic carrier density. The electron and hole lifetime are defined by,

$$\tau_n = (n-n_o)/U_n \quad \text{and} \quad \tau_p = (p-p_o)/U_p$$

where $n_o(p_o)$ is the equilibrium electron (hole) density. Notice that the lifetime under steady-state conditions is a function of the external generation terms if G_n is different from G_p [2].

For the photodiode under consideration we assume, for simplicity, that the optical generation of carriers occurs by band to band transitions, i.e. that the generating source creates free carriers by exciting electrons from the valence band into the conduction band. In this case equal number of electron and holes are created and

$$G = G_n = G_p \tag{1}$$

(for the case of generation from trap levels, see for example Bube [3] and Milnes [4]. The electron and hole recombination terms are then given by

$$U = U_n = U_p = \frac{pn - n_i^2}{\tau_{no}(p+p_1) + \tau_{po}(n+n_1)} \tag{2}$$

and the continuity equations under steady-state conditions become

$$\text{div } J_n = q(U - G) \tag{3}$$

$$\text{div } J_p = -q(U - G) \tag{4}$$

The current densities have a drift component, due to the electric field $\mathscr{E}$, and a diffusion component, due to the gradient of carriers,

$$J_n = q(n \mu_n \mathscr{E} + D_n \text{ grad } n) \tag{5}$$

$$J_p = q(p \mu_p \mathscr{E} - D_p \text{ grad } p) \tag{6}$$

$\mu_n (\mu_p)$ is the electron (hole) mobility and $D_n (D_p)$ the electron (hole) diffusion constant. For nondegenerate semiconductors the following Einstein relation is valid

$$D_{n,p} = \mu_{n,p}(kT/q) \tag{7}$$

k being Boltzmann's constant and T the temperature. Finally the electric field $\mathscr{E}$ is related to the local space charge density by the Poisson relation

$$\text{div } \mathscr{E} = (q/\varepsilon) (p - n + N_D - N_A) \tag{8}$$

where ε is the permittivity of the substrate and $N_D (N_A)$ the density of ionized donor (acceptor) impurities.

The system of equations (1) to (8) is quite complex, even for a one-dimensional structure. In general it is impossible to solve the system by analytical methods; a numerical analysis is required even for simple doping profiles. However, by making a number of simplifying assumptions, one may for some cases derive analytical expressions closely related to the exact numerical solutions.

Since our purpose in this section is to pay attention to the physical parameters influencing the quantum efficiency η of a photodiode and not to compute it exactly for a particular structure, we will consider the simple case of a one-dimensional abrupt junction using classical approximations.

One-dimensional study of an abrupt photodiode

In this section the abrupt photodiode is studied by means of a simple classical model; however more elaborate models exists in the literature (see for example [5]).
Assume the substrate ($x_j < x < L$) of the photodiode (see Fig.1) has a constant density N_D of donor impurities, while the surface region is uniformly doped by a density N_A of acceptor impurities. In order to calculate the distributions of the electric field $\mathcal{E}$ and the electrostatic potential ψ for the one-dimensional structure under consideration, the following classical approximations are used.

The device is divided into a depletion region ($x_p < x < x_n$), surrounding the metallurgical junction ($x = x_j$), and two quasi-neutral regions, respectively (0, x_p) and (x_n, L).
The light incident on the photodiode generates per unit time at a distance x from the surface an equal number of electrons and holes given by

$$G(x) = G_n(x) = G_p(x) = \alpha\, I_o \cdot \lambda \cdot e^{-\alpha x}/hc \qquad (9)$$

Within each quasi-neutral region we assume that the density of excess carriers, generated by the light source or injected as a result of the external applied voltage, remains small compared to the equilibrium majority carrier density, i.e. the case of low injection of excess carriers is considered.

The previous assumptions imply that the majority carrier density within the quasi-neutral surface region (0, x_p) is almost equal to its equilibrium value and that the electrostatic potential ψ may be considered as constant and equal to its equilibrium value which is taken equal to zero.

$$\psi(x) = 0 \qquad \text{for } 0 < x < x_p \qquad (10)$$

This relation implies a zero electric field

$$\mathcal{E}(x) = -d\psi/dx = 0 \qquad \text{for } 0 < x < x_p \qquad (11)$$

In fact under non-equilibrium conditions a small non-zero electric field exists within the quasi-neutral region and even is necessary to insure the drift current of the majority carriers as a result of the external applied voltage V_a. However, for the calculation of the electrostatic potential within an abrupt junction this small electric field may be neglected in a first approximation (quasi-equilibrium approximation).

The quasi-neutral region (x_n, L) may be treated in a similar way :

$$\psi(x) = -(V_a + \psi_b) \qquad (12)$$

$$\mathcal{E}(x) = 0 \qquad \text{for } x_n < x < L \qquad (13)$$

ψ_b is the built-in potential of the abrupt junction given by

$$\psi_b = (kT/q) \ln (n_i^{-2}/N_A N_D) \qquad (14)$$

Within the interval (x_p, x_n) the depletion approximation is used, i.e. the density of free carriers is supposed negligible compared to the net density of ionized impurities which, at room temperature and for nondegenerate semiconductors, is equal to the net density of impurities :

$$|p(x) - n(x)| \ll N_A \qquad \text{for } x_p < x < x_j \qquad (15)$$

$$|p(x) - n(x)| \ll N_D \qquad \text{for } x_j < x < x_n \qquad (16)$$

According to Poisson's equation one obtains for the interval (x_p, x_j)

$$d^2\psi/dx^2 = qN_A/\epsilon \qquad (17)$$

$$d\psi/dx = (qN_A/\epsilon)(x-x_p) \qquad (18)$$

$$\psi = (qN_A/2\epsilon)(x-x_p)^2 \qquad (19)$$

Similarly for the interval (x_j, x_n) we get

$$d^2\psi/dx^2 = -qN_D/\epsilon \qquad (20)$$

$$d\psi/dx = -(qN_D/\epsilon)(x-x_n) \qquad (21)$$

$$\psi = -(qN_D/2\epsilon)(x-x_n)^2 - V_a - \psi_b \qquad (22)$$

The continuity conditions of the electric field and the electrostatic potential at the metallurgical junction x_j impose two relations :

$$N_A(x_j - x_p) = N_D(x_n - x_j)$$

$$(q/2\epsilon)N_D \cdot \left[1 + (N_D/N_A)\right](x_n - x_j)^2 = -V_a - \psi_b$$

from which the boundaries of the depletion region may be calculated

$$x_n = x_j + \left[-2\epsilon N_A(V_a + \psi_b)/q N_D(N_D + N_A)\right]^{1/2} \quad (23)$$

$$x_p = x_j - \left[-2\epsilon N_D(V_a + \psi_b)/q N_A(N_D + N_A)\right]^{1/2} \quad (24)$$

A. <u>Quasi-neutral region</u> $(0, x_p)$

Knowing the distribution of the electric field and of the potential within the device, we now calculate the minority carrier distribution by solving the corresponding continuity equation. The current density of minority carriers within the quasi-neutral region $(0, x_p)$ is essentially a diffusion current, since the electric field in this region is extremely small :

$$J_n(x) = q D_n (dn/dx) \quad (25)$$

After substitution of this expression into the continuity equation (3) one obtains

$$D_n(d^2n/dx^2) = U - \alpha I_0 \lambda e^{-\alpha x}/hc \quad (26)$$

The recombination term U given by (2) may, for the minority carriers within the interval $(0, x_p)$, be approximated by

$$U = (n - n_0)/\tau_{no} \quad (27)$$

since $p(x) \simeq p_0 = N_A \gg n_i \gg n(x)$.

The following boundary conditions are imposed to the continuity equation (26) :

at $x = 0$: $\quad J_n(o) = q s \cdot \left[n(o) - n_0\right] \quad (28)$

at $x = x_p$: $\quad n(x_p) = n_0 \cdot e^{qV_a/kT} \quad (29)$

The first condition accounts for the surface recombination velocity s characterizing the semiconductor surface [6] . The second condition is the classical Boltzmann equation valid under low injection conditions.

Straightforward calculations allow the determination of the minority carrier distribution. In the absence of an external generation source ($I_o = 0$) the electron distribution is

$$n_d(x) = n_o + n_o(e^{qV_a/kT} - 1) \cdot \frac{sL_n \sinh(x/L_n) + D_n \cosh(x/L_n)}{sL_n \sinh(x_p/L_n) + D_n \cosh(x_p/L_n)}$$

where

$$L_n = (D_n \tau_{no})^{1/2}$$

is the diffusion length of electrons. The resulting dark current density of electron is, according to (25),

$$J_{n,d}(x) = (qD_n n_o/L_n)(e^{qV_a/kT} - 1) \frac{sL_n \cosh(x/L_n) + D_n \sinh(x/L_n)}{sL_n \sinh(x_p/L_n) + D_n \cosh(x_p/L_n)}$$

The minority carrier density, when light is incident on the junction, is denoted by

$$n(x) = n_d(x) + n_{ph}(x)$$

where $n_{ph}(x)$ represents the local excess carrier density resulting from the generation source. From the continuity equation one obtains, for $\alpha L_n \neq 1$,

$$n_{ph}(x) = -\frac{\alpha I_o \lambda L_n^2}{D_n(\alpha^2 L_n^2 - 1) hc} \left\{ e^{-\alpha x} + \frac{L_n(s + \alpha D_n) \sinh[(x-x_p)/L_n] - e^{-\alpha x_p}[L_n s \sinh(x/L_n) + D_n \cosh(x/L_n)]}{sL_n \sinh(x_p/L_n) + D_n \cosh(x_p/L_n)} \right\} \quad (30)$$

The case $\alpha L_n = 1$ can be treated by taking the limit of (30) as $\alpha \to 1/L_n$. The resulting current density, due to the diffusion of minority carriers, is denoted by

$$J_n(x) = J_{n,d}(x) + J_{n,ph}(x)$$

$J_{n,ph}(x)$ is the photocurrent resulting from the generation and is given by,

$$J_{n,ph}(x) = \frac{q\,\alpha I_0 \lambda L_n^2}{(\alpha^2 L_n^2 - 1)\,hc} \left\{ \alpha\, e^{-\alpha x} \right.$$

$$\left. - \frac{(s+\alpha D_n)\cosh\left[(x-x_p)/L_n\right] - (e^{-\alpha x_p}/L_n)\left[L_n s \cosh(x/L_n) + D_n \sinh(x/L_n)\right]}{s L_n \sinh(x_p/L_n) + D_n \cosh(x_p/L_n)} \right\} \quad (31)$$

Its particular value at the boundary x_p of the depletion region is

$$J_{n,ph}(x_p) = \frac{q\,\alpha I_0 \lambda L_n^2}{(\alpha^2 L_n^2 - 1)\,hc} \left\{ \alpha\, e^{-\alpha x_p} \right.$$

$$\left. - \frac{(s+\alpha D_n) - (e^{-\alpha x_p}/L_n)\left[L_n s \cosh(x_p/L_n) + D_n \sinh(x_p/L_n)\right]}{s L_n \sinh(x_p/L_n) + D_n \cosh(x_p/L_n)} \right\}$$

For the quasi-neutral region $(0, x_p)$ a partial quantum efficiency η_p may be defined by

$$\eta_p = - J_{n,ph}(x_p)\, hc/q\, I_0 \lambda$$

Its value is

$$\eta_p = \frac{\alpha L_n^2}{1 - \alpha^2 L_n^2} \left\{ \alpha\, e^{-\alpha x_p} \right.$$

$$\left. - \frac{1}{L_n} \cdot \frac{L_n(s+\alpha D_n) - e^{-\alpha x_p}\left[sL_n \cosh(x_p/L_n) + D_n \sinh(x_p/L_n)\right]}{sL_n \sinh(x_p/L_n) + D_n \cosh(x_p/L_n)} \right\}$$

$$(32)$$

The influence of η_p of the different parameters V_a, x_j, τ_{no}, N_A, s will be discussed in later sections. However J_{it} is worthwhile to consider already some limiting cases. For large values of the minority carrier lifetime τ_{no}, and hence of the diffusion length L_n, the partial quantum efficiency becomes,

$$\eta_p = \frac{s(1 - e^{-\alpha x_p}) + \alpha D_n}{\alpha(D_n + s\, x_p)} - e^{-\alpha x_p} \qquad (L_n \to \infty) \qquad (33)$$

Furthermore, for a zero recombination velocity s, the so called ideal case is obtained :

$$\eta_p = 1 - e^{-\alpha x_p} \qquad (L_n \to \infty,\ s=0) \qquad (34)$$

In this case the optimum value of η_p is attained, since all photogenerated minority carriers contribute to the photocurrent.

B. Quasi-neutral region (x_n, L)

The minority carrier distribution and the corresponding current density within the quasi-neutral region (x_n, L) may be treated in a similar way. However the boundary conditions for the continuity equation (4) of the holes are now :

$$p(x_n) = p_o \cdot e^{qV_a/kT}$$

$$p(L) = p_o$$

The last condition implies that an ohmic contact exists at $x = L$. The results are, for $\alpha L_p \neq 1$,

$$p(x) = p_d(x) + p_{ph}(x)$$

$$p_d(x) = p_o + p_o(e^{qV_a/kT} - 1) \cdot \sinh[(x-L)/L_p]/\sinh[(x_n-L)/L_p]$$

$$P_{ph}(x) = -\frac{\alpha I_0 \lambda L_p^2}{D_p(\alpha^2 L_p^2 - 1) hc} \left\{ e^{-\alpha x} \right.$$

$$\left. - \frac{e^{-\alpha L} \sinh\left[(x-x_n)/L_p\right] + e^{-\alpha x_n} \sinh\left[(L-x)/L_p\right]}{\sinh\left[(L-x_n)/L_p\right]} \right.$$

(35)

where $L_p = (D_p \tau_{po})^{1/2}$ is the diffusion length of holes. For the current density one obtains :

$$J_p(x) = J_{p,d}(x) + J_{p,ph}(x)$$

$$J_{p,d}(x) = -(qD_p p_0/L_p)(e^{qV_a/kT} - 1)\cosh\left[(x-L)/L_p\right]/\sinh\left[(x_n-L)/L_p\right]$$

$$J_{p,ph}(x) = -\frac{q\alpha I_0 \lambda L_p^2}{(\alpha^2 L_p^2 - 1) hc} \left\{ \alpha e^{-\alpha x} + \right.$$

$$\left. \frac{e^{-\alpha L}\cosh\left[(x-x_n)/L_p\right] - e^{-\alpha x_n}\cosh\left[(L-x)/L_p\right]}{L_p \sinh\left[(L-x_n)/L_p\right]} \right\}$$

(36)

The partial quantum efficiency η_n of the quasi-neutral region (x_n, L) is defined as

$$\eta_n = -J_{p,ph}(x_n)\, hc/qI_0\lambda$$

Its value is

$$\eta_n = \frac{\alpha L_p^2}{\alpha^2 L_p^2 - 1} \left\{ \alpha e^{-\alpha x_n} + \frac{e^{-\alpha L} - e^{-\alpha x_n} \cosh\left[(L-x_n)/L_p\right]}{L_p \sinh\left[(L-x_n)/L_p\right]} \right\}$$

(37)

For large values of the minority carrier lifetime τ_{po}, and hence of the diffusion length L_p, the partial quantum efficiency becomes

$$\eta_n = e^{-\alpha x_n} \left[1 - \frac{1 - e^{-\alpha(L-x_n)}}{\alpha(L-x_n)} \right] \qquad (L_p \to \infty)$$

(38)

On the other hand, for a thick substrate one obtains

$$\eta_n = \alpha L_p e^{-\alpha x_n}/(\alpha L_p + 1) \qquad (L \to \infty) \qquad (39)$$

Finally the so called ideal case occurs for a semi-infinite substrate and an infinite lifetime of the carriers :

$$\eta_n = e^{-\alpha x_n} \qquad (L, L_p \to \infty)$$

(40)

C. <u>Depletion region (x_p, x_n)</u>

Consider the continuity equation for electrons :

$$dJ_n/dx = q U - q \alpha I_0 \lambda e^{-\alpha x}/hc$$

The total electron current density within the depletion region

$$J_n(x) = J_{n,d}(x) + J_{n,ph}(x)$$

can be obtained by integration of the continuity equation :

$$J_n(x) = J_{n,d}(x_p) + J_{n,ph}(x_p) + q \int_{x_p}^{x} U \, dx + (qI_o \lambda /hc)(e^{-\alpha x} - e^{-\alpha x_p})$$

As mentioned before, the case of low carrier injection is considered. We may them assume that the recombination then $\int U dx$ is practically independent of the external generation source, i.e. that the dark current density is

$$J_{n,d}(x) = J_{n,d}(x_p) + q \int_{x_p}^{x} U \, dx$$

and that the photocurrent density due to the electrons is

$$J_{n,ph}(x) = J_{n,ph}(x_p) + (qI_o \lambda /hc)(e^{-\alpha x} - e^{-\alpha x_p})$$

In other words, we assume that all optically generated carriers effectively contribute to the photocurrent since the additional recombination, due to the generated carriers, is considered as negligible.

The contribution of the entire depletion region to the photocurrent density is

$$J_{DR,ph} = J_{n,ph}(x_n) - J_{n,ph}(x_p) = (qI_o \lambda /hc)(e^{-\alpha x_n} - e^{-\alpha x_p})$$

(41)

and one may define a corresponding partial quantum efficiency :

$$\eta_{DR} = -J_{DR,ph} \, hc/qI_o \lambda = e^{-\alpha x_p} - e^{-\alpha x_n} \quad (42)$$

D. Total_quantum_efficiency

The total photocurrent density of the abrupt junction is given by

$$J_{ph} = J_{n,ph}(x_p) + J_{DR,ph} + J_{p,ph}(x_n) \quad (43)$$

and, according to (31), (36) and (41) is directly proportional to the intensity I_o of the light at the surface (x=0) of the photodiode.

According to the previous theory the total quantum efficiency of the abrupt photodiode is

$$\eta = -J_{ph} \cdot hc/qI_0\lambda = \eta_n + \eta_{DR} + \eta_p \qquad (44)$$

The dependence of η on the junction parameters and on the wavelength λ of the incident light will be discussed in the next sections.

One may notice that, according to (34), (40) and (42), the quantum efficiency of an ideal photodiode, having a semi-infinite substrate and infinite lifetimes, is equal to unity for all wavelengths.

Quantum efficiency of a field induced photodiode

Before discussing the case of an abrupt p-n junction having a metallurgical junction at a finite depth x_j beneath the semiconductor surface, it is worthwhile to consider the simpler case of field induced junctions.

In CCD and CID structures a gate electrode is used to induce an inversion layer at the surface of the semiconductor. The resulting field induced junction can be considered as a shallow abrupt junction with a junction depth x_j equal to zero. The width of the surface inversion layer of the structure (see Fig. 2) may be assumed infinitely thin; the width x_n of the adjacent depletion region depends on the doping of the n-type substrate and on the potential on the external gate electrode. The remaining part (x_n, L) of the semiconductor substrate may be considered as a quasi-neutral region with an ohmic contact at $x = L$. Clearly the theory, developped in the previous sections, applies to the field induced junction for $x_p = x_j = 0$.

The quantum efficiency of the field induced junction has two components:

$$\eta = \eta_{DR} + \eta_n \qquad (45)$$

where according to (42) and (37)

$$\eta_{DR} = 1 - e^{-\alpha x_n} \qquad (46)$$

$$\eta_n = \frac{\alpha L_p^2}{\alpha^2 L_p^2 - 1} \left\{ \alpha e^{-\alpha x_n} + \frac{e^{-\alpha L} - e^{-\alpha x_n} \cosh\left[(L-x_n)/L_p\right]}{L_p \sinh\left[(L-x_n)/L_p\right]} \right\}$$

$$(47)$$

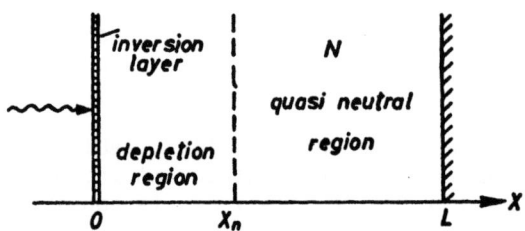

Fig. 2 One-dimensional field induced photodiode structure.

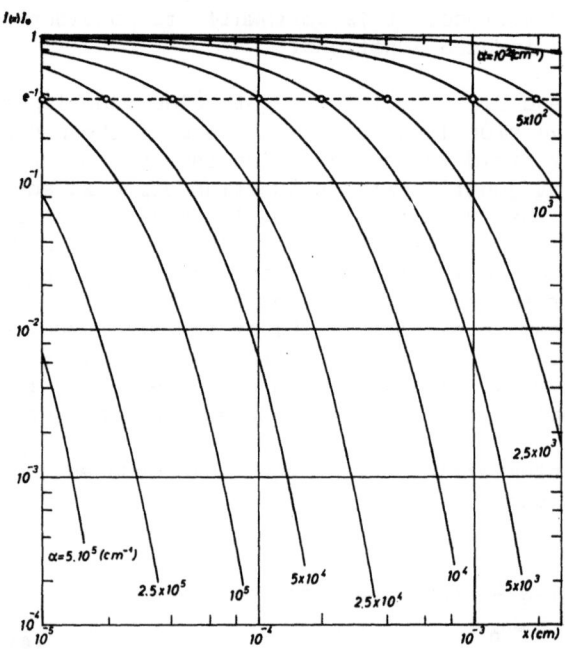

Fig. 3. Ratio $I(x)/I_o$ as a function of the depth x for different values of the absorption coefficient α. The light pentration depth is indicated by "o".

The last result has been used by Brown and Chamberlain [7] in the study of a charged-coupled imaging array. The following limiting cases may be useful

$$L_p \to \infty \quad : \quad \eta_n = e^{-\alpha x_n} \left[1 - \frac{1 - e^{-\alpha(L-x_n)}}{\alpha(L - x_n)} \right] \quad (48)$$

$$L \to \infty \quad : \quad \eta_n = \alpha L_p \, e^{-\alpha x_n} / (\alpha L_p + 1) \quad (49)$$

$$L_p, L \to \infty \quad : \quad \eta_n = e^{-\alpha x_n} \quad (50)$$

The last case corresponds to the ideal induced junction for which $\eta = 1$, according to (46).

To start the discussion of the quantum efficiency, we first introduce the concept of light penetration depth. The light intensity $I(x)$ within the semiconductor decreases according to the law

$$I(x)/I_0 = e^{-\alpha x}$$

Figure 3 represents this ratio as a function of the distance x for different values of the absorption coefficient α. The light penetration depth is defined as the distance from the semiconductor surface at which the light intensity has fallen to 1/e of the surface light intensity; its value $1/\alpha$ is indicated on figure 3. The partial quantum efficiency η_{DR} may easily be deduced from figure 3 and is represented in figure 4 as a function of the absorption coefficient for different values of the depletion width x_n. The wavelength scale, indicated in figure 4, is for a Silicon substrate. Clearly η_{DR} increases for increasing values of α, since more photons are then absorbed within the depletion region, and for increasing values of x_n, since all carriers generated within the depletion region effectively contribute to the photocurrent. Note that the quantum efficiency η_{DR}, and hence η, of field induced junctions, especially in the short wavelength range, is not limited by surface recombination phenomena, since the depletion layer extends up to the semiconductor surface and is shielded by an inversion layer.

We now start the discussion of the other structural parameters influencing the quantum efficiency of a field induced junction. In order to illustrate each effect by an example,

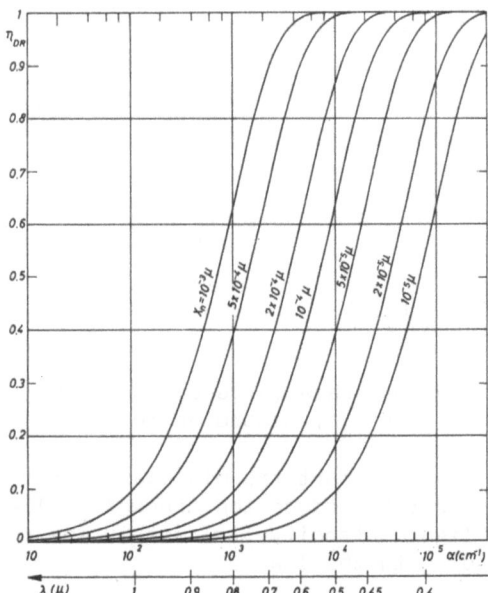

Fig.4. Partial quantum efficiency η_{DR} vs. absorption coefficient α (wavelength scale λ for Si) for different values of the depletion width x_n.

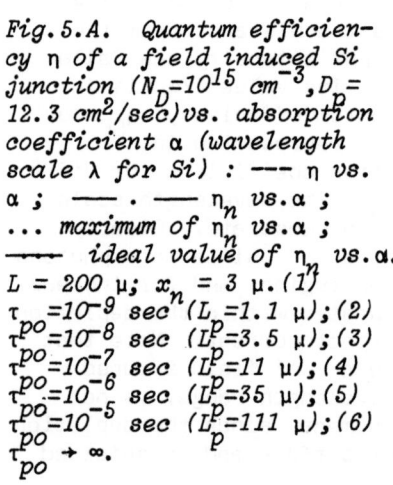
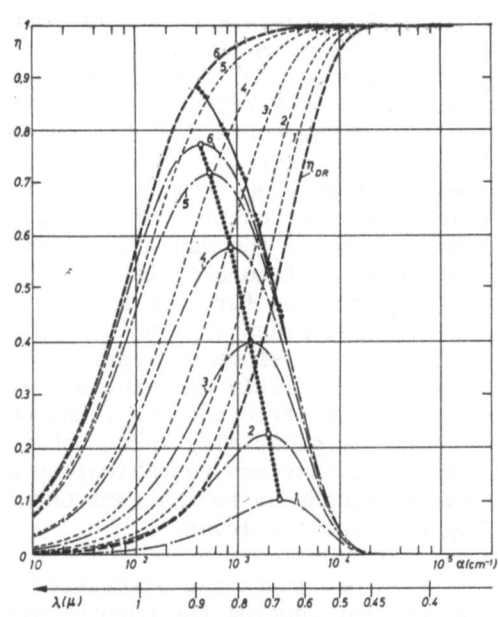

Fig.5.A. Quantum efficiency η of a field induced Si junction ($N_D=10^{15}$ cm^{-3}, $D_p=$ 12.3 cm²/sec)vs. absorption coefficient α (wavelength scale λ for Si) : --- η vs. α ; —·— η_n vs.α ; ... maximum of η_n vs.α ; —— ideal value of η_n vs.α. $L = 200$ µ; $x_n = 3$ µ.(1) $\tau_{po}=10^{-9}$ sec ($L_p=1.1$ µ);(2) $\tau_{po}=10^{-8}$ sec ($L_p=3.5$ µ);(3) $\tau_{po}=10^{-7}$ sec ($L_p=11$ µ);(4) $\tau_{po}=10^{-6}$ sec ($L_p=35$ µ);(5) $\tau_{po}=10^{-5}$ sec ($L_p=111$ µ);(6) $\tau_{po} \to \infty$.

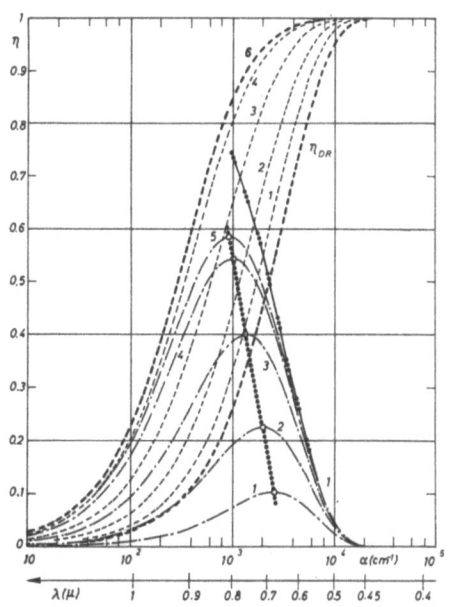

Fig.5.B. Quantum efficiency η of a field induced Si junction ($N_D=10^{15}$ cm^{-3}, $D_p=12.3$ cm^2/sec) vs. absorption coefficient α (wavelength scale λ for Si):
--- η vs. α; —·— η_n vs. α; ... maximum of η_n vs. α; ▬▬▬ ideal value of η_n vs. α. $L = 50$ μ; $x_n=3$ μ. (1) $\tau_{po}=10^{-9}$ sec ($L_p=1.1$ μ); (2) $\tau_{po}=10^{-8}$ sec ($L_p=3.5$ μ); (3) $\tau_{po}=10^{-7}$ sec ($L_p=11$ μ); (4) $\tau_{po}=10^{-6}$ sec ($L_p=35$ μ); (5) $\tau_{po}\to\infty$.

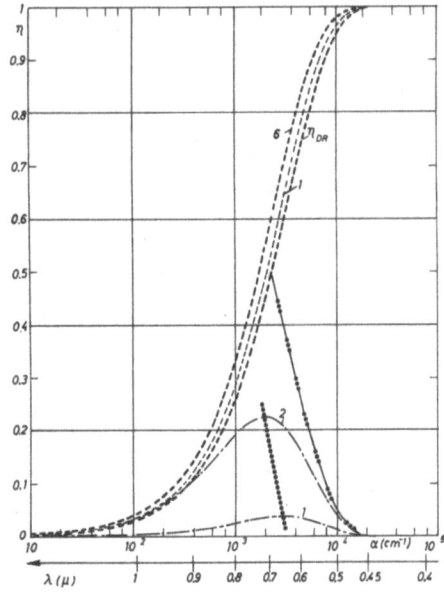

Fig.5.C. Quantum efficiency η of a field induced Si junction ($N_D=10^{15}$ cm^{-3}; $D_p=12.3$ cm^2/sec) vs. absorption coefficient α (wavelength scale λ for Si):
--- η vs. α; —·— η_n vs. α; ... maximum of η_n vs. α; ▬▬▬ ideal value of η_n vs. α. $L = 5$ μ; $x_n = 3$ μ.
(1) $\tau_{po} = 10^{-9}$ sec ($L_p=1.1$ μ); (2) $\tau_{po}\to\infty$.

we consider a Si substrate with a donor concentration of 10^{15} atoms/cm^3; a hole mobility of 475 cm^2/V.sec yields a diffusion constant D_p of 12.3 cm^2/sec at room temperature for the minority carriers.

The influence on the quantum efficiency η of the lifetime τ_{po} and of the doping dependent diffusion constant D_p can be explained in the following way. The light generated minority carriers diffuse through the quasi-neutral region over a mean distance equal to the diffusion length $L_p = \sqrt{D_p \tau_{po}}$, if $L_p < L - x_n$, or equal to the length of the quasi-neutral region $L - x_n$, if $L_p > L - x_n$. Obviously in the last case η_n, and hence η, is independent of the diffusion length L_p and can be computed from (48), valid for L_p tending to infinity.

Consider now the case $L_p < L - x_n$, and assume that the light penetration depth $1/\alpha$ is larger than L_p. Since the minority carriers generated within a diffusion length from the boundary x_n of the depletion region effectively contribute to the photocurrent, any increase of L_p yields an increase of the quantum efficiency η_n, and hence of η. We conclude that, for $1/\alpha > L_p$, η_n increases with L_p and tends to a maximum value, given by (48) and (46), when L_p becomes equal to or larger than the substrate region $L - x_n$.

When the light penetration depth $1/\alpha$ is much smaller than the length $L - x_n$ of the quasi-neutral substrate, an increase of L will not affect the quantum efficiency; the expression (49), valid for L tending to infinity, may be used for η_n. In the last case η_n will increase for increasing values of L_p, if $L_p < 1/\alpha$, but will tend to a maximum value, given by the expression (50) valid for the ideal case, when L_p becomes larger than $1/\alpha$. In the last case the junction collects all minority carriers generated by the photons under consideration.

Finally we mention that, when the diffusion length L_p becomes extremely small, the quantum efficiency η_n becomes negligible according to (47); in this case η is reduced to the partial quantum efficiency η_{DR} of the depletion region which only depends on x_n and α.

The previous effects are illustrated in figure 5 where the quantum efficiency of a field induced Si junction is plotted as a function of the absorption coefficient α for different values of the lifetime τ_{po}, hence of the diffusion length L_p; three different substrate lengths are considered, but the depletion width x_n is constant. The last width determines, for each value of α, the minimum value η_{DR} of the quantum efficiency η. The maximum of η is dictated by the expression (48) and depends on the length L, especially for small values of α. Figure 5 clearly shows that the maximum of η_n, considered as a function of α, shifts from the large to the low absorption range for increasing values of L_p. This is due to the fact that the condition $L_p \gg 1/\alpha$ for saturation of η_n is first fulfilled for large values of α; the ideal value $\eta_n = e^{-\alpha x_n}$ is indicated in figure 5. As a result η becomes equal to unity for large α values.

Figure 5 also indicates that the influence of the substrate length L on the total quantum efficiency is strong when the light penetration depth is large. Figure 6 illustrates this effect more directly by plotting η as a function of the length L of the substrate or of the epitaxial layer, for different values of the absorption coefficient and different values of the lifetime; the depletion width x_n is considered as constant. The diffusion length or, for a given substrate, the lifetime of the minority carriers, determines the maximum attainable efficiency for increasing values of L; this maximum depends on the absorption coefficient and can be computed from the expression (49). Furthermore, for small values of L, the quantum efficiency becomes at all wavelengths nearly independent of the lifetime if the diffusion length L_p remains larger than $L - x_n$. Figure 6 (and also figure 5C) clearly shows that high values of η in the near infrared can be obtained by using a small lifetime and a small substrate length.

When using the field induced junction in the integration mode, the width x_n of the depletion region becomes a function of the integration time. The dependence of the quantum efficiency η on the width x_n is illustrated in figure 7 for different values of the absorption coefficient and two values of L; a constant lifetime is considered. The function η_{DR} increases with x_n, while η_n decreases; however, since all carriers generated within the depletion region effectively contribute to the photocurrent, which is not the case in the quasi-neutral substrate region due to recombination phenomena, the total quantum efficiency is an increasing function of x_n. Furthermore, the influence of x_n on η is wavelength dependent. In general large variations of η will occur for x_n values near the light penetration $1/\alpha$. For $x_n \gg 1/\alpha$, the depletion region absorbs almost all photons and η is nearly equal to η_{DR}. For very small values of the lifetime τ_{po}, the contribution of the substrate is also negligible and η is again nearly equal to η_{DR} (see fig. 5); however in the last case η is equal to η_{DR} for all wavelengths. For $x_n \ll 1/\alpha$ the

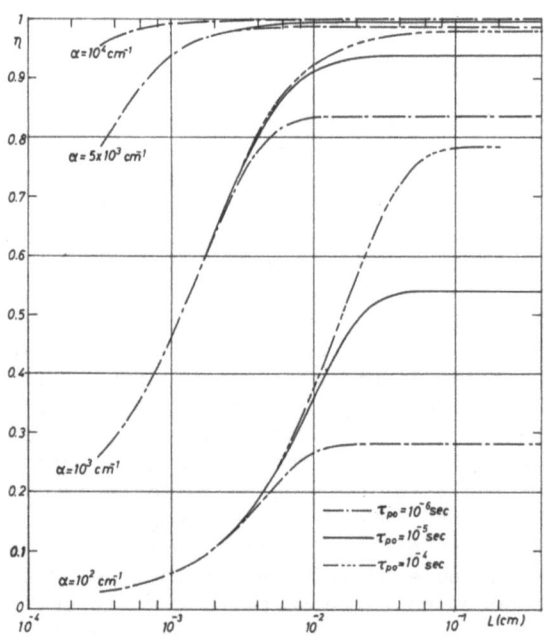

Fig.6. Quantum efficiency η of a field induced Si junction ($N_D=10^{15}cm^{-3}$, $D_p=12.3\ cm^2/sec$) vs. substrate length L for different values of α and different values of τ_{po}; $x_n = 3\ \mu$.

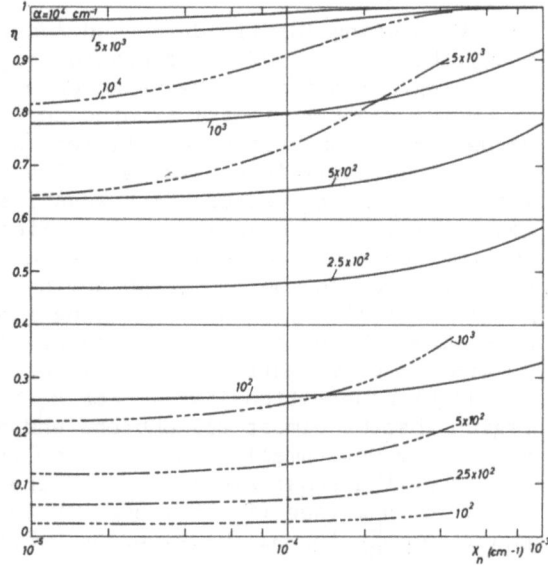

Fig.7. Quantum efficiency η of a field induced Si junction ($N_D=10^{15}cm^{-3}$, $D_p=12.3\ cm^2/sec$) vs. depletion width x_n for different values of α;
—— L=200 μ;
—–—– L = 5 μ;
$\tau_{po} = 10^{-6}$ sec.

photocurrent essentially results from light absorption in the substrate region; in this case η is almost equal to $η_n$ and strongly depends on the length L of this region and (or) on the lifetime $τ_{po}$ of the minority carriers (see fig. 5). The example shown in figure 7 indicates that the variation of η with x_n, especially in the visible range of the spectrum, can be important when the substrate length is small.

As a final result we present in figure 8 the quantum efficiency, as a function of the absorption coefficient with x_n as parameter, for two different values of L. In general, for a thick substrate, the quantum efficiency is high, even in the near infrared range of the spectrum, and is nearly independent of the width x_n of the depletion region. For a thinner substrate the near infrared quantum efficiency becomes smaller, but the dependence of η on x_n becomes important, especially in the visible range of the spectrum. Figure 8 also shows that the partial quantum efficiency $η_n$ strongly decreases for increasing values of x_n, even for a thick substrate; the maximum value of $η_n$ shifts from the heavy to the low absorption range for increasing values of x_n, since the reduction of $η_n$ is most effective for large values of α.

We conclude that the shape of the quantum efficiency versus the absorption coefficient η depends on the parameters N_D, $τ_{po}$, x_n and L of the field induced junction, according to the expressions (45), (46) and (47). The particular values suited for an optimal imaging structure depend of course on the specific application under consideration.

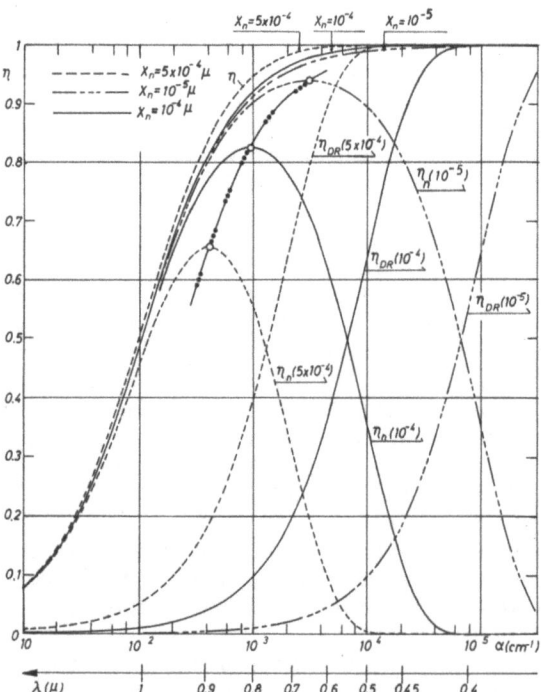

Fig.8.A. Quantum efficiency η of a field induced Si junction ($N_D = 10^{15} cm^{-3}$, $D_p = 12.3\ cm^2/sec$) vs. α (wavelength scale λ for Si) for different values of x_n.

(A) $L = 200\ \mu$;
$\tau_{po} = 10^{-5}\ sec.$

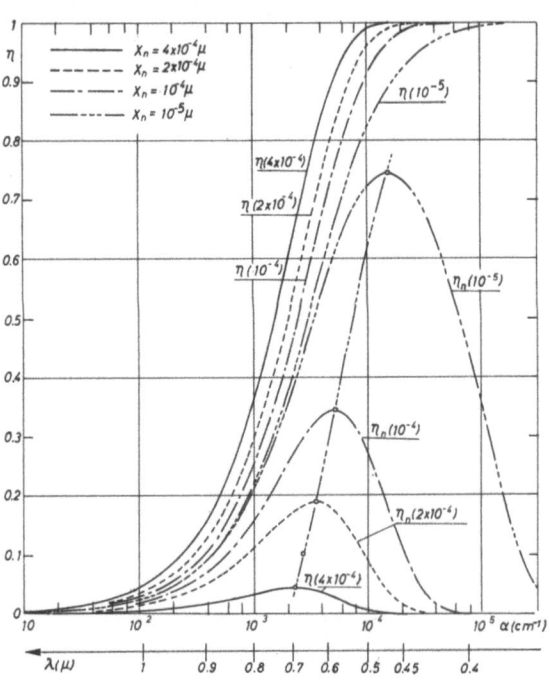

Fig.8.B.

(B) $L = 5\ \mu$
$\tau_{po} = 10^{-5}\ sec.$

Quantum efficiency of an abrupt photodiode

According to the general theory the quantum efficiency of an abrupt photodiode is given by

$$\eta = \eta_p + \eta_{DR} + \eta_n \qquad (51)$$

where η_p is the partial quantum efficiency of the p-type surface quasi-neutral region

$$\eta_p = \frac{\alpha L_n^2}{1 - \alpha^2 L_n^2} \left\{ \alpha e^{-\alpha x_p} - \frac{1}{L_n} \frac{L_n(s + \alpha D_n) - e^{-\alpha x_p}\left[s L_n \cosh(x_p/L_n) + D_n \sinh(x_p/L_n)\right]}{s L_n \sinh(x_p/L_n) + D_n \cosh(x_p/L_n)} \right\} \qquad (52)$$

η_n the partial quantum efficiency of the n-type substrate of length L

$$\eta_n = \frac{\alpha L_p^2}{\alpha^2 L_p^2 - 1} \left\{ \alpha e^{-\alpha x_n} + \frac{e^{-\alpha L} - e^{-\alpha x_n} \cosh\left[(L-x_n)/L_p\right]}{L_p \sinh\left[(L-x_n)/L_p\right]} \right\} \qquad (53)$$

and finally η_{DR} the partial quantum efficiency of the depletion region

$$\eta_{DR} = e^{-\alpha x_p} - e^{-\alpha x_n} \qquad (54)$$

Exactly the same expression for η_n was used in the study of the field induced junction. Hence we know, as far as the substrate region is concerned, that the quantum efficiency η considered as a function of the length $L - x_n$ is nearly independent of the lifetime τ_{po} of the substrate minority carriers for values of $L - x_n$ smaller than the diffusion length; for larger values of $L - x_n$ the quantum efficiency tends to a maximum depending on the lifetime τ_{po} and on the particular wavelength. Furthermore the quantum efficiency increases for increasing values of the diffusion length L_p of the minority carriers until L_p becomes of the same order as the length of the quasi-neutral substrate or until L_p becomes larger than the light penetration depth $1/\alpha$. Finally we know that the partial quantum efficiency η_n decreases for increasing values of the depth x_j of the metallurgical junction beneath the surface and for increasing values of the reverse bias V_a

applied to the junction. From the study of the field induced junction we also know that the maximum value of η_n as a function of α shifts from the heavy to the low absorption range by increasing the value of x_n.

We continue the discussion of the other structural parameters influencing the value of η. In order to illustrate each effect by an example, we consider a Si abrupt photodiode having a surface acceptor concentration of 10^{17} atoms/cm^3 and a substrate donor concentration of 10^{15} atoms/cm^3. The corresponding hole mobility μ_p in the substrate is 475 cm^2/V.sec and the diffusion constant D_p is 12.3 cm^2/sec at room temperature; the figures for the surface region are μ_n = 525 cm^2/V.sec and D_n = 13.6 cm^2/sec.

The partial quantum efficiency of the photodiode, given by (54) may be written as

$$\eta_{DR} = e^{-\alpha x_p}\left[1 - e^{-\alpha(x_n-x_p)}\right]$$

and is equal to the product of the partial quantum efficiency η_{DR} (see fig.4) of a field induced junction, having a depletion width x_n-x_p, by an absorption dependent attenuation factor $e^{-\alpha x_p}$. The last factor makes η_{DR} tending to zero for large values of α, since the light penetration depth becomes small. The width $x_n - x_p$ of the depletion region depends on the doping profile and on the reverse bias V_a applied to the junction [see equ. (23), (24)]. The depth x_p, for a fixed value of V_a, is related to the depth x_j of the metallurgical junction. An example is given in figure 9, where η_{DR} is plotted as a function of the absorption coefficient α for different values of x_j, V_a remaining constant.

The next problem we examine is the dependence of the efficiency η_p, and hence of η, on the surface recombination velocity s. The total quantum efficiency η of a photodiode is plotted as a function of α in figure 10 for different values of s. We see that surface recombination affects the quantum efficiency, especially when the light penetration depth $1/\alpha$ is small. A plot of η as a function of s is shown in figure 11 for different values of the absorption coefficient α. Clearly η goes, for all values of α, from a high to a low value when s increases; the transition between both states occurs, according to (52), for a particular s value given by

$$s_o = (D_n/l_n)\,\mathrm{cth}\,(x_p/L_n) \qquad (55)$$

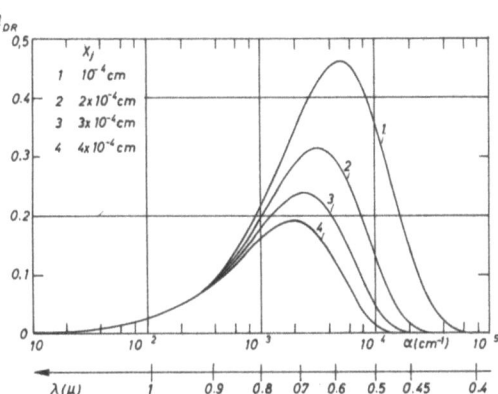

Fig. 9. Partial quantum efficiency η_{DR} of an abrupt Si photodiode ($N_A=10^{17}$ cm^{-3}, $N_D=10^{15}$ cm^{-3}) vs. α (λ scale for Si) for different values of the junction depth x_j; $V_a = -5$ V.

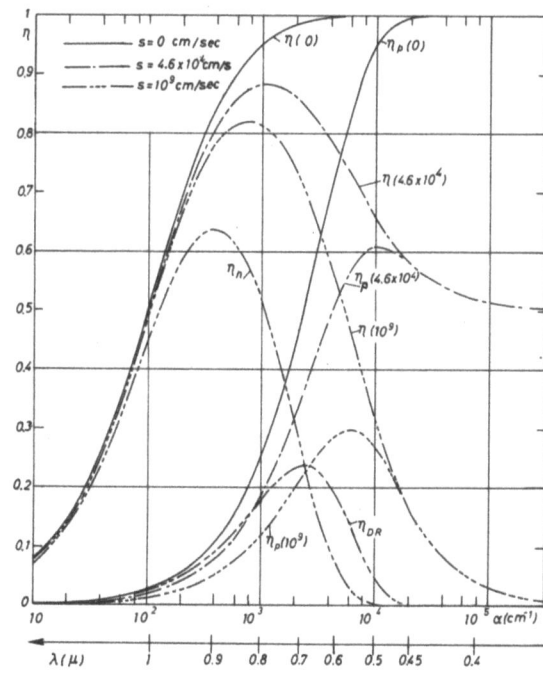

Fig. 10. Quantum efficiency η of an abrupt Si photodiode ($N_A=10^{17}$ cm^{-3}, $N_D=10^{15}$ cm^{-3}, $D_n = 13.6$ cm^2/sec, $D_p = 12.3$ cm^2/sec) vs. α (λ scale for Si) for different values of s; $x_j=3$ μ, $L=200$ μ, $V_a = -5$ V, $\tau_{no} = \tau_{po} = 10^{-5}$ sec, $L_p = 111$ μ, $L_n = 117$μ

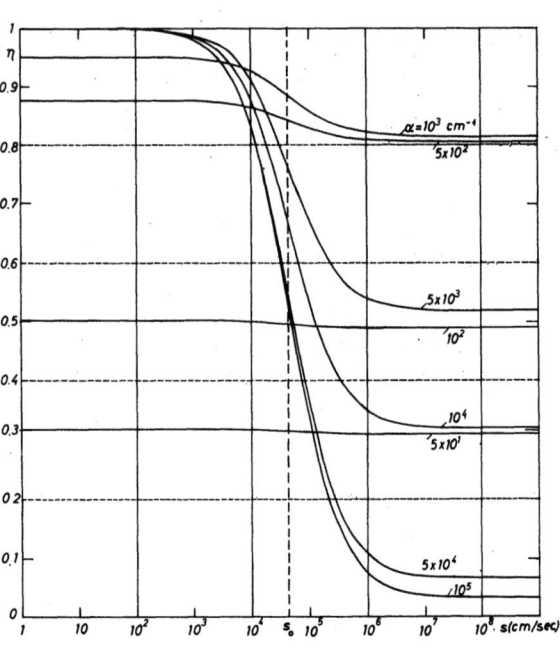

Fig. 11. Quantum efficiency η of an abrupt Si photodiode ($N_A = 10^{17}$ cm^{-3}, $N_D = 10^{15}$ cm^{-3}, $D_p = 13.6$ cm^2/sec, $D_n = 12.3$ cm^2/sec) vs. s for different values of α; $x_j = 3$ μ, $L = 200$ μ, $V_a = -5$ V, $\tau_{no} = \tau_{po} = 10^{-5}$ sec ($L_p = 117$ μ, $L_n = 117$ μ).

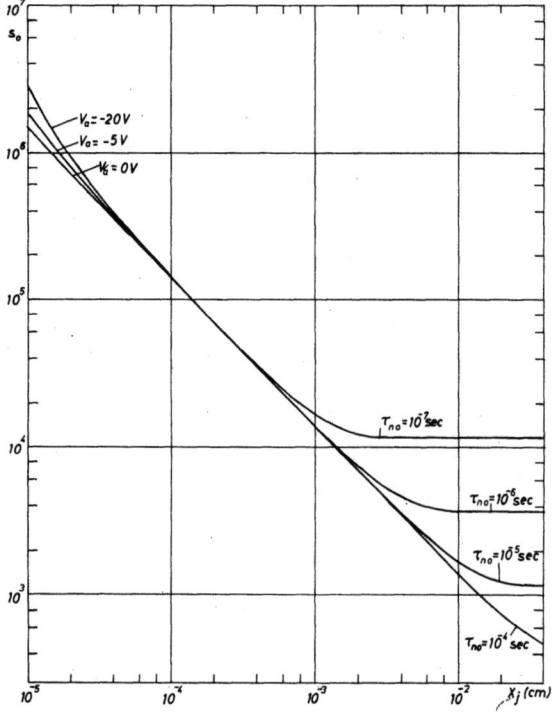

Fig. 12. Transition value s_o vs. x_j for an abrupt Si photodiode ($N_A = 10^{17}$ cm^{-3}, $N_D = 10^{15}$ cm^{-3}, $D_n = 13.6$ cm^2/sec) for different values of V_a and different values of τ_{no}.

Figure 11 also shows that η remains almost constant, for the photodiode under consideration, for values of s ranging from zero to 10^3 cm/sec. Practical photodiodes usually have s values lying within this range.

The transition value x_o is independent of the absorption coefficient; its value as a function of x_j is plotted in figure 12 for different values of the lifetime τ_{no} and different values of the bias voltage V_a applied to the junction. For shallow pn junctions, s_o only depends on the width x_p, and hence on the reverse bias V_a since $s_o \propto D_n/x_p$. For large values of x_j the value of s_o tends to $s_o = D_n/L_n$ and depends, for a given surface doping, on the lifetime τ_{no} of the minority carriers. From figure 12 one may conclude that the particular value s_o is usually larger than the surface recombination velocity encountered in practical photodiodes. Hence, from now on only the case s = 0 will be studied. According to (52) we have for s = 0 :

$$\eta_p = \frac{\alpha L_n^2}{1 - \alpha^2 L_n^2} \left\{ \alpha e^{-\alpha x_p} - \frac{\alpha L_n - e^{-\alpha x_p} sh(x_p/L_n)}{L_n ch(x_p/L_n)} \right\} \quad (56)$$

Consider next the influence of the lifetime τ_{no}, and hence of the diffusion length L_n, of the electrons in the surface region. If L_n is larger than the width x_p of the surface region, almost all light generated minority carriers within this region will diffuse towards the junction and contribute to the photocurrent; in this case η_p becomes independent of L_n and tends to the ideal value obtained from (56) for L_n tending to infinity

$$\eta_p = 1 - e^{-\alpha x_p} \quad \text{(ideal case : } L_n \to \infty \text{ , s = o)}$$

(57)

When L_n becomes smaller than x_p, a reduction of η_p occurs, especially for large absorption coefficients for which the light penetration depth is small.

These results are illustrated in figure 13, where η is plotted vs α for different values of τ_{no}, all other parameters remaining constant. The wavelength dependent decrease of η_p for decreasing values of τ_{no} yields a shift of the maximum of the total quantum efficiency from the heavy to the low absorption range.

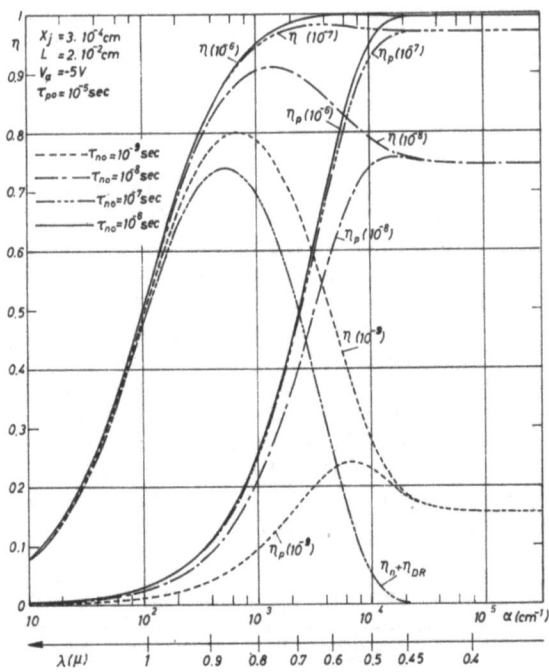

Fig. 13. Quantum efficiency η of an abrupt Si photodiode ($N_A = 10^{17}$ cm^{-3}, $N_D = 10^{15}$ cm^{-3}, $D^n = 13.6$ cm^2/sec, $D^p = 12.3$ cm^2/sec) vs. α(λ scale for Si) for different values of τ_{no}; $s = 0$, $\tau_{po} = 10^{-5}$ sec ($L_p = 111$ μ).

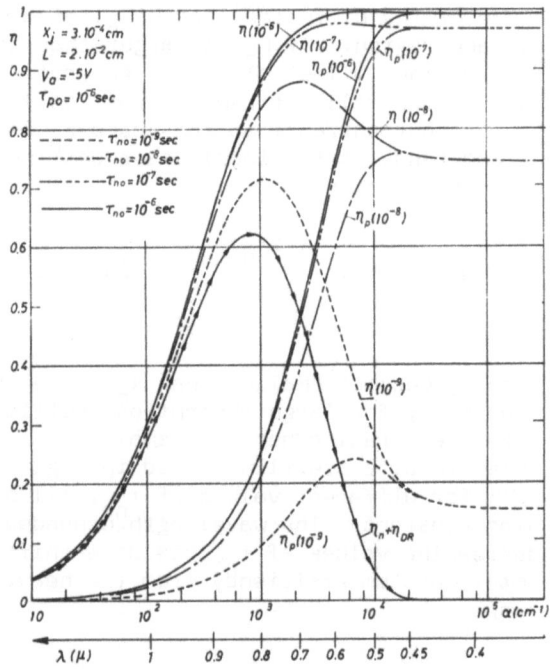

Fig. 14 Quantum efficiency η of an an abrupt Si photodiode ($N_A = 10^{17}$ cm^{-3}, $N_D = 10^{15}$ cm^{-3}, $D^n = 13.6$ cm^2/sec, $D^p = 12.3$ cm^2/sec) vs. α (λ scale for Si) for different values of τ_{no}; $s = 0$, $\tau_{po} = 10^{-6}$ sec ($L_p = 35$ μ).

Fig.15 Quantum efficiency η of an abrupt Si photodiode ($N_A=10^{17}$ cm^{-3}, $N_D=10^{15}$ cm^{-3}, $D^p=13.6$ cm^2/sec, $D^n=12.3$ cm^2/sec) vs. α ($λ$ scale for Si) for different values of $τ_{no}$; $s = 0$, $τ_{po} = 10^{-6}$ sec ($L_p^{po}= 35$ μ).

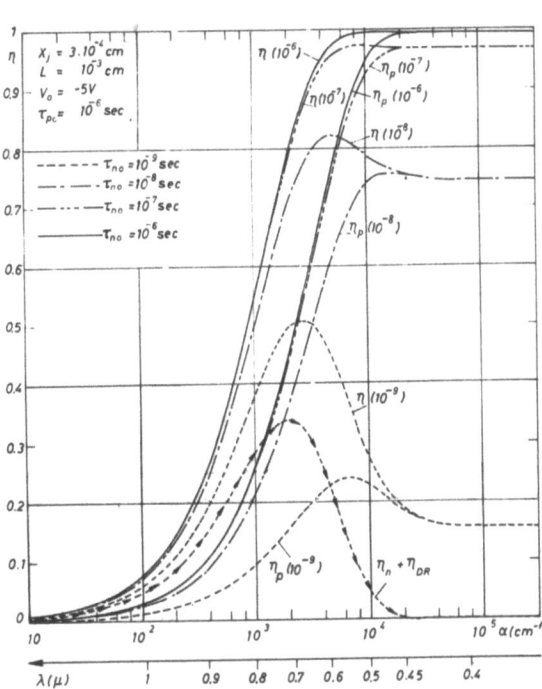

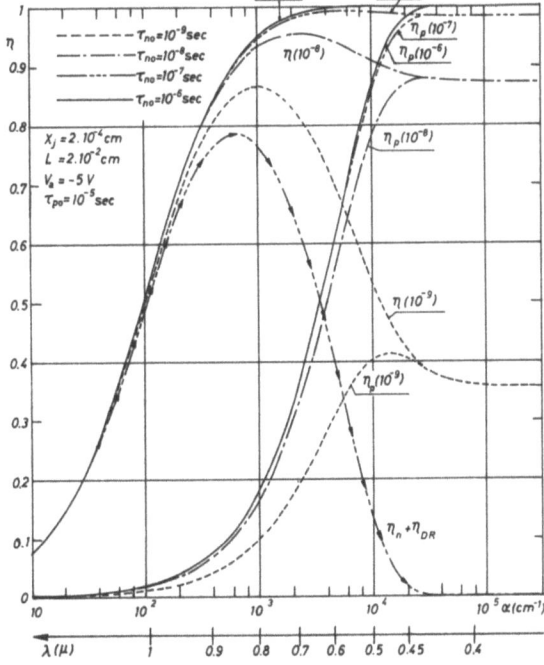

Fig.16 Quantum efficiency η of an abrupt Si photodiode ($N_A = 10^{17}$ cm^{-3}, $N_D = 10^{15}$ cm^{-3}, $D^p = 13.6$ cm^2/sec, $D^n = 12.3$ cm^2/sec) vs. α ($λ$ scale for Si) for different values of $τ_{no}$; $s = 0$, $τ_{po} = 10^{-5}$ sec ($L_p = 111$ μ).

Figure 13 shows that the total quantum efficiency η tends, for large values of α, to a constant value depending on the lifetime τ_{no} of surface minority carriers. For very large values of α the contribution η_{DR} of the depletion region and η_n of the substrate region become negligible, and according to (56) one obtains

$$\eta = \eta_p = 1/\mathrm{ch}(x_p/L_n) \qquad \text{(for } \alpha \to \infty, s=0) \qquad (58)$$

The last formula can be useful, as a rough estimation of η for large values of α, when choosing the parameters of a photodiode.

Figure 13 also shows that η_p is small for small values of α; since the light penetration depth is large, only a few photons are absorbed within the surface region. This implies that the shape of the total quantum efficiency η depends, for the range of small α values, essentially on the parameters influencing η_n an η_{DR}, i.e. x_j, V_a, L_p and L. For example, as illustrated in figure 14, a decrease of the lifetime τ_{po} of the substrate region, reduces the value of η in the near infrared, but does not affect the shape of η for large values of α. As a result the maximum of η as a function of α becomes smaller, but shifts towards the strong absorption range. A similar effect is obtained by a reduction of the length L of the substrate; this is illustrated in figure 15.

A variation of the depth x_j of the metallurgical junction modifies the shape of η for large values of α. This can be seen in figure 16 by comparison with figure 13 and in figure 17 where η is plotted vs. x_j for different values of α and two values of L. A decrease of x_j yields an increase of η_p, especially for large values of α, if $L_p \ll x_p$ [see also equation (58)] since the junction collects then a larger part of the carriers generated within the surface region; however, for $L_p > x_p$, a decrease of η_p occurs when x_p becomes smaller than the light penetration depth $1/\alpha$, since the total number of carriers generated within the surface region decreases. In the last case the decrease of η_p is, however, almost completely compensated by an increase of the partial quantum efficiencies η_{DR} and η_n, especially when L is large. As a result, a decrease of x_j increases the total quantum efficiency η, especially for large values of α, when $L_n < x_p$, and slightly reduces the value of η, especially when L is small, for $L_n > x_p$ and $x_p < 1/\alpha$. Figure 17 clearly shows that a junction depth in the range of one to a few microns allows to get large values for η in the visible range of the light spectrum; low values of η in the near infrared range can be obtained by reducing the length L of the substrate or, as already mentioned,

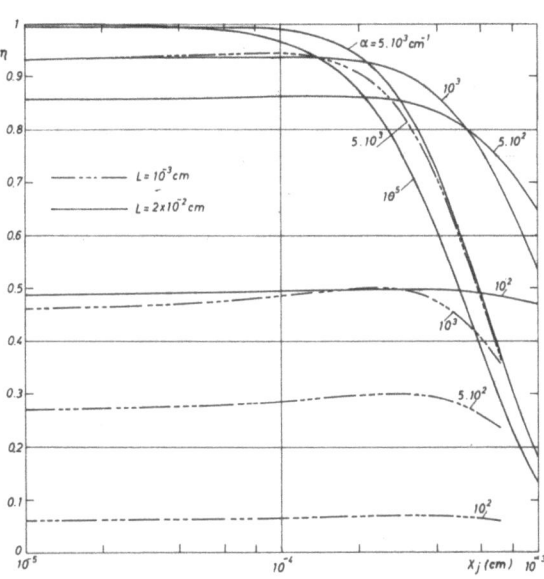

Fig. 17 Quantum efficiency η of an abrupt Si photodiode
($N_A = 10^{17}$ cm^{-3}, $N_D = 10^{15}$ cm^{-3}, $D_n = 13.6$ cm^2/sec, $D_p = 12.3$ cm^2/sec) vs. x_j for different values of α and two values of L; $s = 0$, $V_a = -5$ V, $\tau_{po} = 10^{-5}$ sec ($L_p = 111$ µ), $\tau_{no} = 10^{-8}$ sec ($L_n = 3.7$ µ).

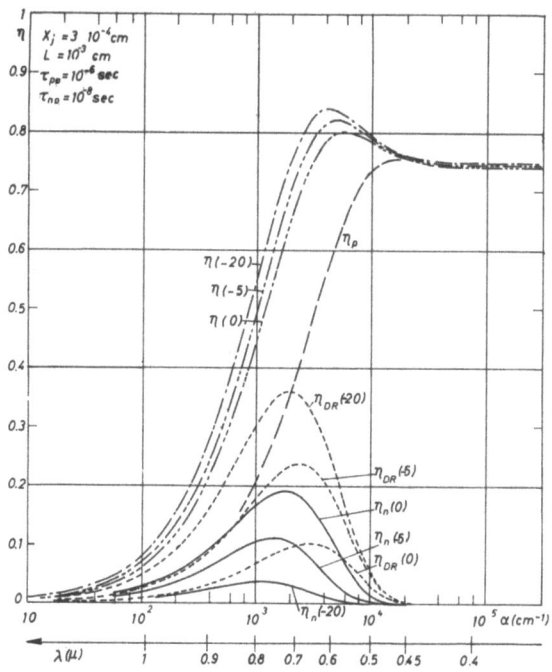

Fig. 18 Quantum efficiency η of an abrupt Si photodiode
($N_A = 10^{17}$ cm^{-3}, $N_D = 10^{15}$ cm^{-3}, $D_n = 13.6$ cm^2/sec, $D_p = 12.3$ cm^2/sec) vs. α (λ scale for Si) for three values of V_a : 0, -5 and -20 V; $s = 0$, $x_j = 3$ µ, $L = 10$ µ, $\tau_{no} = 10^{-8}$ sec ($L_n = 3.7$ µ), $\tau_{po} = 10^{-6}$ sec ($L_p = 35$ µ).

the diffusion length L_p of the substrate minority carriers.

The last parameter to be discussed is the negative bias voltage V_a applied to the junction. The width $x_n - x_p$ of the depletion region depends on the doping profile of the junction and is an increasing function of $|V_a|$. Since all carriers generated within the depletion region effectively contribute to the photocurrent, η_{DR} is an increasing function of $|V_a|$. Figure 18 illustrates this effect by an example. For p^+n junctions the bias dependent broadening of the depletion region essentially occurs at the n side [see equations (23) and (24)]. As a result the width x_p of the surface region and hence the value of η_p are only slightly dependent on the bias voltage; from the discussion of figure 17 we conclude that a slight decrease of x_p affects η_p for large values of α if $L_n < x_p$, and yields then a slight increase of η since η_{DR} and η_n are negligible in this range. Usually the influence of V_a on η may be neglected for large values of α (see fig. 18).

For smaller values of α, we know that η_p slightly decreases with $|V_a|$ for $L_n > x_p$ and $x_p < 1/\alpha$; however, in this range the influence on η_{DR} and η_n of the bias dependent value of x_n becomes dominant for η. From the study of the field induced junction we know that η_n decreases, especially when L is small, for increasing values of x_n (see fig. 8 and also fig. 18). This decrease of η_n (and of $\eta_n + \eta_p$) is however smaller than the corresponding increase of η_{DR}. As a net result the total quantum efficiency η increases with $|V_a|$ for $x_p < 1/\alpha$, especially when the substrate length L is small. This increase of η is wavelength dependent and becomes smaller when the light penetration depth is larger. These effects are clearly illustrated in figure 18.

As a final result we present in figure 19 the so called spectral response $S(\lambda)$ of a photodiode, for different values of the lifetimes and two values of the single silicon dioxide film on top of the semiconductor surface. These curves results from the product, as a function of the wavelength λ, of the total quantum efficiency $\eta(\lambda)$ by the transmittance T of the multilayer structure discussed in a previous section. According to its definition, T is the ratio of the light power I_o transmitted into the substrate (at x = o), to the light power I_{inc}, incident on the oxide layer; taking the definition of η into account, one obtains

$$S(\lambda) = \eta(\lambda) \cdot T(\lambda) = - h c J_{ph}(V_a)/q \lambda I_{inc}(\lambda)$$

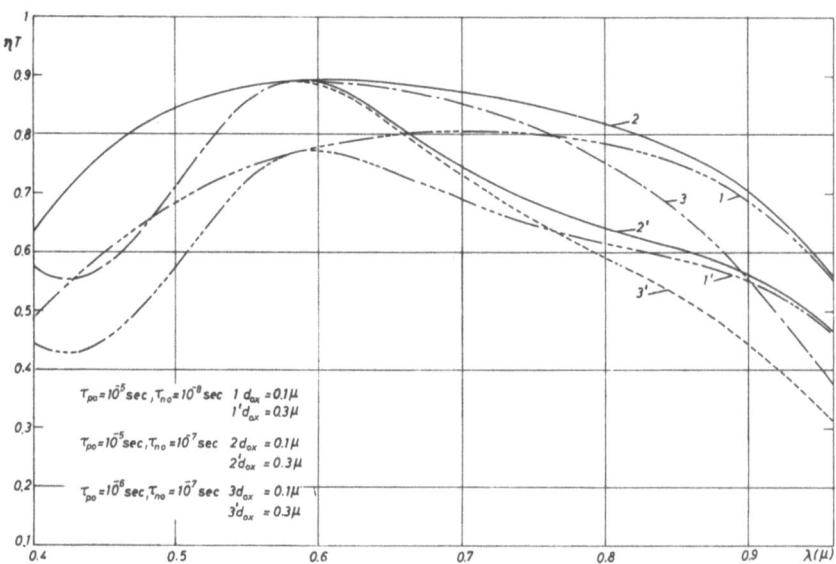

*Fig. 19 Spectral response of an abrupt Si photodiode
($N_A = 10^{17}$ cm^{-3}, $N_D = 10^{15}$ cm^{-3}, $D_n = 13.6$ cm^2/sec,
$D_p = 12.3$ cm^2/sec) vs. λ for different values of
the lifetimes τ_{no}, τ_{po} and two values of the
thickness d_{ox} of the SiO_2 layer; $s = 0$, $x_j = 3$ μ, $L = 200$ μ, $V_a = -5$ V.*

Hence the spectral response, besides its dependence on $\eta(\lambda)$, strongly depends on the single or multilayer structure on top of the photodiode. The photocurrent measured on practical photodiodes is directly related to the spectral response $S(\lambda)$.

Remarks

1. Parallel-illuminated junction

The active area of a photodiode can be larger than the geometrical area of the p-n junction. Indeed, consider for example the region surrounding the geometrical area of the junction in the plane of the metallurgical junction, parallel to the semiconductor surface; the minority carriers generated within a diffusion length from the boundary of the junction will diffuse towards the junction and contribute to the photocurrent. The same conclusion holds for all planes parallel to the semiconductor surface; the surface region yields the largest contribution. The additional photocurrent, due to this lateral generation, could lead especially for small photodiodes, to an apparent quantum efficiency larger than unity since the total active area becomes larger than the geometrical area of the photodiode.

One should notice that the lateral generation strongly depends on the light transmittance of the multilayer structure covering the semiconductor in this region; the presence of a metal film or of a thick oxide may strongly prevent a lateral generation.

A complete study of this effect requires a two-dimensional analysis, since the junction curvature [6] at the boundaries of the geometrical area must be considered. If necessary, one can simulate this effect in a first-order approximation by taking a p-n junction perpendicular to the semiconductor surface. The n region, being in our case the substrate, is submitted to a generation term

$$g(x) = \alpha I_o \lambda \ e^{-\alpha x}/hc$$

where x is the distance from the surface. The resulting photocurrent flows in the direction y parallel to the semiconductor surface (see fig. 20). A detailed study of this case has been presented by Ambroziak [8] . Denote by y_n the width of the depletion region of the lateral junction, and assume an ohmic contact exists at a distance L' from the metallurgical junction. The continuity equation for the holes in the region (y_n, L') is

$$D_p(d^2p/dy^2) - (p - p_o)/\tau_{po} + g(x) = 0$$

assuming the hole current is a diffusion current. The boundary conditions are

$$p(y_n) = p_0 \cdot \exp(qV_a/kT)$$

$$p(L') = p_0$$

The surface recombination effects, for $x = 0$, are neglected. Straightforward calculations yield the hole photocurrent, and allow to define a partial quantum efficiency $\eta_{n,1}$ at the n side of the lateral photodiode; one obtains; at a distance x from the semiconductor surface,

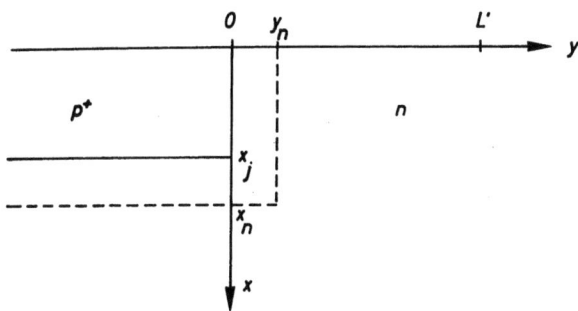

Fig. 20 *Lateral generation in a p-n junction perpendicular to the semiconductor surface.*

$$\eta_{n,1}(x) = \alpha L_p e^{-\alpha x} \tanh\left[(L' - y_n)/2 L_p\right]$$

Clearly $\eta_{n,1}$ attains a maximum value for $x = 0$. For large values of L', the expression for $\eta_{n,1}$ becomes

$$\eta_{n,1}(x) = L_p(\alpha e^{-\alpha x})$$

and indicates that the lateral photodiode collects all minority carriers generated within a diffusion length L_p from the boundary y_n of the depletion region. Finally one must add the contribution of the depletion region $(0, y_n)$; the corresponding partial quantum efficiency is

$$\eta_{DR,1}(x) = y_n (\alpha \, e^{-\alpha x})$$

The minority carriers generated within the p region may be considered as contributing essentially to the photocurrent of the junction parallel to the semiconductor surface; as a result we conclude that the additional quantum efficiency of the total junction in the plane x can be approximated by

$$\eta_1(x) = (L_p + y_n) \alpha \, e^{-\alpha x} \qquad (59)$$

for large values of L'. This expression applies for values of x within the interval (o, x_n) (see fig. 20); integration of (59) yields

$$\int_o^{x_n} \eta_1(x) \, dx = (L_p + y_n)(1 - e^{-\alpha x_n}) = (L_p + x_n - x_j)(1 - e^{-\alpha x_n})$$

(60)

The contribution of the region $x > x_n$ may be neglected, in a first-order approximation. Finally the expression (60) must be multiplied by the length l of the boundary between the principal photodiode and the lateral photodiode. For a photodiode having a geometrical area A, the total quantum efficiency may then be approximated by

$$\eta A + l \, (L_p + x_n - x_j)(1 - e^{-\alpha x_n})$$

η being the quantum efficiency (per unit area) of the principal photodiode. The relative influence of the lateral photodiode depends on the geometry of the photodiode and on the transmittance of the multilayer structure on top of the semiconductor; its dependence, through the factor $1 - \exp(-\alpha x_n)$, on the absorption coefficient α is illustrated in figure 4.

2. Effect of drift field

So far we neglected, for the minority carriers in the quasi-neutral regions, the influence of the electric field on the current flow and on the resulting quantum efficiency; this approximation is allowed for abrupt p-n junction, since the field resulting from the applied voltage remains small in the quasi-neutral regions. However, for junctions with a different impurity profile, a drift field may be present, even in the absence of an external applied voltage.

The profile pf a p-n junction depends on the doping technique used to introduce impurities into the semiconductor substrate. For a two-step diffusion process (predeposition, drive-in) the concentration distribution of acceptor impurities in the surface region may be approximated by a Gaussian distribution [6]

$$N_A(x) = N_s \exp(-x^2/4\,Dt)$$

N_s being the surface concentration, D the diffusivity of the acceptor impurities and t the drive-in diffusion time.

The equilibrium concentration of holes p(x) within the quasi-neutral region (o, x_p) is almost equal to $N_A(x)$. Furthermore since the hole current density under equilibrium conditions is zero, we have

$$N_A(x)\,\mu_p\,\mathcal{E}_o(x) - D_p(dN_A/dx) = 0$$

Hence the equilibrium drift field $\mathcal{E}_o(x)$ is given by

$$\mathcal{E}_o(x) = (kT/q)\left[\,d(\ln N_A)/dx\,\right]$$

and arises as a result of the gradient of doping concentration. For the particular case of a Gaussian distribution one obtains

$$\mathcal{E}_o(x) = -(kT/2\,q\,Dt)\,x$$

The drift field varies linearly with x, and is in such a direction as to enhance the flow of minority carriers towards the depletion region. We conclude that the drift field has the property of partially compensating the effects of surface and bulk recombination by

- sweeping the minority carriers from the surface region, where recombination occurs, towards the depletion region,
- reducing the transit time of minority carriers from their optical generation point to the collecting depletion region, thus reducing the probability of bulk recombination.

As a result, the drift field enhances the quantum efficiency η_p of the surface region, especially for large values of the absorption coefficient. In case of low carrier injection it is usually assumed that the field $\mathcal{E}(x)$ within the quasi-neutral region under non-equilibrium conditions remains almost equal to its equilibrium value $\mathscr{E}_0(x)$. Even then, it is mathematically impossible to develop a closed-form solution of the continuity equation for the variable drift field case; a computer analysis is required. An analytical study of the η_p-enhancing effect of the drift field resulting from a Gaussian distribution is possible and has been made by Gary [9] assuming an infinite lifetime for the minority carriers. The last assumption may be omitted if the average drift field in the surface region is used

$$\bar{\mathscr{E}}_0 = (1/x_p) \int_0^{x_p} \mathscr{E}_0 dx = - (kT/4 q Dt) x_p = \mathscr{E}_0(x_p)/2$$

This simplification in the drift field is equivalent to assuming that the impurity profile is exponential rather than Gaussian. According to Gary, the partial quantum efficiency η_p of the surface region becomes, for the case of a constant drift field $\bar{\mathscr{E}}_0$,

$$\eta_p = \frac{\alpha L_n^2}{(\alpha L_n)^2 - 2 m \alpha L_n^2 - 1} \Bigg\{ (2m - \alpha) e^{-\alpha x_p}$$

$$+ \frac{(M - m) \left[(S - M - m) e^{-\alpha x_p} - (S + \alpha - 2m) e^{(M - m)x_p} \right]}{(S - M - m) e^{-(M + m)x_p} - (S + M - m) e^{(M - m)x_p}} \cdot e^{-(M+m)x_p}$$

$$+ \frac{(M + m) \left[(S + \alpha - 2m) e^{-(M+m)x_p} - (S + M - m) e^{-\alpha x_p} \right]}{(S + M - m) e^{(M - m)x_p} - (S - M - m) e^{-(M + m)x_p}} \cdot e^{(M-m)x_p} \Bigg\}$$

where

$S = s/D_n$
$m = q\bar{\mathscr{E}}_0/2 kT$
$M = (m^2 + 1/L_n^2)^{1/2}$

This result must be compared to the expression (32) valid in the absence of a drift field.

According to Gary [9] the constant drift field approximation yields an overestimate of the quantum efficiency η_p (since the exact field is, in fact, zero at the semiconductor surface), but remains a reasonable approximation to use in the general case.

The η_p-enhancing effect of the constant drift field becomes important when

$$(q/2\ kT)\ |\overline{\mathcal{E}_o}| \gg 1/L_n$$

In particular, for a Gaussian profile one obtains, by taking the average drift field,

$$L_n \gg (8\ Dt/x_p)$$

This condition may be satisfied quite easily in practical photodiodes.

In this case a simpler expression may be used for η_p, since $M = m$,

$$\eta_p = \frac{2m\left[(S + \alpha - 2m)\ e^{-2mx_p} - S\ e^{-\alpha x_p}\right]}{(\alpha - 2m)\left[S - (S-2m)\ e^{-2mx_p}\right]} - e^{-\alpha x_p}$$

In particular, for s = o, one obtains

$$\eta_p = 1 - e^{-\alpha x_p}$$

We conclude that the partial quantum efficiency η_p of the surface region becomes equal to its ideal value for large values of the drift field, when the surface recombination velocity is neglected. We know that the last assumption is valid for practical photodiodes. Hence the partial quantum efficiency η_p of such photodiodes is then equal to its value for an abrupt photodiode having an infinite diffusion length L_n in the surface region [see equ. (57)].

1. W. Shockley and W.T. Read, Jr., Phys. Rev., 87, pp. 835-842, (1952).

2. F. Van de Wiele, Proc. IEEE, 61, pp. 793-794, (1973).

3. R.H. Bube, Photoconductivity of solids, J. Wiley, (1960).

4. A.G. Milnes, Deep impurities in semiconductors, J. Wiley, (1973).

5. J.C. Tandon, D.J. Roulston, S.G. Chamberlain, Solid-State Electron., 15, pp. 669-685, (1972).

6. A.S. Grove, Physics and technology of semiconductor devices, J. Wiley, (1967).

7. R.W. Brown and S.G. Chamberlain, Physica Status Solidi (a) 20, pp. 675-685, (1973).

8. A. Ambroziak, Semiconductor photoelectric devices, London Iliffe Books, (1968).

9. P.A. Gary, Modeling and Optimization of a Silicon Photosensor for a Reading Aid, Stanford University, Technical Report, n° 4822-1, May (1967).

BULK TRAPPING

by

A. Goetzberger

Institut für Angewandte Festkörperphysik der Fraunhofergesell-
schaft, D - 78 Freiburg, Eckerstr. 4, F.R. Germany.

Abstract

Bulk trapping phenomena are caused by impurities or imperfections. The basic processes capture and emission are introduced. Important properties of traps like energy level,
capture cross section and donor or acceptor character are
considered. Trap occupation statistics, recombination rate
and lifetime are derived from basic assumptions. Deep level
centers can be classified as traps or recombination centers
by introducing a classification level. The trap occupancy in
a space charge region is determined by the ratio of emission
rates.

1. Introduction and definitions

In an ideal, perfectsemiconductor there exists a forbidden
gap which is free of states that can be occupied by electrons.
This type of semiconductor is called intrinsic semiconductor.
An extrinsic semiconductor is doped with impurities having a
low ionization energy for electrons and holes to give predominantly n- or p-type conductivity. When the ionization energy of the impurity is higher it will not be ionized at room
temperature and will act as a trap or recombination center.
Stated differently, a doping impurity gives rise to a shallow
energy level or a recombination center or trap with a deep

energy level. It should be stressed at the outset that the
difference between shallow levels, traps and recombination
centers is only a qualitative one. A given impurity level
can act differently dependent on temperature and carrier densities. This will be discussed in detail below. It should
also be noted that energy levels are not only caused by impurities but also by a large variety of crystal defects. This
applies particularly to compound semiconductors. Often impurities form complexes with vacancies or other defects giving
rise to characteristic energy levels or systems of energy
levels.

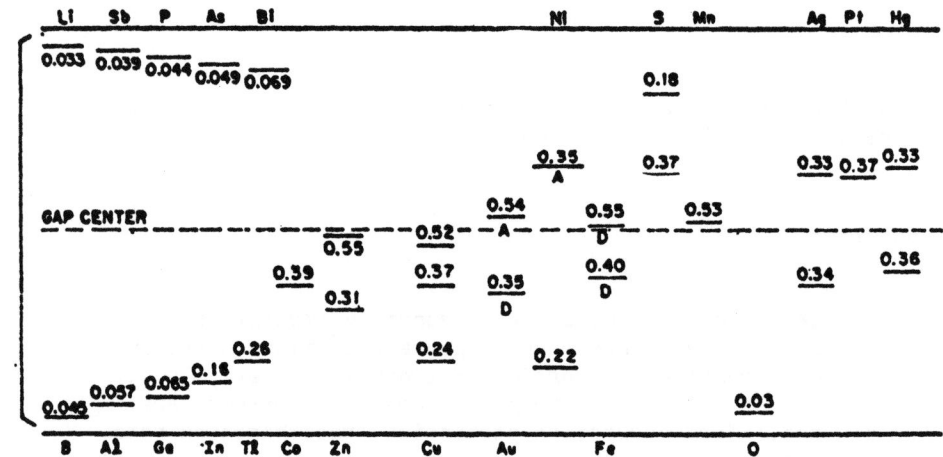

Fig. 1 *Energy levels of impurities in silicon.*

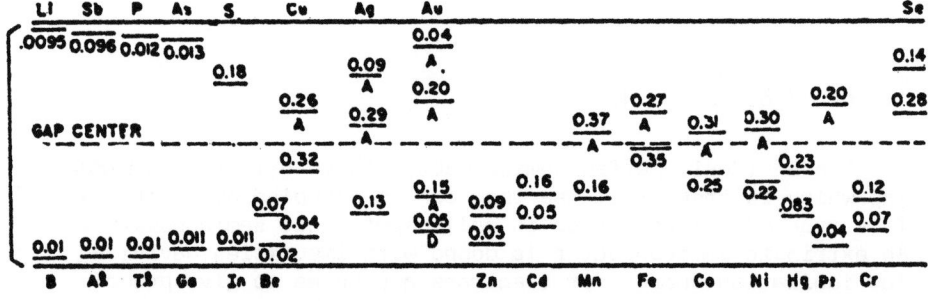

Fig. 2 *Energy levels of impurities in germanium.*

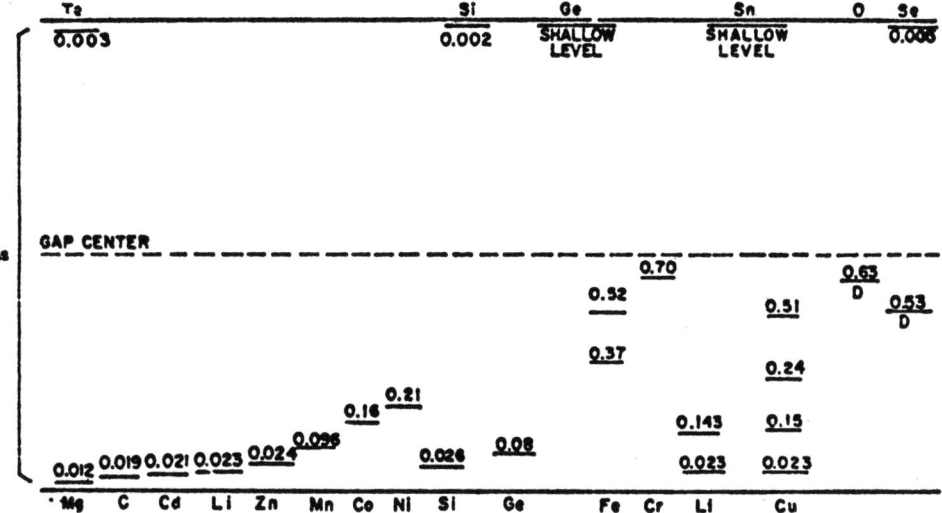

Fig. 3 Energy levels of impurities in GaAs.

Fig. 1a-c[1] gives a summary of energy levels due to various impurities in Si, Ge and GaAs. It will be noted that most deep centers are associated with more than one energy level. Before the trapping behavior can be analyzed, a few fundamental relations have to be introduced.

Following the basic concepts of Shockley and Read[2] we define four basic processes :

a) Electron capture from the conduction band.
b) Electron emission from the trap into the conduction band.
c) Hole capture from the valence band.
d) Hole emission into the valence band.

Process a is described by a capture rate $c_n = \sigma_n V_{th} [cm^3 \, sec^{-1}]$ where σ_n = electron capture cross section, V_{th} = average thermal velocity of electrons. Similarly process c is described by $c_p = \sigma_p V_{th}$. Process b is dominated by the emission rate $e_r [sec^{-1}]$ and also d by $e_p [sec^{-1}]$.

The rate at which electrons and holes are captured and emitted depends on the density of free carriers and the occupation of the traps. In thermal equilibrium both are determined by the same Fermi energy E_F. Energy levels are further characterized by being donor- or acceptor-like. A donor is neutral when occupied by an electron and positive when ionized, an acceptor level can capture an electron and thus become negatively charged. Doubly charged donors or acceptors are also possible.

If a total number of donors N_D is contained in a unit volume of semiconductor a certain number N_D^+ will be charged positively and $N_D = N_D - N_D^+$ will be neutral. In thermal equilibrium this number is given by a Fermi-Dirac-Distribution :

$$N_D^+ = \frac{N_D}{1 + g \exp \frac{E_F - E_D}{kT}} = N_D (1 - f_T) \qquad (1)$$

where g is the spin degeneracy factor and f_T is the fraction of traps occupied by electrons. E_F = Fermi Level, E_D = donor ionization energy.

The negatively charged acceptors are given by a similar relation

$$N_A^- = \frac{N_A}{1 + \frac{1}{g} \exp \frac{E_A - E_F}{kT}} = N_A f_T \qquad (2)$$

The factor g is mostly assumed to be 2 but other values have also been suggested in the literature. g can be different for different traps and is generally not very well known. From equ. (1) it follows that the contribution of g can also be added to the trap energy as $kT \ln(g)$ and since it is usually very small compared to E_D it is very hard to obtain quantitatively. Only when temperature is changed the influence of g is expected to enter. Fortunately in practical cases this question plays a very minor role.

Now we can express the rate equations sketched in Fig. 4

a : $r_n = N_T c_n (1 - f_T) n$ (3)

b : $g_n = N_T e_n f_T$ (4)

c : $r_p = N_T c_p f_T p$ (5)

d : $g_p = N_T e_p (1 - f_T)$ (6)

where $r_{n,p}$ = capture rate, $g_{n,p}$ = generation rate, N_T = density of traps (N_T is used when the difference between donor and acceptor states is not relevant).

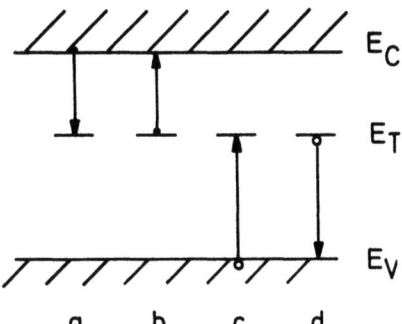

*Fig. 4 The four basic capture and emission processes; E_c is conduction band edge, E_T Energy of trap level.
• = Electron, o = Hole.*

Capture is proportional to the density of electrons and the density of unoccupied traps. Emission is only proportional to the density of occupied traps. Thus $c_{n,p}$ and $e_{n,p}$ have different dimensions.

One impurity or defect can cause more than one level. Inspection of Figs. 1a, b and c shows that most deep traps are indeed multilevel centers. The following discussion of traps will necessarily have to be brief and many interesting properties of traps have to be omitted. For further details the reader is referred to the literature. An excellent monograph was recently published by A.G. Milnes [3].

2. Trap Depth.

The depth of a shallow trap can be easily estimated using the effective mass concept : The Coulomb attraction of a positive charge in a medium with the dielectric constant of the semiconductor is considered to act on an electron with effective mass m^*. Now the ground state of the hydrogenic system is calculated

$$E_T = \frac{(m^*/m_0)}{\epsilon^2} E_H \qquad (7)$$

where m_0 = mass of the free electrons, ϵ = dielectric permittivity of the semiconductor, E_H = ground state energy of hydrogen atom. This simple estimate gives values remarkably close to the experimental ones. For silicon for instance E_n = 0.04 eV compared to a range of 0.033 to 0.069 eV for various donors. The effective mass approximation works so well because the wave function of the bound electron extends over many lattice sites so that the discontinuity of the lattice does not enter.

This approach fails for deep levels which have a very constricted wave function with high binding energy. The theory has to take the nature of the impurity and its incorporation into the lattice into account and therefore becomes very complicated. Even then the results are not very satisfactory. The energy levels of deep impurities are therefore still largely an empirical affair.

3. Capture Cross Sections.

Besides energy level capture cross sections are a basic atomic property of a given impurity. Experimentally they are obtained by measuring the capture rate c. The rate of change of electron density n in the conduction band into empty donor levels is from (3)

$$\frac{dn}{dt} = r = - c_n \, n \, N_o \qquad (8)$$

If the electrons all have the same velocity, the capture rate would be simply

$$c_n = \sigma_n V_{th} \qquad (9)$$

where σ_n is the capture cross section for electrons. In reality the electrons have a thermal distribution of velocities and furthermore the capture cross section may depend on electron energy. These effects are averaged for practical reasons by dividing the experimentally determined capture rates by the mean thermal velocity V_{th} of electrons or holes :

$$\sigma_{n,p} = c_{n,p}/V_{th}$$

The experimental values of σ cover a considerable range, from 10^{-12} to 10^{-22} cm^2. Most of these capture cross sections are only accurate within an order of magnitude. If better values are desired the measurements become very elaborate and time consuming.

The differences of capture cross sections can be qualitatively understood by taking into account that dependent on occupancy a level may be either Coulomb attractive, neutral or repulsive. These three cases are illustrated in Fig. 5.

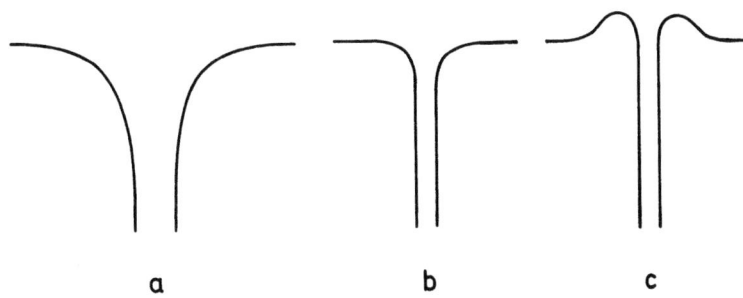

Fig. 5 Field distribution for a, a Coulomb attractive b, a neutral c, a repulsive center.

For the attractive center the capture cross section can be estimated by a similar procedure as in the case of the energy level (equ. 7)[4,5]. Assuming that a free carrier is captured if it comes sufficiently close to a charged center that the binding energy is equal to the thermal energy kT the equation

$$kT = \frac{q^2}{r\varepsilon} \qquad (10)$$

results, q = elementary charge, r = distance from attractive center. The capture cross section is then

$$\sigma = r^2\pi = \frac{q^2\pi}{kT\varepsilon} \qquad (11)$$

or about 10^{-12} cm^2 at room temperature. This is close enough to the large experimental capture cross sections. Nevertheless it is theoretically very difficult to account for the large size of the attractive and neutral capture cross sections. The major problem is transfer of the energy (of the order of one half the gap for deep traps) to the lattice which can only occur by phonons. Since the phonon energies are much smaller than the energy of the free carrier, capture has to involve multiple phonon processes. This would make carrier trapping a very unlikely process and therefore lead to small capture cross sections. Lax[6] has proposed a model which partially explains the problem of so called "giant cross sections". In this cascade model it is assumed that capture initially occurs into excited states with large orbits. The energy is then transmitted in steps to phonons as the electron makes successive transitions into the ground state.

4. Trap Occupation and Dynamics.

If an excess density if minority carriers is introduced into a semiconductor containing deep levels we have to distinguish between trapping and recombination processes. An electron may be captured into a trap (process a in Fig. 4) and then, after some time - the trapping time - reemitted into the conduction band (process b). On the other hand while the electron is trapped a hole may be captured from the valence band thus rendering the center again neutral. An excess electron may be captured and reemitted several times before disappearing through recombination as shown in Fig. 6.

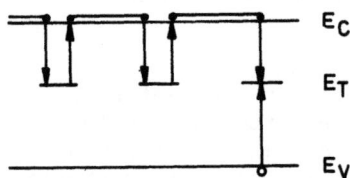

Fig. 6 Typical distribution of a trap level. An electron is trapped and reemitted several times into the conduction band before it finally recombines with a hole.

In the following calculation the principle of detailed balance will be used to derive important relations between trap parameters.[2]

The concentration of electrons in the conduction band is described by the difference of a and b of Fig. 4 (equ. (3) and (4)).

$$\frac{dn}{dt} = - c_n \, n \, N_T \, (1 - f_T) + e_n \, N_T \, f_T \qquad (12)$$

Exactly the same relation exists between reactions c and d involving holes :

$$\frac{dp}{dt} = - c_p \, p \, N_T \, f_T + e_p \, N_T \, (1 - f_T) \qquad (13)$$

The principle of detailed balance now states that in steady state each of the rates (12) and (13) is zero since the concentration of electrons as well as holes does not change. Therefore

$$c_{ne} \, n(1 - f_T) = c_{ne} \, f_T \qquad (14)$$

Where the subscript e denotes equilibrium rates (14) has to holed for arbitrary positions of the Fermi level. In particular we can choose the Fermi level such that the traps are half filled ($f_T = 1/2$). The electron density for this case is called n_1. Then we obtain a general relation between c_{ne} and e_{ne} :

$$\frac{e_{ne}}{c_{ne}} = n_1 \qquad (15)$$

and also

$$\frac{e_{pe}}{c_{pe}} = p_1 \qquad (16)$$

Physically n_1 and p_1 are the carrier densities when the Fermi level is at the trap energy as is seen from equ. (1).[x] Thus if c_{ne} (or σ_n) and the trap depth E_T are known e_{ne} can be calculated and vice versa.

We can also postulate a carrier density n^*, p^* such that the capture rate of holes is equal to the capture rate of electrons. Then by the principle of detailed balance the emission rates have to have the same value also.

Equating, process a and d and b and c in (12) and (13) we find the following relation

$$\frac{c_n}{c_p} = \frac{p^*}{n_1} = \frac{p_1}{n^*} \qquad (17)$$

---

[x] Strictly speaking this is only true when the degeneracy factor g is one but as was pointed out before the influence of g is very minor.

This relation is not as useful as (15) and (16) because n^*, p^* are not directly related to the trap energy. This means that in practice σ_n, σ_p and E_T have to be determined experimentally in order to describe a trap level completely.

We shall now consider the dynamics of recombination. In steady state there is constant recombination requiring that $\frac{dn}{dt} = \frac{dp}{dt}$. Equating (12) and (13) and using (15) and (16) f_T can be computed. If this expression is reinserted into (12) and (13) the steady state recombination rate $U = \frac{dn}{dt}$ is obtained:

$$U = - \frac{np - n_i^2}{\frac{n + n_1}{c_p N_T} + \frac{p + p_1}{c_n N_T}} \qquad (18)$$

In the last expression $n_1 p_1 = n_i^2$ has been used. It is practical to define limiting electron and hole lifetimes τ_n and τ_p by

$$\tau_n = \frac{1}{c_n N_T} \; ; \; \tau_p = \frac{1}{c_p N_T} \qquad (19)$$

Then (14) becomes

$$U = \frac{dn}{dt} = \frac{dp}{dt} = - \frac{np - n_i^2}{\tau_p(n+n_1) + \tau_n(p+p_1)} \qquad (20)$$

Let us now consider an n-type semiconductor with electron and hole density n_o, p_o with $n_o \gg p_o$. If the hole density has been raised to $p + \delta p$ the electron density adjusts to $n + \delta n$ with $\delta n = \delta p$ to preserve charge neutrality. With these values (20) reduces to

$$\frac{dp}{dt} = - \frac{n_o \delta_p}{\tau_p(n_o+n_1) + \tau_n p_1} \qquad (21)$$

Since for n-type $n_o \gg n_1$, p_1, we get

$$\frac{dp}{dt} \approx \frac{\delta p}{\tau_p} \qquad (22)$$

Quite generally the lifetime is defined as

$$\tau_{n,p} = \frac{\delta_{n,p}}{U} \qquad (23)$$

Thus the above definition (19) is justified. For small deviations from equilibrium and strong doping hole lifetime in n-type semiconductor is therefore τ_n and electron lifetime in p-type semiconductor is τ_p.

5. Multilevel Traps.

For multilevel centers a few general rules can be recognized. Fig. 7 shows a generalized multilevel center. In order to understand trap behavior it is important to realize that the occupancy of the trap changes by one unit of charge when the Fermi level crosses that level.

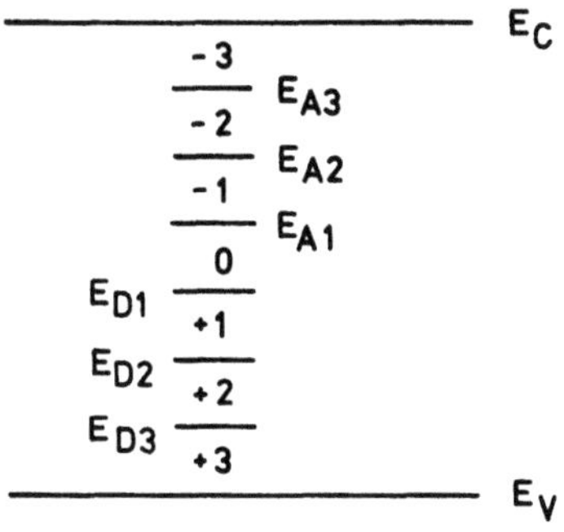

Fig. 7 Multilevel trap. E_{A1} ... are successive acceptor levels, E_{D1} ... successive donor levels. Numbers between levels give charge state of trap when Fermi level is in this range.

In Fig. 7 E_{A1}, E_{A2} ... are successive acceptor levels and E_{D1}, E_{D2} ... are successive donor levels. When the Fermi level is between E_{D1} and E_{A1} the trap is neutral. When it crosses E_{A1} one electron is captured and the charge is -1. Then follows -2 and so on. The same is true for positive charge when the Fermi level moves towards the valence band. This is indicated by the numbers between the levels in Fig. 7. The arrangement of levels must be as shown in Fig. 7 since it requires more and more energy to add successive charges.

It is possible to imagine that an energy level falls within the conduction or valence band. Such a level is of no consequence because it requires more energy to fill it than the production of free pairs by direct band to band transition. Therefore the latter process will be preferred.

From Fig. 7 it can also be understood that a shallow donor (E_{D1} close to the conduction band E_c) can never have acceptor levels as well. At most it may have deeper multiply charged donor levels.

When the occupation of such multilevel centers is calculated it is necessary to take into account that they are not independent. A single impurity atom can only be in one state. The mathematical treatment of such systems is very involved[7] and will not be pursued further here.

6. Classification of Traps or Recombination Centers.

Another question of importance is whether a deep level will act as a trap or a recombination center. Let us imagine a doubly charged donor. Such a level is likely to act as an electron trap since it will be attractive to electrons but repulsive to holes even after having captured an electron. Recombination centers on the other hand are likely to be singly charged and attractive to minority carriers. In this case capture rate of minority carriers is high because of the Coulomb attraction. Capture rate of majority carriers is also high because their density is high.

Trapping behavior also depends strongly on temperature. A relatively shallow level within the thermal activation energy from the band cannot bind a carrier at room temperature. If temperature is lowered, however, it may become a very active trap.

After these qualitative considerations a more quantitative treatment of trapping vs. recombination behavior will be given. This problem is of greatest interest when carrier densities deviate from equilibrium. n and p are then described by quasi Fermi levels E_{Fn} and E_{Fp} according to the following definitions :

$$n = n_i \exp\left(\frac{E_{Fn} - E_i}{kT}\right) \; ; \; p = n_i \exp\left(\frac{E_i - E_{Fp}}{kT}\right)$$

Where n_i = intrinsic carrier density; E_i = intrinsic energy level (n_i is chosen as reference density in this case because then the relations are particularly simple).

According to Stöckmann[8] it is possible to define a demarkation level E_T which defines the behavior of the trap. Considering Fig. 4 and equations (3) - (6) we can postulate that for an imperfection to act as an electron trap two conditions have to be fulfilled :

1. In the empty state the probability to capture an electron from the conduction band must be larger than that for emission of a hole to the valence band.

2. In the occupied state the probability for thermal excitation into the conduction band must be larger than the probability for its recombination with a free hole. Therefore the conditions for electron traps are :

$$r_n \gg g_p \quad ; \quad g_n \gg r_p \tag{24}$$

For hole traps

$$r_p \gg g_n \quad ; \quad g_p \gg r_n \tag{25}$$

For recombination centers we get :

$$r_n \gg g_p \quad ; \quad g_n \ll r_p \tag{26}$$

and for generation centers

$$r_n \ll g_p \quad ; \quad g_n \gg r_p \tag{27}$$

Expressing the inequalities (24) by (3) - (6) and (15), (16) we obtain for (24a) :

$$r_n = c_n n(1-f_T)N_T = \sigma_n V_{th} n_i \exp \frac{E_{Fn} - E_i}{kT}$$

$$\times (1-f_T) N_T \gg g_p = e_p(1-f_T)N_T =$$

$$= \sigma_p V_{th} p_1 (1-f_T) N_T = \sigma_p V_{th} n_i \exp \frac{E_i - E_T}{kT} (1 - f_T) N_T$$

This reduces to

$$\sigma_n \exp \frac{E_{Fn} - E_i}{kT} \gg \sigma_p \exp \frac{E_i - E_T}{kT} \tag{28}$$

Taking the logarithm of (28)

$$kT \ln \frac{\sigma_n}{\sigma_p} + E_{Fn} - E_i > E_i - E_T \qquad (29)$$

$$E_{Fn} > 2E_i - E_T - kT \ln \frac{\sigma_n}{\sigma_p} \equiv E_T^* \qquad (30)$$

Similarly for (24b):

$$E_{Fp} > 2E_i - E_T - kT \ln \frac{\sigma_n}{\sigma_p} \quad E_T^* \qquad (31)$$

The important conclusion is that the same demarkation level is sufficient to classify relations (24) - (27). Once E_T is known the property of the trap can be derived according to the following rules. The centers are

electron traps, if $\qquad E_{Fn}, E_{Fp} > E_T^*$

hole traps, if $\qquad E_T^* > E_{Fn}, E_{Fp}$

recombination centers, if $E_{Fn} > E_T^* > E_{Fp}$

generation centers, if $\quad E_{Fp} > E_T^* > E_{Fn}$

The classification level E_T^* is in the mirror image position of the trap level E_T with respect to the intrinsic level E_i if the kT terms can be neglected, e.i. if the capture cross sections are equal:

$$E_i - E_T = E_T - E_i \qquad (32)$$

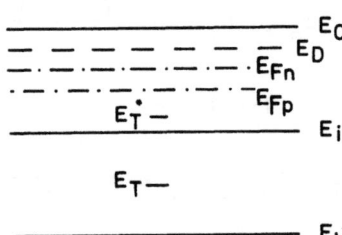

Fig. 8 Classification of traps. Shown in an n-type semiconductor with shallow donors E_D and a trap at E_T. Trap classification level is E_T^*. In this figure E_T^* E_{Fn}, E_{Fp}. Therefore E_T is an electron trap.

Fig. 8 gives an example for an electron trap : in an n-type semiconductor with donor levels close to the conduction band the electron Fermi level E_{Fn} is in the upper half of the band. If additional holes are brought into the semiconductor the hole Fermi level E_{Fp} moves downward. A trap at E_T in the lower half of the band has its classification level at E_T^*. If the conditions are as in Fig. 8 the trap is an electron trap. If more holes are added E_{Fp} moves even lower and after it crosses E_T^* E_T will be a recombination center.

7. Traps in Space Charge Regions.

When traps are located within a space charge region e.g. a pn junction or surface space region their occupancy is only determined by emission rates since capture is negligible due to the low carrier concentration in such a region. In steady state equilibrium exists between electron emission and hole emission (process b and c in Fig. 4) (Equ. (4) and (6))

$$N_T e_n f_T = N_T e_p (1 - f_T) \qquad (33)$$

Thus

$$f_T = \frac{e_p}{e_n + e_p} = \frac{1}{e_n/e_p - 1} \qquad (34)$$

In a space charge region the traps emit holes and electrons alternatively generating reverse current. Equ. (34) relates the occupancy to the ratio of the emission rates. If we make use of relations (15) and (16) and of the definition of n_1 and p_1 in terms of the trap level E_T we find:

$$\frac{e_n}{e_p} = \frac{c_n}{c_p} \exp \frac{2(E_T - E_i)}{kT} \qquad (35)$$

Thus we can state that if a trap is around the middle of the gap its occupancy will be determined by the ratio of the capture cross sections. As we move the traps away from midgap the exponential factor very rapidly takes over forcing traps in the upper half of the gap to be empty and in the lower half to be full of electrons. From the influence on the total charge in the space region of a junction one can thus also determine the donor or acceptor character of a trap.

Further, one of the most accurate means for determining capture cross sections consists in observing capture and emission rates in p-n junctions after a sudden change of applied voltage.[9,10]

References

1. S.M. Sze, Physics of Semiconductor Devices, Wiley, New York (1969).

2. W. Shockley and W.T. Read, Phys. Rev., 87, 835 (1952).

3. A.G. Milnes, Deep Impurities in Semiconductors, Wiley, New York (1973).

4. A. Rose, Concepts in Photoconductivity and Allied Problems, Interscience Publishers, New York (1963).

5. R.H. Bube, Photoconductivity of Solids, Wiley, New York (1960).

6. M. Lax, Phys. Rev., 119, 1502 (1960).

7. W. Shockley and J.T. Last, Phys. Rev. 107, 392 (1957).

8. F. Stöckmann, Phys. Stat. Sol. (a) 20, 217 (1973).

9. C.T. Sah, L. Forbes, I. Rosier, and A.F. Tasch Jr., Solid State Electron, 13, 759 (1970).

10. D.V. Lang, J. Appl. Phys., 45, 3023 (1974).

Surface Trapping

A. Goetzberger

Institut für Angewandte Festkörperphysik der Fraunhofer -
Gesellschaft 78 Freiburg, Eckerstrasse 4

Abstract

A survey of surface trapping phenomena related to surface states is given. First the electrical behavior of the surface state free surface is treated. Then the various properties of surface states are introduced. They influence capacitance, loss factor and surface potential. The conditions at real surfaces may further alter those parameters. Surface state distribution v.s. energy has a characteristic shape. The origin of surface states is considered for intrinsic and extrinsic surfaces. Finally the influence of surface states in devices is investigated. In junction devices recombination and generation play a major role. In charge coupled devices trapping by surface states is very important.

1. Introduction

Just like in the bulk allowed states may occur at the surface of a semiconductor within the energy gap. These states are called surface states. Owing to the two-dimensional nature of the surface, their density is measured per unit area in contrast to bulk states which are measured per unit volume. A third type of states, similar to surface states, occur at the interface between two adjacent materials. These states are called interface states. Very often they are also simply called surface states.

If the lattice of a semiconductor is terminated abruptly this perturbation of the crystal symmetry leads to discrete levels whose properties depend on the position of the cut in the unit cell[1,2]. This results in a very high density of electronic states, roughly one state per surface atom. This number has been confirmed many times experimentally on clean surfaces in a high vacuum. The density of surface states drops drastically if it is possible to saturate the dangling bands at the surface by chemical reaction. The best way to observe this is to thermally oxidize silicon. Since the greatest body of scientific and technical work was carried out on silicon, the $Si-SiO_2$ interface will be treated predominantly in this review.

2. Surface without Surface States.

Before interface states can be introduced, the ideal semiconductor surface has to be described mathematically. Fig.1 contains the pertinent quantities and definitions. The electrostatic potential varies near the surface. It becomes ψ_s at the surface and is then called surface potential. All the other definitions are self-explanatory. The energy bands can be bent up or down by an external electric field leading to a voltage-variable capacitance of the semiconductor surface. Since one of the most obvious effects of surface and interface states is a change of the surface capacitance C_D and its field dependence, it is necessary to establish this dependence first for the ideal semiconductor surface.

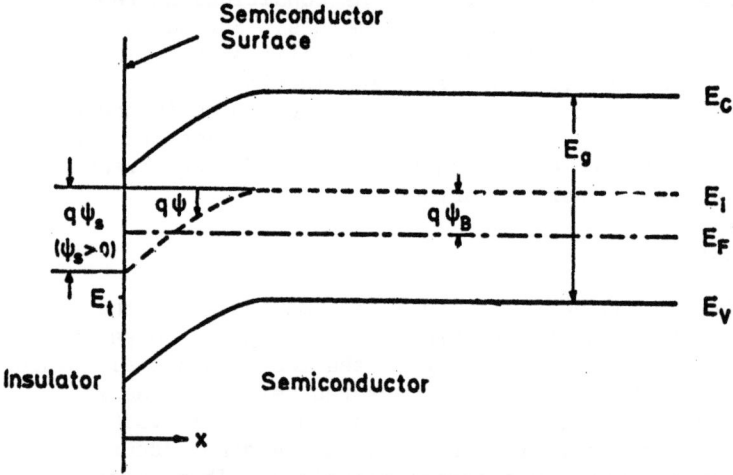

Fig. 1. Energy band diagram at the surface of a p-type semiconductor. Energy terms given in this sketch are converted to potentials by dividing by the electronic charge. It should be noted that the surface potential ψ_s is positive when the bands are bent downwards.

In practical devices, the field is applied by means of a metal electrode separated by a thin insulator film from the semiconductor. The voltage dependence of such an MOS (metal-oxide-semiconductor) structure can be obtained by integration of Poisson's equation. It is given by the oxide capacitance C_{ox} and the space charge capacitance C_D in series :

$$C = \frac{C_{ox} C_D}{C_{ox} + C_D} \tag{1}$$

C_D is further given by (for a p-type semiconductor)

$$C_D = \frac{\varepsilon_s}{\lambda_p} \frac{1 - e^{-q\psi_s/kT} + (n_o/p_o)(e^{q\psi_s/kT} - 1)}{G(\psi_s, n_o/p_o)} \tag{2}$$

where ε_s is the dielectric permittivity of the semiconductor. λ_p the extrinsic Debye length of the semiconductor, k the Boltzmann constant, and n_o, p_o the bulk densities of electrons and holes, respectively. q is the electronic charge.

$$G(\psi_s, n_o/p_o) = \left[e^{-q\psi_s/kT} + \frac{q\psi_s}{kT} - 1 + \frac{n_o}{p_o}(e^{q\psi_s/kT} - q\frac{\psi_s}{kT} - 1) \right]^{1/2} \tag{3}$$

The relation between surface potential ψ_s and voltage V applied to the metal is obtained by realising that V is divided into the voltage drop across the oxide V_{ox} and ψ_s.

$$V = V_{ox} + \psi_s \tag{4}$$

where V_{ox} is given by

$$V_{ox} = \frac{\varepsilon_s}{\lambda_p C_{ox}} \frac{2 kT}{q} G(\psi_s, n_o/p_o) \tag{5}$$

Three ranges can be distinguished for the surface potential depending on carrier concentrations at the surface (Fig. 2).

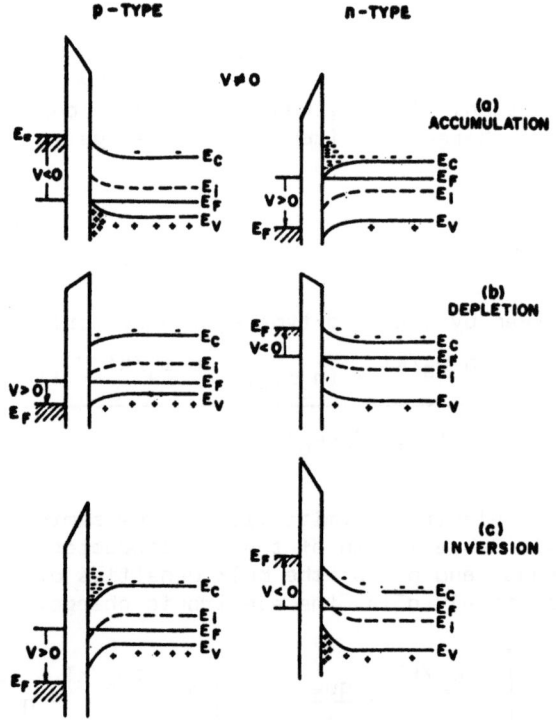

Fig. 2. Energy band diagrams for ideal MOS capacitors for both n-type and p-type semiconductors and a series of bias voltages showing accumulation (a) depletion, (b) and inversion, (c) layers respectively.

<u>p-type silicon</u>

1. $\psi_s < 0$ enhanced hole concentration : accumulation.

2. $2\psi_B > \psi_s > 0$ depleted hole concentration : depletion.

3. $\psi_s > 2\psi_B$ enhanced electron concentration : inversion or n-channel.

Where ψ_B is the potential difference between the Fermi potential E_F and the intrinsic potential E_i in the bulk.

n-type silicon

1. $\psi_s > 0$ enhanced electron concentration
accumulation.

2. $2\psi_B < \psi_s < 0$ depleted electron concentration depletion.
depletion.

3. $\psi_s < 2\psi_B$ enhanced hole concentration
inversion or p-channel.

Total differential capacitance of the MOS capacitor is a series combination of two capacities : oxide capacity and semiconductor space charge capacity. Only the semiconductor capacity depends on voltage.

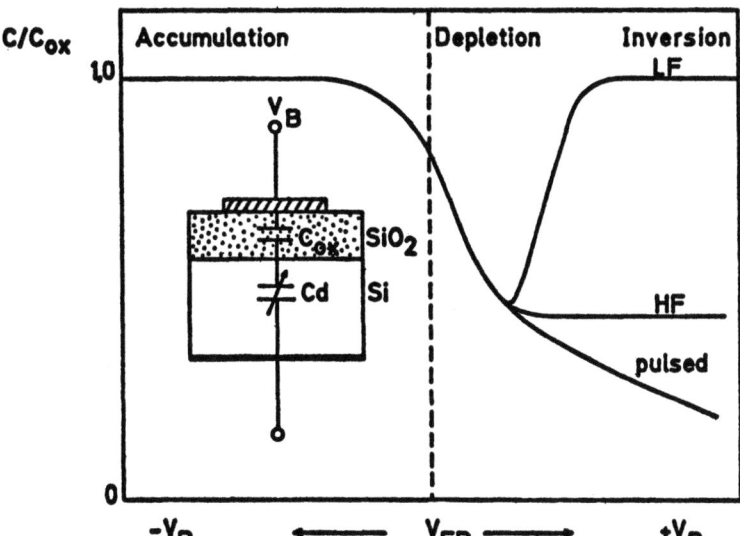

Fig. 3. *Typical capacitance-voltage curves (c-v curves) for an MOS capacitor on n-type semiconductor measured at low frequency (LF), at high frequency (HF), and under pulsed conditions. For a p-type semiconductor the sides are reversed. The insert demonstrates the capacitance distribution accross the element.*

Typical capacitance-voltage curves (c-v curves) for p-type semiconductor material are shown in Fig. 3. For n-type material, the sides are reversed. Capacity is normalized with respect to oxide capacity. In the same way as for carrier concentrations, three regions can be defined.
For negative bias in the accumulation region, the space charge capacity of the semiconductor is large. Total capacitance is close to the oxide capacity. Toward a positive bias voltage, the depleted surface zone acts as a dielectric in series with the oxide. The total capacity drops. The capacitance curve passes through a minimum and increases again when an inversion layer is formed. The increase is dependent on the ability of the carrier concentration in the channel to follow the applied signal. This is only possible at low frequencies (f ≈ 10 Hz) where recombination-generation rates of minority carriers can keep up with small signal variation .

For the high-frequency curve, recombination and generation cannot follow the small variation and thus contribute to differential capacity. An inversion layer, however, is formed in accord with dc-bias voltage. Total capacitance saturates at the value for which strong inversion starts to appear and where the width of the depletion zone is limited.

For a pulsed c-v measurement, the minority carrier channel is not in equilibrium with bias voltage. The depletion zone then increases and differential capacitance decreases continuously with applied voltage until breakdown in the semiconductor increases the carrier concentration.

3. Electrical properties of interface states.

The electrical effects of interface states can be qualitatively understood as follows :

A. <u>Capacitance</u> : An interface state constitutes an additional allowed state at the interface. It therefore adds a capacitance of one elementary charge per state. This capacitance is a sharply peaked function of the surface potential and thus of applied voltage. The peak occurs for the voltage for which the Fermi level crosses the interface state level.

B. <u>Conductance</u> : Capture and emission of carriers from interface states are not infinitely fast, but are associated with a time delay. This time delay can be expressed by an RC representation of the interface state. This time constant contributes to ohmic losses. Note that the ideal MOS structure does not show any losses. If such losses are present they are most likely due to interface states.

C. <u>Surface potential</u> : The previously mentioned capacitance and conductance are ac effects. In addition to these effects, interface states cause a dc effect. The charge stored in interface states modifies the electric surface field. More applied voltage is needed to change the surface potential by a given amount when interface states are present than in the ideal case. The effect is seen as a streched-out of the capacitance-voltage curves.

The admittance of an interface state can be calculated when only one type of carrier, electrons or holes interact with the states. Then Shockley Read [3] recombination statistics can be applied and the small signal approximation is made to obtain the small signal interface state admittance [4] Y_s :

$$Y_s = j\omega \frac{q^2}{kT} \frac{N_s f_o (1 - f_o)}{(1 + j\omega f_o/c_n n_{so})} \qquad (6)$$

where $j = \sqrt{-1}$ is the imaginary unit, ω is the angular frequency, N_s is the interface state density, f_o is the dc Fermi level, c_n the electron capture rate, and n_{so} the time averaged free electron density at the interface. In the case of hole capture $c_n n_{so}$ has to be replaced by $c_p p_{so}$.

Equation (6) describes the admittance of a series RC network with a capacitance C_s and a time constant τ :

$$C_s = q^2 N_s f_o (1 - f_o)/kT \qquad (7)$$

$$\tau = \frac{f_o}{c_n n_{so}} = \frac{f_o}{c_n n_o} e^{-\frac{q\psi_s}{kT}} \qquad (8)$$

In practical interfaces, a single level state is very rare and can only be realised by using special techniques [5]. Normal interface states are always continuously distributed in energy. Taking into account a distribution of interface states[6,7] yields a modified expression for the admittance

$$Y_s = \frac{q N_{ss}}{2 \tau_m} \ln(1 + \omega^2 \tau_m^2) + j \frac{q N_{ss}}{\tau_m} \times \qquad (9)$$

$$\times \text{arc tan} (\omega \tau_m)$$

where N_{ss} is the interface state density and

$$\tau_m = \frac{1}{c_n n_{so}} \qquad (10)$$

Equation (9) results in a dispersion of the time constants, i.e. a parallel network of RC terms. The experimentally observed dispersion is much bigger than predicted by eq. (9). This can be explained by the effect of interface charge fluctuations. Parasitic charge of not clearly established origin which is located close to the interface must be assumed to be statistically distributed across the interface and to give rise to built-in fluctuations of the surface potential [6]. The situation is explained in fig. 4. According to eq. (10) small changes of ψ_s result in sizable differences of time constants. Charge in interface states will also add to the fluctuations.

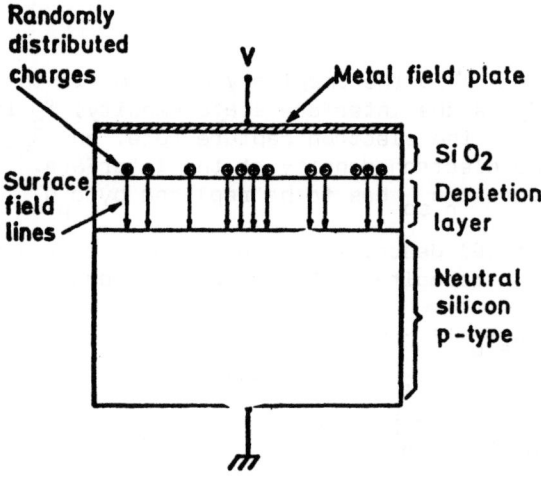

Fig. 4 Randomly distributed surface charges at interface but still in the oxide give rise to fluctuations of surface potential.

The first equivalent circuit suggested by Nicollian and Goetzberger [6] to explain the effects of charge fluctuations is shown in fig.5. This equivalent circuit leads to the following relations for capacitance and conductance of the MIS capacitor.

$$C_p = C_D(\bar{\psi}_s) + q\, N_{ss} (2\Pi\sigma_s^2)^{-1/2} \int_{-\infty}^{+\infty} \exp\left(\frac{q}{kT}\left[\psi_s + \psi_o + \frac{q}{2kT}(\frac{\psi_s - \bar{\psi}_s}{\sigma_s})^2\right]\right) \times$$
$$\times \arctan \exp\left(\frac{q}{kT}(\psi_s - \psi_o)\right) d\psi_s \qquad (11)$$

$$G_p/\omega = \frac{1}{2} q\, N_{ss} (2\Pi\sigma_s^2)^{-1/2} \int_{-\infty}^{+\infty} \exp\left(\frac{q}{kT}\left[\psi_s + \psi_o + \frac{q}{2kT}(\frac{\psi_s - \bar{\psi}_s}{\sigma_s})^2\right]\right) \times$$
$$\times \ln\left\{1 + \exp\left(\frac{2q}{kT}(\psi_s - \psi_o)\right) d\psi_s\right\} \qquad (12)$$

where $\bar{\psi}_s$ is the average surface potential, and σ_s is the standard deviation of the surface potential.

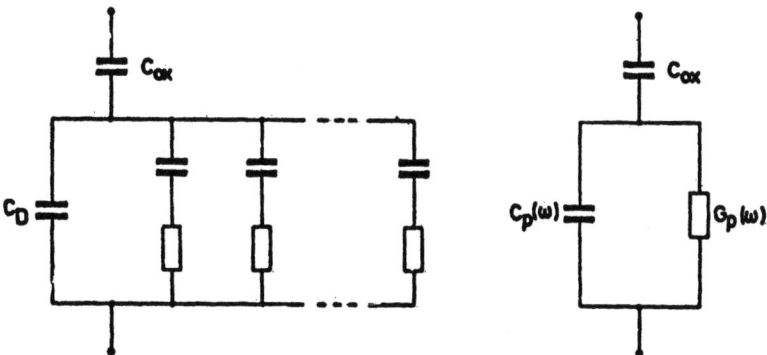

Fig. 5. Equivalent circuit taking into account charge fluctuations in the oxide, an equipotential plane is assumed in the interface for the ac signal.

4. Experimental Data on Surface States.

Many different techniques for measuring surface states have been introduced in the past. They are too diverse and complicated to be described here. The interested reader is referred to the literature. Only the most important results will be introduced in this subsection. The surface state density distribution across the gap has the following general features. (Fig.6)[8]. The density of states is lowest in the middle of the gap and increases towards the band edges, more strongly towards the valence than towards the conduction band. This general shape is independant of the absolute value of surface state density. Modern technology permits very low values of surface state densities, typically less than $10^{10}/cm^2$ in the middle of the gap. For trapping phenomena the capture cross sections are also of great importance because they determine trapping and detrapping rates. Capture cross sections depend on technological details of oxide growth, of crystal orientation, and on energetic positions of the surface states. Very few measurements of such cross sections have been made. Fig. 7. contains a compilation of such data. It is seen that for electron capture the cross sections are dropping towards the band edge. The reasons for the distribution of states as well as the behavior of the capture cross sections are not well known.

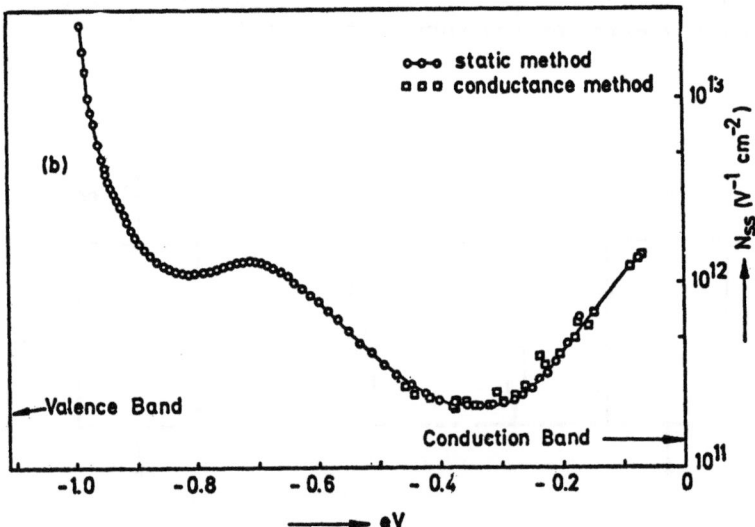

Fig. 6. Distribution of surface states in forbidden gap for thermally oxidized silicon. After Ref. 8.

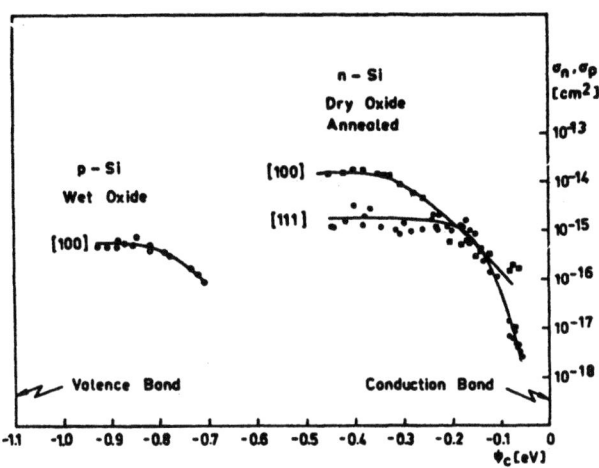

Fig. 7. Capture cross sections for surface states. σ_n is plotted for n-type silicon near the conduction band for two different orientations and for σ_p in p-type material near the valence band.

5. Origin of Interface States.

The origin of surface states is much less well understood than that of bulk states. Apparently a number of mechanismus can result in surface or interface states.

We distinguish between intrinsic and extrinsic interface states. By intrinsic interface, we mean a clean interface between SiO_2-Si which only contains silicon and oxide, in contrast to an extrinsic interface which is contaminated by impurities other than Si and O.

5.a) Intrinsic Interface.

For the ideally oxidised SiO_2-Si interface, it can be imagined that all dangling bonds which cause a high interface state density in the case of a free silicon surface are saturated by bonds with oxygen. [9,10]. The strong binding energy of oxygen to silicon shifts the energy of the bonds to lower energies away from the window of the silicon band gap where they are measured electrically [11]. It is therefore plausible that the electrically measured SiO_2-Si interface density is very low, far less than the number atoms or bonds in the interface.

However, it is easily visualized in the schematic arrangement of the bonds in the two-dimensional drawing of fig.8 which can be used to represent the three-dimensional tetrahedral configuration, that the wide lattice of the SiO_2 does not match the lattice of the silicon. An ideal contact of the two components cannot be achieved. There will always be a fraction of unsaturated bonds and imperfections of the arrangement in the vicinity of the interface which will cause a remaining density of interface states. Since the chemical affinity of oxygen to silicon is very large this fraction is small.

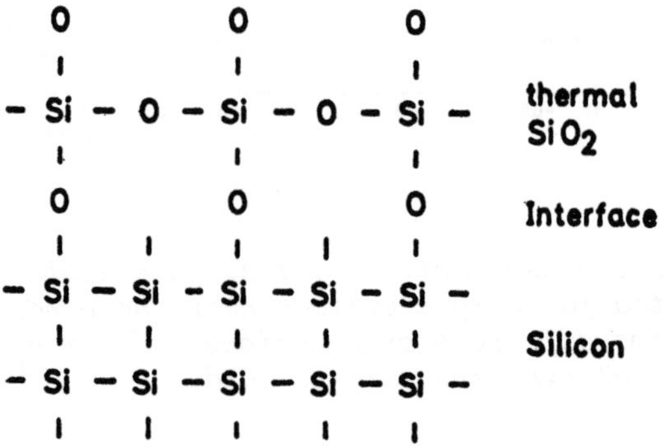

```
      O           O           O
      |           |           |
  - Si - O -  Si  - O -  Si -      thermal
      |           |           |     SiO2
      O           O           O
      |     |     |     |     |    Interface
  - Si - Si - Si - Si - Si -
      |     |     |     |     |    Silicon
  - Si - Si - Si - Si - Si -
      |     |     |     |     |
```

Fig. 8. Schematic drawing of the configuration of silicon and oxygen atoms in the interface to illustrate the lattice mismatch.

The intrinsic surface states depend crucially on oxidation technology for thermally oxidized silicon. Dominant parameters which control the interface properties are the technological parameters for the oxide growth. The oxide growth proceeds by three consecutive reactions[12]:

i) Transfer of oxygen into the oxide already formed.
ii) Diffusion of oxygen through the oxide and
iii) Reaction of oxygen with silicon at the interface to form SiO_2.

At high temperature ≥ 1000°C the oxydation is diffusion controlled which leads to a square root law for oxide growth. At low temperatures ≤ 1000°C the growth is controlled by surface reaction which leads to a linear growth rate. Both types of reactions lead to a thin region of SiO_2 containing partially ionized silicon near the SiO_2 interface where diffusing oxygen reacts to form new SiO_2. There is a thin region with a reduced oxygen concentration which has recently been experimentally identified[13]. The concentration of this partially ionised or excess silicon depends upon the final high-temperature oxidation or annealing condition. The connection of interface states and charge with the oxydation temperature is explained by Deal's[12] oxidation triangle shown in fig.9.

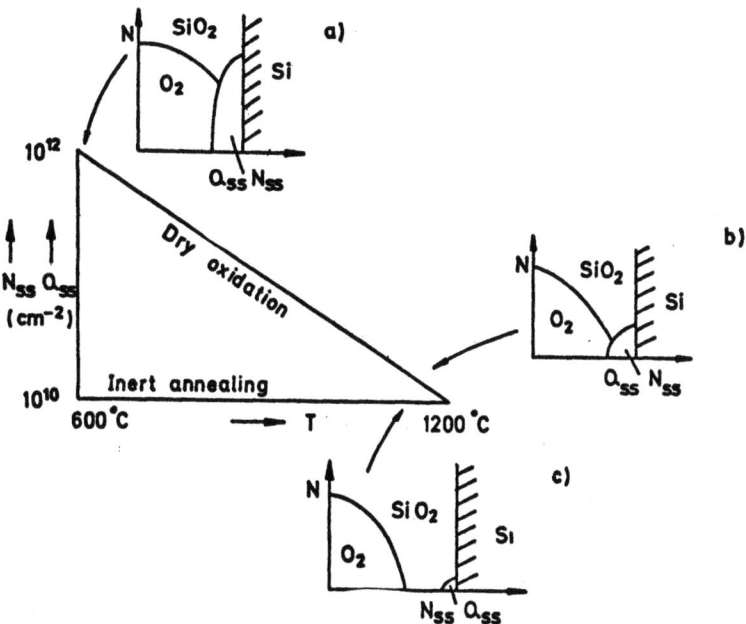

Fig. 9. *Oxidation triangle after Deal [12]. The inserts a), b), c) demonstrate the partial content of chemical components, interface states and charge near the interface.*

A high interface state density is obtained by low-temperature oxidation because there are many unsatisfied silicon bonds produced by the slow reaction rate (Insert a) in fig. 9. A low interface state density is obtained at high temperature because the reaction there is high enough to dominantly produce SiO_2 (Insert b) in fig. 9. The same low interface state density can be achieved by annealing a low-temperature oxide having excess oxygen (Insert c) in 9. The annealing, however, must not be too long to avoid depletion of oxygen.

The proportionality of the interface state density to the oxide charge led to the suggestion of the "charge model" to interprete interface states [14]. This model explains fast inter face states by the binding of mobile charge carriers in the semiconductor to charge centers in the oxide in the immediate vicinity of the interface by Coulomb attraction. A positive charge center in the oxide will give rise to a bound donor state analogous to that found in bulk silicon. A negative charge center will cause an acceptor state.

The random distribution of the depth of the charge centers in the oxide and the partial overlap of their potential will smear out the distribution of the bound states. Centers located deep in the oxide form states closer to the band edges. Overlapping charges will produce deep states. We expect a distribution of interface states with a high density near the band edges and a drop-off towards midgap when the energy position is deeper than 0.1 eV.

5.b) Extrinsic Interface

Impurities may enhance or compensate the oxide charge as well as the number of fast interface states. The property of the impurity is dependent on the electrical and chemical nature of the impurity and its mobility in the interface region. Many elements have a tendency to be gettered in the disturbed region near the interface so that they accumulate in this region even if they are only present in a small concentration. Their effect on interface charge or state density is therefore enhanced.

All the alkali impurities cause a positive space charge in the oxide in the vicinity of the interface. Normally not all of the impurity atoms are electrically active, indicating again different types of incorporation of the impurity into the host material. An increase of the interface state density proportional to the fixed oxide charge has also been found. This result supports the charge model discussed for intrinsic interfaces. However, the increased interface state density could also be caused by additional strain in the interface so that these effects are still not conclusive for the model.

A pronounced effect on interface states is observed for oxidation or annealing in presence of water of hydrogen [10,15]. These effects can be described by a chemical reaction of unsaturated silicon bonds with water at elevated temperatures under formation of silanol groups [15]

$$-\overset{|}{\underset{|}{Si}}{}^\bullet + H_2O \rightleftharpoons -\overset{|}{\underset{|}{Si}}-OH + H$$

A similar reaction is obtained for hydrogen

$$-\overset{|}{\underset{|}{Si}}{}^\bullet + H_2 \rightleftharpoons -\overset{|}{\underset{|}{Si}}-H + H$$

At temperatures below 500 °C, annealing in an ambient containing water or hydrogen leads to a reduction of the interface state density. At high temperatures, the reaction is reversed and the interface state density is increased again [15]. By varying the partial water pressure, the reaction can be weighted to one side or the other. In presence of hydrogen or water, the annealing triangle is not valid any more because the reaction of hydrogen with oxide charge centers and interface states is different. The oxide charge decreases continuously during annealing even at high temperatures while the hydrogen bond of the above reaction splits again and causes an increase of the interface state density.

The behavior of other elements is even less well understood. Chlorine itself has no effect on interface state density or oxide charge [16]. However, it can be used to bind and neutralize sodium ions which cause an oxide charge [17,18]. Very important for this behavior of chlorine is its preferential incorporation into the disturbed layer in the vicinity of the interface [18].

Gold is a particularly significant impurity in silicon devices because of it pronounced effects on carrier life time. Gold is also gettered in the interface region during oxide growth[19]. It seems to be incorporated in the oxide in the form of a neutral atom. However, it causes some positive space charge and a large interface state density, larger than the number of gold atoms in the interface[19]. This behavior cannot be explained at present.

6. Recombination and Trapping via Surface States.

6.a) Surface Recombination

Recombination and trapping by surface states is very similar to the same process in the bulk. A major difference arises from the fact that the recombination occurs in the plane of the surface. For small deviations from equilibrium in the bulk the total number of carriers recombining at the surface is obtained by the same reasoning as has been appled before for bulk recombination. This number of carriers is in general terms per cm^2 and sec.

$$U_s = - \frac{\sigma_n \sigma_p V_{th} N_{ss}(n_s p_s - n_i^2)}{\sigma_n(n_s + n_i \exp{(E_T-E_I)/kT)} + \sigma_p(p_s + n_i \exp{(E_i-E_T)/kT})} \quad (13)$$

This equation is valid for homogeneously doped semiconductors and also for the surface region of p-n junctions. For the quantative calculation of surface recombination a difficulty arises because surface states are very often distributed continuously across the gap. Thus a single trap level E_T is not applicable.

A simpler situation results in the surface region of a p-n junction :

For reverse bias, carrier densities are very low such that n_s, $p_s \ll n_i$. Equation (13) then can be simplified to give

$$U_s = - \frac{\sigma_n \sigma_p V_{th} n_i N_{ss}}{\sigma_n \exp{(E_T-E_i)/kT} + \sigma_p \exp{(E_i-E_T)/kT}} \quad (14)$$

A further simplification results by realizing that the most efficient centers for recombination are those at midgap. We can therefore set $E_T = E_i$. In addition, usually $\sigma_n = \sigma_p$ is assumed. This assumption is not realistic because all measurements show that σ_n and σ_p differ greatly, sometimes by an order of magnitude. Nevertheless, the error made by setting $\sigma_n = \sigma_p$ is at most a factor of 2.

Using this simplifications we obtain

$$U_s = - \frac{1}{2} \sigma V_{th} N_{ss} n_i = - \frac{1}{2} S_o n_i \quad (15)$$

where S_o is the recombination velocity. The minus sign indicates that carriers are generated rather than recombined. The recombination velocity is directly proportional to the interface density near midgap within a range of a few kT. Interface states further away from midgap do not contribute much to the recombination rate.

When carriers are injected (forward bias) one or both of the densities n_s, p_s are above equilibrium. For this case, we obtain

$$U_s = S_o \frac{p_s n_s - n_i^2}{n_s + p_s + 2 n_i} \qquad (16)$$

In the neutral part of the base region of a transistor n_s and p_s are much larger than n_i. We obtain

$$U_s = S_o \frac{p_s n_s}{n_s + p_s} \qquad (17)$$

It should be noted that in the case of recombination, the same recombination velocity S_g is applicable as for generation although the currents involved are different by many orders of magnitude.

Surface recombination velocities achievable today are between 1 and 10 cm/s/ A rough calculation shows that this value is low enough not to affect device performance. Nonetheless, passivation of junctions is still one of the critical processes that have to be watched carefully.

6.b) Trapping Phenomena in Charge Coupled Devices.

One type of device that is still much affected by interface states are charge coupled devices [20]. Because charge is moved along the interface interaction with interface states is much more intense than in other classes of devices. Interface states, among other factors limit the number of stages of a CCD.

An important concept for CCDs is the charge transfer efficiency which is defined as the fraction of charge that is lost at each transfer. It has to be made small to make possible efficient CCD action. If a packet of charge follows an empty well while propagating down a CCD it is obvious that a certain number of interface states will empty during transfer of the empty well and will be filled when the charge packet arrives. The slower traps will release their charge during the next cycle thus constantly depleting the charge packet. For this mode of operation, the transfer inefficiency is [21]

$$\varepsilon_p = \frac{kTqN_{ss}}{C_o V_s} \cdot \ln(m+1) \qquad (18)$$

where m is the number of phases of the device. For an interface state density of 2×10^{10} cm^{-2} eV^{-1} a transfer inefficiency of 2.3×10^{-3} was found [21]. Loss of charge by this mechanism can be significantly reduced by circulating a background charge even in the empty wells to keep interface states constantly filled. This requires definition of a zero at the level of the background charge (fat zero). But even with a background charge, interface states will cause losses due to a number of second order effects. Tompsett has studied these effects in detail [21]. These effects are :

i) Edge effect : A large charge packet occupies a larger area under the electrode than a small one because of the curved shape of the potential walls. Therefore a different area of interface becomes active in trapping.
ii) Charge captured during transfer.
iii) Variable transfer time : A large charge packet has a larger transfer time than a small one thus permitting more interface states to be filled.
iiii) Variable trap occupation : The same difference between large and small charge packets applies to interface states under the electrodes.

Numerical calculations indicate that edge effect and charge capture during transfer are the most important contributions [21]. Operation with fat zero lowers ε_p by about one order of magnitude.

Besides transfer inefficiency, interface states also contribute noise to the CCD output signal [21]. Interface states having time constants of the order of the transfer time again contribute most to this noise. Interface states with smaller time constants do not empty at all while the faster interface states emit their charge fast enough to catch the original charge packet. When charge captured from one packet is re-emitted into the following packet noise in succeeding packets is correlated. Considering these facts, the signal to noise ratio caused by interface state trapping can be calculated to [21]

$$S/N = \frac{A_s C_o^2 V_s^2}{2 nqTN_{ss}(1 + \gamma_g P_s)\ln 2} \qquad (19)$$

where A_s is the area under the transfer electrode occupied by signal charge. V_s is the change in interface potential due to signal charge and n is the number of transfers. The quantity γ_g is the ratio of the gap region which is never occupied by static charge to the area under the transfer electrode.

Noise due to interface states is of course only one of the possible noise sources. Measurements on bulk CCDs, however, show significantly lower noise. In these devices, the charge cannot reach the interface. This result indicates that an interface state density of the order of 10^{10} cm^{-2} still has a deleterious effect on the performance of CCDs.

References.

1   I. Tamm Physik. Z. Sowjetunion, 1, 733, (1932).

2   W. Shockley, Physic. Rev. 56, 317, (1939).

3   W. Shockley and W.T. Read, Phys. Rev. 87, 835, (1952).

4   See, for example, A. Goetzberger and S.M. Sze, Metal-Insulator-Semiconductor Physics, in Appl. Solid State Science 1, Academic Press, New York, (1969).

5   W. Fahrner and A. Goetzberger, J. Appl. Phys. 44, 725, (1973).

6   E. H. Nicollian and A. Goetzberger, BSTJ. 46, 1055,(1966).

7   K. Lehovec Appl. Phys. Letters 8, 48, (1966).

8   K. Ziegler and E. Klausmann, Appl. Phys. Letters, 26, 400, (1975).

9   D.E. Eastman, et al Phys. Rev. Lett. 28, 1378 (1972).

10 L.F. Wagner, et al Phys. Rev. Lett. 28, 1381 (1972).

11 Yudurain and J. Rubio, Phys. Rev. Lett., 26, 138 (1972).

12 B.E. Deal et al. J. Electrochem. Soc., 121, 198C (1974).

13 T.W. Sigmon, W.K.C. Chu, E. Luganijo, and J.W. Mayer, Appl. Phys. Lett. 24, 105 (1974).

14 A. Goetzberger, V. Heine and E.H. Nicollian, Appl. Phys. Lett. 12, 95 (1968).

15 F. Mantello and P. Balk, J. Electrochem. Soc. 118, 1463, (1971).

16 M. Schulz, E. Klausmann, and A. Huurle, Proc. Conf. Comp. Semicond. Interfaces (1975).

17 R.J. Kriegler 12th Annual Proc on Reliability Phys. Las Vegas 1974.

18 R.J. Kriegler and Denki Kasagu J. Electrochem. Soc. Japan 41, 466 (1973).

19 P.F. Schmidt, L.P. Adda, J. Appl. Phys. __45__, 1826 (1974).

20 W.S. Boyle and G.E. Smith, BSTJ, __49__, 87, (1970).

21 M.F. Tompsett, IEEE-Trans. __ED-20__, 45 (1973).

Section III

DIODE AND TRANSISTOR ARRAYS

CHARGE STORAGE OPERATION OF SILICON PHOTODETECTORS

Gene P. Weckler

Vice President, Director of Engineering, Reticon Corp.,
910 Benicia Avenue, Sunnyvale, CA. 94086

INTRODUCTION

Although most of the "sound and fury" surrounding the development of the solid-state image sensor has taken place since 1970, serious developmental efforts were going on as early as 1960.

In 1960 a patent was issued to F. W. Reynolds at Bell Telephone Laboratories describing a characteristic of the p-n junction which later became known in the imaging field by the generic term "Charge-Storage Operation". Although the physical principles were adequately described, the technology required to successfully implement the concept was not yet available and those who took an early look at the practicality of this concept discarded it. However, during the early 1960's the necessary technology was being developed.

This basic concept of charge storage is fundamental to all solid-state image sensors. Today's solid-state image-sensing devices differ not in how they detect the absorption of a photon, but in how this information is transferred to the external user.

GENERAL DESCRIPTION OF SOLID-STATE IMAGE SENSING

Electronic imaging devices perform the task of converting a pattern of incident illumination falling upon the surface of the sensor into a voltage waveform which is an ordered, sequential reproduction of the incident radiation pattern. This voltage waveform is referred to as a video signal. The mechanism of conversion of the spatial pattern to a time-varying voltage waveform is referred to as scanning.

In a solid-state array image sensor, the scanning function is accomplished by the application of one or more electrical signals in such a way as to interrogate at one time one and only one photosensitive element in the array. These scanning signals are applied so as to provide an ordered sampling of all the picture elements in the array.

The basic mechanism for converting the photons of the incident light pattern to an electrical signal is associated either with the depletion region of a p-n junction or the depletion region induced by applying the appropriate voltage to a MOS capacitor. In the discussion to follow the term photodiode will be used to mean either a diffused or an induced depletion region.

In the normal photoconductive or photovoltaic mode the output from a photodiode depends on the rate of photon absorption, however, in an image sensor, the photodiodes are sampled in a periodic manner. If the photodiodes are operated in either the normal photoconductive or photovoltaic mode, the active properties of the diode are used only during the sampling time. It was apparent that a photon flux integration mode was necessary to allow each photodiode to be active 100 percent of the time yet maintaining a short readout time, thereby providing an effective gain in responsivity proportional to the increase in active time. The photon flux integration mode thus became charge-storage operation.

CHARGE-STORAGE OPERATION OF A PHOTODIODE

Charge-storage operation is based on the principle that if a p-n junction is reverse biased, then open circuited, the charge stored on the depletion-layer capacitance decays at a rate proportional to the incident illumination level.

In the dark, only generation-recombination current (dark current) is available to discharge the depletion-layer capacitance and, since both depletion-layer capacitance and generation-recombination current are directly proportional to area, the time constant is independent of area. In the dark, time constants of several minutes may be attained in large-area properly processed silicon planar junctions; however, for most standard MOS processes and for small-geometry photodiodes a time constant of several seconds is more typical.

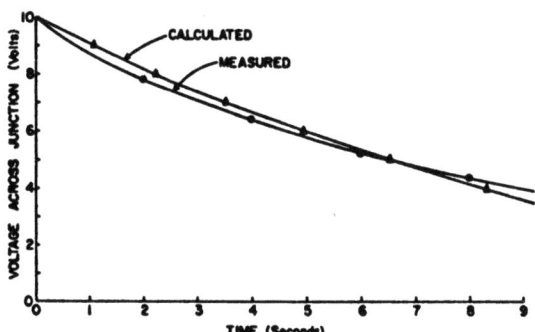

Fig. 1. Voltage decay of a silicon planar p-n junction in the dark.

Figure 1 shows a typical unilluminated decay characteristic for a typical high-quality p-n junction. The calculated curve was obtained by equating capacitive displacement current to the generation-recombination current.

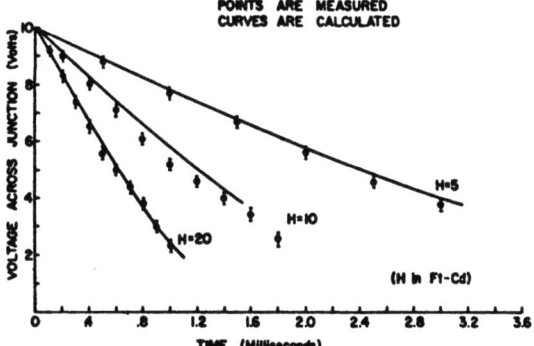

Fig. 2. Voltage decay of an illuminated silicon planar p-n junction.

Figure 2 shows decay characteristics of the same diode for several illumination levels. The photon-generated current is directly proportional to the illumination level and therefore the amount of charge removed in a given interval of time is directly proportional to the integral of illumination taken over that interval. Thus, by monitoring the charge required periodically to re-establish the initial-voltage condition, one may obtain a signal proportional to the incident illumination. The advantages of this mode of operation are:

1. Improved responsivity resulting from integration of the incident illumination.

2. Electronic control of responsivity by varying the integration time, hence a wide dynamic range.

PRACTICAL STRUCTURES FOR OBTAINING CHARGE-STORAGE OPERATION

Three functional elements are necessary to realize charge-storage operation:

1. A charge-storage element.

2. A current generator whose output is proportional to the incident illumination level, and

3. A nearly ideal switch.

Two practical structures possessing the above characteristics will be described. The first one uses a MOS field-effect transistor as the switch and the second uses a diode switch.

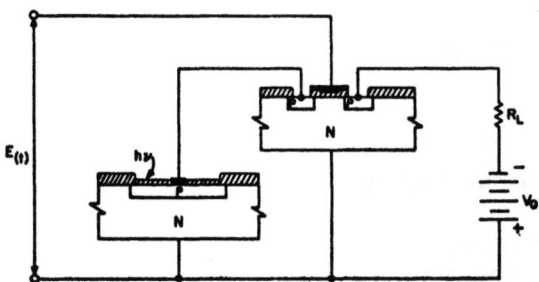

Fig. 3. Practical structure using an MOS switch for charging the photodiode.

Figure 3 shows how a p-n photodiode and a p-channel MOST can be combined to accomplish this task. With no voltage on the gate, the photodiode is shunted only by the source diode of the MOST which is reverse-biased at the same potential as the photodiode. It is apparent that the photodiode may actually be the source diode without affecting

the operation of the device. When the gate is made sufficiently
negative to invert the region under the gate, a high conductance exists
between source and drain, allowing recharging of the source (photodiode)
junction capacitance. The use of the source diode as the photosensitive
junction provided a structure which was very easy to integrate into
very densely packed arrays. A version of this structure is used in
most solid-state image sensors.

Storage-mode operation has also been realized using two p-n
junction diodes, furthermore, these two diodes may actually be the
emitter-base and base-collector junctions of a transistor.

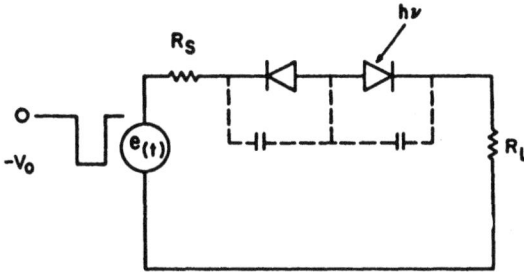

Fig. 4. Practical structure using
a p-n junction as the switch.

Figure 4 shows how two diodes may be connected to operate in the
storage mode. The junction capacitances are shown externally to
facilitate the discussion to follow. The diode on the right is a photodiode and functions as a charge-storage element and as a current
source proportional to incident illumination. The diode on the left is a
small-area, low-leakage diode and performs the functions of a switch.

When a negative voltage is applied by the pulse generator, the
diode on the left is forward biased and the diode on the right becomes
reverse biased to nearly the full amplitude of the pulse. Upon
termination of the pulse, the device on the left, having a smaller
depletion-layer capacitance than does the photodiode, shares the
charge initially stored on the photodiode and both junctions become
reverse-biased to almost the peak value of the pulse. The resulting
polarities across the junction are shown in Figure 5.

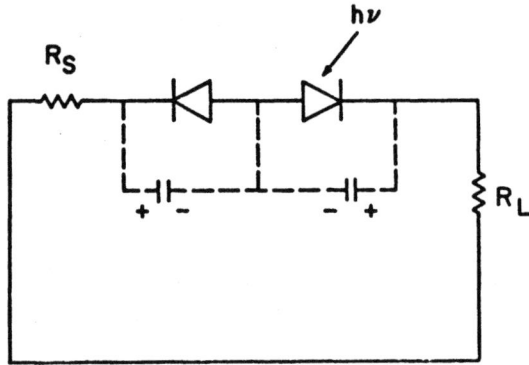

Fig. 5. Polarity of junction voltages during integration time.

The photodiode is now charged to a voltage nearly equal to the pulse voltage and is isolated from the external circuit by the diode on the left side since it, too, is reverse biased. The rate of decay of stored charge depends on the magnitude of the incident illumination. For zero illumination, the decay rate is determined primarily by the generation-recombination current of the photodiode. As the illumination level is increased, the rate of decay increases. The two diodes in this circuit may be replaced by a single phototransistor.

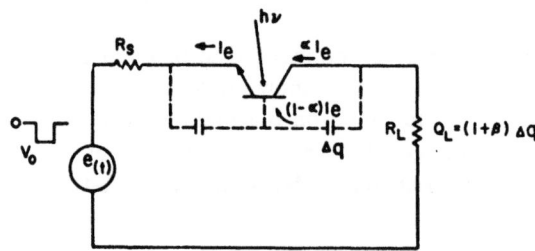

Fig. 6. Storage mode operation of a phototransistor.

Figure 6 shows how the two diodes may be replaced by a phototransistor with a floating base connected in series with a resistive load and a voltage pulse source. The application of a negative pulse will forward-bias the emitter base junction and reverse-bias the base collector junction. The depletion layer capacitance of the collector junction becomes charged to about the peak value of the pulse. During the charging interval, the transistor is conventionally biased and a collector current beta times the base-collector junction-capacitance (C_{bc}) charging current flows through the load in addition to the C_{bc} charging current.

Upon termination of the pulse, the emitter junction shares the charge initially stored on the collector-junction capacitance and both junctions become reverse biased so that the collector junction is isolated by the reverse-biased emitter-base junction from the external circuit.

In a double-diffused planar phototransistor the emitter-base junction leakage is considerably smaller than the base-collector leakage for the same reverse-bias conditions. The emitter-base junction not only has considerably less area than does the base-collector junction but it also has a much larger grading constant, both of which factors tend to reduce its leakage well below that of the collector-base junction

ANALYSIS OF THE STORAGE MODE

The voltage decay across a p-n junction under the condition of zero incident illumination will first be discussed and later the case of an illuminated junction assuming the linearly graded junction approximation.

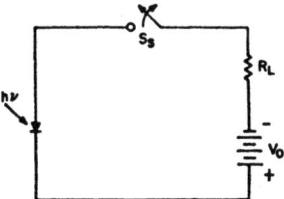

An idealized circuit for analyzing storage mode operation of a *p-n* junction diode.

Let us consider the circuit shown in Figure 7. At time zero, the switch opens and the voltage across the diode begins to decrease. Since there is no illumination incident on the diode, the only current available to discharge the space-charge capacitance is the generation-recombination current in the space-charge region which is given by

$$I_{gr} = \frac{Aqn_i}{2\tau_0} W \qquad (1)$$

where
  I_{gr} is the generation-recombination current;
  W is the space-charge width (meters);
  A is the junction area (square meters);
  q is the electronic charge;
  n_i is the intrinsic concentration; and
  τ_0 is the effective lifetime in the space-charge region.

Substituting for the voltage dependence of the space-charge width gives the generation-recombination current as a function of voltage across the junction, that is,

$$I_{gr} = \frac{Aqn_i}{2\tau_0}\left(\frac{12\epsilon}{qa}\right)^{1/3} V^{1/3} \qquad (2)$$

where
  a is the net doping gradient at the junction; and
  ϵ is the permittivity of silicon.

For an open-circuit junction it is seen that the generation-recombination current must equal the capacitive displacement current, and the resulting balance expression is

$$C_{(v)} \frac{d}{dt} V_{(t)} = -I_{gr}. \qquad (3)$$

The junction capacitance is a function of applied voltage and for a graded junction is given by

$$C(v) = A\left(\frac{qa\epsilon^2}{12}\right)^{1/3} V^{-1/3}. \qquad (4)$$

Substituting (2) and (4) into (3) yields the first-order separable equation

$$V^{-2/3}\frac{dV}{dt} = -\frac{3n_i}{2\tau_0}\left(\frac{144q}{\epsilon a^2}\right)^{1/3}. \quad (5)$$

Integrating (5) and employing the boundary condition that at $t = 0$ the voltage across the junction is V_0, yields the expression for junction voltage as a function of time:

$$V(t) = \left[V_0^{1/3} - \frac{n_i}{6\tau_0}\left(\frac{144q}{\epsilon a^2}\right)^{1/3} t\right]^3. \quad (6)$$

It is seen that this expression is independent of junction area. The effective lifetime in the space-charge region may be determined experimentally using the expression

$$I_rC = \frac{qn_i\epsilon}{2\tau_0} A^2 \quad (7)$$

which is valid if excess surface currents are negligible, and avalanche multiplication is not significant at the value of applied voltage used for the measurement of junction capacitance and generation-recombination current.

Now consider the diode in Figure 7 to be illuminated so that photocurrent adds to the diode dark current, causing a more rapid discharge of the junction capacitance. The photocurrent may be considered as a current source of magnitude,

$$I_p = I_0AH, \quad (8)$$

in shunt with the junction capacitance, where

I_o is the basic photosensitivity for the p-n structure expressed in amperes per unit area per ft-cd of illumination;

H is the illumination level in ft. cd; and

A is the photosensitive area in m^2.

Now the current balance equation for $t \geq 0$ can be written as

$$C_{(v)} \frac{d}{dt} V_{(t)} = -(I_{gr} + I_p). \tag{9}$$

Substituting the expressions for the capacitance, generation-recombination current, and photocurrent yields the differential equation:

$$V^{-2/3} \frac{dV}{dt} + I_0 H \left(\frac{12}{qa\epsilon^2}\right)^{1/3} V^{-1/3} + \frac{n_i}{2\tau_0} \left(\frac{144q}{\epsilon a^2}\right)^{1/3} = 0. \tag{10}$$

An explicit solution of this equation for junction voltage does not exist. However, it is possible to form a separable equation that yields the solution for time in terms of voltage:

$$t = \frac{6\tau_0}{n_i} \left(\frac{\epsilon a^2}{144q}\right)^{1/3} (V_0^{1/3} - V^{1/3})$$

$$+ \frac{\tau_0^2 a I_0 H}{3 n_i^2 q} \ln \left[\frac{\frac{n_i}{2\tau_0} \left(\frac{12q^2\epsilon}{a}\right)^{1/3} V^{1/3} + I_0 H}{\frac{n_i}{2\tau_0} \left(\frac{12q^2\epsilon}{a}\right)^{1/3} V_0^{1/3} + I_0 H}\right]. \tag{11}$$

The first term of the right-hand side of this expression is the solution obtained in (6) for the case of zero photo-generator current. If the second term in (11) were expanded in an infinite series it would be found that the first term of the expansion would identically cancel the first term in (11), leaving an infinite series involving both generation-recombination current and photocurrent. Such an approach, although rigorous, lends little insight into the physical processes taking place.

An alternate approach is to assume the photocurrent to be predominant, hence neglecting the generation-recombination current in (9). This assumption is justified for sufficiently short integration times or high illumination levels. For example, the p-n junction described by Figure 1 required more than 100 ms for the generation-recombination current to reduce junction voltage by 1 percent of its initial value at room temperature, whereas a light level of 100 ft cd resulted in complete discharge in less than a millisecond. The simplified differential equation for current balance results:

$$V^{-1/3}\frac{dV}{dt} = -I_0 H \left(\frac{12}{qa\epsilon^3}\right)^{1/3}. \qquad (12)$$

This equation is separable and yields the expression for voltage as a function of time and illumination level:

$$V_{(t)} = \left[V_0^{2/3} - \frac{2}{3} I_0 H \left(\frac{12}{qa\epsilon^3}\right)^{1/3} t\right]^{3/2}. \qquad (13)$$

This expression is plotted in Figure 2, assuming an initial voltage of 10 V and a grading constant of $3 \times 10^{29} \text{m}^{-4}$ as used in Figure 1. The value of I_0 for the diode structure was determined to be approximately 0.048 Å/m²ft.cd.

SUMMARY

This paper describes storage-mode operation in much the same sequence as it was developed. Storage mode operation has been fundamental to the development of solid-state image sensors. While the size of the individual elements in the sensor was being decreased so that more elements could be realized in a given area, it was charge-storage operation that provided a still usable responsivity.

REFERENCES

Gene P. Weckler: "Operation of p-n Junction Photodetectors in a Photon Flux Integrating Mode"; IEEE Journal of Solid-State Circuits, V. SC-2, No. 3, September 1967.

Gene P. Weckler: "Charge Storage Lights the Way for Solid-State Image Sensors"; Electronics, May 1, 1967.

R. H. Dyck and Gene P. Weckler: "Integrated Arrays of Silicon Photodetectors for Image Sensing"; IEEE Transactions on Electron Devices, V. ED-15, No. 4, April, 1968.

XY ADDRESSING

P. Jespers
Université Catholique de Louvain
Microelectronics Laboratories
Bâtiment Maxwell
1348 Louvain-la-Neuve, Belgium.

Abstract

XY addressing of photosensing arrays is dependent of the type of photosensor used. After a brief survey of various sensors, the organization schemes of linear and two dimensional arrays are described.

1. Photosensors

1.1. Continuous mode operation :

A light beam shining on a semiconductor creates a non equilibrium situation which can be detected in several ways :
- enhanced conductivity.
- increased saturation current of a reverse biased PN junction.
- photovoltaic e.m.f.   etc ...

One of the most widely used photosensors is the photodiode. Assuming the efficiency of converting photons in electrons in a PN junction is η (see section II.2 on quantum efficiency), the number of electrons released by ϕ photons falling on a unit area diode per second is $\eta\phi$. The released charge per second, or photocurrentdensity J_L, thus is $q\eta\phi$. Each photon having an energy $h\nu$, the incident power P, expressed on Watt cm^{-2}, is equal to $\phi h\nu$. Eliminating ϕ between the two above expressions yields :

$$J_L = \frac{qnP}{h\nu} \quad (A.cm^{-2}) \qquad (1)$$

For a diode with an area A, the photocurrent I_L thus is given by :

$$I_L = \frac{qAnP}{h\nu} \quad \text{or} \quad \frac{qAn\lambda P}{hc} \qquad (2)$$

with $\nu = c/\lambda$

(c : velocity of light, λ : wavelength).

The current I_L usually is very small under most common incident light levels; it ranges from a few picoamperes to microamperes. Sensitive detectors thus are necessary in order sense images electrically and extremely large gain bandwidth products are necessary in order to read out high definition arrays continuously. Fortunately a lot of gain can be obtained if the charge storage mode operation is used instead of the continous mode.

1.2. Charge storage mode operation of photodiodes

In the charge storage mode operation, the photodiode is preloaded first under a fixed reverse bias of a few Volts. Then the circuit is opened so that the junction behaves like a capacitor which discharges smoothly under the influence of the light falling on the junction and also of the junction leakage current itself. While the photodiode "integrates" the light, it provides a means to evaluate the total irradiant energy received. The charge lossed during the light integration time indeed is proportional to the light power received by the diode multiplied by the duration of light integration.

The voltage drop across the diode also is significant for the total energy received. But, since the depletion capacitance varies with reverse bias, the relationship between irradiant energy and reverse bias modification is nonlinear. The problem was dealt with in the literature [1,2] , and will be briefly reviewed hereunder.

Let us call Q the stored charge and C(V) the differential voltage dependent capacitance of the photodiode. Knowing that :

$$C(V) = \frac{dQ}{dV} \qquad (3)$$

the following equation must be integrated in order to find the voltage variation across the reverse biased junction :

$$I = \frac{dQ}{dt} = C(V)\frac{dV}{dt} \qquad (4)$$

Considering a constant value I_L for I, given by (2), the following result is obtained:

$$-\frac{1}{I_L}\int_{V_i}^{V} C(V)\,dV = t \qquad (5)$$

(V_i represents the initial voltage across the diode).

Among expressions used for C(V), a convenient one is the following:

$$C(V) = A \cdot K \cdot V^{-m} \qquad (6)$$

where A represents the diode area
K is a constant depending on the semiconductor parameters
and m is taken equal to 1/2 for an abrupt junction and 1/3 for a linear junction.

After performing the integration contained in equation (5), and taking in account (6), we obtain

$$\frac{V}{V_i} = \left[1 - \frac{I_L}{C_i V_i}(1-m)t\right]^{\frac{1}{1-m}}$$

where C_i denotes the value of C(V) for V equal to V_i.

Knowing that the factor $I_L/C_i V_i$ represents the reciprocal of the time T which is needed in order to fully discharge C_i (supposed to be constant) under a photocurrent I_L, we rewrite the above equation as follows:

$$\frac{V}{V_i} = \left[1 - (1-m)\frac{t}{T}\right]^{\frac{1}{1-m}} \qquad (7)$$

$$\text{with } T = \frac{C_i V_i}{I_L}$$

A graph representing V/V_i is given in fig.1, which shows the departure from linearity resulting from the voltage dependance of C(V) during the discharge process. It should be noted that (7) is not valid anymore when V approaches zero volt, for expression (6) is an uncorrect representation of C(V) near the origin. A better approximation would be:

$$C(V) = C_o(1 - \frac{V}{\phi})^{-m} \qquad (8)$$

where C_o represents the zero volt junction capacitance and ϕ the built in potential of the depletion region.

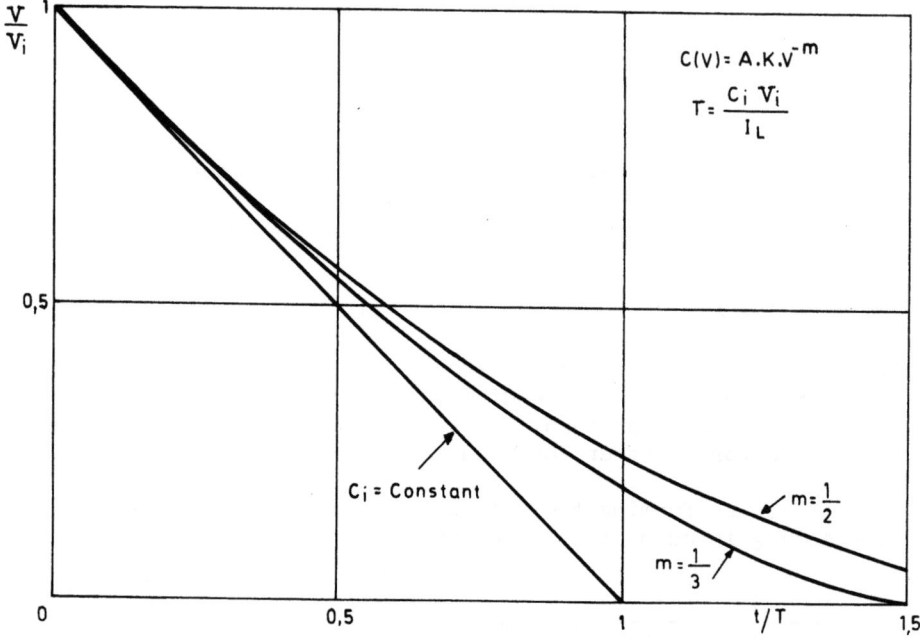

Fig.1. Under constant illumination, the reverse voltage across the diode decays non linearly because of the voltage dependance of the junction capacitance.

1.3. Comparison between continuous and charge storage modes

In order to compare the performances of photodiodes in the continuous and charge storage modes, it is necessary to consider first the manner in which the information is being sampled.

In the continuous mode, it is sufficient to measure the photocurrent diode I_L directly. In the storage mode, several possibilities exist. The most classical method was proposed by G. Wecker [3] and may be called the voltage sampling method as is illustrated in fig.2. In the first case (a), the diode is reverse biased during the so called sampling time T_S by

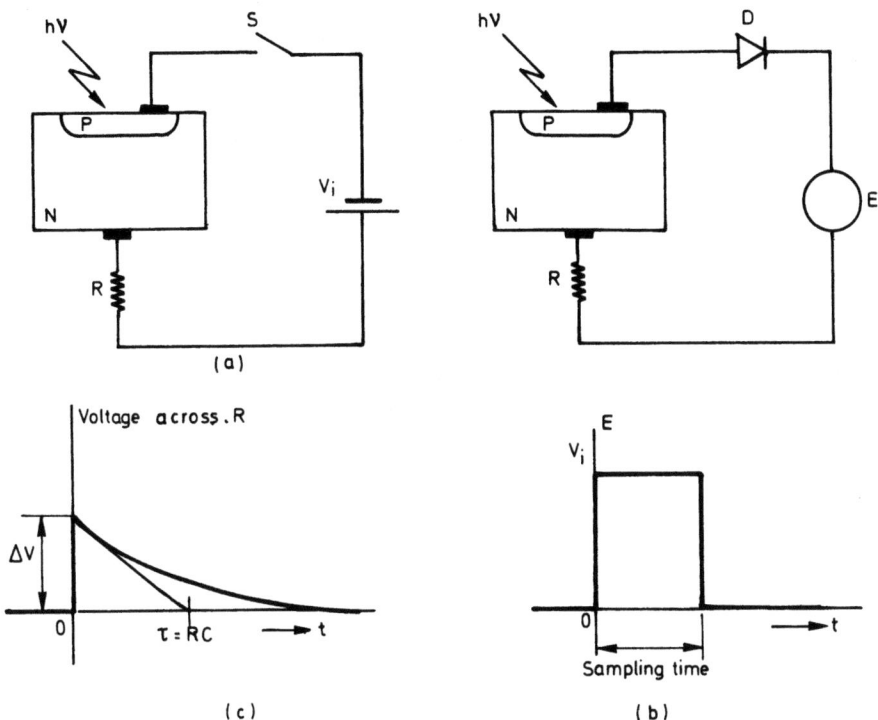

Fig. 2. Charge storage operation of photodiodes. In (a) a constant voltage supply V_i and a series switch are used. In (b), the supply is pulsed and the switch is replaced by a diode. The output signal in both cases is represented in (c).

closing the switch S. At the very time the switch is closed, the voltage drop ΔV across the diode resulting from previous light integration, produces an equal voltage step across the series resistor R. The corresponding instantaneous peak current I_o is given by $\Delta V/R$. Since :

$$\Delta V = \frac{I_L T_i}{C_i}$$

(assuming C_i is constant, and T_i represents the integration time), the current I_o may be written as :

$$I_o = \frac{I_L}{RC_i} T_i = I_L \frac{T_i}{\tau} \qquad (9)$$

where τ represents the circuit time constant RC_i.

A gain factor G can now be defined as the ratio of the charge storage peakcurrent I_o to the current I_L which is obtained in the continuous mode;

$$G = \frac{I_o}{I_L} = \frac{T_i}{\tau} \qquad (10)$$

For instance, considering a typical junction capacitance of 1 pF and a resistor R of 1 kΩ , yields a time constant τ of 1 ns, which leads to a gain factor G of 10^6 if an integration time T_i of 1 ms is considered.

The second circuit, represented under (b) in fig.2., operates in a similar manner. Instead of a series switch, a diode D is used and the constant voltage supply V_i is replaced by a pulse generator delivering short pulses of amplitude V_i and duration T_s. When the pulse generator produces a positive output pulse, the diode D is forward biased and recharging of the junction capacitance occurs like in the previous case. When the voltage of the pulse generator is returned to zero, the diode is blocked and light integration starts.

It will be shown later (section III 4 on phototransistors) that the dynamic impedance of the series diode D plays an important role with respect to the recharging process. This impedance indeed varies like the reciprocal of the current flowing through the diode, causing excessive reloading time. Consequently, in many cases, full recharge of the diode cannot be achieved with the usual sampling times. The voltage drop across the diode thus does not reach its steady state value, nor does the current drop to zero at the end of the sampling time. It is convenient therefore to introduce another gain factor G' which is defined as the ratio of the average current I_{av} observed during the sampling time to the continuous photocurrent I_L. Let us call ΔQ the charge lossed during the light integration time T_i :

$$I_L = \frac{\Delta Q}{T_i} \qquad (11)$$

The average reloading current is given by :

$$I_{av} = \frac{\Delta Q}{T_s} \qquad (12)$$

Thus :

$$G' = \frac{T_i}{T_s} \qquad (13)$$

Let us consider for instance a photodiode mosaic comprising 400 x 500 diodes. Because reading out occurs sequentially, each diode is sampled during only 1/200,000 of the total frame time. The new gain factor G', given by the ratio of the frame time to the sampling time, is thus equal to 200,000.

1.4. FET phototransistors

The implementation of the switch in series with the photodiode represented in fig. 2 (a) can be achieved by means of a FET transistor like shown in fig.3 (a). The photodiode and the source electrode of the FET transistor naturally need not to be different; the combination of both leads to the FET phototransistor shown in fig.3 (b).

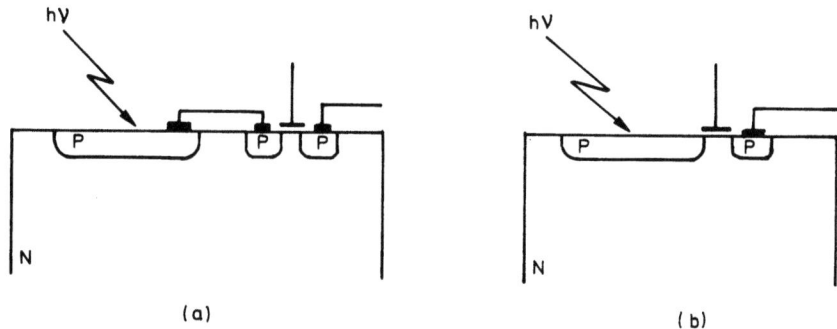

Fig.3. A photodiode and a FET switch may be combined in order to form a photo field effect transistor.

1.5. Bipolar phototransistors

The series diode D represented in fig 2 (b), also may be integrated within the photosensitive area, resulting in a structure which is a bipolar phototransistor (fig.4). Contrarily to the FET phototransistor, the bipolar phototransistor provides additional gain.

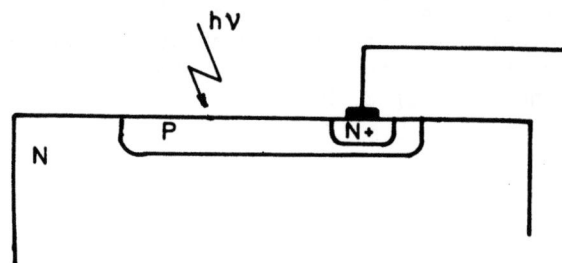

Fig.4. The introduction of the series diode D of fig.2 (b) into the photodiode itself leads to a bipolar phototransistor structure.

In order to understand the difference, let us consider the charge storage operation of the phototransistor shown in fig.4. During the sampling time, the photodiode junction - the collector of the phototransistor in the present case - is depleted by removing majority carriers from both sides of the junction. The resulting displacement current which is injected in the series diode - now called the emitter junction - in turn causes a huge current injection from emitter into the base, for the emitter efficiency factor is large. This is the consequence of the higher doping level of the emitter with respect to the base doping. Transistor action thus takes place, introducing additional gain into the device. Consequently a small fraction only of the current flowing through the transistor (the reciprocal of the current gain h_{fe}) is used for the purpose of collector recharging.

It would seem natural to say that bipolar phototransistor exhibit gain factors equal to G or G', times the current gain of the phototransistor. However this is not realistic, since voltage sampling does not exploit the current gain at all, for the initial voltage drop across the series resistor is still the same as before. The only difference is the lengthening of time constant τ, which now is multiplied by h_{fe}; introducing an appreciable release of the specifications to be imposed to the detector amplifier (e.g. in the example considered in 1.3, a time constant of 0.1 μs is obtained if a current gain of 100 is achieved).

In order to take full advantage of the increased gain factor, charge sampling should be considered instead of voltage sampling. Such a procedure however is not recommandable for it introduces unwanted current gain sensitivity. In the case of a phototransistor array for instance, the unmatched current gains produce unacceptable Fixed-Pattern Noise (F.P.N).

1.6. Charge storage operation of photo MIS capacitors

Besides photodiodes, the second most currently used photosensing element is known as the photo-MIS (Metal-Insulator-Semiconductor) capacitor; a typical example is represented in fig. 5. MIS capacitors are used in the charge storage operation mode.

Basically an MIS capacitor is a field induced junction. The transparent gate electrode usually is made of a thin layer of polysilicon or, more recently, metal oxide [4]. This electrode induces a wide depletion layer at the surface, within the silicon. During light integration, mobile minority carriers generated within the semiconductor are swept to the surface by means of the electrical field existing in the depletion region. They form gradually an inversion layer. Assuming that the contribution from thermal generation can be kept small compared to the light generation term, it is obvious that the charge contained in the inversion layer bears the same information as the charge lossed in a photodiode. A detailed analysis of this charge collection mechanism can be found in [5].

Reading out photo MIS capacitors may be performed in several manners. First, one may transfer the charges parallel

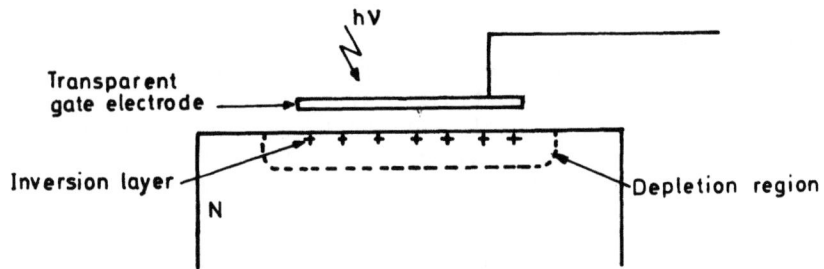

Fig. 5. Cross section view of a photo M.I.S. capacitor.

to the surface, creating a Charge Transfer Device (C.T.D.). This class of devices will be considered in the next section as separate topic. Another means to collect the information consists to inject the carriers previously stored in the inversion layer into the bulk. In the substrate the injected minority carriers recombine giving rise to a measurable recombination current. Charge Injection Devices (C.I.D.) being non linear capacitors, no net DC current can flow through them.

Thus, in order to detect useful light generated signals, one must be able to separate the displacement current caused by the applied gate voltage steps from the actual recombination current. Integration of the output current provides an excellent means to cancel out these switching signals, since they are necessarily equal and of opposite polarity, while charges generated by light during the integration time cause a non zero output signal, once sampling is achieved.

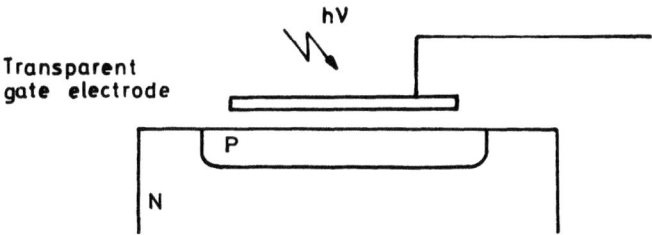

Fig.6. An early version of a C.I.D. structure

In the early work on CID sensors [6], a combination of an MIS capacitor and a diode was investigated (fig.6). This device basically may be considered as a diode whose potential excursions are controlled by means of the transparent gate electrode. Such a device also may be regarded as a third version of the circuit represented in fig. 2 with a series switching capacitor instead of a series switch S or a series diode D. It does not provide superior characteristics with respect to the simple MIS capacitor considered above, moreover additional processing steps are required for fabrication. A CID version of this device was investigated however [7] and is known as the Bucket Brigade Device (BBD). BBD sensors will be discussed in the sections devoted to charge transfer devices.

When a field induced junction is considered, one of the major problems is the relatively long times required for recombination of the injected carriers. One is faced with two conflicting requirements, one is the need for long inversion layer formation times in order to allow low level illumination performances, the other is fast recombination in order to prevent image smearing out. The solution [8] of this problem consists in the introduction of an epi-junction beneath the surface, approximatively 10 µ deep, providing efficient carrier collection without modification of the lifetime experienced in the epi-layer itself (fig.7). An additional advantage of this underneath junction is the reduction of minority carrier collection area obtained. Without the epi-junction, charges generated deep in the bulk could indeed be collected at the surface far from the site overhanging the generation area.

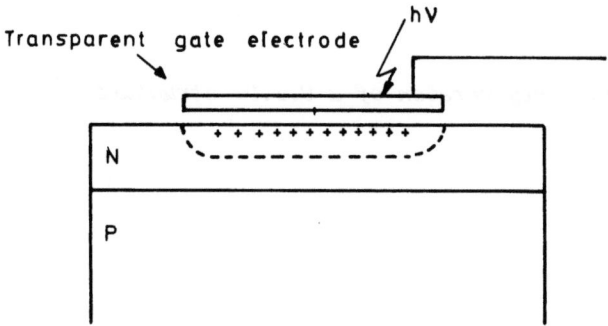

Fig.7. The introduction of an epi-junction in C.I.D. sensors enables to reduce the recombination time of the injected minority carriers and improves crosstalk between adjacent cells.

2. Linear photosensor arrays

Linear arrays of photosensors are currently available. They are formed by a line of individual photosensors with a self scanning addressing network. Ring counters or address decoders are the most widely used scanning networks. In the first case a dynamic shift-register is used with a single bit being transferred from input to output terminals under the control of external clock signals. Each time the bit signal reaches the end of the register, a new "one" is generated at the input terminal, originating a new scan. In the second case, scanning occurs in a similar manner accordingly to addresses generated by a separate counter.

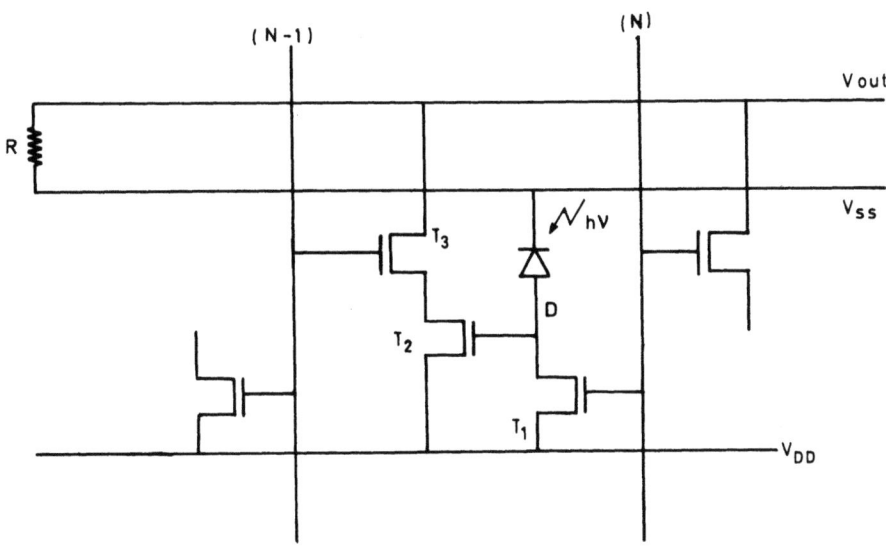

Fig.8. Representation of the usual voltage and charge sampling circuit used in linear array of photodiodes.

One of the early versions [9] of linear photodiode arrays is shown in fig.8. It was designed in order to allow voltage sampling as well as charge sampling. Reloading of each photosensitive diode occurs by means of a series switch T_1 which is controlled by a corresponding signal delivered by the scan generator. Each time, a diode is recharged, the transistor T_3 of the next cell is also turned on. The common output resistor R thus is connected to the source of transistor T_2, providing a mean to sense the voltage drop across the next diode which will be recharged at the following clock signal. This procedure is called the "voltage sampling mode". Charge sampling is somewhat simpler for it does not use the transistor pair T_2 and T_3, but operates along the lines already described before.

In the voltage sampling mode, transistor T_2 operates like a source follower. Non uniformities of the threshold voltage of this transistor may introduce therefore fixed-pattern noise to an extend of 0.1 to 0.2 Volts. In the charge sampling mode another source of FPN however exists also which is dependant on the gate to source and drain capacitances of the series transistor T_1. This problem will be analyzed in a later section.

3. XY addressing

There are many similarities existing between solid state memories and XY addressing networks designed for photosensing mosaïcs. Photoarrays usually operate in a word organization structure providing parallel or serial video output. Only very small arrays intended for data capture have parallel output terminals, in most cases a parallel to serial output register is used providing full multiplexing of the video information. Other mosaïcs operate accordingly to a bit selection scheme.

In the word organisation structure shown in fig 9 (a), the output signal is read line by line under control of the row selection register. A whole line time is available for addressing each element along the line. In the case of fig. 9 (b), the substrate is used as a common output terminal, each cell being addressed by the coïncidence of appropriate row and column scanning voltages.

3.1. Word organization

In a word organized mosaïc, each cell is provided with one individual switch. For instance, fig. 10 represents a bipolar phototransistor array [10] in which the emitters play the role of series diodes like in fig. 2 (b). When a row is turned on, the video information stored in each cell along this row becomes available on the corresponding column bus. The

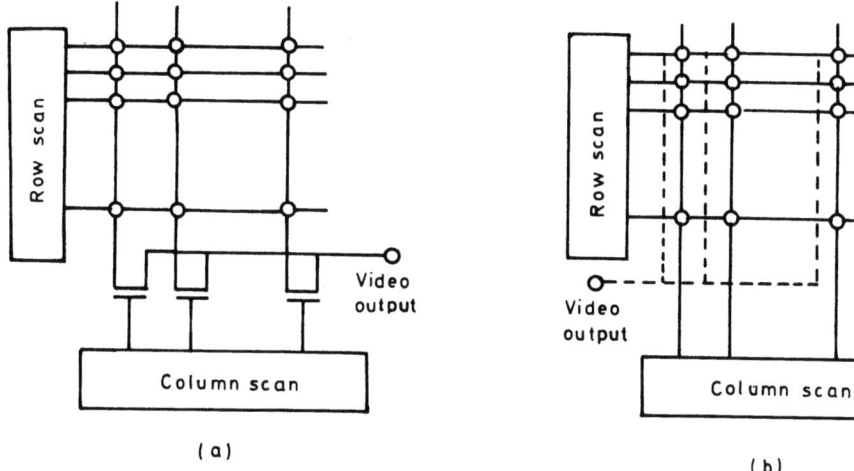

Fig.9. Two dimensional arrays may be organized in a word (a) or bit (b) structure.

reverse biased emitter junctions of all other phototransistors not belonging to the selected row prevent mixing of the row video signals. Reading out occurs sequentially by charge sharing between each column and one common load.

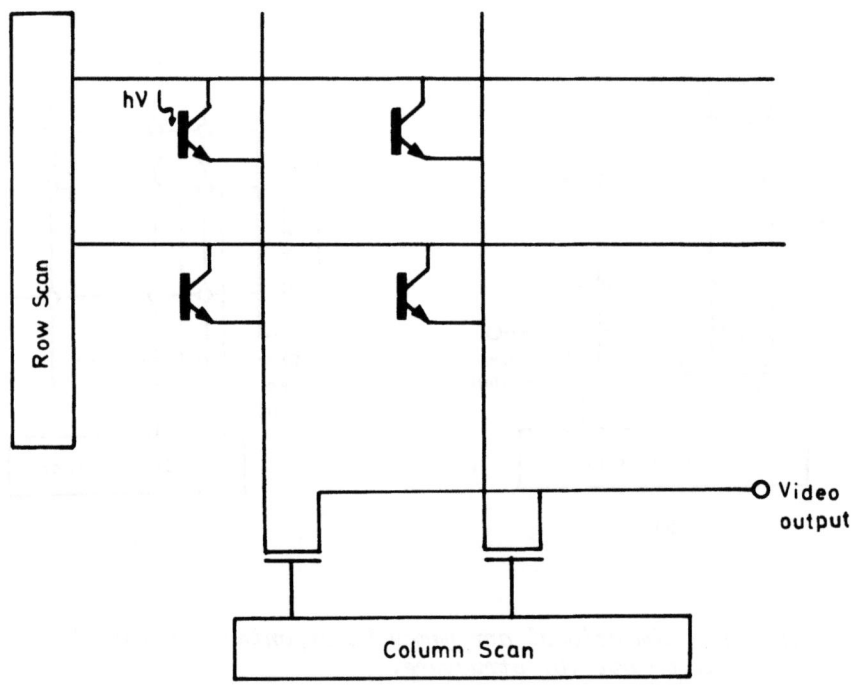

Fig.10. A typical XY addressed phototransistor array [10]

A similar organization is shown in fig. 11 for a FET photo-sensor array [11]. Like in the previous case, each column is used as an intermediate video storage medium.

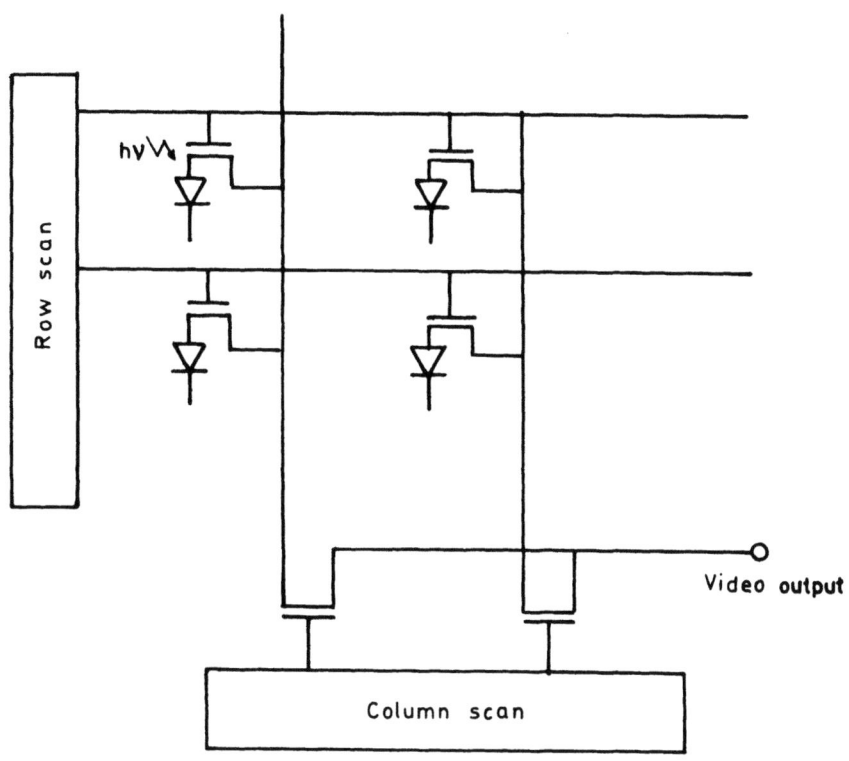

Fig.11. A typical XY addressed MOS photodiode array [11].

A word organized CID structure finally is shown in fig.12, in which the columns are implemented as buried stripes collecting the vertically injected charges. Again intermediate storage is provided through the collector column stray capacitance (see section   ).

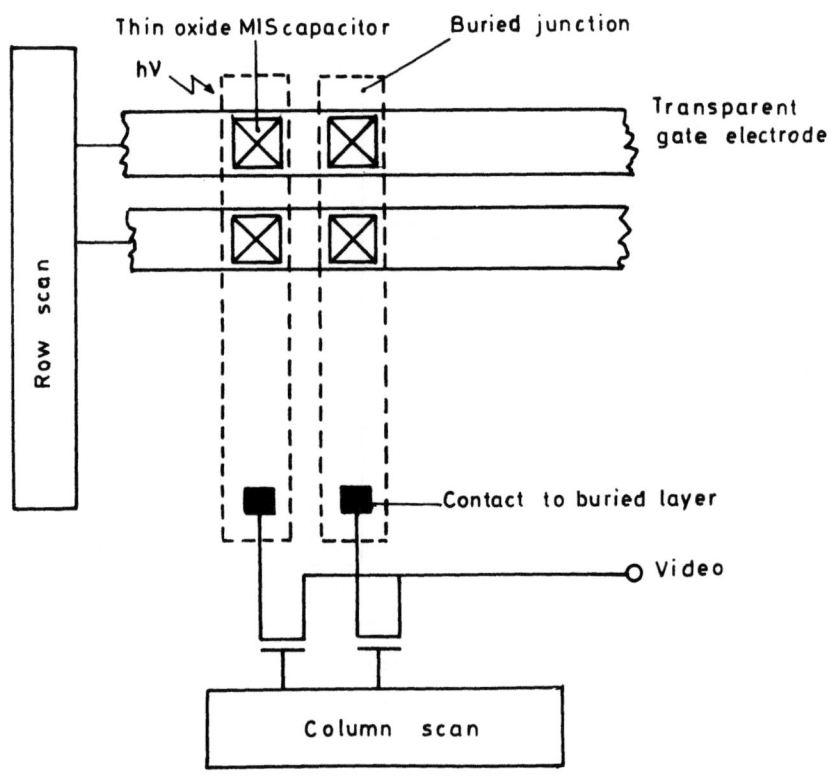

Fig. 12. A word organized CID structure using polysilicon gates and buried collectors for XY addressing.

3.2. Bit organization

In a bit organized mosaic, the coincidence between line and column signals is detected by a double switch. E.g. the bipolar matrix [12] shown in fig. 13 comprises two series switches, one is the emitter junction of the bipolar transistor, the other the series FET transistor. Intermediate storage of the video signal is not required of course, the output signal being obtained sequentially at the substrate signal.

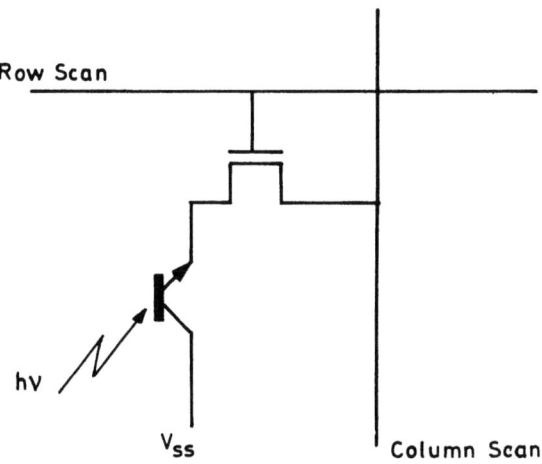

Fig. 13. A bit organized bipolar photoarray using an additional MOS transistor per cell for switching [12].

Two series FET transistors may also be used in order to select a given cell in a bit oriented structure.

In a recent paper, an interesting bit organized two dimensional CID was proposed [13] which is shown in fig. 14. It consists in a double photo MIS device with a diffused junction in between MIS capacitors. Its functioning is based on the fact that charge injection into the bulk occurs only when the gate voltages of both capacitors are low together. In all other circumstances no injection takes place, because either one gate electrode remains high and charge transfer occurs along the surface from the selected capacitor towards the non selected one, of the same cell, or both remain high and nothing changes.

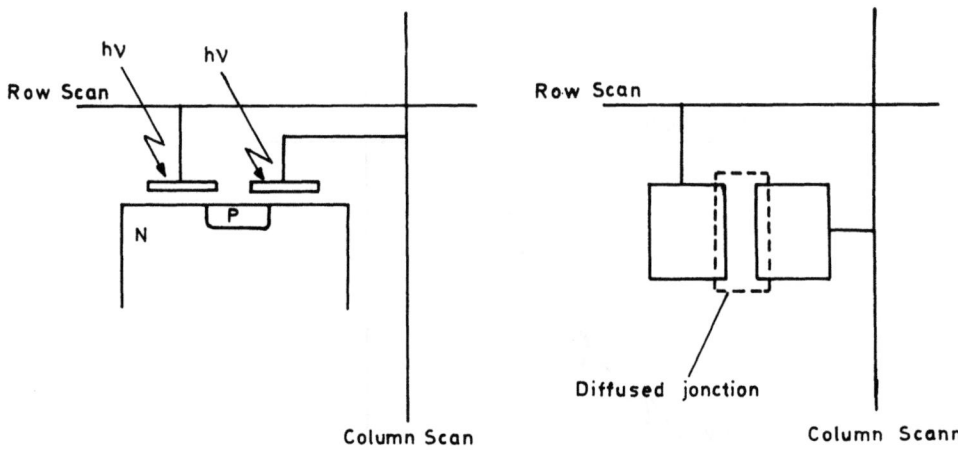

Fig.14 A bit organized CID photoarray using two MIS capacitors for photosensor [13]

BIBLIOGRAPHY

[1] Photosensitivity and Scanning of Silicon Image Detector Arrays, S.G. Chamberlain, IEEE Journal of Solid State Circuits 4, Dec. 1969, pp. 333-342.

[2] Question concerning a p-n Photodiode in the Integrating Mode, R. Van Overstraeten, G. Declerck, L. Schreurs. IEEE Journal of Solid State Circuits 6, Dec. 1971, p. 419.

[3] Operation of P-N Junction Photodetectors in a Photon flux Integration Mode, G.P. Weckler. IEEE Journal of Solid State Circuits, 2, Sept 1967; p.65-73.

[4] Transparent Metal Oxide Electrode CID Imager Array.
D. Brown, M. Ghezzo, M. Garfinkel.
IEEE Solid State Circuit Conference, Dig. Techn. Pap.
Philadelphia, 1975, pp. 34-35.

[5] Quantum Efficiency of a Silicon Gate Charge-Coupled Optical Imaging Array,
R.W. Brown, S.G. Chamberlain,
Physi Stat. Sol. (a) 20, 1973, pp. 675-685.

[6] Multielement Self Scanned Mosaïc Sensors.
P.K. Weimer, W.S. Pike, G. Sadasiv, F.V. Shallcross, L Meray-Horvath,
IEEE Spectrum 6, 1969, p. 52-64.

[7] Integrated MOS and Bipolar Analog Delay Lines using Bucket-Brigade Capacitor Storage.
F.L.J. Sangster.
IEEE Solid State Circuit Conf., Dig. Techn. Pap, Philadelphia , 1970, p. 74-75.

[8] Operational characteristics of CID Imager.
G.J. Michon and H.K. Burke.
IEEE Solid State Circuit Conf., Dig. Techn. Pap., Philadelphia, 1974, p. 26-27.

[9] Fixed-Patter, Noise in Photomatrices.
Peter W. Fry, P.J.W. Noble, R.J. Rycroft.
IEEE Journal of Solid State Circuits, 5, oct. 1970, pp. 250-254.

[10] Transfer Functions of Imaging Mosaïcs Utilizing the Charge Storage Phenomena of Transistor Structures,
I. Tepper, R. Anders, D.H. Mc Cann,
IEEE Transactions on Electron Devices, 15, April 1968, pp. 226-237.

[11] MOS Electronics for a Reading Aid for the Blind.
James D. Plummer
Technical Report n° 4828-6, Solid State Electronics Laboratories, Stanford University, August 1971.

[12] Integrated Arrays of Silicon Photodetectors for Image Sensing,
R.H Dyck, G.P. Weckler,
IEEE Transactions on Electron Devices, 15, April 1968, pp. 196-201.

[13] Charge Injection Imaging.
G.J. Michon, H.K. Burke
IEEE Solid State Circuit Conf., Dig. Techn. Pap.,
Philadelphia, 1973, p. 133-139.

PHOTODIODE SENSOR ARRAYS

Marvin H. White

Westinghouse Electric Corporation,
Advanced Technology Laboratory,
Baltimore, Maryland U.S.A.

ABSTRACT. A discussion of photodiode sensor arrays which includes the basic photodiode sensor, photodiode address and scan circuits, on-chip MOS electrometer amplifiers, off-chip analog signal processing, and image performance of photodiode arrays. The signal-to-noise (S/N) ratio is formulated at the photodiode sensor for Nyquist, shot, and equivalent first-stage current noises. CMOS and varactor-boot strapped PMOS scan registers are presented and their associated address and reset MOS switches. Examples of photodiode sensor arrays are line arrays, or matrix area arrays, and special-purpose arrays such as circular configuration.

1. SPECTRAL RESPONSE

The spectral response of a photodiode approaches an ideal photon detector in the 400 nm to 800 nm wavelength band[1] as illustrated in figure 1. A photodiode with a deep diffusion, 12 μm P$^+$/P$^-$/N junction, has improved response in the infrared because the long-wavelength photogenerated carriers are collected with improved efficiency; however, there is some sacrifice in the short-wavelength or blue response. Conversely, a shallow diffusion, 2 μm P$^+$/N junction has improved response in the blue since the short-wavelength carriers near the silicon surface are collected more efficiently than with the deeper diffused diode structure. The deep diffusion has several disadvantages: (1) a restriction on the achievable pitch of the array due to lateral diffusion and (2) increased noise due to the additional sidewall capacitance of the diffusion. The short-wavelength fall-off in

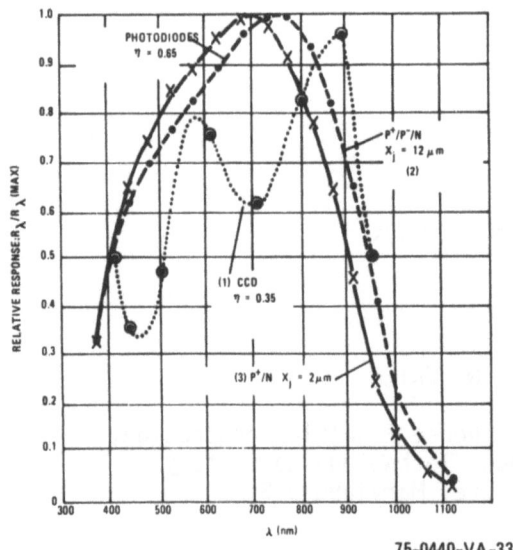

Figure 1. Photodiode Relative Spectral Response vs Wavelength[1]

spectral response ($\lambda < 400$ nm) is a result of surface recombination of photogenerated carriers while the long-wavelength fall-off ($\lambda > 800$ nm) up to the band-gap limitation is due to bulk recombination of photogenerated carriers. Typically, a shallow-diffused photodiode with a 2 μm SiO_2 protective overcoat layer has a quantum efficiency $\eta \simeq 0.70$ in the 400 nm to 800 nm wavelength band. The advantage of diffused photodiodes over other sensors is the lack of spectral streaking from element-to-element (i.e., spectral response variations versus wavelength) which is important for multispectral scanning.

1.1 Responsivity

The responsivity concept has been discussed in the chapter "Design of Solid-State Imaging Arrays" and a particular example was taken for an effective irradiance in the 400 nm to 800 nm wavelength band referenced to a 6000°K blackbody temperature. For the photodiode of figure 1, the responsivity becomes

$$R_D = A \frac{\int_{400}^{800} R_\lambda H_\lambda d\lambda}{\int_{400}^{800} H_\lambda d\lambda} = \frac{0.212 \text{fC}}{\mu J/m^2} \left[\frac{A}{1 \text{ mil}^2}\right] \left[\frac{\eta}{0.70}\right]$$
$$= \frac{1325 e^-}{(\mu J/m^2)} \left[\frac{A}{1 \text{ mil}^2}\right] \left[\frac{\eta}{0.70}\right] \quad (1)$$

for an effective quantum efficiency η and area A. The signal charge collected on a photodiode operating in the charge-storage integration mode[2] in an exposure time τ is

$$Q_S = R_D H(\text{eff.}) \tau = R_D E(\text{eff.}) \quad (2)$$

where H(eff.) is the denominator in equation (1). The maximum signal charge Q_S(sat.) is related to the reset voltage V_R across the photodiode and the photodiode storage capacitance C,

$$Q_S(\text{sat.}) = V_R C \equiv R_D E_{\text{sat.}}(\text{eff.}) \quad (3)$$

which defines a saturation exposure density. Thus, to handle large exposure densities we need a sizeable capacitance C; however, we shall see this limits the minimum detectable signal by the kTC noise.

The signal in general will consist of a leakage current component I_L which accumulates over the exposure time to provide a "signal" charge

$$Q_S' = R_D E(\text{eff.}) + I_L \tau = (I_{ph} + I_L) \tau \quad (4)$$

where $I_{ph} = R_D E(\text{eff.})$ is the photocurrent collected. In general, the leakage current is typically less than 0.1 pA for a 1 mil^2 area (i.e., < 15 nA/cm^2 leakage current). The leakage or dark current is a strong function of temperature and it may be written as,[3]

$$I_L = q n_i \left(\frac{A_B W_j}{\tau_{g-r}} + A_S s\right) \quad (5)$$

where n_i is the intrinsic carrier density $n_i \sim \exp(-E_G/2kT)$ with a value n_i (300°K) = 1.45 x 10^{10} cm^{-3}. A_B and A_S are the areas associated with the bulk and surface components of leakage, W_j the width of the space charge region, s the surface recombination velocity and τ_{g-r} the bulk generation recombination lifetime with typical values of $s < 10$ cm/sec and $\tau_{g-r} > 100$ μsec. Special gettering techniques with phosphorous for bulk gettering and hydrogen for surface-state annealing have yielded $s < 1$ cm/sec and $\tau_{g-r} > 1$ msec. In practice, the leakage current is sufficiently low such that small variations in temperature [e.g. $\Delta T \simeq 0.1°C$] do not affect the low light level performance of the photodiode arrays; however, anomalous leakage current spots in the array with extreme temperature sensitivity affect array performance through "streaking" [e.g. "white" lines in the hard copy image] which is highly objectionable to the observer. These anomalous "streakers" [note: streakers usually prefer to remain anomalous] have been attributed to metal-ion precipitates of the silicon surface which give rise to generation recombination centers with unusually high emission rates as a function of temperature. Rapid "quenching" of silicon photodiode arrays from high temperatures "freezes" these centers in the bulk and prevents precipitation at the surface.

2. NOISE EQUIVALENT SIGNAL (N.E.S.)

The noise equivalent signal (N.E.S.) is defined as the input exposure density E(eff.) which will make the signal-to-noise (S/N) = 1 at a specified position in the signal flow path. A conventional reference point is the storage node of the photodiode, although other points may be selected with the same end result for N.E.S. Before we formulate the N.E.S. the various noise sources must be examined in the photodiode array.

2.1 Johnson-Nyquist Noise[4] (kTC)

An electronic switch or a conductor has a finite resistance to the flow of charge carriers. This electrical resistance defines the fluctuation of charge across a capacitor C. To illustrate this effect we consider a series RC circuit as shown in figure 2 with an ideal switch on a resistance with associated noise voltage generator v(t). The differential equation which describes the charge across the capacitor may be written as,

$$\frac{dQ}{dt} + \frac{Q}{RC} = \frac{v(t)}{R} \tag{6}$$

which can be integrated to yield,

$$Q(t) = \frac{1}{R} e^{-t/RC} \int_0^t v(\tau) e^{\tau/RC} d\tau \tag{7}$$

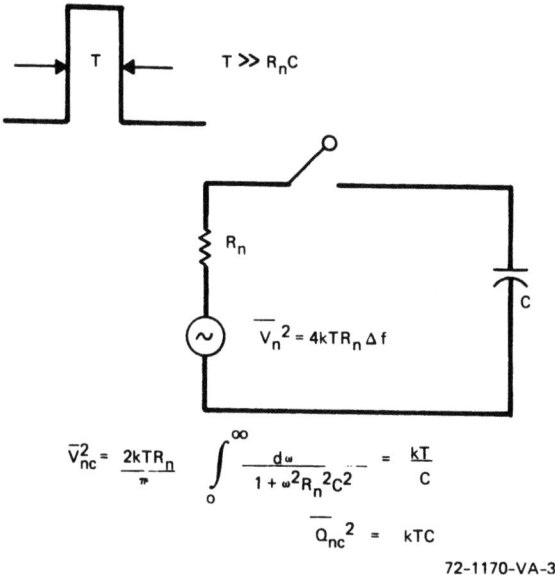

Figure 2. Johnson-Nyquist Thermal Noise

The fluctuation in Q(t) may be obtained as follows:

$$\overline{q_n^2} = \overline{\Delta Q(t)^2} = \overline{\left[Q(t) - \overline{Q(t)}\right]^2}$$
$$= \frac{1}{R^2} e^{-2t/RC} \int_0^t \int_0^\tau \langle v(\tau_1)v(\tau_2) \rangle e^{(\tau_1+\tau_2)/RC} d\tau_1 d\tau_2 \qquad (8)$$

For the "white" noise Johnson-Nyquist power spectra the thermal fluctuations of charge carriers in the resistor give

$$\langle v(\tau_1)v(\tau_2) \rangle = 2kTR \, \delta\langle(\tau_1 - \tau_2)\rangle \qquad (9)$$

Substitution of equation (9) into (8) yields

$$\begin{aligned}\overline{q_n^2} &= kTC \, (1-e^{-2t/RC}) \\ &= kTC \text{ for } t \gg 1/2 \text{ RC} \\ &= 2kTt/R \text{ for } t \ll 1/2 \text{ RC}\end{aligned} \qquad (10)$$

Examination of equation (10) reveals the noise charge fluctuations approach zero as $t \ll RC$. [The above results are analogous to the theory of Brownion Motion for which the particle displacement $\overline{x_n^2} \sim \overline{q_n^2}$ and the diffusion coefficient $D = kT/R$.] In general, the address or reset time intervals (i.e., the time the switch is closed) are sufficiently long such that $t \gg 1/2RC$. For two independent switch closures and measurements the total charge uncertainty is obtained by addition of the fluctuations,

$$\overline{q_n^2} \text{ (2 independent measurements)} = 2kTC \tag{11}$$

We should note the probability density for the Johnson-Nyquist noise has a Gaussian distribution. An intuitive derivation of equation (10) may be obtained if we note the fluctuation energy by the equipartition theorem is $kT/2$ which we can equate to $\overline{q_n^2}/C$ the electrostatic energy of a capacitor.

2.2 "Shot" Noise ($e\overline{I}\tau$)

The fluctuation in the intensity of a charge carrier stream as it flows from one point to another is the so-called "shot" effect. The fluctuation in this stream is random, similar to the thermal fluctuations of the Nyquist noise; however, the probability density has a Poisson distribution. The spectral noise current density may be written as,

$$\overline{i_n^2} = 2e\overline{I}\,\Delta f \tag{12}$$

where $\overline{I}$ is the average current flow. We will apply the "shot" effect to the collection of photo and thermally generated charge carriers in a reverse-biased photodiode. If the photodiode is reverse-biased for an exposure time τ, then the Fourier transform of equation (12) multiplied by a square wave transform yields,

$$\overline{q_n^2} = \int_0^\infty \left(\frac{\overline{i_n^2}}{\Delta f}\right) \tau^2 \left(\frac{\sin \pi f \tau}{\pi f \tau}\right)^2 df = e\overline{I}\tau \tag{13}$$

with the exposure time interval acting as a filter for the "white" noise spectrum of the "shot" process. The average current $\overline{I} = I_{ph} + I_L$ as used in equation (4) may be substituted into equation (13)

$$\overline{q_n^2} = \underset{\text{photon}}{eR_D E(\text{eff.})} + \underset{\text{leakage}}{eI_L \tau} \tag{14}$$

2.3 Equivalent Noise Current in the Preamplifier

We may write a general expression for the equivalent noise current in the preamplifier due to surface or "1/f" noise in the MOS electrometer, if one is used in conjunction with the photodiode, or any equivalent noise source which may be placed at the input of the preamplifier. This noise current spectral density may be transformed to the photodiode storage node by the expression,

$$\overline{q_n^2} = \left(\frac{C}{g_m}\right)^2 \int_0^\infty \frac{\overline{i_n^2}}{\Delta f} |T(f)|^2 \, df \tag{15}$$

where C is the storage node capacitance, g_m the transconductance of the MOS electrometer, and $T(f)$ the transfer function of the analog signal processor. Equation (15) may be written in a slightly different form as,

$$\overline{q_n^2} = \left(\frac{\overline{i_{no}^2}}{\Delta f}\right) B_{eff} \left(\frac{C}{g_m}\right)^2 \tag{16}$$

where $\overline{i_{no}^2}/\Delta f$ is determined at a characteristic corner frequency of the noise spectrum. The effective bandwidth is,

$$B_{eff} = \frac{1}{2\pi} \int_0^\infty \frac{\overline{i_n^2}/\Delta f}{\overline{i_{no}^2}\Delta f} |T(\omega)|^2 \, d\omega \tag{17}$$

If we consider a special case with a "white" noise spectral density and the transfer function of equation (31) in the chapter "Design of Solid-State Imaging Arrays," then the effective bandwidth becomes,

$$B_{eff} = \frac{2}{\pi} \int_0^\infty \frac{\sin^2 \frac{\omega \tau_0}{2}}{1 + \omega^2/\omega_0^2} \, d\omega = \frac{\omega_0}{2} \left(1 - e^{-\omega_0 \tau_0}\right) \tag{18}$$

where ω_0 is the bandwidth of the preamplifier and τ_0 is the time difference between clamping to the reset reference and the sample determination. If we use equation (12) as the noise current spectral density, then the noise charge becomes,

$$\overline{q_n^2} = e\overline{I_A}\omega_0 \left(1 - e^{-\omega_0 \tau}\right) \left(\frac{C}{g_m}\right)^2 \tag{19}$$

where $\bar{I}_A$ is the average current flow into the amplifier. If we consider thermal noise in the form of surface or "1/f" noise in the electrometer amplifier, then the noise charge may be written in the form $\overline{q_n^2} = kTC_{ST}$, where C_{ST} is an effective storage trap capacitance. Notice equation (13) and the example expressed by equation (19) illustrate the basic feature of the electrometer, namely, a low C/g_m ratio is desired to minimize off-chip noise contribution to the N.E.S.

2.4 Formulation of Noise Equivalent Signal (N.E.S.)

If we consider the various noise contributions of sections (2.1, 2.2, 2.3) and the signal charge described by equation (2), then the noise equivalent signal (N.E.S.) becomes,

$$\text{N.E.S.} = \frac{1}{R_D} \left[\beta kTC + e\bar{I}_L \tau + \left(\frac{\overline{i_{no}^2}}{\Delta f}\right) B_{eff} \left(\frac{C}{g_m}\right)^2 \right]^{\frac{1}{2}} \quad (20)$$

$$\underbrace{}_{\substack{\text{Johnson} \\ \text{Nyquist}}} \quad \underbrace{\phantom{e\bar{I}_L\tau}}_{\text{Shot}} \quad \underbrace{\phantom{\left(\frac{i_{no}^2}{\Delta f}\right)}}_{\substack{\text{Preamplifier} \\ \text{Noise Current}}}$$

where β = 1 or 2 dependent upon a single or double address switch closure and $\overline{i_{no}^2}/\Delta f$ is determined at the input of the off-chip preamplifier. Alternately, equation (20) may be written with $\overline{v_{no}^2} = \overline{i_{no}^2}/g_m^2$ to reference the noise current to the input of the on-chip electrometer. In the case where surface or "1/f" noise is present in the electrometer we have,

$$\frac{\overline{i_{no}^2/g_m^2}}{\Delta f} = 4kTR(\omega) \quad (21)$$

$$\sim \frac{N_{ST}}{\omega C_A}$$

where N_{ST} is the surface trap density and C_A is the area of the electrometer gate region.[5,6] The total storage capacitance $C = C_p + C_A$, where C_p is the parasitic capacitance of interconnect or line capacitance and photodiode capacitance. If we combine equations (20) and (21), then the N.E.S. may be minimized with respect to the gate area of the electrometer amplifier.

Figure 3 illustrates the measured N.E.S. ($\mu J/m^2$) of a photodiode sensor chip[7] as a function of input exposure density E ($\mu J/m^2$). This particular chip has an N.E.S. limited by the Johnson-Nyquist noise of the storage capacitance C = 0.75 pF and the reflected current noise in the electrometer due to the high C/g_m ratio [g_m = 130 μmhos at V_R = 8V]. The photodiode sensor has an area of 18 μm x 22 μm as discussed in the chapter "Design of Solid-State Imaging Arrays" Section 4, and with an effective quantum efficiency of η = 0.70 the responsivity is 0.13fC/$\mu J/m^2$.

If we consider only the Johnson-Nyquist noise, then the calculated N.E.S. becomes

$$\text{N.E.S.} \begin{pmatrix} \text{J-N noise} \\ \text{limited} \end{pmatrix} \simeq \frac{\sqrt{2kTC}}{R} = 0.60 \ \frac{\mu J}{m^2} \qquad (22)$$

which is close to the measured values at low exposure densities. The radiation "shot" noise contribution does not disturb the N.E.S. until the exposure density is on the order of the above noise,

$$\begin{aligned} E \text{ (radiation shot)} &\simeq R \ \frac{(\text{N.E.S.})^2}{q} \\ &= \frac{2kTC}{qR} \simeq 300 \ \frac{\mu J}{m^2} \end{aligned} \qquad (23)$$

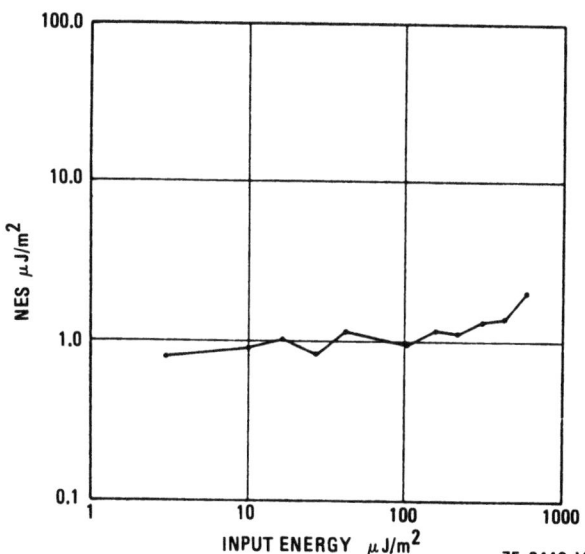

Figure 3. Noise Equivalent Signal (N.E.S.) as a Function of Exposure Density E ($\mu J/m^2$) [After reference (7)]

which is illustrated in figure 3. Figure 3 was taken with an exposure time of $\tau = 1$ msec and referenced to a 6000°K blackbody in the 400 nm to 800 nm wavelength band.

3. CMOS SCAN REGISTERS AND ADDRESS CIRCUITRY

Complementary MOS-FET (CMOS) shift registers may be used to scan or commutate photodiode sensors with the advantages of low voltage operation and low power dissipation. The photodiodes operate in the charge storage integration mode in which the photodiode is periodically reversed-biased and reset to a known reference voltage V_R. Figure 4 illustrates a CMOS shift-register composite circuit to control the amplifier address and reset switches. A "one" is input to the first stage of the shift register (SR1) and shifted through successive stages with a 2ϕ register clock (ϕ_A, ϕ_B). The "one" appears at interstage points as gate pulses to the amplifier address switches. In this example the shift register output addresses a pair of photodiodes with a corresponding pair of output busses (i.e., OUT-A, OUT-B). The CMOS shift register is of the dynamic type with the clocks 180 degrees out of phase from each other. The nested reset pulse is generated by an AND gate whose inputs are the external clock ϕ_C and the SR output at the particular stage of interest.

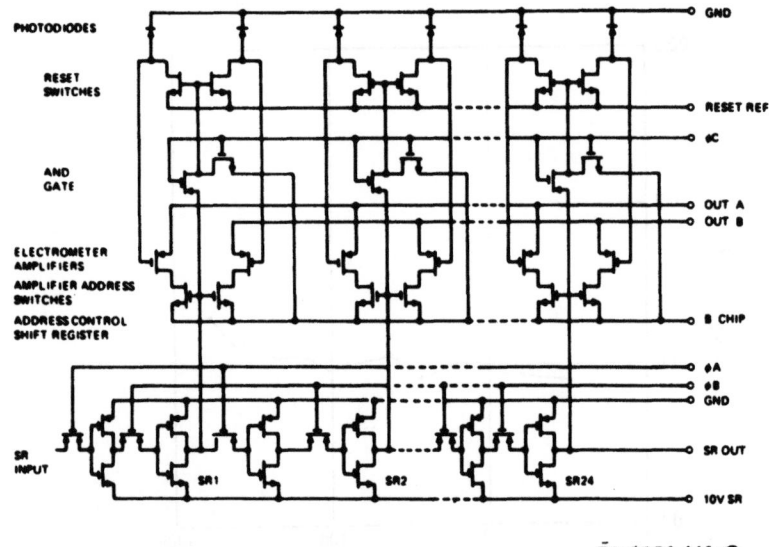

Figure 4. CMOS Shift Register Composite Circuit [After reference (7,8)]

Figure 5 illustrates the photodiode sensor and amplifier combination, which is an enlargement from figure 4. Q_1 is the reset switch, Q_2 the electrometer amplifier, and Q_3 the address switch controlled by the SR output. In this particular linear array each photodiode has an associated p-channel electrometer amplifier. Figure 6 illustrates the sensor timing and signal output waveforms. The output waveform may be divided into three distinct time periods:

(1) READ SIGNAL of nth integration time

(2) RESET photodiode for $n+1^{th}$ integration time

(3) READ initial condition of $n+1^{th}$ integration time and subtract from step (1)

The "READ SIGNAL" voltage is a function of the initial conditions (voltage) in the nth integration time, leakage (I_L) and signal (I_S) current, and electrometer amplifier gain (g_m). The output voltage e_o may be related to the charge Q_G on the gate of the electrometer by

$$e_o = \frac{R_f g_m Q_G}{C} = \frac{R_f g_m}{C} (I_L + I_S) \tau \qquad (24)$$

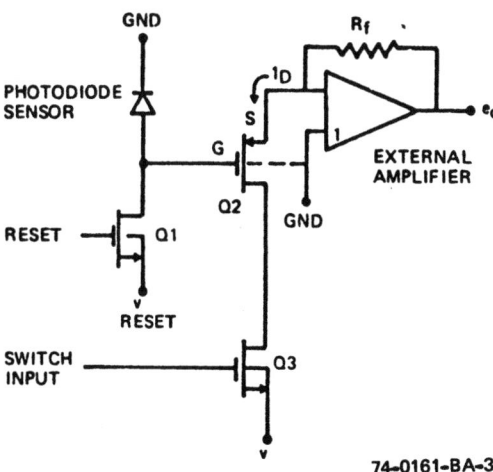

Figure 5. Equivalent Circuit of Photodiode Sensor and Amplifier [After reference (7,8)]

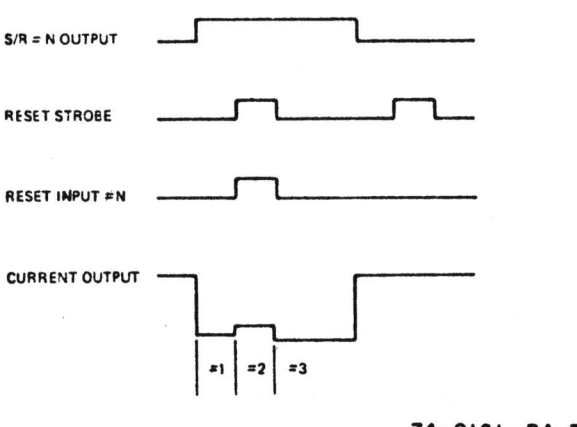

Figure 6. Sensor Timing and Output [After reference (7,8)]

where C is the capacitance from the gate electrode of Q_2 to ground and τ the integration time. The transconductance g_m of the electrometer is related to the physical properties of the structure by the expression,[3]

$$g_m = \frac{\bar{\mu}_p}{X_0} \left(\frac{W}{L}\right) K_0 \epsilon_0 (V_{GS} - V_T) \qquad (25)$$

where $\bar{\mu}_p$ is the hole mobility, X_0 the oxide thickness, $K_0\epsilon_0$ the dielectric constant of the oxide, (W/L) the width-to-length ratio, V_{GS} the gate-to-source voltage, and V_T the device threshold voltage. In the reset interval (#2) the electrometer output is a measure of the reset voltage level (i.e., $V_{GS} = V_R$). In the (#3) interval the electrometer output is a measure of the initial condition for the start of the integration or exposure time. The initial condition interval (#3) differs from the reset interval (#2) by the addition of feedthrough signal voltage

$$\Delta V = \frac{C_{GD}}{C_{GD}+C} V_R \qquad (26)$$

where C_{GD} is the gate-to-drain feedthrough capacitance of the reset switch Q_1. The difference between (#3) and (#1) intervals is a measure of the accumulated signal and leakage currents over the exposure interval as expressed by equation (24).

4. PMOS SCAN REGISTERS AND ADDRESS CIRCUITRY

P-channel MOS-FET (PMOS) shift registers, which use the so-called "bootstrapping" or varactor principle,[9] may be employed to scan photodiode sensors with the advantage of extremely low power dissipation. Figure 7 illustrates a varactor PMOS shift register with the storage node "bootstrapped" by the varactor. This particular shift register AND's the clock pulse and the data pulse to provide a scan pulse to actuate the address switch and transfer the stored photocharge from the addressed photodiode to the bus line B_1. Figure 8 illustrates the clock waveforms, data, and address pulses provided by the scan register. The address pulses are provided for each 1/2 stage of the shift register and the shift register only dissipates power in the stage with the data. Thus, the entire shift register has a power dissipation equivalent to a single stage under continuous operation. Figure 7 with a common bus line for many photodiodes is an example of a single amplifier per row of photodiodes in contrast with the single amplifier per photodiode of figure 4.

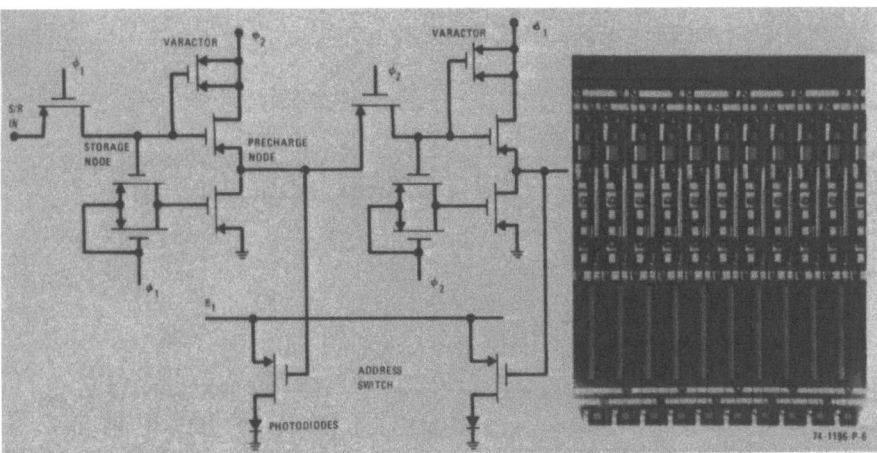

Figure 7. PMOS Shift Register Composite Circuit with Varactor Bootstrapping of Storage Node

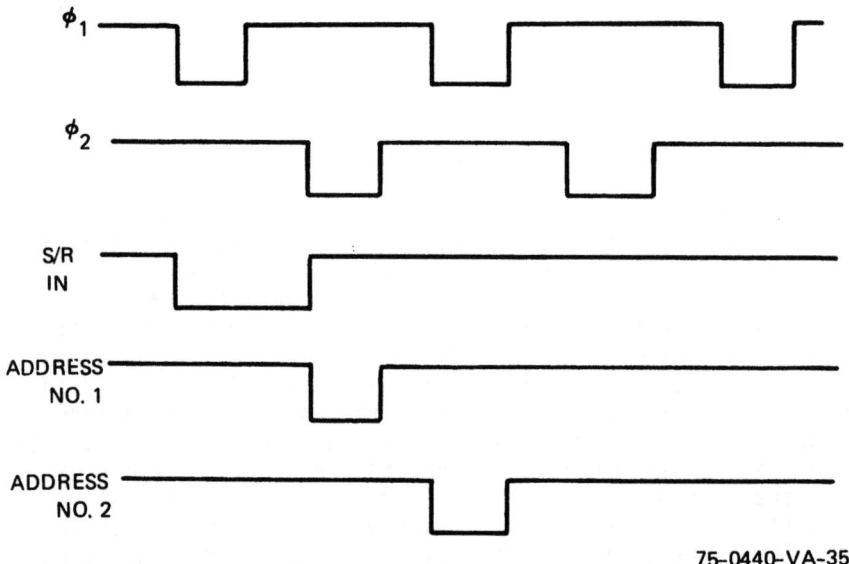

Figure 8. Clock and Address Waveforms for Circuit of Figure 7

The varactor principle used in figure 7 may be explained with reference to figure 9. The voltage "bootstrapped" to the storage node by the varactor may be written as

$$\Delta V = \frac{C_b}{C_b + C_s} V_c$$

where C_b is the varactor capacitance of figure 9, C_s the total storage node capacitance, and V_c the clock voltage. The capacitance C_b has two values:

$$C_b = C_o WL' \quad \text{DATA} = 0$$
$$ = C_o WL \quad \text{DATA} = -V_c$$

where $L \gg L'$. When the storage node = $-V_c$ the inversion layer under the varactor causes the capacitance C_b to become high effectively coupling a large fraction of the clock voltage V_c to the data node where it adds to the data. Thus, the clocks (ϕ_1, ϕ_2) are coupled "full-strength" (i.e., no threshold-voltage loss) to the address switch resulting in high speed operation.

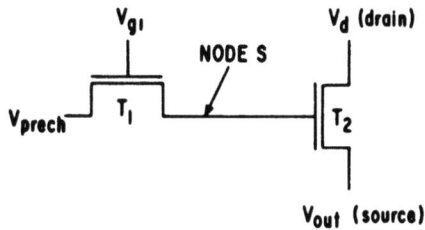

Storage node S in dynamic MOS circuit.

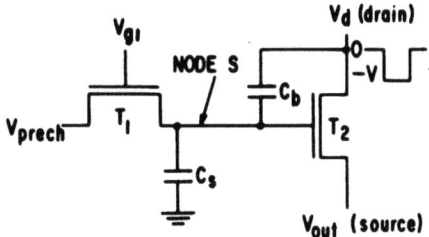

Circuit incorporating bootstrap capacitor C_b.

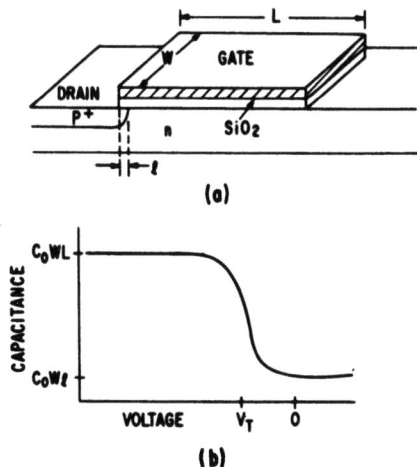

(a) Schematic of MOS varactor. (b) Capacitance as function of voltage for MOS varactor.

75-0440-VA-36

Figure 9. MOS Varactor Principle [After reference (9)]

Another example of a varactor PMOS shift register type is shown in figure 10 where the precharge node is "bootstrapped" with the varactor.10 This shift register provides a different output scan waveform in which DATA and $\overline{\text{DATA}}$ are used to switch the sensor from one voltage (V_R) to another (V_W) and provide a latching feature. Figure 11 illustrates the characteristic waveforms of data (address No. 1) and data (address No. 2). The usefulness of this circuit lies in the validity of the DATA and $\overline{\text{DATA}}$ in the absence of the clock in contrast with figure 8. Thus, a knife-edge scan may be provided with the circuit of figure 10.

5. PHOTODIODE LINE ARRAYS

Photodiode line arrays have the general equivalent circuit shown in figure 12 with a series address switch controlled by a shift register as discussed in sections 3 and 4. The video output may be common to a row of photodiodes as discussed in section 4, or each photodiode may have its own electrometer amplifier and amplifier address switch. The video output shown in figure 12 may be connected to an on-chip electrometer amplifier and reset switch or this function may be performed off-chip as shown in figure 13. The disadvantage of the method shown in figure 13(b) is the inability to remove the reset noise across the capacitance C, while the method shown in figure 13(a) limits the noise to the

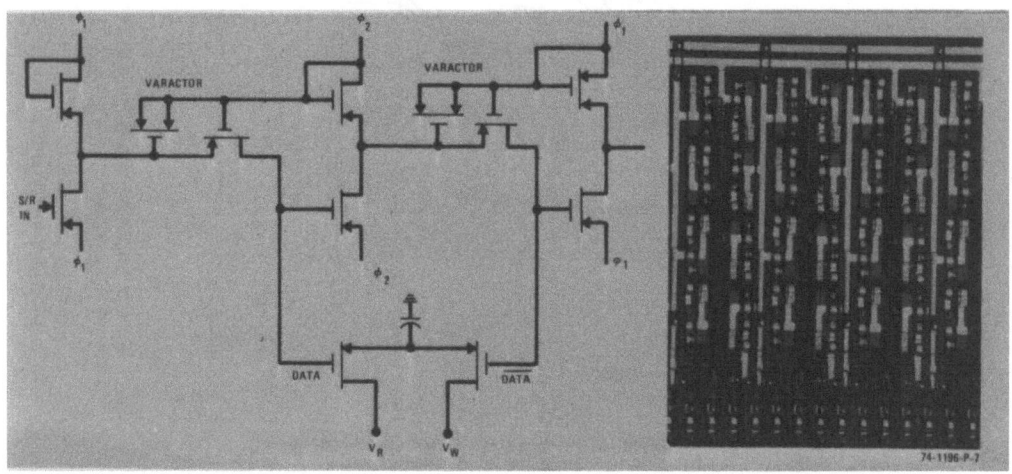

Figure 10. PMOS Shift Register Composite Circuit with Varactor Bootstrapping of Precharge Node

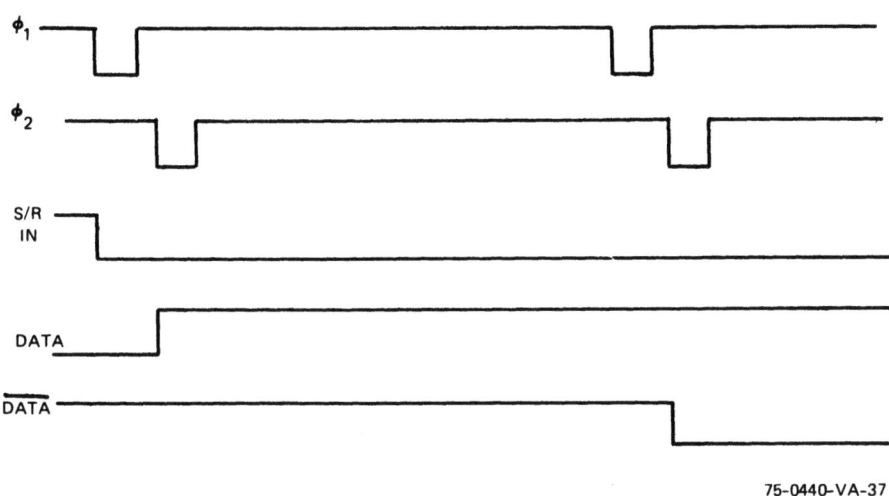

Figure 11. Clock and Address Waveforms for Circuit of Figure 10

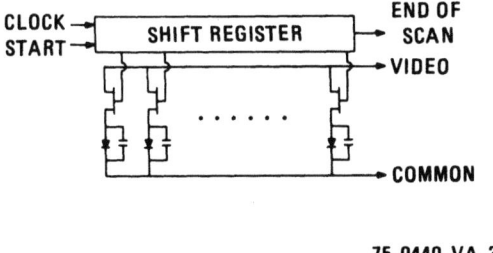

Figure 12. Schematic Diagram of Typical Photodiode Line Scanner

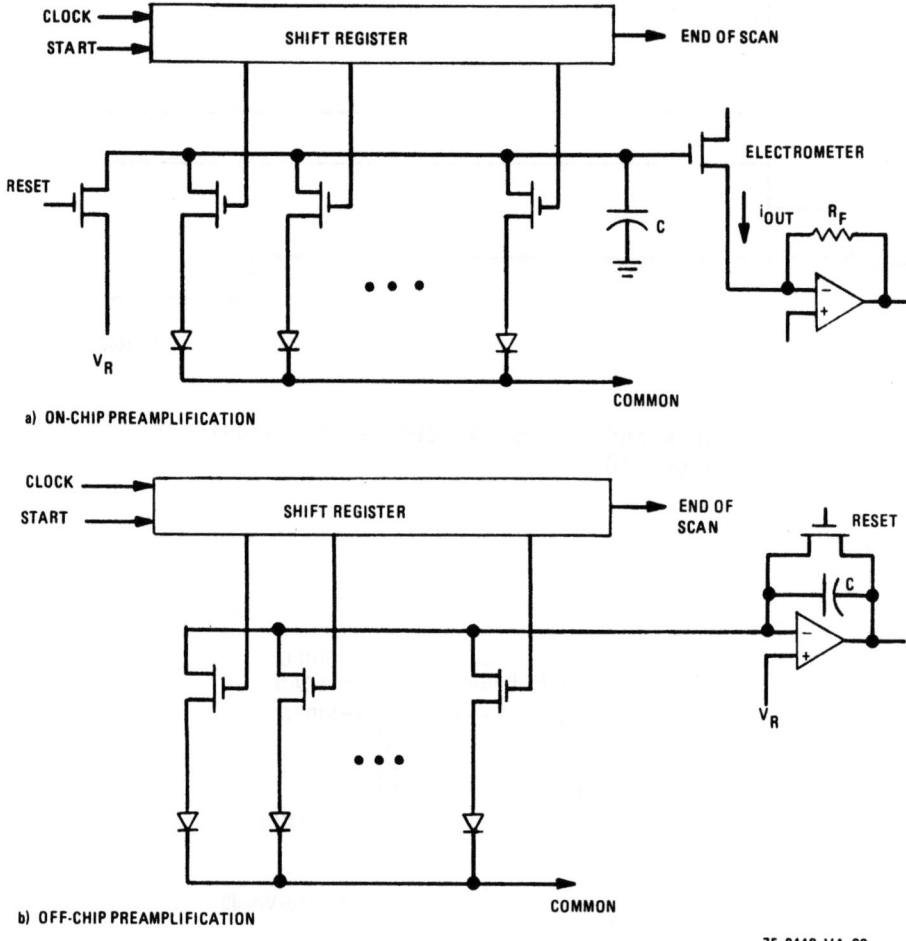

Figure 13. Methods of Preamplification for Photodiode Line Arrays Sharing a Common Preamplifier

diode capacitance C_D, since $C_D \ll C$. A disadvantage of the method shown in figure 13(a) is the need to provide $C \gg C_D$ and hence off-chip noise contributions reflect back to the gate of the electrometer with C/g_m as discussed in section 2.3. Figure 14 illustrates a 256-element photodiode array on 1 mil centers with a 1 mil wide aperture and figure 15 illustrates a 1872-element photodiode array on 15 μm centers (pitch P = 15 μm) with a 16 μm aperture. (See section 4.2 in the chapter entitled "Design of Solid-State Imaging Arrays.") The background in figures 14 and 15 illustrate the light shield and the aperture window of the array.

Figures 14 and 15 are examples of zero-P design; however, to improve the figure-of-merit of the photodiode line array and achieve equal M.T.F.'s at the Nyquist sampling limit [see section 4.1 of the chapter "Design of Solid-State Imaging Arrays"], a bilinear photodiode array may be selected. Figure 16 illustrates such a photodiode array with a 2-P offset in the along-track direction where P = 0.6 mil in the across-track direction. In order to form long line arrays the "butted" assembly technique shown in figure 17 is employed in which the edge of the chip must be placed to within 0.3 mil of the end diodes to maintain image contiguity. Figure 18 illustrates a butted assembly, linear

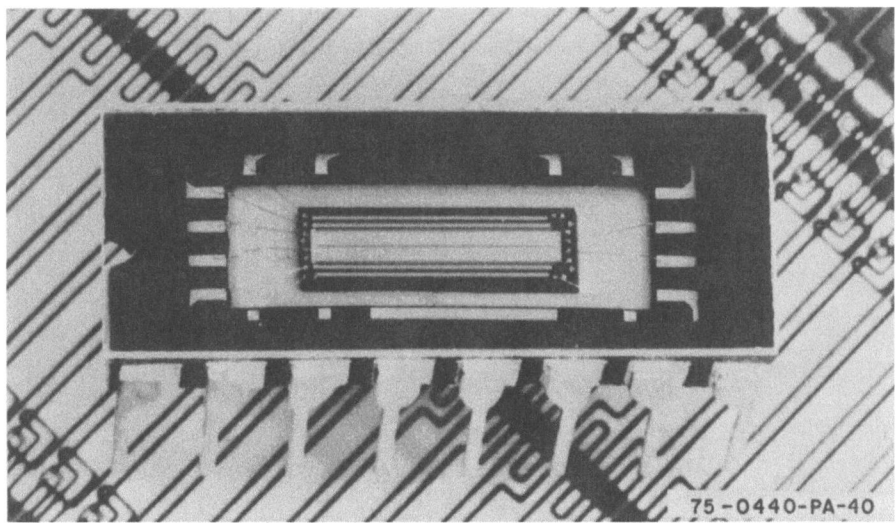

Figure 14. RL-256 Element Photodiode Line Array with Pitch P=1 mil and Aperture Width 1 mil (i.e., $\Delta x = \Delta y = P = 1$ mil) [Courtesy of Gene Weckler, Reticon Corp.]

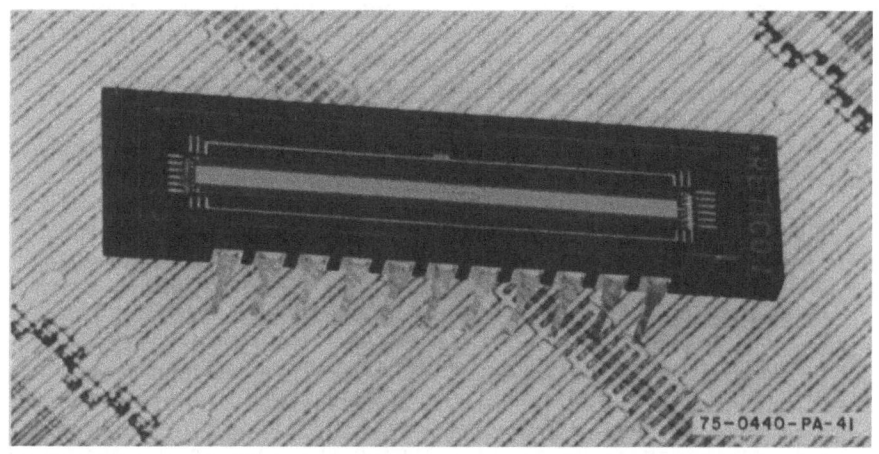

Figure 15. RL-1872 Element Photodiode Line Array with Pitch P=15 μm and Aperture 16 μm (i.e., Δx = P = 15 μm, Δy = 16 μm) [Courtesy of Gene Weckler, Reticon Corp.]

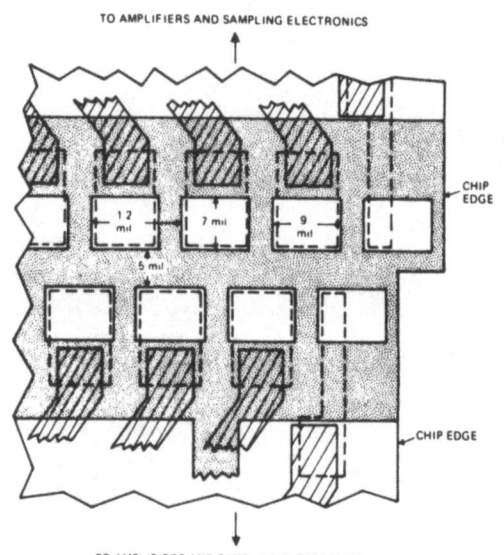

Figure 16. Photodiode Bilinear 2P Array and Chip Edge Geometry [After reference (7)]

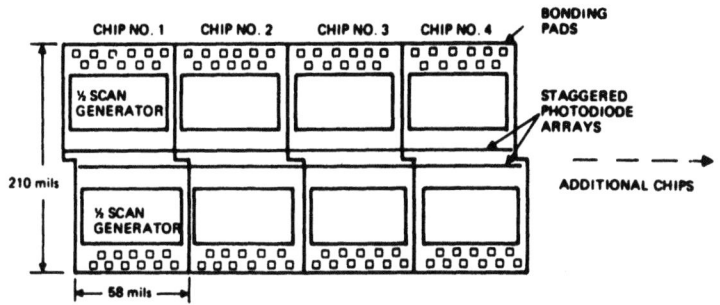

Figure 17. Butted Assembly of Chips with Staggered Photodiode Arrays [After Reference (7)]

array of 1728 photodiodes and associated signal processing for Advanced Earth Resource Observation Applications for NASA/Goddard Space Flight Center.[8,11] Figure 19 is an example of high resolution aerial imaging with such photodiode line arrays. Photodiode line arrays may be used for page reading, film scanning, satellite "push-broom" remote sensing, pattern recognition, spectroscopy, etc.

6. PHOTODIODE AREA ARRAYS

Figure 20 illustrates a schematic diagram of a photodiode matrix array with horizontal and vertical PMOS shift registers to perform the X-Y selection of the photodiode sensor. Each unit cell consists of MOS-FET which serves to "AND" the horizontal and vertical shift register outputs for address of the photodiode sensor. Figure 21 illustrates a 50 x 50 matrix photodiode array with diodes on 4 mil centers in X and Y directions. Figure 22 illustrates a self-scanned array (SSA) of 12 x 38 photodiodes in an Optical Character Recognition (OCR) Wand System for 2-dimensional character imaging. The OCR Wand is passed over the data, either left-to-right or right-to-left, and the characters are

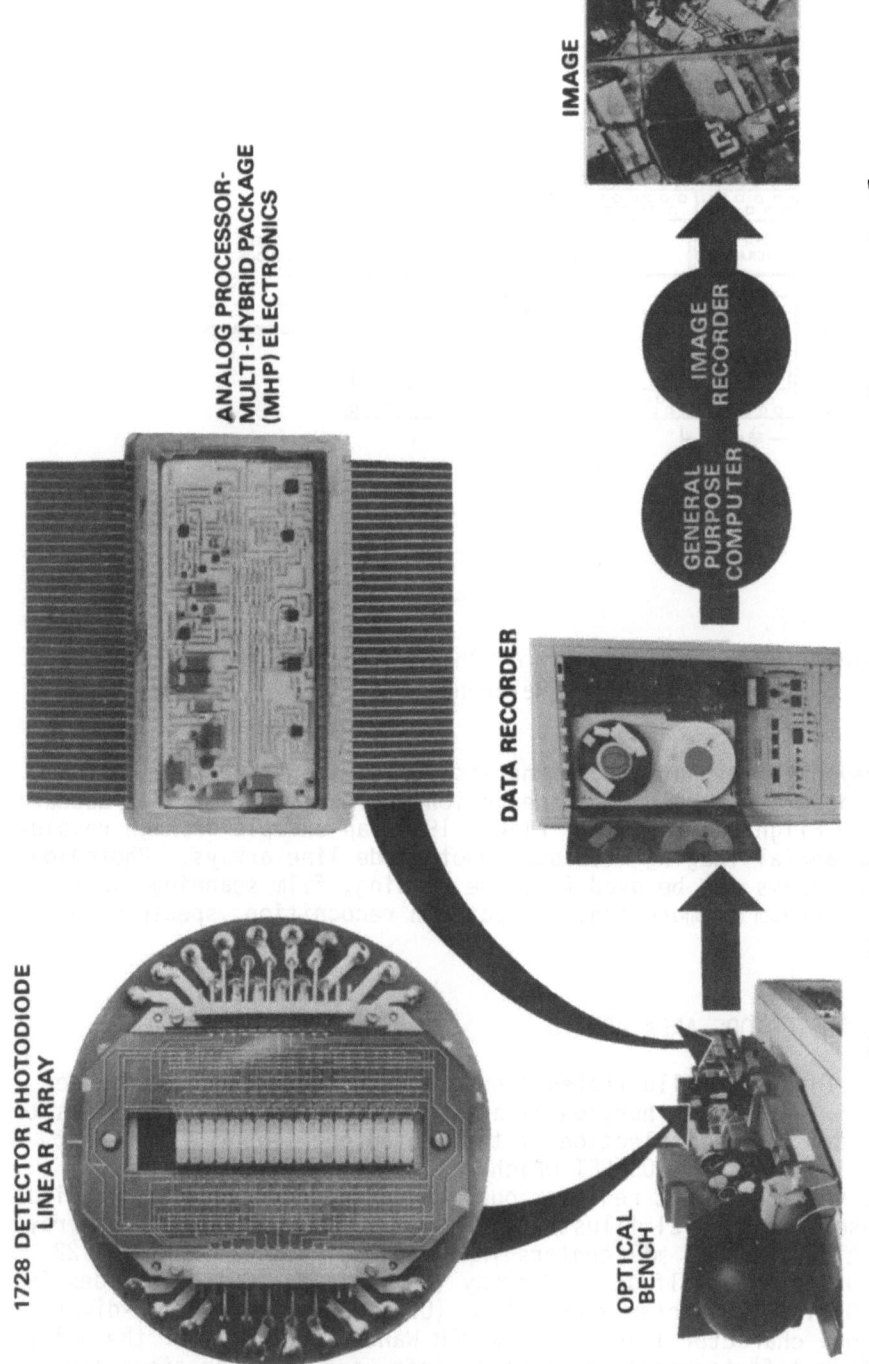

Figure 18. Breadboard Linear Photodiode Array [Reference (11)]

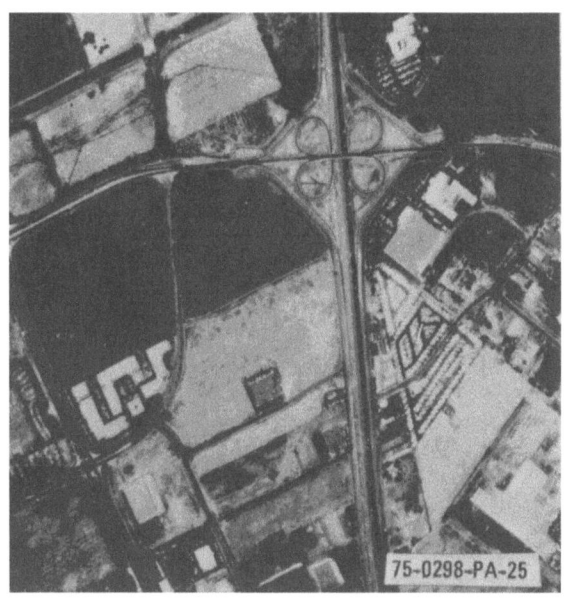

Figure 19. High Resolution Aerial Imaging with Photodiode Line Arrays (reference 8) [Clover Leaf Expressway Inter-Change] Taken with a Scene Translator

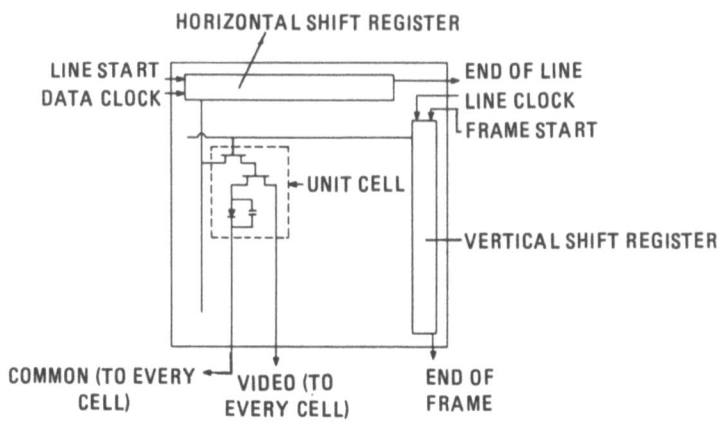

Figure 20. Schematic Diagram of Photodiode Matrix Array (Courtesy of Gene Weckler, Reticon Corp.)

Figure 21. RA 50 x 50 Matrix Photodiode Array (Courtesy of Gene Weckler, Reticon Corp.)

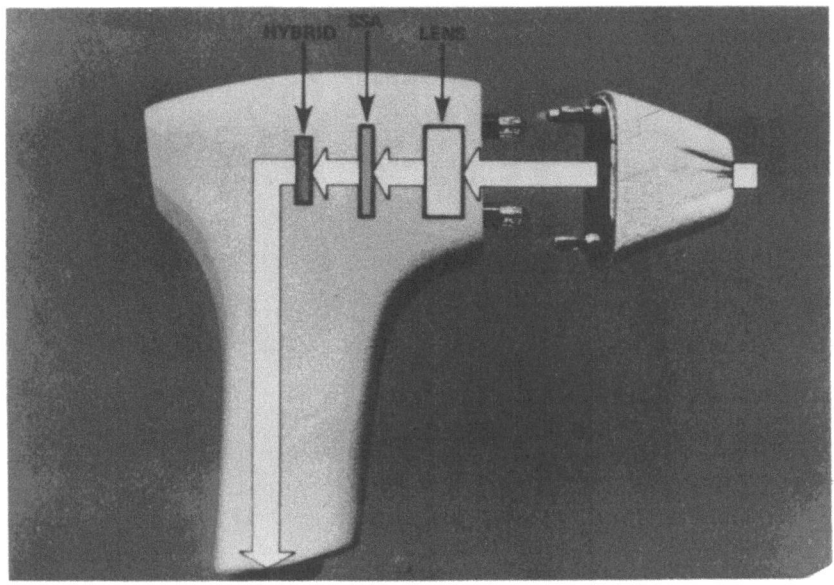

Figure 22. Hand-Held (OCR) Wand 12 x 38 Photodiode Self-Scanned Array (SSA) for Character Recognition (Courtesy of Dick Van Tyne, Recognition Equipment, Inc.)

recognized, edited and transmitted to a data recording system. Decentralized data collection can be performed for point of sale, inventory recording, production control, postal data, airline tickets, etc. Figure 23 illustrates a set of numerics made from a C.R.T. screen with the video processed from the matrix array to modulate the Z-axis of the C.R.T.

Figures 22 and 23 illustrate the use of low resolution photodiode matrix arrays for pattern recognition.

7. SPECIAL PURPOSE PHOTODIODE ARRAYS

Figure 24 illustrates a circular 64-element photodiode array for such applications as tracking, alignment, automatic focusing, etc. The 64 photodiodes are equally spaced on a 2-mm diameter circle with a 0.1 mm x 0.1 mm element size as shown in figure 25. The light sensing area is an annulus with an outer radius of 1.05 mm and an inner radius of 0.95 mm. The elements are scanned by a PMOS shift register/ring counter with a two-phase clock drive as shown in figure 26. The time required to scan a complete circle is $64/f_c$, where f_c is the clock frequency.

Figure 23. Numeric Characters Imaged on a 12 x 38 Photodiode Matrix Array (Courtesy of Dick Van Tyne, Recognition Equipment, Inc.)

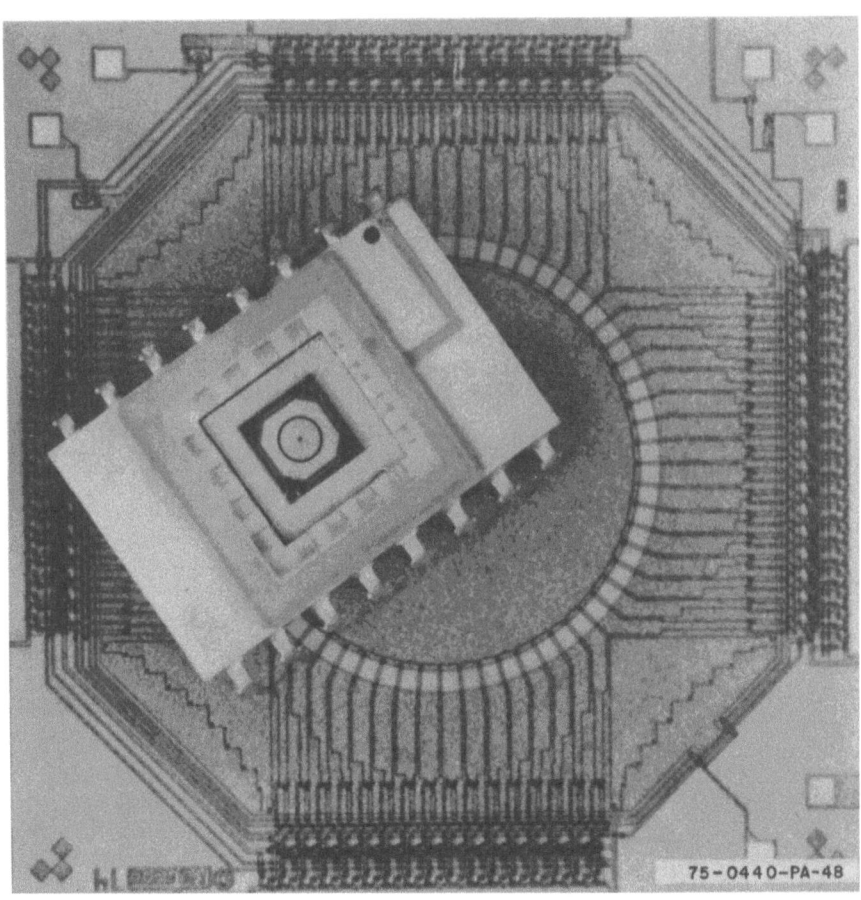

Figure 24. RO-64 Self-Scanned Circular Photodiode Array
(Courtesy of Gene Weckler, Reticon Corp.)

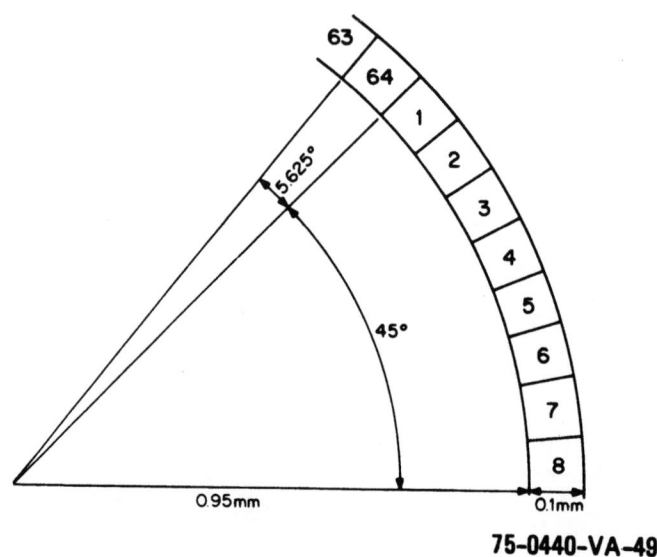

Figure 25. Circular Photodiode Geometry

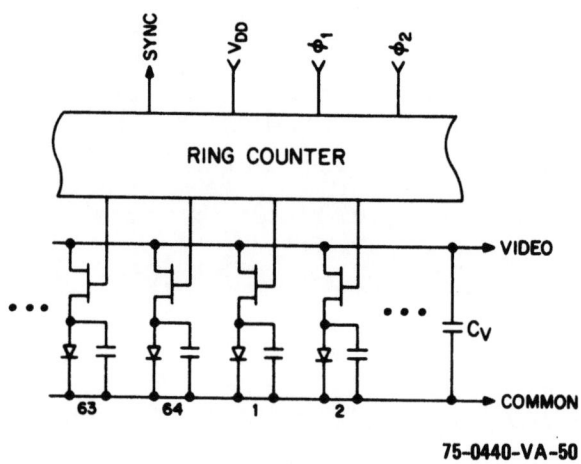

Figure 26. Ring Counter Driven Circular Array

REFERENCES

1. M.H. White, D.R. Lampe, F.C. Blaha, and I.A. Mack, "Characterization of Surface Channel CCD Image Arrays at Low Light Levels," IEEE J. of Solid-State Circuits, SC-9, 1, 1974.

2. G.P. Weckler, "Operation of P-N Junction Photodetectors in a Photon-Flux Integrating Mode," IEEE J. of Solid-State Circuits, SC-2, 65, 1967.

3. A.S. Grove, Physics and Technology of Semiconductor Devices, John Wiley & Sons, Inc., 1967.

4. D.F. Barbe, "Imaging Devices Using the Charge-Coupled Concept," Proc. IEEE, 63, 38, 1975.

5. S. Christensson, I. Lundstrom, C. Svensson, "Low Frequency Noise in MOS Transistors," Journal of Solid-State Electronics, Vol. 11, 797, 1968.

6. E.A. Leventhal, "Derivation of 1/f Noise in Silicon Inversion Layers from Carrier Motion in a Surface Band," Journal of Solid-State Electronics, Vol. 11, 621, 1968.

7. Advanced Scanners and Imaging Sensors for Earth Observations, NASA SP-335 AdHoc Advanced Imagers and Scanners Hocking Group, Sponsored by NASA-GSFC, Dec. 1972.

8. M.H. White and D.R. Lampe, "Noise Considerations in Solid-State Imagers," Intercon, New York City, N.Y., 1974.

9. R.E. Joynson, J.L. Mundy, J.F. Burgess, C. Neugobauer, "Eliminating Threshold Losses in MOS Circuits by Bootstrapping Using Varactor Coupling," IEEE J. of Solid-State Circuits, SC-7, 217, 1972.

10. J.J. Tiemann, W.E. Engeler, R.D. Baertsch, "A Surface Charge Correlator," IEEE of Solid-State Circuits, SC-9, 403, 1974.

11. L.L. Thompson and R.A. Tracy, "Advanced Solid-State System for Remote Sensing from Satellite," Symposium on Management and Utilization of Remote Sensing Data (American Society of Photogammetry), Sioux Falls, N.D., Nov. 1973.

PHOTOTRANSISTOR ARRAYS

P. Jespers
Université Catholique de Louvain
Microelectronics Laboratories
Bâtiment Maxwell
1348 Louvain-la-Neuve, Belgium.

Abstract

The performances of phototransistor arrays are analyzed considering the limitations resulting from the variable dynamic impedance of the series junction emitter. The influence of current gain non uniformities across the array also are taken into account. An experimental 400 . 500 elements array is described.

INTRODUCTION

The charge storage mode operation of phototransistors was described already in the section on XY addressing. In this chapter we will concentrate our analysis on the limitations mainly of phototransistor arrays. We will assume that phototransistor as well as photodiode sensors may adequately be described by the model shown in fig. 1. We recall that C_c represents the reverse biased collector junction capacitance which is discharged by means of the photocurrent I_L, and that the series diode D stands for the emitter junction. Resistor R represents the resistor R_e multiplied by the current gain h_{fe} when a phototransistor is considered, and resumes to the actual series resistor in case two separate opposite diodes are considered.

The main drawback of phototransistors used in the storage mode of operation results from the wide impedance variations of the series diode D. Under high level conditions, the current flowing in the circuit during the sampling time usually is large

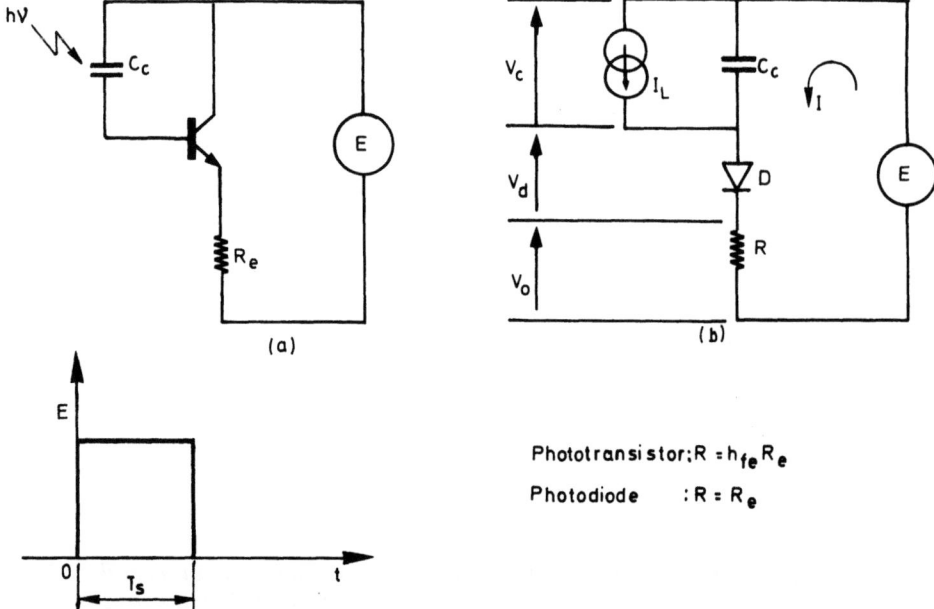

*Fig 1. Voltage sampling mode of phototransistor sensors.
(a) actual circuit. (b) equivalent circuit.*

enough in order to make the dynamic impedance of D almost negligible with respect to R. But when low light levels are considered, this is no more true and rather long sampling times are required in order to achieve the full loading of C_c. It will be shown that the response curve of phototransistors under these circumstances exhibits a high degree of non linearity and that an unwanted memory effect takes place, causing smearing out of images. A tradeoff between these effects exists, as will also be shown in the next section.

Influence of the series diode on the performances of phototransistors
---

In order to start our discussion, let us assume that the diode D may be described by the well known expression :

$$I = I_s \exp \frac{V_d}{V_T}$$

where : I_s represents the saturation current.
V_d the forward voltage drop across the diode.
and V_T is equal to kT/q.

We neglect first the transition capacitance of the series diode
in order to simplify the analytical expressions and understand
clearly the basic limitations of phototransistors.

When the integration time is started, the actual voltage
drop across the collector capacitance is equal to the final
voltage attained at the end of the sampling period. The series
diode consequently is reverse biased and equivalent to an open
circuit. When the sampling time is over and the integration time
is initiated, the light falling on the junction slowly discharges
the collector junction. At the end of the integration time,
the series diode suddenly is forward biased again and reloading
of C_c takes place. The current flowing through the photosensor
equals :

$$I = C_c \frac{dV_c}{dt} + I_L \qquad (1)$$

whereas the total voltage drop across the sensor and its series
resistor resumes to :

$$E = V_o + V_d + V_c \qquad (2)$$

calling respectively :

V_o : the voltage drop RI across the series resistor.
V_d : the voltage drop across the forward biased diode.
V_c : the voltage drop across the reverse biased photojunction.

After differentiation of (2) and elimination of dV_c/dt in (1),
the following expression is obtained :

$$I = -RC_c \frac{dI}{dt} - C_c \frac{dV_d}{dt} + I_L$$

The derivative of V_d may be replaced by :

$$\frac{dV_d}{dt} = \frac{V_T}{I} \frac{dI}{dt}$$

since :

$$V_d = V_T \ln \frac{I}{I_s}$$

The following differential equation now is obtained :

$$(\tau + \frac{C_c V_T}{I}) \frac{dI}{I - I_L} = - dt \qquad (3)$$

with time constant τ equal to the product RC_c.

With the reasonable assumption that the current I is always much larger than I_L during the sampling period, expression (3) may be simplified, as follows :

$$(1 + \frac{V_T}{RI}) \frac{dI}{I} = - \frac{dt}{\tau} \qquad (4)$$

The above expression clearly indicates that a breakpoint between high level and low level illumination exists, and that it is set by the condition that the voltage drop across the series resistor be respectively larger or smaller than V_T. Indeed, as long as :

$$RI > V_T$$

one observes a purely exponential decay of the current I, whereas the opposite condition corresponds to an hyperbolic much slower recharging process. In the first instance, the dynamic impedance of the series diode does not influence the overal time constant; in the second, it becomes the dominating element. In practice thus the recharging process always slows down progressively while the current decreases. The effect is clearly evidenced in the graphical representation of the following equation obtained by integration of (4) :

$$\ln \frac{I}{I_o} + \frac{V_T}{RI_o} \cdot (1 - \frac{I_o}{I}) = - \frac{t}{\tau} \qquad (5)$$

In this equation I_o represents the initial current flowing through the series diode. Several curves representing $\frac{I}{I_o}$ versus the normalized time t/τ are shown in fig.2 considering four different values of the reduced output voltage :

$$v_o = \frac{RI_o}{V_T}$$

When v_o is very large, the classical exponential solution is found which corresponds to complete elimination of the impedance offered by the series diode. Considering v_o equal to 0.1 yields the opposite case with an initial voltage drop RI_o of only 2.6 mV. A very long sampling time is required in this last case in order to restore the full charge across C_c. For instance, to obtain a current as small as one tenth of I_o, a charging time equal to 92.3 τ would be needed. This time is reduced to 11.3 τ when v_o is equal to one, but still much time is needed in order to achieve the last part of the recharging process. Even when v_o is equal to 4, thus when RI_o is close to 100 mV, full recharge of C_c is only possible after very long sampling times.

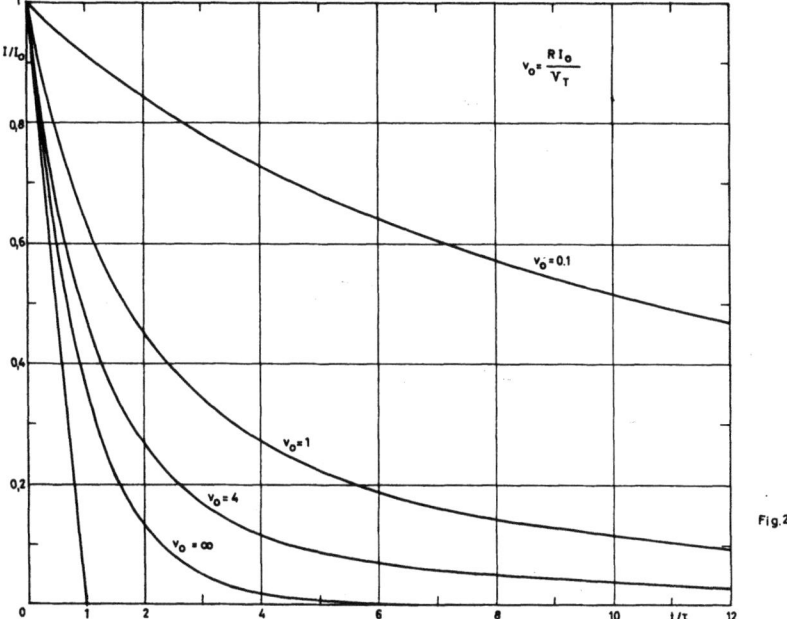

Fig. 2. Reduced output current versus reduced time during the sampling period.

It is legitimate therefore to consider that, in practice, sampling times may not always be sufficiently long in order to reach steady state conditions. This is the overruling situation under low light level conditions that was analyzed by S. Brugler [1,2]. Some of the most interesting results of this work are reviewed hereunder.

Let us call "f" the actual current ratio I/I_o at the end of the sampling time T_s, accordingly to fig. 3. Integrating again equation (4) with the condition :

$$RI = RfI_o \qquad , \text{ for } t = T_s$$

yields the following result :

$$\ln f + \frac{V_T}{RI_o}(1 - \frac{1}{f}) = \frac{T_s}{\tau} \tag{6}$$

or :

$$v_o(\mathscr{V} + \ln f) = \frac{1-f}{f} \tag{7}$$

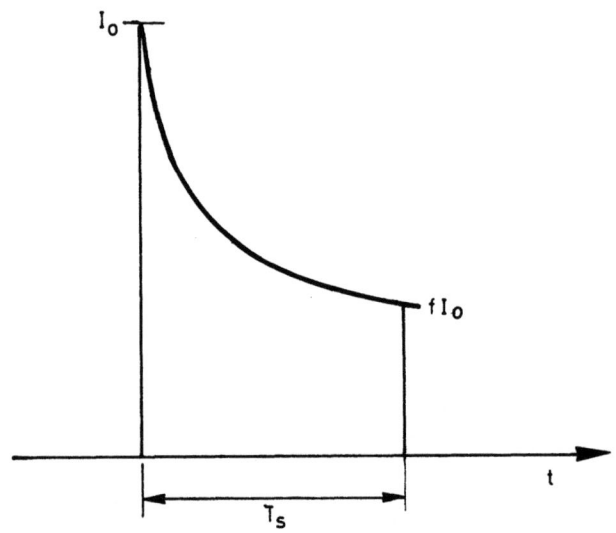

Fig.3.   Incomplete recharge of the collector depletion capacitance.

with $\vartheta = \dfrac{T_s}{\tau}$

The charge restored during the sampling time of course must equal the charge lossed during the light integration time. When a steady state periodical situation is attained, the corresponding boundary condition is given by,

$$\int_0^{T_s} I \, dt = I_L T_i \qquad (8)$$

based on the assumption of a constant light generation current I_L. Since the current I cannot be solved with respect of t, the first term of the above expression is evaluated as follows :

$$\int_{fI_o}^{I_o} t \, dI + f \, I_o \, T_s$$

where t is given by equation (5). After some manipulations, the following result is obtained :

$$\dfrac{RI_o}{V_T} (1 - f) - \ln f = \dfrac{I_p T_i}{V_T C_c} \qquad (9)$$

In this equation :

$$\frac{I_p}{C_c} \cdot T_i$$

represents the total voltage change Δv_c across the back biased diode during the integration time. Equation (9) therefore may be rewritten as follows :

$$v_o(1 - f) - \ln f = \Delta v_c \qquad (10)$$

where both, v_o and Δv_c are reduced voltages, e.g. :

$$\Delta v_c = \frac{\Delta V_c}{V_T}$$

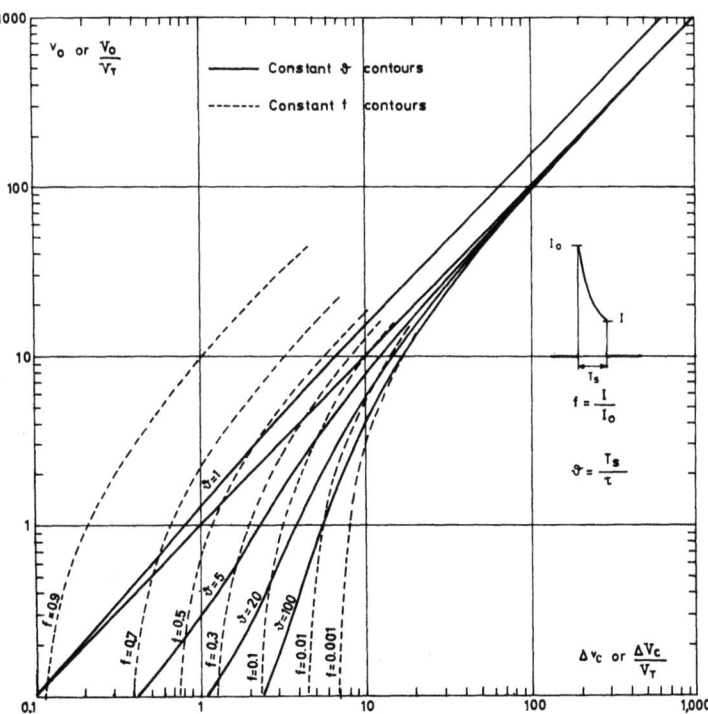

Fig.4. Reduced input-output curves of the phototransistor used in the charge storage mode.

The system formed by equations (7) and (10) will be used in this section in order to predict the behavior of phototransistors under low light illumination conditions. First we consider the relationship between the input variable Δv_c and the output signal v_o. Despite of the fact that no direct relation between these quantities is readily available, the problem can be handled by considering a fixed sampling time, or a constant ϑ , and by changing f from 0 to 1. In this manner, all possible situations regarding fractional recharge are reviewed, and it is possible to compute the corresponding values of v_o and Δv_c. The graph representing v_o versus Δv_c is shown in fig.4. The parameter ϑ has been given four different values ranging from 1 to 100. A rather surprizing conclusion results from this graph, namely that increased sampling times deteriorate the linearity at low light levels. For instance a ΔV_c of 100 mV (or $\Delta v_c = 4$) leads to an output voltage V_o of only 10 mV considering a sampling time as long as a hundred times τ . If the sampling time is decreased the following results are obtained :

T_s in units of τ	V_o mV	Attenuation dB
20	29	− 10.8
5	60	− 4.4
1	156	+ 3.9

In order to understand these results, let us consider first the case which corresponds to a gain factor of approximatively 3.9 dB, namely when T_s equal to τ . Obviously, the apparent gain is a fictious one, caused by the shortening of the sampling time. Indeed when T_s is equal to τ , the fraction of recharge current that is still flowing through the transistor at the end of the sampling time, represents at least 36% of the initial current I_o if an ideal diode is considered. The series impedance of a real diode increases this fraction to the extend of 45%. This means that at the onset of the following sampling pulse, a relatively large recharge current must necessarily flow through the series diode, whatever the actual illumination conditions during the last integration period was. For instance, if no light at all impinged on the photosensor during the last integration period, the series junction still will be forward biased during a few sampling times in order to achieve full recharging of C_c. Consequently, low level illuminations still can be detected, and the linearity is good. Long sampling times however, achieving quasi complete recharging of C_c, tend to make the dynamic impedance of the series diode D much higher. Hence poor linearity is experienced, as shown in fig. 4.

The obvious drawback of short sampling times of course is the memory effect which is the inevitable consequence of relatively high values of f. This effect is illustrated in fig.5

which represents the output voltage, immediately after a dark
scene follows a period of constant illumination. The pulses
shown correspond to successive recharging times with a sampling
time equal to τ . Two situations are investigated, the upper
one corresponds to high level preillumination (V_o = 1 V), the
lower one to low level (V_o = 10 mV). Naturally the effect is
more pronounced in the latter case.

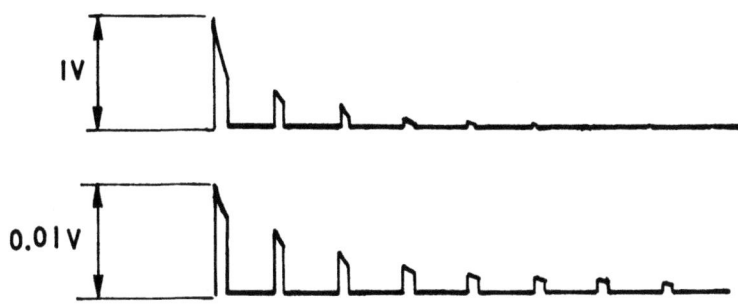

*Fig. 5. Comparison of high and low level responses when a
dark scene follows immediately a period of constant
illumination.*

A tradeoff between linearity and memory effect is easily
recognized in fig. 4, considering constant "f" contours in the
input-output signal plane. These curves, which are directly
obtained from equation (10), have all more or less the same
shape. They have a slope equal to one under large signal conditions and are responsible of the fictious gain mentioned a-
bove. Furthermore they exhibit an asymptotoc behavior under
low level conditions, when :

$$\Delta v_c = - \ln f \qquad (11)$$

Clearly this indicates that low level illumination cannot be achieved without a minimum memory effect, whatever the actual sampling time may be. For instance aΔv_C of 1 means that f cannot be made smaller than 0.368, even if the sampling time was infinitely long. Usual sampling times with increased f, thus cause smearing of the image. E.g. considering again Δv_C of 1 and a sampling time equal to τ, yields a factor "f" of 0.6.

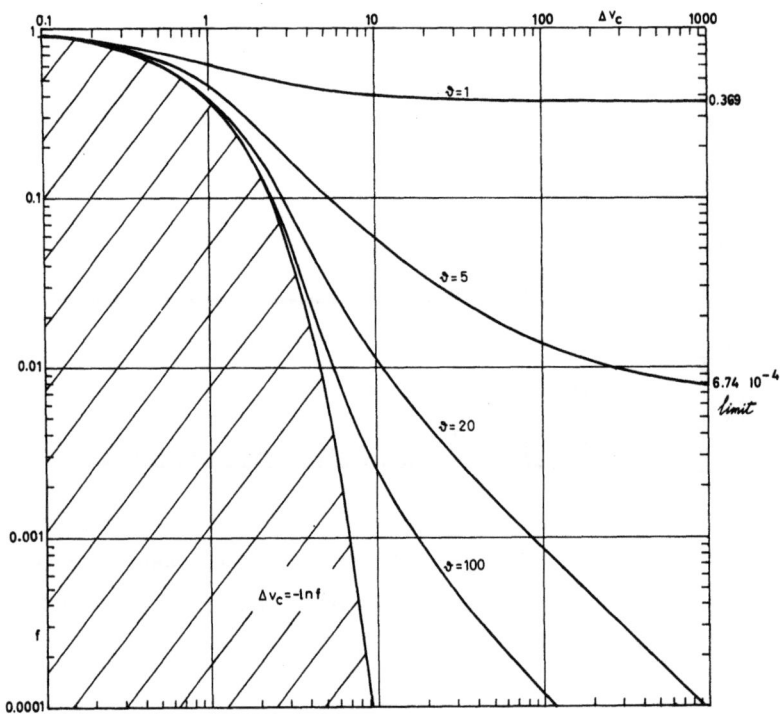

Fig. 6. *Influence of the dynamic range Δv_c and relative sampling time θ on the fractional recharge factor f.*

A somewhat clearer picture of this memory effect may be found in fig. 6 which represents "f" versus Δv_c for different values of the parameter θ. In this picture the forbidden domain defined by (11) is indicated showing that collector voltage changes under the 50 mV (Δv_c = 2) are not advisable regarding smearing out effects, and that increased sampling times do not provide any advantage but less linearity of the response curve.

The same figure however shows that high level performances can be achieved reasonably well. A value of θ between 2 and

5 seems to be acceptable in this respect and, in the same time, a good compromise regarding non linearity.

Effect of the emitter capacitance on the performances of phototransistors.

In the preceding section, we have considered real DC characteristics of the series diode, but the emitter capacitance has been neglected completely. We will now reintroduce this capacitance which will be supposed to be voltage independant, like the collector capacitance as well.

Instead of an open circuit, the diode D now represents a capacitive feedthrough. Charge sharing consequently occurs between C_c and C_e, when switching occurs at the end of the sampling time. The actual base node voltage V can be computed on the basis of the charge continuity equation.

Let us call Q_t the total charge stored in the transistor at the end of the sampling period (*) :

$$Q_t = V_e C_e - V_c C_c \qquad (12)$$

The emitter voltage drop V_e is still close to a few tenths of a Volt, for the complete recharge of C_c is almost impossible. This can be verified easily, if we notice that the voltage change across the emitter junction throughout the entire sampling time is given by

$$V_T \ln (f)^{-1}$$

and that the vertical scale of fig. 6 thus represents a linear plot of this voltage change. Since θ usually is not taken larger than 5, the actual voltage change does not exceed 130 mV indeed.

After E has returned to zero, the new charge balance is given by,

$$Q_t = V(C_e + C_c) \qquad (13)$$

Equalizing (12) and (13) yields the actual base node voltage, which roughly is given by,

$$V = - \frac{C_c}{C_e + C_c} E$$

(*) The charge stored in the neutral region of the transistor is neglected in equation (12), for the current flowing through the transistor is always very small.

after neglecting the contribution of the charge stored in the emitter.

When the integration time resumes, a new charge transfer occurs in the opposite direction which will reestablish exactly the same initial conditions if the junction has not been discharged during the integration time. Otherwise, the emitter junction will be driven more heavily into conduction.

In the foregoing analysis we have considered that all charge transfers take place during a time supposed to be short in comparison with the output pulse time constant. In fact one assumes that the rise time of the signal delivered by the pulse generator E is short compared to $h_{fe} \cdot \tau$, but long with respect to the time constant :

$$R \frac{C_e C_c}{C_e + C_c}$$

This last time constant corresponds to the series combination of R, C_e and C_c, which governs the transient behavior of the blocked transistor operating in the integration mode. Usually there may be one or two orders of magnitude difference between these time constants so that this assumption can be made valid. However, should the rise time of the pulse generator be extremely small, e.g. of the order of a ns, then two additional large current spikes of reverse polarity will be superimposed on the output signal. It would be easy however to filter out these spurious signals.

A typical base-emitter voltage time response showing the effect of light integration is shown in fig.7. These curves were obtained from a computer simulation [3] which took in consideration the variations of the emitter and collector depletion capacitances versus the actual voltage drops. Notice the quasi constant forward voltage drop across the emitter junction during the sampling time, as stated before.

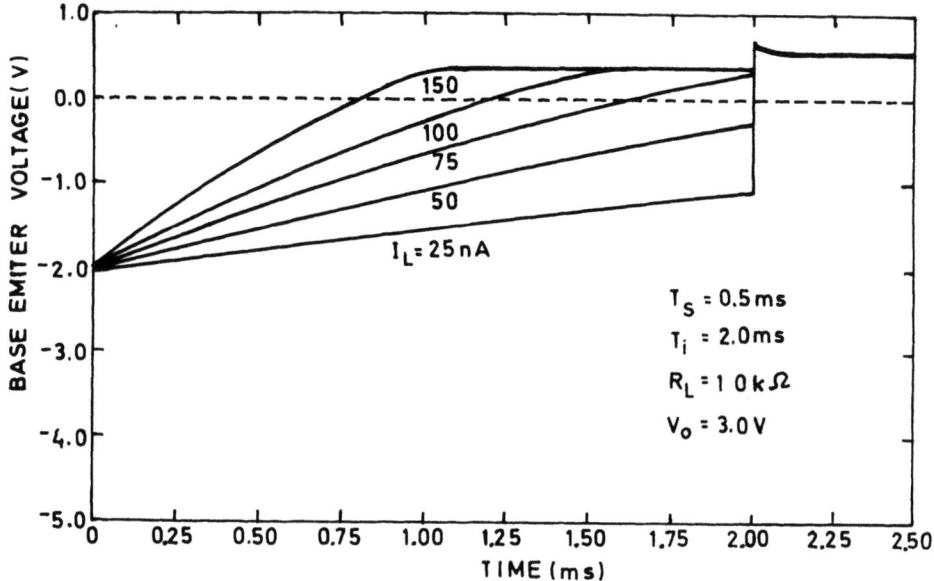

Fig. 7. Base-emitter voltage response for several illumination levels.

Influence of the current gain h_{fe} on the performances of the phototransistor.

Voltage sampling of the output signal should be theoretically insensitive to the actual current gain of the phototransistor. In the ideal case indeed, the output voltage V_o is an exact replica of the voltage change V_c observed at the collector terminal, whether a photodiode are phototransistor is considered. But imperfections are responsible for the non linear responsivity and unwanted memory effects stated before; they also introduce an unwanted current gain sensitivity. The problem has been analyzed again by S. Brugler [1,2] who found the following expression of the sensitivity factor :

$$S = \frac{dV_o}{V_o} / \frac{dh_{fe}}{h_{fe}} = \frac{\theta f}{1 - f} \cdot \frac{v_o}{1 - v_o}$$

A plot of S versus Δv_c is shown in fig.8 for three different values of the reduced sampling time θ . It is apparent that low level operation is very sensitive to current gain. An interesting effect may be noticed which results from the interlacing of the S curves near the origin. Thanks to this effect, there is an optimum sampling time with regard to current sensitivity. For instance in the range of 100 mV, a θ of 5 yields an improvement of 30 or 42% respectively over values of θ , ranging from 1 to 100.

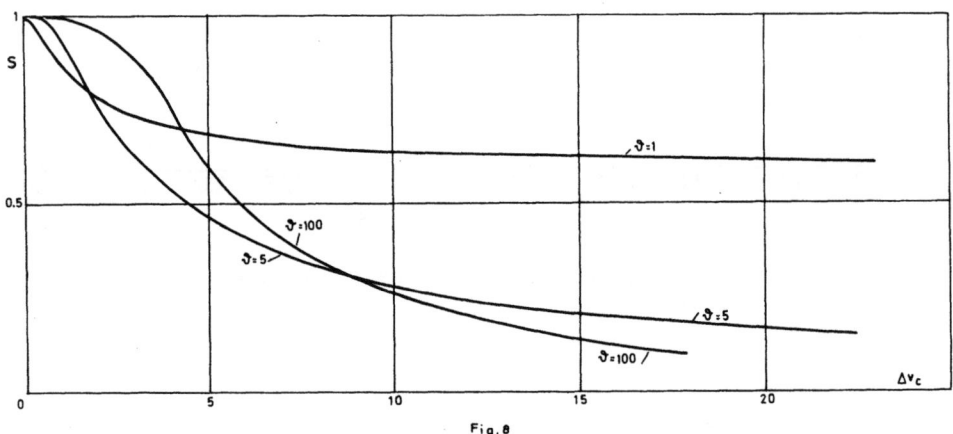

Fig. 8. Sensitivity factor S of phototransistors operating in the charge storage mode versus reduced dynamic range Δv_c and relative sampling time θ .

The problem of current gain sensitivity was reported in several other papers [4,5] and should be considered as one of the main drawbacks with respect to fixed pattern noise.

A typical phototransistor array

During the 1960's several bipolar phototransistor arrays were fabricated by Westinghouse [6] and incorporated into a camera. Much fundamental work was also performed also at Stanford University in an attempt to build a reading aid for the blind [7], the so called Optacon.

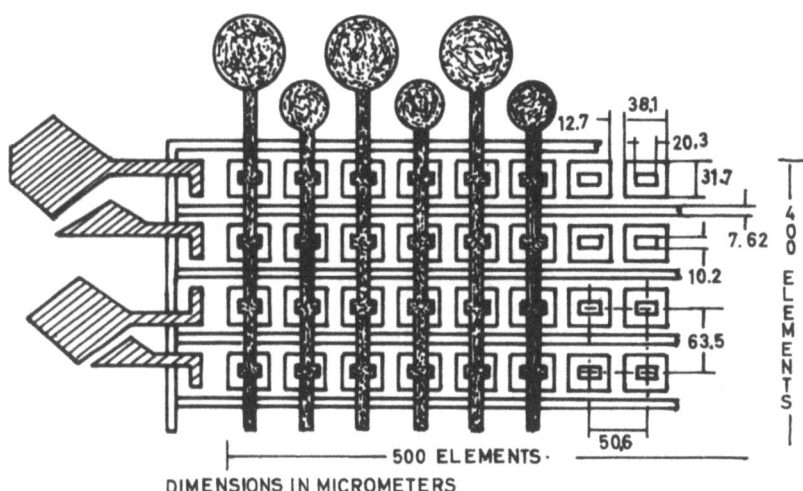

Fig.9. Schematic of the 400 . 500 Silicon Phototransistor matrix developed by Westinghouse [8] .

A 400 × 500 phototransistor array has been described [8], which is shown in fig. 9. The total area of the chip used is approximatively 1 inch square. Spacings between cells is 63,5 μ and 50,6 μ in both directions. XY addressing is achieved by means of a common collector isolation in one direction, and emitter metallization in the other direction. Each emitter falls in the middle of photosensitive area, without contributing to photosensitivity for the metallization stripe shields the emitter completely from light.

The technology used was precautions in order to minimize surface damage. It starts by etch removal of 50 μ from around the entire surface diameter of a high quality raw Si ingot prior to sawing. It was reported that failure to follow this procedure could lead to the propagation of any original surface damage throughout the entire wafer. After sawing the wafer, another 50 μ etch of the planar sides is performed prior to mechanical lapping, polish and final chemical polish. The diffusion process then is initiated by subdiffusion of low resistivity collector stripes which are introduced in order to prevent collector pinchoff between base and substrate. This subdiffusion is approximatively 5 to 6 μ deep and the reported resistance per square ranges from 80 to 100 Ω/square, resulting from a compromise between redistribution during the epi growth, reduction of defects due to doping, and collector stripe resistance. Then a 1 Ωcm epi layer, 4.5 μ thick, is grown followed by a 7000 Å oxydation. The next step includes boron wall diffusion (thickness 5 μ, resistance 50-100 Ω/square), boron base diffusion (thickness 1.25 to 1.5 μ, resistance 300 Ω/square) and the phosphorous emitter diffusion (10 Ω/square). P glass passivation layer, slow cooling and annealing were used like in ordinary bipolar technology, and the metallization finally was achieved by filament evaporation rather than by E beam evaporation in order to avoid X-ray damage at the Si-O_2 interface. The actual h_{fe}'s reported range from 10 to 50.

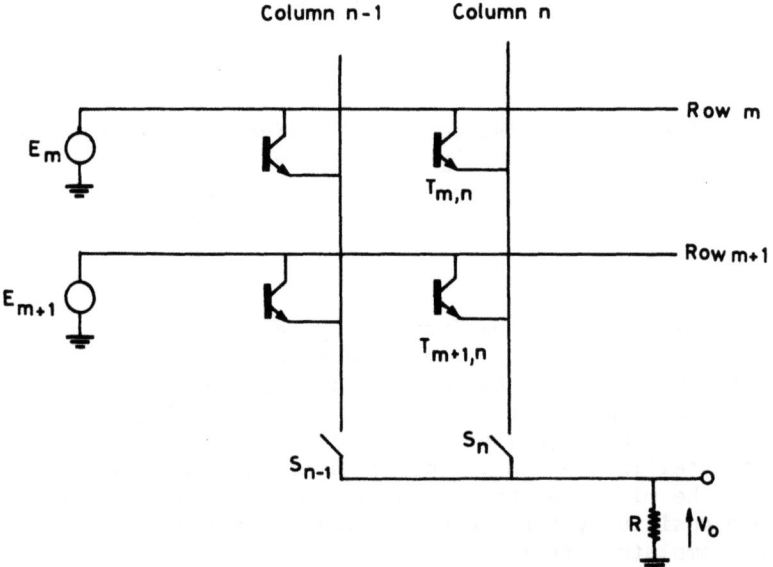

Fig.10. Phototransistor array organization using isolated collector columns and emitter metallization for XY addressing [9]

The array organization will be reviewed hereunder considering an M rows and N columns mozaïc. The circuit diagram is shown in fig. 10, following the presentation that was given in [9]. Since all the emitters along a column are tied together, one should notice that each one is capacitively coupled to ground through series combination of emitter and collector depletion capacitances of the (M-1) remaining transistors. Hence, all emitters are loaded by means of total capacitances, called C_{PT}, whose values are given by,

$$C_{PT} = (M-1) \frac{C_e C_c}{C_e + C_c}$$

Assuming the array is sensed sequentially by means of the column switches S_1 to S_n and line per line, we consider the behavior of a transistor $T_{m,n}$ throughout one frame time. The starting point of our analysis will be the end of the sampling time of $T_{m,n}$ so that any prior charge distribution on this transistor has been erased. The row m is still energized, the switch S_n opens, followed by closure of S_{n+1} etc, till S_n is reached.

1) Reading out of $T_{m,n+1}$, till $T_{m,n}$

No electrical signal are felt by $T_{m,n}$, whose emitter connections is no more tied to the output load resistor R.

2) Reading out of $T_{m+1,1}$ till $T_{m+1,n-1}$

Two changes occur which influence $T_{m,n}$. First E_m is grounded, producing a charge sharing within $T_{m,n}$ between C_c and the series combination of C_e and C_{PT}. A small negative voltage is induced thus on the emitter column through C_e, but it is almost negligible since C_e is much smaller than C_{PT}. Second E_{m+1} is energized, producing a current flowing from the (m+1)th row towards the mth row. The series combination of the transistors $T_{m+1,n}$ and $T_{m,n}$, now is forward biased as far as $T_{m+1,n}$ and the inverse $T_{m,n}$ transistor are concerned. The resulting positive voltage step appearing on the nth column may reach an appreciable amplitude for its results from capacitive divider action formed by $h_{fe} C_c$ of $T_{m+1,n}$ and C_{PT}. The amplification of C_c here represents a severe drawback. Careful attention must be given to this problem. Fortunately, $T_{m,n}$ is blocked, for a large negative base voltage has been imposed at the end of the sampling period, and the unwanted positive voltage step induced by the emitter column can be kept smaller thanks to the proper design of the phototransistor, namely by keeping C_e much smaller than C_c. The problem however becomes more difficult near the end of the integration time.

3) Reading out of $T_{m+1,n}$

The emitters tied to column n are grounded by means of the output load resistor R. Charge sharing takes place restoring the initial situation.

4) Reading out of $T_{m+1,n+1}$ till $T_{m+1,n}$

The column n is left open again and stays at zero volt, unaffected by further scanning of the row (m+1).

5) Reading out of the remaining transistors till $T_{m-1,n}$

The situation is in every respect similar to that of § 2 and following.

6) Reading out of $T_{m,1}$ till $T_{m,n-1}$

As soon E_m is turned on again, $T_{m,n}$ conducts for its emitter is grounded by means of C_{pT}. The latter capacitance stores the output information till the column switch S_n is closed, enabling reading out of the output signal.

Other phototransistor arrays

The 400 x 500 elements array which was considered above, is probably the largest one ever built. With the introduction of MOS photodiode arrays and CCD's, the interest in bipolar transistor arrays has dropped progressively. The potential interest in photosensitive elements that provide some local gain, undoubtedly explains the early interest of many investigators in phototransistor arrays. But, it became soon evident that several drawbacks among which the rather poor dynamic range and increased difficulties experienced with switching an isolated cell in a very large mozaïc, overwhelmed the advantages of local current gain. Moreover this current gain cannot be exploited otherwise than in the voltage sampling mode. The charge sampling mode would be highly sensitive to current gain variations.

Phototransistor arrays seem to be more suited for smaller size arrays. A typical example is the mozaïc used in the Optacon reading aid for the blind. The research programs initiated at Stanford University by the development of this array, have led to a thorough analysis of many of the above mentionned problems. The description of 24 by 6 photomatrix can be found in the work of P. Salsbury [7]. The only difference with respect to the larger array discussed above is a slightly different sampling procedure, the clock generator signal E being applied to the common emitter metallization and the reading out occuring by means of isolated collector rows.

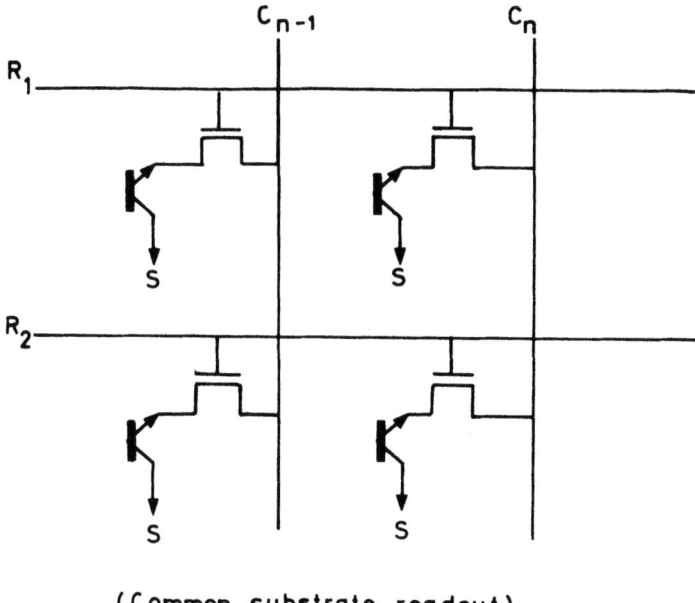

(Common substrate readout)

Fig.11. Phototransistor array organization using an additional MOS transistor for XY addressing [10] .

   A somewhat different approach which needs to be mentioned here is the contribution of G. Weckler [10] who built an array of bipolar phototransistor with only one common collector terminal (fig.11). XY addressing therefore requires an additional switching element which in the present case is a MOS transistor. Although the epitaxial technology is avoided in this manner, other difficulties arise resulting from the implementation of bipolar and unipolar devices on the same chip. A single cell layout is shown in fig. 12.

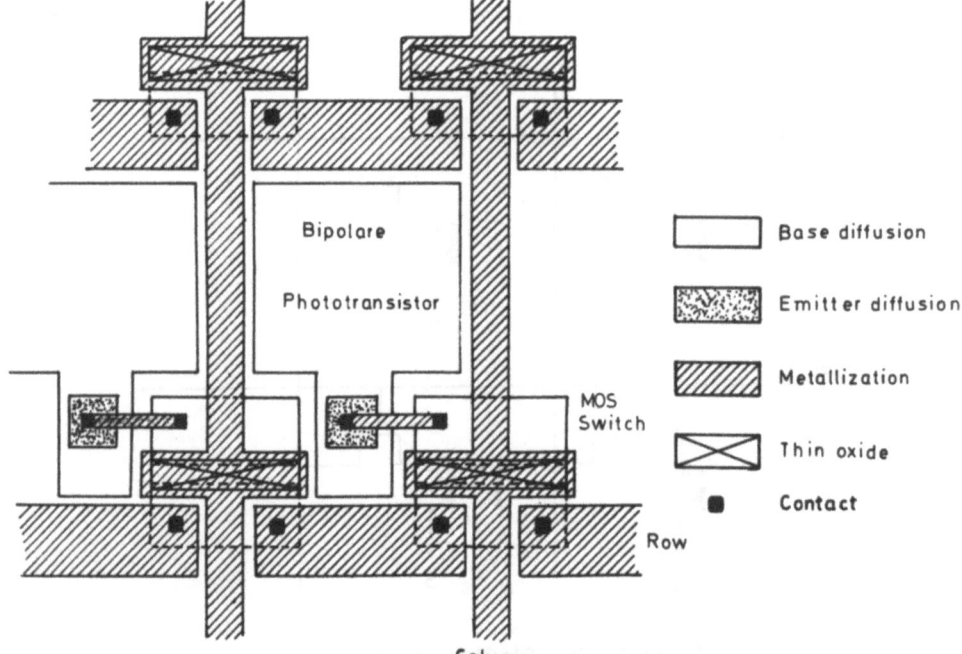

Fig. 12. Schematic of the elementary MOS-phototransistor cell developped by Fairschild [10] .

BIBLIOGRAPHY

[1] Low-Light-Level Limitations of Silicon Junction Photodiodes,
Joseph S. Brugler.
Technical Report N° 4824-1, May 1968, Solid State Electronics
Laboratory of Stanford University, Calif.

[2] Low-Light-Level Properties of the Phototransistor Charge-
Storage Mode.
J.S. Brugler
IEEE Journal of Solid State Circuits 4, June 1969, pp. 136-144.

[3] Charge Storage Mode Operation of Phototransistors in a
Reading Aid, Richard C. Joy.
Technical Report N° 4825-1, March 1968, Solid State Electro-
nics, Laboratory of Stanford University, Calif.

[4] Solid State Imaging - Methods of Approach.
William. F. List, IEEE Transactions on Electron Devices,
Vol ED-15, n°4, April 1968, pp: 256-261.

[5] Image Sensors for Solid State Cameras.
P.K. Weimer, Advances in Electronics and Electron Physics,
Academic Press, Vol 37, 1975, p. 197-198.

[6] Developmental Solid State Imaging System.
R.A. Anders, D.E. Callahan , W.F. List, D.H. Mc. Cann,
M.A. Schuster,
I.E.E.E. Transactions on Electron Devices 15,1968, pp. 191-196.

[7] A Monolithic Image Sensor for a Reading Aid for the Blind.
Phillip J. Salsbury.
Technical Report N° 4828-1, July 1969, Solid State Electronics
Laboratory, Stanford University, Calif.

[8] Solid State Array Cameras.
G. Strull, W.F. List, E.L. Irwin, and D.L. Farnsworth.
Applied Optics, Vol 11, May 1972, pp. 1032-1037.

[9] Transfer Functions of Imaging Mosaics Utilizing the Charge
Storage Phenomena of Transistor Structures.
Irwin Tepper, Roland A. Anders, David H. Mc Cann.
I.E.E.E. Transactions on Electron Devices 15, n° 4,
April 1968, pp. 226-237.

[10] Integrated Arrays of Silicon Photodetectors for Image
Sensing.
R.H. Dyck, G.P. Weckler.
I.E.E.E. Transactions on Electron Devices 15, n°4, April 1968,
pp. 196-201.

Section IV

CTD ARRAYS

CHARGE INTEGRATION AND STORAGE IN MOS PHOTOSENSORS

D. F. Barbe

Naval Research Laboratory,
Washington, D. C. 20375

ABSTRACT. The integration of photon-generated charge and the
storage of charge in surface channel and buried channel MOS
photosensors are discussed. The relationship between minimum
potential and gate voltage as a function of integrated charge
is derived for surface channel and buried channel sensors. The
full-well charge density for polyphase and two phase sensors is
derived and comparisons are given for the important sensor
designs.

1. INTEGRATION AND STORAGE IN MOS CELLS

With an image on an array of MOS integrating cells, the density
of photon-generated carriers in each cell, N_i, is proportional
to the photon flux on the cell, Φ_i, and the integration time,
T_{integ}; i.e., $N_i = \eta \, \Phi_i \, T_{integ}$, where η is the number of carriers
collected in a cell per photon incident on the cell. The purpose
of this chapter is to discuss the behavior of MOS integrating
cells (surface channel and buried channel) as the cell fills
with carriers. Since the upper limit of the dynamic range is
determined by the saturation level, it is important to determine
N_{FULL} for each cell design. A large value of N_{FULL} is especially
important in order to achieve large intrascene dynamic ranges.
For example suppose an array were designed to have $N_{FULL} = 10^{11} cm^{-2}$ and $\eta = 0.6$, and suppose that the average photon flux
from the scene were 10^{12} photons $cm^{-2} sec^{-1}$. Then for an integration time of 1/60 sec., the average value of N would be
$10^{10} cm^{-2}$. A local bright spot in the scene subtending one
element having 100 times the intensity of the average scene

brightness would cause a cell to overfill, and charge would spill into 9 adjacent cells (blooming). If the array had been designed to have $N_{FULL} = 10^{12} cm^{-2}$, a cell could have accomodated all of the carriers generated by the localized bright spot, and there would have been no blooming.

The length of time a charge packet can be stored in an MOS cell is determined by the thermally generated current density - usually called the dark current density, J_d. There are three components of dark current in MOS cells: (1) thermal generation current at the $Si-SiO_2$ interface, (2) thermal generation in the depleted volume, and (3) thermal generation in the neutral bulk within a diffusion length of the depleted volume [1]. Components (1) and (2) have the temperature dependence $J_{d_{1,2}} \sim \exp(E_g/2kT)$, where E_g is the energy gap of silicon, k is Boltzmann's constant and T is the absolute temperature. Component (3) has the temperature dependence $J_{d_3} \sim \exp(-E_g/kT)$. The total dark current is the sum of the three components. A typical value of J_d at 300°K is 5 nA/cm^2. The dark current is a strong function of temperature and can be reduced to insignificant levels by cooling to about -50°C. Using the rule of thumb that the useful storage time is the time required for a cell to fill to 10% of N_{FULL} due to thermal generation, then the storage time is $T_s = 0.1e\ N_{FULL}/J_d$. For $N_{FULL} = 5 \times 10^{11} cm^{-2}$ and $J_d = 5\ nA/cm^2$, $T_s \simeq 1.6$ sec. Typically some elements in an array have generation rates substantially larger than the mean in which case the array performance may be limited by dark current non-uniformities rather than by the mean dark current.

2. SURFACE CHANNEL (SC) MOS INTEGRATION CELLS

When a voltage is applied to the conducting electrode of an MOS capacitor with respect to the substrate, the energy bands in the semiconductor bend. If the applied field is in the direction to repel majority carrier from the surface of the semiconductor, i.e., from the semiconductor-insulator interface, the bands in p-type silicon will bend as shown in Fig. 1(A). The effective voltage across the capacitor, $V_G - V_{FB}$, will be divided between the semiconductor, as indicated by the band bending in the semiconductor, and the oxide, as indicated by the "tilt" of the energy bands in the oxide. The electron potential at the semiconductor-insulator interface under the electrode is lower than that in the bulk of the semiconductor by amount ϕ_{so}; thus a "potential well" of depth ϕ_{so} is formed at the semiconductor-insulator interface under the electrode. If charge is collected in the potential well as a result of photon absorption, injection from an input diffusion, or thermal generation, the potential across the insulator and the semiconductor will be redistributed

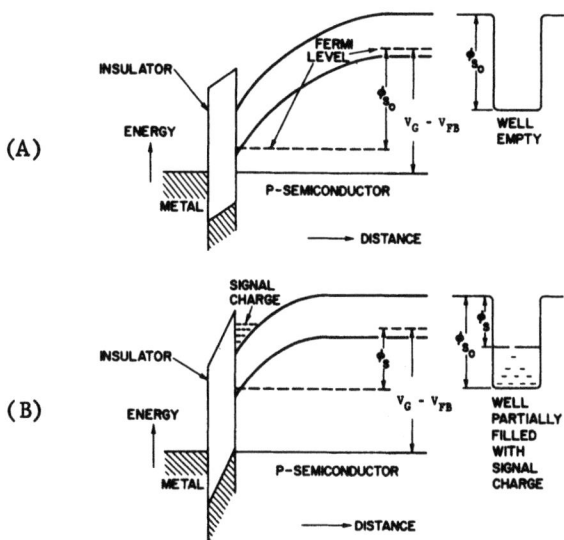

Fig. 1 Energy band diagrams showing
(A) an empty potential well and
(B) a partially filled well.

as shown in Fig. 1(B). Thus, the potential well has been partially filled to ϕ_s. The equation relating the surface potential or potential-well depth to the factors affecting it is discussed in the following paragraphs.

The basic electrostatic design equation for a surface-channel CCD relating surface potential ϕ_s to doping density N_A, oxide thickness d, and gate voltage V_G is obtained by solving Kirchhoff's voltage equation for an MOS capacitor:

$$V_G - V_{FB} = \phi_s + V_{ox} \qquad (1)$$

where V_{FB} is the flatband voltage and V_{ox} is the voltage drop across the oxide. But

$$V_{ox} = \frac{1}{C_{ox}} \text{ [mobile charge density + fixed charge density]}$$

$$= \frac{1}{C_{ox}} [\quad eN \quad + \quad N_A eW \quad], \qquad (2)$$

where $C_{ox} = \epsilon_{ox}/d$, $e = 1.6 \times 10^{-19}$ C, and N is the number of mobile electrons per unit area.

Using the depletion approximation, the width of the depletion region, W, is

$$W = \left(\frac{2\epsilon_s \phi_s}{eN_A}\right)^{\frac{1}{2}}, \qquad (3)$$

where ϵ_s is the dielectric constant of the semiconductor.

Combining (1), (2), and (3) gives

$$V_G - V_{FB} - \frac{eN}{C_{ox}} = \phi_s + \frac{1}{C_{ox}}(2eN_A\epsilon_s\phi_s)^{\frac{1}{2}} \qquad (4)$$

Solving (4) for ϕ_s gives

$$\phi_s = V_G - V_{FB} - \frac{eN}{C_{ox}} + \frac{eN_A\epsilon_s}{C_{ox}^2} - \frac{1}{C_{ox}} \times$$

$$\left[2eN_A\epsilon_s\left(V_G - V_{FB} - \frac{eN}{C_{ox}}\right) + \left(\frac{eN_A\epsilon_s}{C_{ox}}\right)^2\right]^{\frac{1}{2}} \qquad (5)$$

Equation (5) shows how barriers can be controlled by proper choice of gate voltage, doping density, and oxide thickness [2]. For example, (5) shows that ϕ_s decreases when N_A is increased and when C_{ox} is decreased. This is illustrated in Fig. 2, which shows qualitatively the two methods of channel confinement (channel stops)-high doping or thick oxide. These two methods are also used to build in charge flow directionally along the channel in two-phase CCD's.

2.1 Polyphase CCD integration cells

The surface potential of a polyphase CCD during integration is shown in Fig. 3. Region (2) is where photon-generated carriers are being collected, and regions (1) and (3) are barrier regions. When the well is full, the surface potential in region (2) is equal to that in regions (1) and (3); i.e., $\phi_{s_1} = \phi_{s_2} = \phi_{s_3} = \phi_{FULL}$. Equation (4) is used to write the equations describing regions (1), (2), and (3):

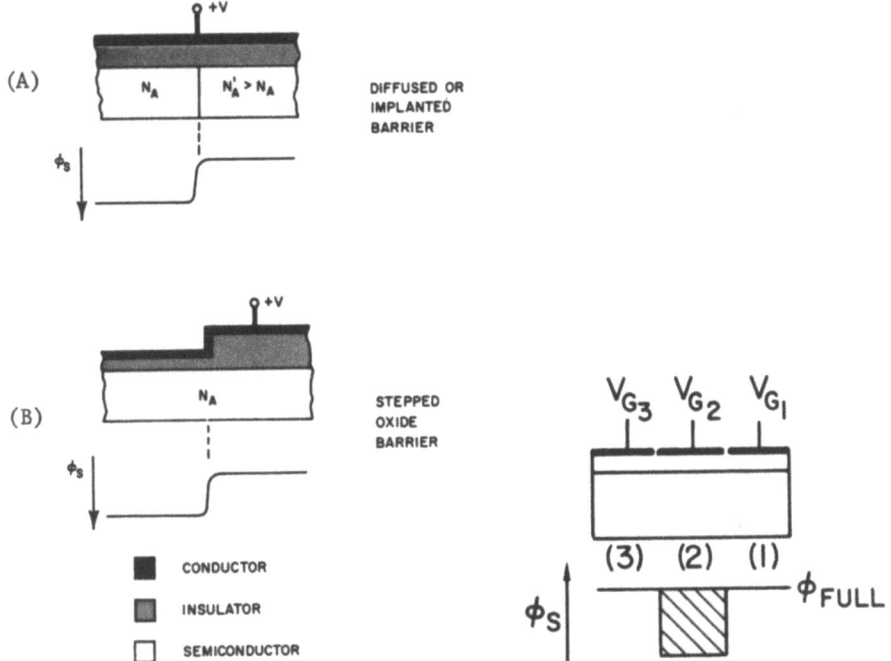

Fig. 2 Potential barrier formation by (A) nonuniform doping of substrate and (B) nonuniform oxide thickness.

Fig. 3 Full well condition for a 3-phase CCD cell.

$$V_{G_1} - V_{FB} - 0 = \phi_{FULL} + K \phi_{FULL}^{\frac{1}{2}}, \tag{6}$$

and

$$V_{G_2} - V_{FB} - eN_{FULL}/C_{ox} = \phi_{FULL} + K \phi_{FULL}^{\frac{1}{2}}, \tag{7}$$

where $K = (2eN_A\epsilon_s)^{\frac{1}{2}}/C_{ox}$. Subtracting Equation (6) from Equation (7) gives

$$N_{FULL} = C_{ox} \Delta V_G/e, \tag{8}$$

where $\Delta V_G = V_{G_2} - V_{G_1}$.

Equations (5) and (8) are combined to give the curves shown

in Fig. 4 of ϕ_S vs. $V_G - V_{FB}$ with N/N_{FULL} as a parameter for a typical polyphase CCD.

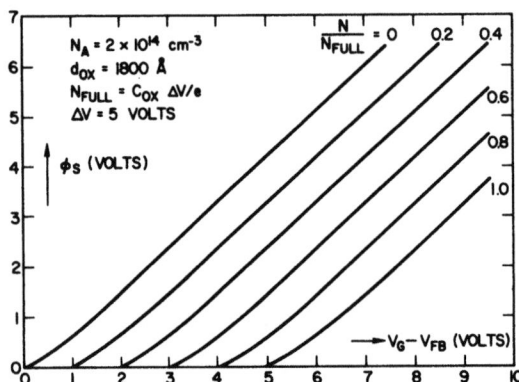

Fig. 4 Surface potential vs gate voltage for a surface-channel CCD cell.

2.2 Two-phase CCD integration cells

For a two-phase CCD, the full-well capacity is determined by the barrier created by either (A) an oxide step or (B) nonuniform substrate doping. Consider Fig. 5 which shows potential profiles for (A) stepped oxide and (B) implanted barrier two-phase CCD integrating cells. Region (1) is that having the thinner oxide or lower substrate doping and region (2) is that having the thicker oxide or higher doping. The thick oxide or high doping provides the barrier adjacent to the potential well so that charge can be collected in the well. When the well is full, the surface potential is equal in regions (1) and (2); i.e., $\phi_{S_1} = \phi_{S_2} = \phi_{FULL}$. Using Equation (4), the equations describing regions (1) and (2) are:

$$V_G - V_{FB_1} - eN_{FULL}/C_{ox_1} = \phi_{FULL} + K_1 \phi_{FULL}^{\frac{1}{2}}, \tag{9}$$

and

$$V_G - V_{FB_2} - 0 = \phi_{FULL} + K_2 \phi_{FULL}^{\frac{1}{2}}, \tag{10}$$

where K was defined previously in Equations (6) and (7).

Solving these equations for N_{FULL} gives

$$N_{FULL} = C_{ox_1} \Delta V_{eff}/e, \qquad (11)$$

where

$$\Delta V_{eff} = -\Delta V + \Delta K \left[(V_2 + K_2^2/4)^{\frac{1}{2}} - K_2/2 \right], \qquad (12)$$

and

$$\Delta V = V_2 - V_1 = (V_{G_2} - V_{FB_2}) - (V_{G_1} - V_{FB_1}),$$

$$\Delta K = K_2 - K_1, \text{ and}$$

$$K = (2eN_A \epsilon_s)^{\frac{1}{2}}/C_{ox}.$$

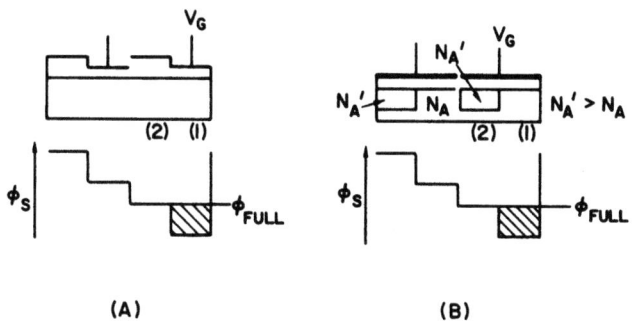

Fig. 5 Full-well conditions for
(A) stepped oxide two phase CCD cell and
(B) implanted barrier two phase CCD cell.

3. BURIED CHANNEL MOS INTEGRATION CELLS

A buried channel CCD is one in which the surface of the semiconductor is doped opposite to that of the substrate as shown in Fig. 6. It is assumed that the buried channel is depleted of charge by the application of a large positive voltage to the buried-channel layer and that minority carrier density, N, (cm^{-2}) is collected in the MOS capacitor.

The profile of potential versus distance into silicon for a buried-channel CCD element is shown in Fig. 6. As for the surface-channel CCD, the basic electrostatic design equation relating the minimum electron potential in the channel, ϕ_{min}; to the doping density of the substrate, N_A; doping density of the channel, N_D; oxide thickness d; thickness of the donor layer, t; and gate voltage V_G is calculated from Kirchhoff's equation [2]:

$$\phi_s + V_G - V_{FB} + V_{ox} = \phi_j + \phi_c .\qquad(13)$$

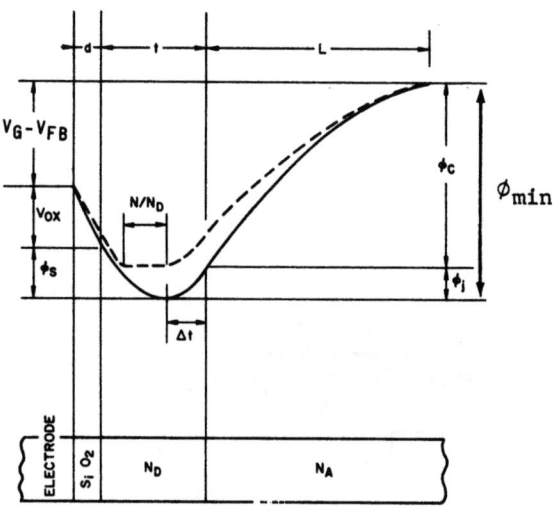

Fig. 6 Energy band diagram for a buried channel CCD.

Since $V_{ox} = \frac{1}{C_{ox}}$ (mobile charge density + fixed charge density), it follows that

$$V_{ox} = \frac{d}{\epsilon_{ox}} eN_D (t - \Delta t - N/N_D). \qquad(14)$$

Using the depletion approximation, it follows that:

$$\phi_s = \frac{eN_D(t-\Delta t-N/N_D)^2}{2\epsilon_s} , \qquad(15)$$

$$\phi_c = \frac{eN_A L^2}{2\epsilon_s} , \qquad(16)$$

and

$$\phi_j = \frac{eN_D(\Delta t)^2}{2\epsilon_s} \quad . \tag{17}$$

Also,

$$\phi_{min} = \phi_c + \phi_j \quad , \tag{18}$$

and

$$\Delta t = (N_A/N_D) L \quad . \tag{19}$$

Combining these equations gives

$$V_G - V_{FB} + V_I - \frac{eN}{C_{eff}} = \frac{N_D}{N_D + N_A} \phi_{min} + \left(\frac{d}{\epsilon_{ox}} + \frac{t}{\epsilon_s}\right)\left(\frac{2e\epsilon_s N_D N_A}{N_D + N_A}\right)^{\frac{1}{2}} \phi_{min}^{\frac{1}{2}} , \tag{20}$$

where

$$V_I = eN_D t \left(\frac{d}{\epsilon_{ox}} + \frac{t}{2\epsilon_s}\right) , \tag{21}$$

$$\Delta t = \left(\frac{2\epsilon_s}{e} \frac{N_A}{N_D(N_D+N_A)} \phi_{min}\right)^{\frac{1}{2}} \tag{22}$$

$$C_{eff}^{-1} = \frac{d}{\epsilon_{ox}} + \frac{t}{\epsilon_s}\left(1 - \frac{\Delta t}{t} - \frac{1}{2} \frac{N}{N_D t}\right), \tag{23}$$

and $N/N_D \leq t - \Delta t$.

Equation (20) has the same form as Equation (4), and it is noted that if t is replaced by zero and $N_D/(N_D + N_A) \phi_{min}$ is replaced by ϕ_s in Equation (20), this equation reduces to the surface-channel equation; i.e., Equation (4).

Equations (20-23) are the basic electrostatic equations for buried-channel CCD's [3]. Figure 7 shows curves of ϕ_{min} vs.

$V_G - V_{FB}$ with N as a parameter for a shallow buried-channel (SBC) device; i.e., a device in which the channel is formed by ion-implantation. Figure 8 shows similar curves for a deep buried-channel (DBC) device; i.e., a device in which the channel is formed by epitaxy. For a SBC CCD, $N_D \gg N_A$; therefore, $\Delta t/t \ll 1$. Then for the SBC case

$$C_{eff}^{-1} \simeq \frac{d}{\epsilon_{ox}} + \frac{t}{\epsilon_s} (1 - \tfrac{1}{2} \frac{N}{N_D}) . \qquad (24)$$

For a DBC CCD Δt, is not negligible, and C_{eff} is given by Equation (23); however, C_{eff} is a slowly varying function of ϕ_{min} and for practical purposes C_{eff} can be treated as a constant in Equation (23).

3.1 Polyphase CCD integration cells

The interpretation of C_{eff} in Equation (20) is the same as the interpretation of C_{ox} in Equation (4), i.e., for a polyphase CCD, the full-well carrier density is

$$N_{FULL} = C_{eff} \Delta V_G / e. \qquad (25)$$

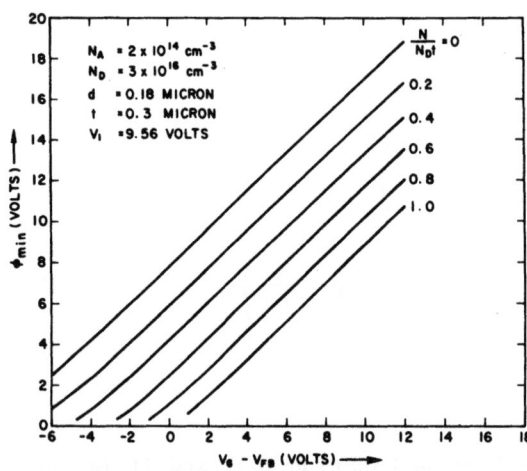

Fig. 7 Minimum potential vs. gate voltage for a shallow buried channel CCD cell.

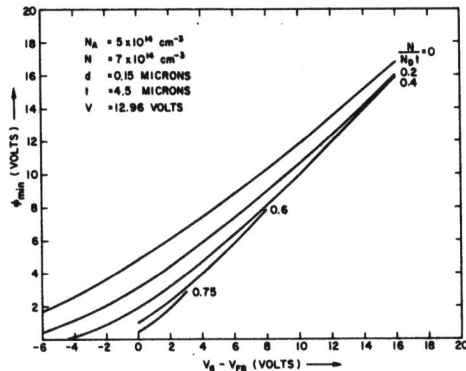

Fig. 8 Minimum potential vs. gate voltage for a deep buried channel CCD cell.

3.2 Two-phase CCD integration cells

A two-phase buried channel CCD can be achieved by the stepped oxide or the implanted barrier methods. In the implanted barrier method, a barrier results when the buried channel region having doping density N_D is partially compensated with doping density N_A' as shown in Fig. 9. The full-well carrier density of a two-phase buried channel CCD can be calculated in the same way as it was calculated for a two-phase surface channel CCD. The result is:

$$N_{FULL} = R_1 \, C_{eff_1} \, \Delta V_{eff}/e, \qquad (26)$$

where

$$\Delta V_{eff} = -\Delta (V/R) + \Delta K \left[(V_2/R_2 + K_2^2/4)^{\frac{1}{2}} - K_2/2 \right], \qquad (27)$$

and

$$R = N_D/(N_D + N_A),$$

$$K = (d/\epsilon_{ox} + t/\epsilon_s) \left[2e\epsilon_s(N_D + N_A) \, N_A/N_D \right]^{\frac{1}{2}}$$

$$V = V_G - V_{FB} + V_I.$$

Consider the implanted barrier case with

Region (1)	Region (2)
$N_{A_1} = 2 \times 10^{14}$	
$N_{D_1} = 3 \times 10^{16}$	$N_{D_2} = N_D - N_A' = 2 \times 10^{16}$
$d = 0.18$ μm	
$t = 0.3$ μm	
$V_{I_1} = 9.56$	$V_{I_1} = 6.37$

Using these values in Eqs. (26) and (27) gives $N_{FULL} = 3 \times 10^{11}$ cm^{-2}.

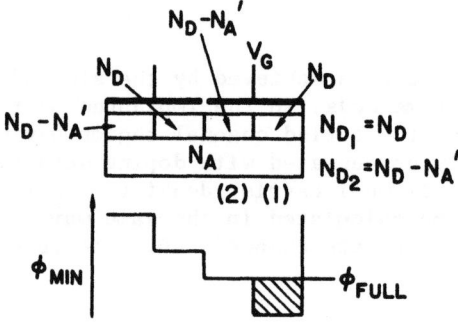

Fig. 9 Full well condition for a implanted barrier two phase buried channel CCD cell.

Table I summarizes the calculated values of N_{FULL} for several types of CCD's. The parameters used in these calculations are those used in devices which have been fabricated and operated; therefore, the values of N_{FULL} given in the table are representative of what is achieved in actual practice.

Table I
N_{FULL} for Typical CCD Designs

Device Type	Polyphase			Two Phase		
				Implanted Barrier		Stepped Oxide
	SC	SBC	DBC	SC	SBC	SC
N_{FULL} (CM^{-2} × 10^{11})	12	9	1.5	8	3	2
Parameters Used to Calculate N_{FULL}	$N_A = 2 \times 10^{14}$	$N_A = 2 \times 10^{14}$	$N_A = 5 \times 10^{14}$	$N_A = 2 \times 10^{14}$	$N_A = 2 \times 10^{14}$	$N_A = 2 \times 10^{14}$
	$d = 0.18\mu$	$N_D = 3 \times 10^{16}$	$N_D = 7 \times 10^{14}$	$N_A' = 3 \times 10^{16}$	$N_D = 3 \times 10^{16}$	$d_1 = 0.1\mu$
	$\Delta V = 10$	$d = 0.18\mu$	$d = 0.15\mu$	$d = 0.18\mu$	$N_A' = 10^{16}$	$d_2 = 0.24\mu$
		$t = 0.3\mu$	$t = 4.5\mu$	$V_G = 10$	$d = 0.18\mu$	$V_G = 10$
		$\Delta V = 10$	$\Delta V = 5$		$t = 0.3\mu$	

4. REFERENCES

1. G. F. Amelio, etc.al., "Charge-Coupled Imaging Devices: Design Considerations," IEEE Trans. Electron Devices, ED-18, pp. 986-992, 1971.

2. D. F. Barbe, "Imaging Devices Using the Charge-Coupled Concept," Proc. IEEE, 63, pp. 38-67, 1975.

3. D. F. Barbe and N. S. Saks, "A Tradeoff Analysis of Transfer Speed versus Charge-Handling Capacity for CCD's," in Proc. 1974 CCD Applications Conf., pp. 114-122, 1974.

INTRODUCTION TO CHARGE-COUPLED DEVICES

C. H. Séquin

Bell Laboratories
Murray Hill, New Jersey 07974, U. S. A.

ABSTRACT. The basic concepts of charge transfer devices will
be reviewed. In particular, the principles of charge-coupled
devices (CCD's) including surface and bulk channel structures and
a variety of different electrode arrangements will be discussed.
Some of the important parameters characterizing the performance
of CCD's will be introduced.

FOREWORD

This lecture intends to give an introduction to the wide
variety of charge-coupled devices for the reader who is unfamil-
iar with the basic concepts. Compared to the extent of the subject, this lecture has been kept rather short since all of this
material is covered in a much more extensive and rigorous manner
in a recently published book Charge Transfer Devices, from which
substantial material and several figures have been adopted with
the explicit permission of Academic Press, Inc. [44].

1. HISTORICAL BACKGROUND

The first integrated charge transfer devices were built in
the late 1960's in the form of bucket-brigade circuits [1], even
though the basic concept of such circuits dates back to
Wiener [2]. He proposed the use of capacitors separated by
telegraph-type repeaters to store and shift analog information.
Later such circuits were implemented with discrete components
using bulky buffer amplifiers with vacuum tubes [3] to realize
analog sampled data delay lines in which the total delay time

could be varied by altering the driving clock frequency of the circuit. And ultimately such circuits using simple MOS switches between the storage capacitors were built with the silicon integrated circuit technology [1,4].

In 1969, while searching for an electrical analog to magnetic bubble memory, Boyle and Smith [5] conceived the idea of charge-coupled devices. In the simplest implementation CCD's consist of closely spaced capacitors on an isolated surface of a semiconductor. When pulsed in the proper sequence, these electrodes generate moving potential wells which can transport packets of minority carriers. This simple but powerful concept proved to be so stimulating, that almost overnight a host of variations of the electrode structure and dozens of potential applications were conceived. It only took a few days to verify the concept experimentally on an existing structure with closely spaced test capacitors [6]. A few months later the first completely designed CCD was demonstrated as a delay-line and as a simple line imaging device [7]. The idea then caught on rapidly since the scientific community was ready, mentally and technologically, to accept and develop the new ideas.

In retrospect it appears that many workers had been pursuing a route that eventually might also have led to charge-coupled devices. Further development of the integrated MOS bucket-brigade circuits might also have brought forth the more functional viewpoint of moving potential wells. And surface charge transistors [8], in which the source and drain of an ordinary MOSFET are replaced with the inversion layer under an MOS electrode, can also form a charge transfer device, when strung together in series and operated with the proper pulse sequence.

On the practical side there was a well-developed MOS technology awaiting the new concepts. With only small changes in the processing sequence the new designs could be fabricated on existing lines. Starting with only eight elements [7], larger and larger devices were soon built, growing in size at a rate of more than an order of magnitude per year.

Though CCD's were originally conceived as a replacement for binary magnetic bubbles, it became evident immediately that due to their analog charge handling capabilities they could be used in applications other than the storage of digital information. CCD's have been used to build analog delay lines and many related devices such as transversal filters or Fourier correlators, and they have begun to play an important role in signal processing. However, most of the initial enthusiasm for CCD's was generated by their potential use as solid-state image sensors, owing to lack of a practical alternative in the early 1970's. Using the photoelectric properties of silicon, the information is placed into a CCD by optical means. Incident photons generate minority carriers, which are integrated in the individual potential wells under the transfer electrodes. Subsequently, the information is read out electrically in serial form.

For the first time, charge-coupled devices allowed the implementation of solid-state image sensors which were free of the striation problems of earlier approaches and which thereby demonstrated that the time for the realization of solid-state image sensors with medium to high resolution capabilities and good uniformity had finally arrived.

2. CHARGE-COUPLED DEVICE PRINCIPLE

2.1 The MOS Capacitor

Charge-coupled devices are composed of closely spaced electrodes each of which forms an MOS capacitor. Some aspects of the physics of an MOS capacitor will be reviewed to prepare for the understanding of CCD's. Figure 1 shows a row of MOS capacitors formed by a metal electrode deposited on a thermally oxidized p-type silicon substrate. If at time $t = t_o$ suddenly a positive voltage V_2 is applied to one of the metal electrodes (Fig. 1a), the majority carriers underneath, holes in this case, are repelled, and a potential well forms in the silicon substrate which is at first depleted of free carriers. Subsequently minority carriers, electrons, thermally generated in or near this potential well, will accumulate in this potential well within a layer of less than 10 nm of the interface. The interface potential in the presence of some minority charge Q_{sig} is then given by

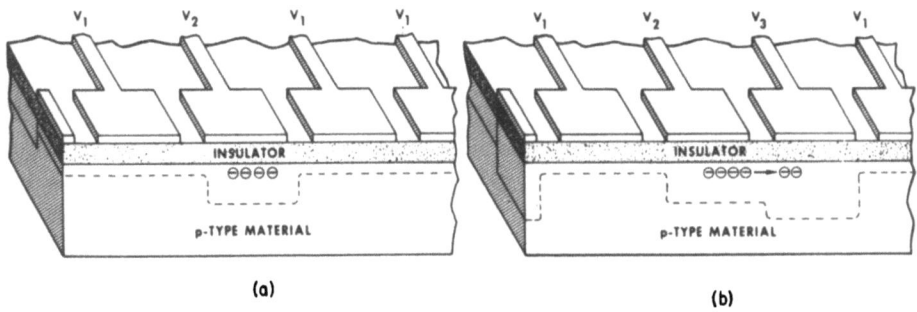

Fig. 1   Basic cell of a 3-phase CCD formed by three MOS capacitors shown (a) during integration and storage and (b) during charge transfer.

$$\varphi_s = V_G' + V_o - (2V_G'V_o + V_o^2)^{1/2} \qquad (1)$$

with

$$V_G' = (V_G - V_{FB}) + Q_s/C_o$$

$$V_o = qN_A\varepsilon_o\varepsilon_s/C_{ox}^2$$

$$C_{ox} = \varepsilon_o\varepsilon_{ox}/d_{ox}$$

where V_G is the voltage applied to the gate electrode; Q_s is the mobile charge in the inversion layer; C_{ox} is the oxide capacitance given by the thickness of the oxide layer d_{ox} and its permittivity ε_o; ε_s is the dielectric constant of the substrate; and N_A its doping density in acceptors/cm^3. As time progresses and the collected charge in the well increases, the interface potential is thus reduced. At the same time the depletion region the width of which is given by

$$X_d = (2\varepsilon_o\varepsilon_s\varphi_s/qN_A)^{1/2} \qquad (2)$$

shrinks, and correspondingly the electrode-to-substrate capacitance increases.

This thermal relaxation time or storage time of the MOS capacitor depends on the quality of the bulk material and on the integrity of the interface and can be several hundred seconds. For time intervals short compared to this relaxation time the MOS capacitor can serve as a storage element for analog information represented by the amount of charge in the well. This charge can be injected electrically or can be generated by photoelectric processes in the silicon. In the latter case the amount of charge contained in the well is a function of the integrated light flux, and the device can thus be used for image sensing purposes.

2.2 Charge Transfer

If two electrodes are placed so closely that their depletion regions overlap and their potential wells merge or "couple" any mobile minority charge will accumulate at the location with the highest interface potential. Thus it is possible to transfer charge in a controlled manner from one electrode to an adjacent electrode. In this context it is convenient to think of the potential well as a bucket and of the minority charge as a fluid that partially fills this container. The potential distribution

of the empty well, plotted with positive values in the downward
direction, can be used to depict schematically the size of the
bucket, and the potential profile in the presence of charge is
used to show the fluid surface of the partially filled well. The
area between the two surfaces represents the amount of charge as
a liquid sitting at the bottom of the bucket, whereas in the real
device the charge resides at the interface. This model has proven
very useful in depicting qualitatively the processes in a CCD.
In Fig. 1 and in the following cross-sectional drawing of various
electrode structures the dotted line represents the potential
profile in empty devices.

With two adjacent electrodes simultaneously turned on to the
same potential the charge packet will distribute uniformly underneath the two electrodes. When the first electrode is turned off,
the charge will be pushed completely to the second one (Fig. 1b).
To continue this charge transfer the following electrode is pulsed
to a high potential, and so on.

These electrodes can be lumped together in a periodical
manner into phase systems, and several packets of charge can then
be transferred simultaneously. With the described isotropic
electrodes, at least three phases are required to define the
direction of transfer. Only one neighbor to an electrode holding
charge at a certain moment must be turned "on", and the third
electrode in each "cell" must be held at a low potential to act
as a blocking electrode that prevents the backward flow of
charge. It was this single-level 3-phase principle which was
originally proposed [5] and which led to the first realization of
a charge-coupled device [7].

Although a 3-phase CCD consists in principle of an array of
MOS capacitors which could all be formed with a single mask level,
a real device is typically much more complicated. At least four,
and normally more, mask levels are required because of the following reasons:

(1) The minority carriers have to be confined laterally in the
direction perpendicular to the charge transfer, and thus
transfer channels have to be outlined. This can be done
either by a channel stopping diffusion or, in the case of an
n-type substrate, by a thick field oxide.

(2) The devices normally require input and output diodes in order
to be useful.

(3) The topologically unavoidable crossings in the 3-phase
electrode structure may necessitate diffused crossunders and
contact windows.

The narrow gaps between the transfer electrodes of the described 3-phase CCD are unattractive features. They are frequently subject to metal shorts and thus reduce the yield of these devices. In addition, the interface potential under the bare gaps is poorly controlled. A change in the charging condition of the outer surface can produce potential bumps which inhibit the complete transfer of charge and thus affect the performance of the device in an undesirable way. Other electrode structures have therefore been developed, which result in a completely covered transfer channel, but at the expense of at least one additional mask level.

3. TRANSFER ELECTRODE STRUCTURES

3.1 Nondirectional Electrodes

The first devices with a completely sealed transfer channel used four phases of overlapping electrodes (Fig. 2) formed in two levels of metallization separated by a deposited insulator [8,9]. An insulator of high quality has to be used to avoid interlevel shorts through pinholes and to obtain high yield. Good results have been obtained [10,11] with a thermally grown oxide on polysilicon electrodes in the first level and with aluminum electrodes in the second level (Fig. 3). Alternatively, aluminum can also be used to form the electrodes which are subsequently anodized to form an insulator of the required integrity. The attractiveness of this double level aluminum metallization [12] lies in the high conductivity of aluminum and in the fact that no high temperature steps are required to form the electrode structure.

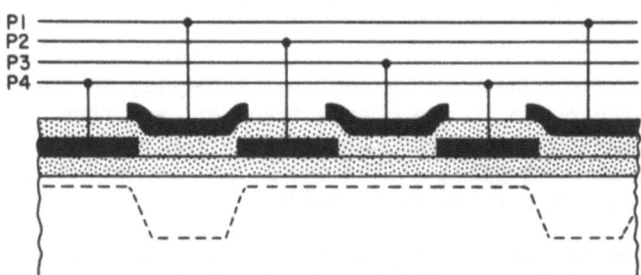

Fig. 2   Four-phase CCD electrode structure employing two metal levels with an intermediate deposited insulator. (Ref. 44)

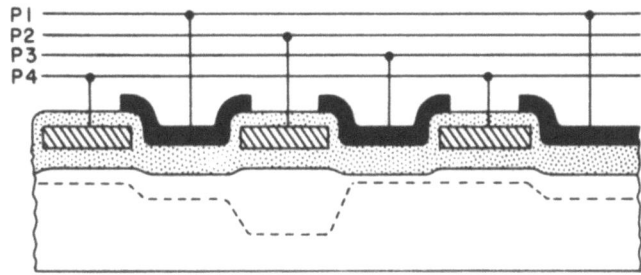

Fig. 3 Four-phase CCD electrode structure employing combination of oxidized polysilicon and metal electrodes. (Ref. 44)

In principle the number of electrodes per CCD cell could readily be increased. However, in most cases this becomes impractical because of the topological problems involved in connecting the electrodes to the many bus lines. With the opposite goal in mind, to shrink the cell size and thus to increase the packing density, the 2-level metallization scheme has also been used to make 3-phase devices. Each phase line then feeds electrodes which alternately lie in the first and second level of metallization. The double level aluminum metallization is especially useful for the implementation of such a 3-phase approach, since both levels can be used to form bus bars of high conductivity (Fig. 4).

Two other approaches have been used to build 3-phase charge-coupled image sensors with a completely covered transfer channel. Using three levels of polysilicon [13] a structure can be built in which each of the three phases is formed in a separate level of interconnection (Fig. 5). The features within each level can thus be made really large though the overall minimum cell

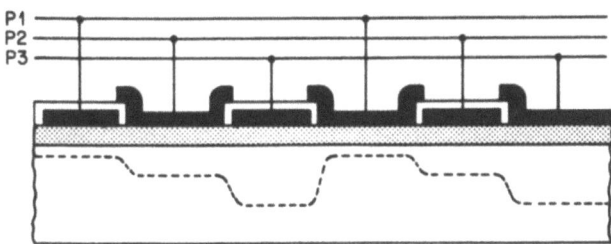

Fig. 4 Three-phase CCD electrode structure formed with two levels of anodized aluminum. (Ref. 44)

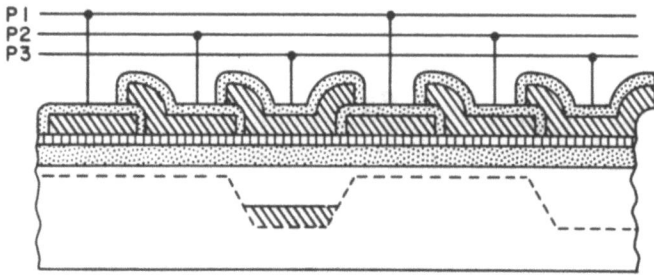

Fig. 5 Three-phase CCD electrode structure employing three levels of overlapping polysilicon electrodes. (Ref. 44)

dimension is still very compact. A major source of trouble can thereby be avoided. Intralevel shorts produced by mask defects, by flaws in the photoresist or by dust in the etching process no longer render a device completely inoperable and are less likely to cause any harm at all. Interlevel shorts, on the other hand, are not more problematic than in 4-phase structures and can be kept at low values when an insulator of high integrity, such as a thermally grown oxide on polysilicon electrodes, is used. A drawback of the 3-level approach is the rather long processing sequence to deposit, dope, define and oxidize the three levels of polysilicon. This approach has also disadvantages when it is applied to image sensors with circuit side illumination where the light has to pass through the electrode structure in order to reach the silicon substrate. Absorption in the polysilicon electrodes results in a significant loss in the blue response, and the many oxide-silicon interfaces cause losses due to reflections and optical interferences [14].

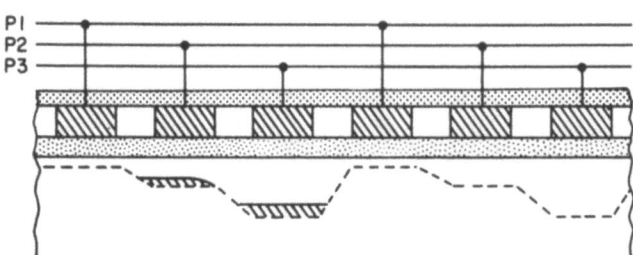

Fig. 6 Three-phase CCD electrode structure in which the electrodes are formed by selective doping of a single layer of high resistivity polysilicon. (Ref. 44)

These losses are minimized as far as possible in a single-level approach (Fig. 6) where the whole active area is covered with a single layer of high resistivity polysilicon in which the electrodes are formed by localized selective doping [15]. The necessary crossovers, address bus lines, light shields and bonding pads are formed in an additional level of true metal.

3.2 Directional Electrodes

In devices with symmetrical electrodes the phase relation between three or more clock systems determines the direction of charge transfer. On the other hand, two clocks are sufficient if the directionality [16] is built into the electrodes. If the potential profile shows an asymmetrical step under each electrode, the signal charge will accumulate in the deeper "storage" part of the well. The charge packets will then remain isolated from one another by the built-in potential "barriers" even when both electrode systems are held at the same potential. Charge transfer will occur only if the two phases are offset by a large enough voltage so that the signal charge can flow from the storage part of one electrode across the potential barrier of its neighbor.

A step in the interface potential can be produced in many different ways. The two most common approaches are a step in the thickness of the gate oxide and a localized change of the doping of the substrate near the interface. Other schemes using different dielectrics, differences in work function or fixed charges in the gate insulator have also been proposed. Even a normal 4-phase device can be operated in a 2-phase mode, by pulsing neighboring electrodes in pairs, suitably biased with a dc-voltage offset.

In Fig. 7 a 2-phase device is shown which is produced by connecting adjacent electrodes in a 2-level structure in pairs [10,11]. If the gate oxide difference under the two metal levels is not producing a sufficiently high barrier, it can be enhanced by an ion-implanted barrier which is self-aligned with the first-level electrodes [17,18]. Other ways to produce stepped-oxide 2-phase CCD's using only a single level of metal rely on the breaks that occur in an evaporated metal at sharp oxide steps. In one scheme the formation of a break is enhanced by forming a deliberate undercut in a dual dielectric structure [19]. In another approach (Fig. 8) the metal is evaporated at an oblique angle onto a castellated oxide to produce step coverage on one edge but a complete break on the other [20]. However, these latter single-level metal structures run into topological limitations in the fabrication of more complicated devices. Furthermore, the open breaks may cause yield or reliability problems.

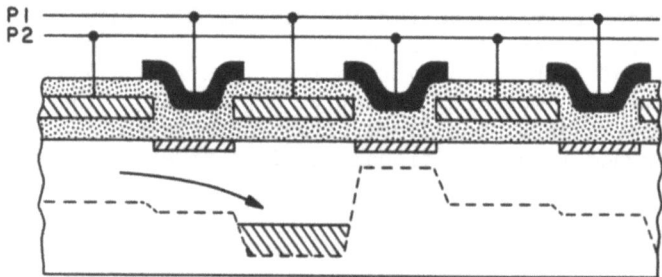

Fig. 7 Two-phase CCD electrode structure employing overlapping electrodes connected in pairs and an optional implant to enhance the potential step. (Ref. 44)

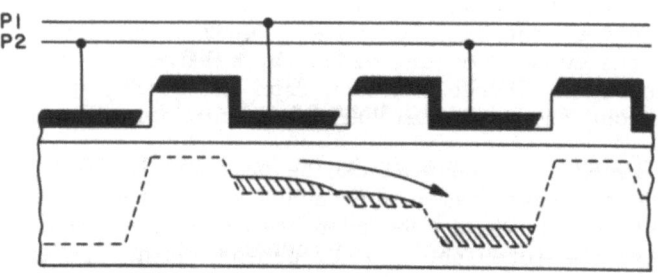

Fig. 8 Two-phase CCD electrode structure employing stepped electrodes formed by oblique evaporation. (Ref. 44)

A potential step can also be produced with a localized ion-implanted region placed under one side of the electrode [21]. However, if the barrier is not properly aligned with the edge of the transfer electrodes (Fig. 9), a potential pocket is formed which retains some charge at all times. In each transfer the signal charge is added to the carriers in this pocket, and in the subsequent transfer across the implanted barrier it has to be skimmed off again from this reservoir of carriers. The amount of signal charge transferred will thus vary with any modulation of the barrier height or width [22].

In that respect the above device operates in a manner similar to an integrated MOSFET bucket-brigade device (BBD) [1,4] which is shown in Fig. 10. In this latter device the storage areas under each electrode are formed by a conductive region of opposite polarity to that of the substrate. The areas between these reverse-biased diodes, having the original substrate doping, form

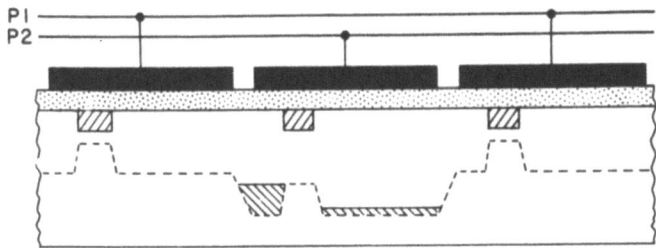

Fig. 9   Two-phase CCD electrode structure employing ion-implanted barriers. (Ref. 44)

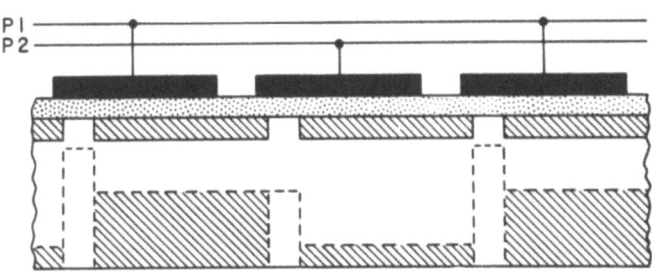

Fig. 10   MOS bucket-brigade electrode structure. (Ref. 44)

the barrier regions. Alternatively, and the way in which these devices were originally conceived, this region can be understood as a MOSFET channel. A BBD is thus an integrated chain of MOSFET's with deliberately enhanced gate to drain capacitances. The two sets of MOSFET's are coupled with their gates to the two phase lines and act as switches which allow the signal charge to flow in a 2-phase manner from capacitor to capacitor. Because of the channel length variation with varying source-drain voltage these devices have limited transfer efficiency (see Section 5) and have therefore not gained too much significance for high resolution image sensors. However, by using special channel geometries or shield electrodes the transfer efficiency of BBD's can be improved substantially [4,23].

To avoid the bare gaps between the transfer electrodes, 2-phase devices with implanted barriers have also been built with a single continuous sheet of high resistivity polysilicon, in which the transfer electrodes are formed by selectively doped areas (Fig. 11). This approach is particularly useful in bulk channel devices (see Section 4) where the stronger fringe fields

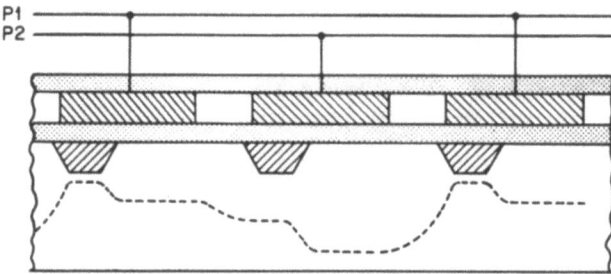

Fig. 11 Two-phase CCD electrode structure using a single layer of selectively doped polysilicon and ion-implanted barriers.

alleviate the registration problems [24] and thus prevent the formation of potential pockets which can retain charge and thereby degrade transfer efficiency.

Many variations of the electrode structures presented in this already rather lengthy list have been proposed, or implemented and are still being invented. Metal electrodes can be replaced by polysilicon electrodes to obtain transparency required for image sensors. Oxidized polysilicon could be replaced with anodized aluminum where speed is the main goal and necessitates high conductivity in the electrodes. Undercutting schemes are used with 2-level electrode structures to reduce the amount of overlap between the two levels, and additional implants can be used to enhance the barrier heights produced by thick-thin oxide steps.

Electrodes placed over oxide steps or using implanted barriers for directionality can be combined in the same device. This has been realized in an offset-mask technique which has been developed to produce high density 2-phase electrode structures. An effective barrier width as small as half the minimum design feature F can be obtained by offsetting one mask against another by half a feature dimension. A first mask is used to form thick-thin oxide steps each of length F. The polysilicon electrodes are then placed offset by $F/2$ over the thick-thin oxide steps. This forms one phase of directional electrodes. The first polysilicon electrode and the original thick oxide then jointly serve as a mask for a low energy ion implant which forms barriers of width $F/2$ which are self-aligned with the edges of the polysilicon electrodes. After implantation, the exposed gate oxide is stripped to the substrate, a new uniform thin gate oxide is grown, and the second electrode level is deposited (Fig. 12). In these devices both electrode phases are formed in a separate level of

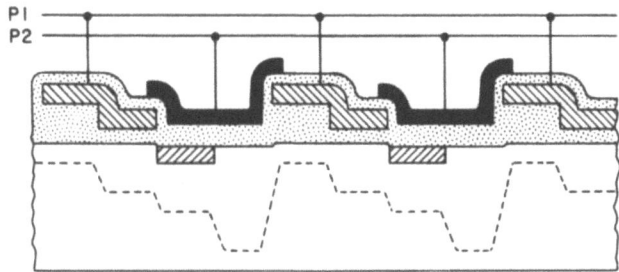

Fig. 12 Two-phase CCD electrode structure formed with offset mask technique using ion-implanted barriers under second level electrodes. (Ref. 44)

interconnection. This results in the same advantages as discussed earlier for the 3-phase 3-level electrode structure: less critical mask feature dimensions and higher yield. The second level is especially uncritical since it could take the form of a continuous strip of metal. This electrode structure looks attractive for the formation of high-density memories, but may be less suitable for image sensors because of the different optical properties of the various MOS structures.

In general, a 2-phase structure can be operated with only one active phase and with the other set of electrodes held at an intermediate constant potential. The active phase is pulsed about the dc phase with a sufficiently large amplitude so that it drains out all the charge from the dc phase in its "on" condition and dumps all the charge in its "off" condition. This principle of operation can lead to a uniphase device [26]. An implanted barrier which defines the directionality is combined with thick-thin oxide areas in such a manner that the thick oxide areas act as a quasi-static phase under which the interface potential varies only a small amount as the single electrode is pulsed (Fig. 13). Two or three implants are used to shift the thresholds in the various areas to the correct value, which allows the potential in the thin areas to be pulsed symmetrically about the thick oxide potential.

The uniphase structure looks attractive from an operating point of view and also for the formation of large memory planes or area image sensors, in which the whole active area could be covered by a continuous sheet of metal. However, because of the many processing steps required to build such a structure, the tight tolerances on the implants and the low signal handling compared to the applied pulse potentials, it remains questionable, whether this structure will gain commercial significance.

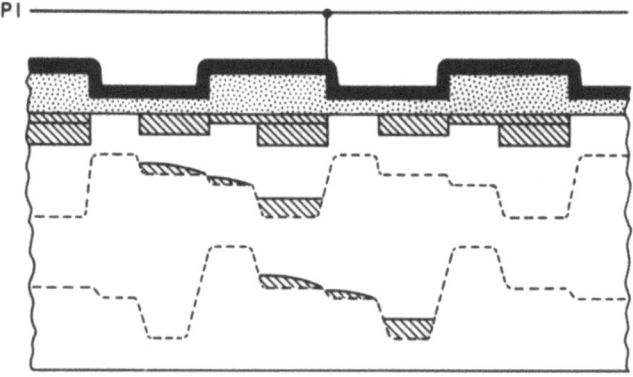

Fig. 13 Uniphase CCD electrode structure employing a combination of thick-thin oxide steps and various implanted regions. (Ref. 44)

4. SURFACE AND BULK CHANNEL CCD'S

The interaction of the signal charge packets with interface states imposes certain lower limits on transfer inefficiency and transfer noise (see Section 5). To prevent this interaction the effective transfer channel can be located away from the Si-SiO$_2$ interface deeper into the bulk of the device [27,28]. This can be achieved with the use of an epitaxial or ion-implanted silicon layer of opposite polarity to that of the substrate. This layer is in electrical contact with the output diode (Fig. 14), which drains out all mobile carriers when suitably reverse biased. The potential minimum under each transfer electrode will then be formed inside this layer, but generally away from the Si-SiO$_2$ interface. Its exact location depends upon the doping profiles, the applied voltages and the amount of signal charge present (Fig. 15). The clock pulses applied to the transfer electrodes modulate the channel potential to produce moving potential wells just as in the case of a surface channel charge-coupled device (SCCD). In these bulk channel charge-coupled devices (BCCD) which have also been called "buried channel CCD's" [27,29] or "peristaltic CCD's" [28] the signal charge moves completely inside the bulk of the silicon. The carriers thus never come in contact with the interface states. However, similar trapping and reemission action now takes place with the traps in the bulk of silicon, but the effective density of these traps which interact with the signal is normally much lower than typical densities of interface states [30].

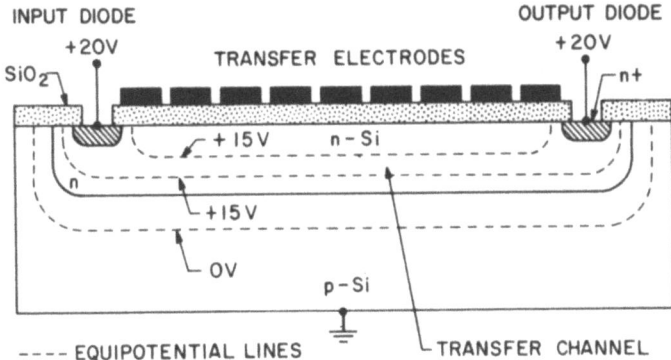

Fig. 14 Longitudinal cross section through a bulk channel CCD showing the channel layer in contact with the reverse biased input-output diodes and equipotential lines which indicate the actual position of the transfer channel.

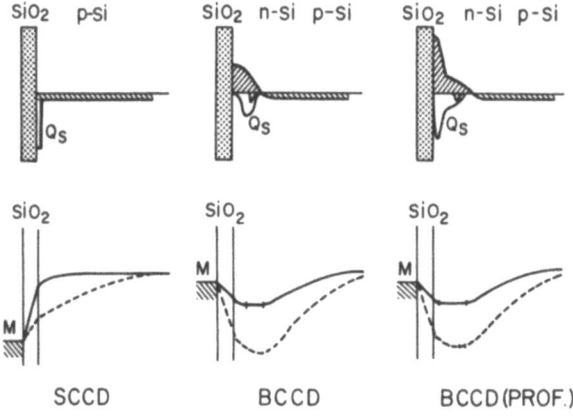

Fig. 15 Schematic doping profiles and charge distribution (top) and potential distribution (bottom) perpendicular to the semiconductor surface in a surface channel CCD, a bulk channel CCD, and a BCCD with a profiled channel. The dashed lines represent the nearly empty well and the solid lines the profile in the presence of a substantial charge packet.

The region of depleted silicon between the signal charge and the gate oxide increases the effective thickness of the gate insulator. The fringe fields thus are considerably stronger than in an SCCD of comparable geometry and will speed up the transfer of charge (see Section 5). Devices with a 5 μm thick epilayer have been operated at clock rates above 100 MHz [31]. On the other hand, a thicker effective gate insulator reduces the capacitance per area and thus the signal handling capability. To obtain maximum signal handling, a shallow implant concentrated near the Si-SiO$_2$ interface is required. However, the best features of both devices can be combined. The combination of a lightly doped epilayer and a shallow concentrated implant generates a profiled channel [32] which carries large packets close to the surface but has a potential minima which lies deep in the bulk in the presence of small packets (Fig. 15). This structure thus combines a large signal handling capability with strong fringe fields which speed up the transfer of the last fraction of charge.

The choice between a surface channel or one of the bulk channel devices will mainly depend on the performance required in a particular application. The lower transfer inefficiency and noise and the higher operating speed of a BCCD are bought at the expense of a more complicated processing sequence.

5. PERFORMANCE PARAMETERS

5.1 Signal Handling Capability

To first order the amount of charge Q_S which can be stored in each individual potential well depends on the "size of the bucket"

$$Q_s = A \cdot C_{ox} \cdot V_p, \tag{1}$$

i.e., its active area A and its "depth" given by $C_{ox} \cdot V_P$ where C_{ox} is the gate oxide capacitance per unit area and V_P is the pulse potential applied to the electrode with respect to the neighbor electrodes. The amount of signal current I_s which can flow through a 3-phase device pulsed at an element or clock rate f_c then becomes

$$I_s = A \cdot C_{ox} \cdot V_P \cdot f_c. \tag{2}$$

Devices with more than three symmetrical electrodes per unit cell can be operated in a special mode where more than only one electrode is turned on at all times. The maximum signal current I_s can then be

$$I_s = (p-2) \cdot A \cdot C_{ox} \cdot V_P \cdot f_c \tag{3}$$

where p is the number of phases. This mode has some significance in 4-phase devices where it can effectively double the signal handling capability.

In devices with symmetrical electrodes the clock amplitudes V_P can, at least in principle, be adjusted within reasonable limits to match the required signal. At the lower limit, minimum pulse amplitudes of typically about 1-2 V are required to overcome anomalies in the potential distribution at the edges of the electrodes or threshold voltage differences between adjacent electrodes. An upper limit for V_P is reached when the fields at the oxide-silicon interface become strong enough to produce avalanche breakdown. The generated carriers would then flood the potential well and lower the interface potential. In the presence of a full charge packet the main potential drop occurs across the gate oxide, which also has a limited breakdown strength of about 5×10^6 V cm^{-2}, or about 50 V for a 100 nm thickness. The maximum number of electrons that can be stored at the Si-SiO$_2$ interface is thus on the order of 10^{13} cm^{-2}.

In 2-phase devices the maximum signal current and the necessary operating pulse potentials are more or less predetermined by the built-in potential step or barrier, although in most devices the step height also scales to a certain extent with the applied clock potential. For electrodes with a built-in barrier $\varphi_b(V_P)$ and a storage part of area A_{st} with a capacitance per unit area C_{st} the signal handling capability is

$$Q_s = \varphi_b(V_P) \cdot A_{st} \cdot C_{st}. \tag{4}$$

Since $\varphi_b(V_P)$ normally increases with applied electrode voltage, the device will have the highest signal handling capability when operated with overlapping clock pulses, also called push clocks [33], so that one electrode is fully turned on before the other one turns off. This mode of operation is also more advantageous than the nonoverlapping or drop clock mode with regard to obtaining best transfer efficiency (see Section 5.2).

If 2-phase devices are operated with clock amplitudes which do not exceed the built-in barrier height $\varphi_b(V_P)$, then not all charge can transfer from one well to the next, and the device then operates in the incomplete charge transfer mode or bucket-brigade mode. The amount of charge retained behind the barrier in the previous well is not absolutely constant but is modulated by changes in the barrier height or width, which depend on the amount of signal charge transferred. This mode has thus serious drawbacks with respect to the transfer efficiency that can be obtained.

5.2 Transfer Efficiency

In a suitably designed 3-phase CCD, which is operated with sufficiently large clock pulses, eventually all the charge would flow from one well to the next, if enough time is allowed for this transfer. The main fraction of a reasonably large charge packet will transfer within a very short time on the order of only a few ns under the influence of the self-induced electrostatic forces. But once the charge density remaining under the sender electrode has fallen to about 10^{10} electrons/cm^2 the final stage of the charge transfer is governed by thermal diffusion. The time constant with which the last few percent of charge flow from the sender electrode will thus increase with the square of the electrode length. For a 10 μm long electrode it is about 30 ns.

If insufficient time is allowed for the transfer, the sender well cannot empty to a negligible fraction, and a small signal dependent amount of charge is left behind. This effect is cumulative and can amount to substantial signal degradation after many transfers. Assuming that a constant fraction ε of the signal is left behind, an isolated charge packet which has undergone n transfers will be reduced by a factor

$$Q_{out} = Q_{in}(1-\varepsilon)^n \approx Q_{in}(1-n\varepsilon) \tag{5}$$

and the subsequent, originally empty packet will have accumulated some residual charge

$$Q_{res} = Q_{in}(1-\varepsilon)^n n\varepsilon \approx Q_{in} \cdot n\varepsilon. \tag{6}$$

The approximations hold as long as the "inefficiency product" nε is sufficiently small compared to unity, so that second order terms can be neglected. The degradation of a single charge

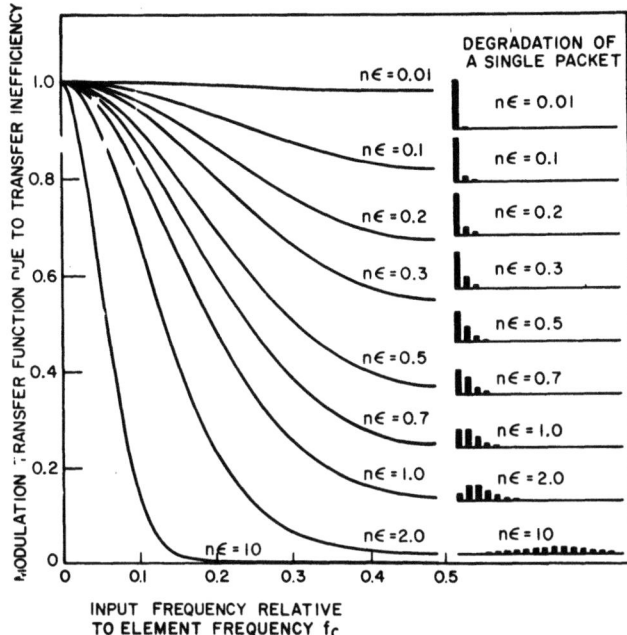

Fig. 16 Effect of incomplete transfer on the modulation transfer function and on the appearance of an isolated charge packet for different values of the inefficiency product $n\epsilon$. (Ref. 43)

packet for increasing inefficiency products $n\epsilon$ is illustrated on the right-hand side of Fig. 16.

The curves on the left-hand side represent the signal degradation in the frequency domain calculated from the same linear model. Charge transfer devices are analog sampled data systems, and the maximum signal frequency f which can be properly represented is equal to half the clock frequency f_c. The transfer functions have thus been normalized with respect to f_c. The stronger roll-off towards higher frequencies with higher inefficiency products $n\epsilon$ is evident. For imaging devices $n\epsilon$ products in excess of 0.3 would lead to a noticeable loss of resolution near the Nyquist limit.

As an introduction we had considered the limitations imposed on transfer efficiency by the clock rate and the thermal diffusion time constant. Obviously shorter electrodes will permit higher speed of operation. However, in practice it is just as effective to use substrate materials of high resistivity and thicker gate insulators. Then the fringe fields will reach under the electrodes and the carriers will move much more rapidly under the influence of these drift fields [34-36]. This effect becomes especially important in bulk channel devices [37]. If it were only for the limitations discussed so far, transfer inefficiencies below 10^{-6} could readily be obtained at MHz clock rates.

In bucket-brigade devices, or in 2-phase CCD's operating in the incomplete transfer mode because of insufficient clock pulse amplitudes, the speed of the transfer is limited by the bottleneck represented by the MOSFET channel or by the barrier region. Furthermore, the length of this channel or the height of the barrier is modulated by the signal charge transferred. Thereby the charge retained in the previous well is also a function of the signal packet size. Typically the transfer inefficiency is on the order of 10^{-3} unless special care is taken to reduce these modulation effects.

In devices with electrode separations of a few micrometers such as the original 3-phase CCD, or in 2-phase CCD's with improperly aligned implanted barriers, spurious potential bumps can form which also retain part of the signal charge. Normally these undesired barriers are narrow and largely dominated by fringe effects. The modulation effects due to the signal charge can then be especially strong, and the device will show a correspondingly poor transfer efficiency [38,39].

In CCD's of proper design operating in the complete transfer mode the dominant mechanism producing transfer inefficiency is due to the interaction with interface states [40,41]. A signal charge packet coming in contact with empty interface states will supply the necessary charge to fill these states almost instantaneously. As the charge packet moves on, these interface states release the captured charges again but with considerably slower time constants which depend on the energy levels of the states in the band gap. Many carriers will thus be released too late so that they can no longer join their own charge packet but will be added to one of the following packets, and they will thus produce a small tail following the last full charge packet. The loss experienced by the first full packet of a sequence increases with the time elapsed since the last transfer of charge through the device (Fig. 17a), since all interface states which have emptied in the meantime have to be refilled.

This effect can be mitigated by operating the device with some bias charge also called "background charge" or "fat zeros" (Fig. 17b). In this mode most of the interface states are kept filled during most of the time and ideally each signal packet would lose into the interface states the same amount of charge which it would gain from the previous packet. In reality this

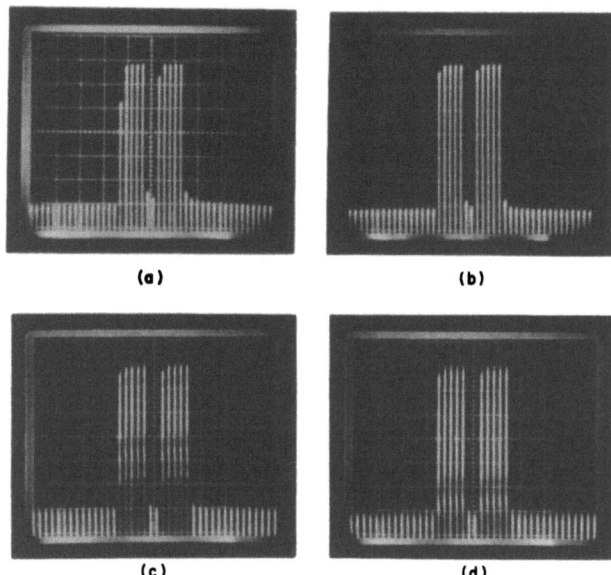

Fig. 17 Output signals observed on a 256-element 3-level 3-phase CCD: (a,b) on an SCCD and (c,d) on a BCCD, both without and with a 10% bias charge. (Refs. 13 and 30)

cancellation is not perfect since the size of a charge packet increases somewhat with its charge content. A full charge packet will thus reach a larger number of interface states at the edges of the potential well than a bias charge packet does. This remaining residual edge effect limits the transfer efficiency in wide channel devices to the range of 10^{-4} to 10^{-5}. In principle the same effects take place in bulk channel devices owing to interaction with the traps in the silicon bulk. However, because of the lower effective trap density, transfer inefficiencies below 10^{-5} can be obtained in BCCD's (Fig. 17c,d) [13].

5.3 Dark Current

In the potential wells of a CCD, as in any solid-state image sensor, thermally generated carriers are added to those produced by the incident light. The two most important sources in a potential well under an MOS electrode are the generation of carriers via midband states located at the Si-SiO_2 interface and in the depleted silicon bulk. This dark current background will thus appear superposed on the video signal. Since the generation rate

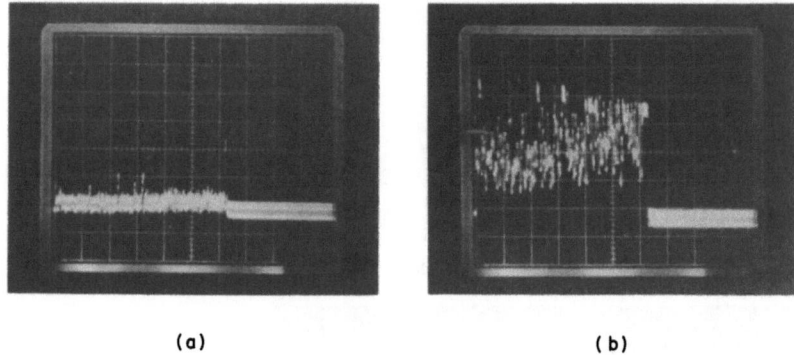

(a)   (b)

Fig. 18   Integrated dark current distribution in a linear CCD for integration times of (a) 0.1 s and (b) 1 s, respectively, showing noisy background and several individual dark current spikes. (Ref. 44)

is not uniform over the whole device, the granularity of this background will typically set a lower limit to the amount of light required to produce an image of a certain quality. Of course, the dark current contribution increases with longer integration times (Fig. 18), and it can be reduced by cooling the device.

In present-day devices the average dark current is in the range from 1 to 10 nA cm^{-2}. At this rate a potential well under an electrode pulsed to 10 V would be completely filled in about 10 s. These numbers include already the more complicated processing sequence required to make a real device. Much longer storage times can be obtained on isolated MOS capacitors.

In addition to the dark current accumulated in the integration sites, some consideration has also to be given to the contribution of the readout registers. In principle all packets which pass through the same registers at the same transfer rate will get the same dark current contribution, and nonuniformities will thus average out. Furthermore, the readout time is normally not longer than the integration time, and thus if the dark current is low enough to give adequate levels in the unaveraged integration sites, the contributions in the readout registers can typically be neglected. In surface channel devices the latter dark current component even has a beneficial aspect since it provides an automatic bias charge which improves transfer efficiency.

5.4 Noise

The noise in the video signal can, of course, never be lower than the fluctuations given by the Poisson distribution of the randomly arriving photons. The variance of this noise is given by the square root of the number of collected carriers and thus gets lower as the light flux is reduced. The thermal noise at the input of the preamplifier and other system noise, however, are constant and will thus become dominant in low light level applications.

In addition to these general considerations, the signal in a CCD is subject to transfer noise. The amount of charge left behind in each transfer has not only a systematic dependence on the signal, which results in transfer inefficiency, but also shows statistical fluctuations introducing additional noise [40,42]. These fluctuations are related to the absolute amount of charge which is captured and subsequently reemitted in each transfer. In a well-designed device, operating in the complete charge transfer mode, the transfer noise will then be determined mainly by the density of interface states or bulk traps. For this reason surface channel devices will exhibit more transfer noise than bulk channel devices. Furthermore, in SCCD's the size of the charge packet and therefore the transfer noise are fairly constant, while in BCCD's the size of the charge packet varies with its charge content. At low signal levels the signal packets interact with fewer bulk traps, and the BCCD has therefore much less transfer noise. This is also the reason that BCCD's, when operated at small signal levels with no additional bias charge, still exhibit good transfer efficiency [30].

6. CONCLUSION

A wide variety of designs of charge transfer devices are available and their operation is well understood. For high-resolution image sensors considerations of good transfer efficiency and low noise are dominant. For this reason devices such as 2-, 3-, and 4-phase CCD's operating in the complete charge transfer mode are used almost exclusively at present. For low light level applications bulk channel devices are preferred. Other important considerations are long storage times, or low and uniform dark current. This latter property is strongly technology dependent. A great deal of effort is currently being spent to develop processes which meet all of the above specifications and which produce defect-free devices with high yield.

REFERENCES

[1] Sangster, F. L. J., "The Bucket Brigade Delay Line, A Shift Register for Analogue Signals," Philips Tech. Review 31, 97-110 (1970).

[2] Wiener, N., in "Cybernetics," p. 144, John Wiley & Sons, Inc., N. Y. (1948).

[3] Hannan, W. J., Schanne, J. F., Waywood, D. J., "Automatic Correction of Timing Errors in Magnetic Tape Recorders," IEEE Trans. on Military Electronics 9, 246-254 (1965).

[4] Sangster, F. L. J., "Integrated Bucket Brigade Delay Line Using MOS Tetrodes," Philips Tech. Review 31, 266 (1970).

[5] Boyle, W. S., and Smith, G. E., "Charge Coupled Semiconductor Devices," Bell Syst. Tech. Jour. 49, 587-593 (1970).

[6] Amelio, G. F., Tompsett, M. F., and Smith, G. E., "Experimental Verification of the Charge Coupled Device Concept," Bell Syst. Tech. Jour. 49, 593-600 (1970).

[7] Tompsett, M. F., Amelio, G. F., Smith, G. E., "Charge Coupled 8-Bit Shift Register," Appl. Physics Lett. 17, 111-115 (1970).

[8] Engeler, W. E., Tiemann, J. J., and Baertsch, R. D., "Surface Charge Transport in Silicon," Appl. Physics Lett. 17, 469-472 (1970).

[9] Engeler, W. F., Tiemann, J. J., and Baertsch, R. D., "The Surface-Charge Transistor," Appl. Physics Lett. 17, 469-472 (1971).

[10] Kosonocky, W. F., and Carnes, J. E., "Charge Coupled Digital Circuits," IEEE Jour. of Solid-State Circuits SC-6, 314-322 (1971).

[11] Kosonocky, W. F., and Carnes, J. E., "Two Phase Charge Coupled Devices with Overlapping Polysilicon and Aluminum Gates," RCA Review 34, 164-202 (1973).

[12] Collins, D. R., Shortes, S. R., McMahon, W. R., Bracken, R. C., and Penn, T. C., "Charge Coupled Devices Fabricated Using Aluminum-Anodized Aluminum-Aluminum Double-Level Metallization," Jour. Electrochem. Soc. 120, 521-526 (1973).

[13] Bertram, W. J., Mohsen, A. M., Morris, F. J., Sealer, D. A., Sequin, C. H., and Tompsett, M. F., "A Three Level Metallization Three-Phase CCD," IEEE Trans. on Electron Devices ED-21, 758-767 (1974).

[14] Sequin, C. H., Morris, F. J., Shankoff, T. A., Tompsett, M. F., and Zimany, E. J., "Charge-Coupled Area Image Sensor Using Three Levels of Polysilicon," IEEE Trans. on Electron Devices ED-21, 712-720 (1974).

[15] Kim, C-K., and Snow, E. H., "p-Channel Charge Coupled Devices with Resistive Gate Structure," Appl. Physics Lett. 20, 514-515 (1972).

[16] Kahng, D., "Charge Coupled Devices," U.S. Patent No. 3,700,932 (1972).

[17] Erb, D. M., Kotyczka, W., Su, S. C., Wang, C., and Clough, G., "An Overlapping Electrode Buried Channel CCD," IEDM, Washington, D. C., Tech. Digest, 24-26 (1973).

[18] Kim, C-K., "Two-Phase Charge Coupled Linear Imaging Devices with Self-Aligned Barrier," IEDM, Washington, D. C., Tech. Digest, 55-58 (1974).

[19] Berglund, C. N., Powell, R. J., Nicollian, E. H., and Clemens, J. T., "Two-Phase Stepped Oxide CCD Shift Register Using Undercut Isolation," Appl. Physics Lett. 20, 413-414 (1972).

[20] Baker, I. M., and Beynon, J. D. E., "Charge Coupled Devices with Submicron Electrode Separations," Electronics Lett. 9, 48-49 (1973).

[21] Krambeck, R. H., Walden, R. H., and Pickar, K. A., "A Doped Surface Two-Phase CCD," Bell Syst. Tech. Jour. 51, 1849-1866 (1972).

[22] Krambeck, R. H., Strain, R. J., Smith, G. E., and Pickar, K. A., "Conductively Connected Charge Coupled Device," IEEE Trans. on Electron Devices ED-21, 70-72 (1974).

[23] Berglund, C. N., and Thornber, K. K., "A Fundamental Comparison of Incomplete Charge Transfer in Charge Transfer Devices," Bell Syst. Tech. Jour. 52, 147-182 (1973).

[24] Walsh, L., and Dyck, R. H., "A New Charge-Coupled Area Imaging Device," CCD Appl. Conf., San Diego, Proc., 21-22 (1973).

[25] Bower, R. W., Zimmerman, T. A., and Mohsen, A. M., "A High Density Overlapping Gate Charge Coupled Device Array," IEDM, Washington, D. C., Tech. Digest, 30-32 (1973).

[26] Amelio, G. F., "A Little 'Bit' on Charge Coupled Devices," IEDM, Washington, D. C., abstract 1.3 (1971).

[27] Walden, R. H., Krambeck, R. H., Strain, R. J., McKenna, J., Schryer, N. L., and Smith, G. E., "The Buried Channel Charge Coupled Device," Bell Syst. Tech. Jour. $\underline{51}$, 1635-1640 (1972).

[28] Esser, L. J. M., "Peristaltic Charge-Coupled Device: A New Type of Charge-Transfer Device," Electronics Lett. $\underline{8}$, 620-621 (1972).

[29] Kim, C-K., Early, J. M., and Amelio, G. F., "Buried Channel Charge Coupled Devices," NEREM, Boston, Record of Tech. Papers, Part 1, 161-164 (1972).

[30] Mohsen, A. M., and Tompsett, M. F., "The Effects of Bulk Traps on the Performance of Bulk Channel Charge Coupled Devices," IEEE Trans. on Electron Devices $\underline{ED-21}$, 701-712 (1974).

[31] Esser, L. J. M., Collet, M. G., van Santen, J. G., "The Peristaltic Charge Coupled Device," IEDM, Washington, D. C., Tech. Digest, 17-20 (1973).

[32] Esser, L. J. M., "The Peristaltic Charge Coupled Device for High Speed Charge Transfer," ISSCC, Philadelphia, Digest of Tech. Papers, 28-29 (1974).

[33] Mohsen, A. M., McGill, T. C., Mead, C. A., "Charge Transfer in Overlapping Gate Charge Coupled Devices," IEEE Jour. of Solid-State Circuits $\underline{SC-8}$, 191-207 (1973).

[34] Carnes, J. E., Kosonocky, W. F., and Ramberg, E. G., "Drift-Aiding Fringing Fields in Charge-Coupled Devices," IEEE Jour. of Solid-State Circuits $\underline{SC-6}$, 322-326 (1971).

[35] Engeler, W. E., Tiemann, J. J., and Baertsch, R. D., "Surface Charge Transport in a Multielement Charge Transfer Structure," Jour. of Appl. Physics $\underline{43}$, 2277-2285 (1972).

[36] Lee, H. S., and Heller, L. G., "Charge Control Method of Charge Coupled Device Transfer Analysis," IEEE Trans. on Electron Devices $\underline{ED-19}$, 1270-1279 (1972).

[37] Collet, M. G., and Vliegenthart, A. C., "Calculations on Potential and Charge Distributions in the Peristaltic Charge Coupled Device," Philips Res. Reports 29, 25-44 (1974).

[38] Amelio, G. F., "Computer Modelling of Charge Coupled Device Characteristics," Bell Syst. Tech. Jour. 51, 705-730 (1972).

[39] Séquin, C. H., "Experimental Investigation of a Linear 500-Element 3-Phase Charge-Coupled Device," Bell Syst. Tech. Jour. 53, 581-610 (1974).

[40] Tompsett, M. F., "The Quantitative Effects of Interface States on the Performance of Charge Coupled Devices," IEEE Trans. on Electron Devices ED-20, 45-55 (1973).

[41] Carnes, J. E., and Kosonocky, W. F., "Fast Interface State Losses in Charge Coupled Devices," Appl. Physics Lett. 20, 261-263 (1972).

[42] Carnes, J. E., and Kosonocky, W. F., "Noise Sources in Charge Coupled Devices," RCA Review 33, 327-343 (1972).

[43] Séquin, C. H., "Interlacing in Charge Coupled Imaging Devices," IEEE Trans. on Electron Devices ED-20, 535-541 (1973).

[44] Séquin, C. H.; and Tompsett, M. F., "Charge Transfer Devices," Suppl. #8 to "Advances in Electronics and Electron Physics," Academic Press, Inc., New York (1975).

ORGANIZATIONS OF CHARGE-COUPLED IMAGE SENSORS

C. H. Séquin

Bell Laboratories
Murray Hill, New Jersey 07974, U. S. A.

ABSTRACT. The basic readout organizations of solid-state image sensors using charge transfer readout registers will be reviewed, and some of their advantages and shortcomings will be discussed.

1. THE BASIC CONCEPT

Most early attempts to make a solid-state image sensor used an array of sensor elements which were addressed by a decoder or a shift register and which read out the photogenerated charges into a common bus (Fig. 1). These early approaches produced unsatisfactory results because of uniformity problems [1]. The readout of different sensor elements produced different switching transients which thus superposed a fixed pattern noise onto the video signal in the form of many striations. When the individual sensor elements were provided with internal gain to produce signals which are large compared to the capacitive pulse pickup, these striations were even more distinct because of gain nonuniformities. This fixed pattern noise cannot be eliminated by filtering. There is also another inherent limitation. In a uniform device operated at low light levels the signal-to-noise ratio S/N will be mainly determined by the thermal noise in the input stage of the preamplifier. The ratio S/N decreases as the square root of an increasing capacitance at the detection node. This represents a serious disadvantage in high resolution devices which have many sensor elements attached to a long readout bus line.

In image sensors with charge transfer readout registers the signal charge packet themselves, rather than the addressing bits,

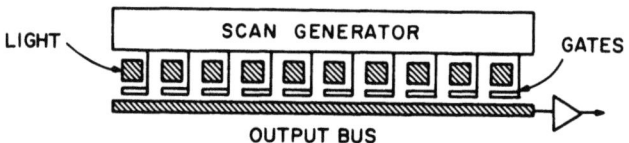

Fig. 1 Organization of shift register-addressed array of photodiodes.

are moved through the device. All the charge packets from the individual sensor elements are transferred to a single detection node, which can be implemented with a little output diode or with a small floating gate, both of which can have a capacitance of less than 0.1 pF. The ultimate limit of sensitivity given by the thermal input noise of the preamplifier can thus be considerably lower.

The size of the charge packets is not altered by pulse pick-up during their transfer to the output diode. Furthermore, all charge packets are pushed to the detection node by the same clock pulses applied to the same electrodes. The effect of pulse pick-up is thus the same for every time slot and every emerging signal packet. Therefore, pattern noise due to capacitive coupling of the clock pulses is virtually nonexistent.

On the other hand, the readout registers require good transfer efficiency so that the integrity of the individual charge packets is preserved. Since, in general, not all charge packets undergo the same number of transfers, finite transfer inefficiency will result in a resolution shading in which the smearing increases with the distance from the detection node. A localized spot of high transfer inefficiency, which may be produced by a defect such as a broken electrode, causes a new kind of fixed pattern distortion. All packets transferring through that spot will experience a stronger smearing, and the display will show an abrupt step to lower resolution at that point. Considerations of the overall inefficiency products $n\varepsilon$ for the various readout registers is of prime concern and will be discussed in more detail as we review the various readout organizations of one-dimensional arrays or linear sensors, and of two-dimensional arrays or area image sensors.

2. LINEAR IMAGE SENSORS

Linear sensors consist of a single row of photosensitive elements and can thus be used to monitor positions in one dimension or to record an optical spectrum. Two-dimensional pictures can be obtained with a linear sensor, when the second dimension is scanned by mechanical means, such as a movable mirror, or, alternatively, when the object to be scanned is moved in front of the sensor. Devices with a single row of resolution elements can have a relatively small active area and a simple readout organization, and high spatial resolution along the axis of the device can easily be obtained. By mechanically scanning an image with such a device, perpendicular to its axis, pictures with several million resolution spots can be produced.

2.1 Integration in a Transfer Channel

In the simplest implementation an isolated charge transfer channel (Fig. 2) can be used as a linear sensor [2]. The clock pulses are stopped with one set of electrodes held at a high potential, thus forming the integrating potential wells underneath. Light incident from the circuit side enters the silicon bulk either through transparent electrodes or through the gaps between metal electrodes and generates hole-electron pairs. The minority carrier, electrons for an n-channel device, will end up in the nearest potential well, while the holes are discarded in the bulk. After a suitable integration time the amount of charge collected in the individual potential wells is representative of the integrated light flux incident in the vicinity of this well. The transfer electrodes are now suitably pulsed to produce readout by shifting the charge packets to the output diode where the video signal appears in serial form.

If light is continuously incident on the device, minority carriers will still be generated during readout, and additional smeared information will be superposed on the original charge pattern. In short devices, when the readout period can be made sufficiently short compared to the integration time, this smeared component may be tolerable, particularly since, for a static video input pattern, the contribution of charge added to all packets is the same, and the smeared component will thus constitute a constant

Fig. 2 Charge-coupled device integrating the video information in the transfer channel.

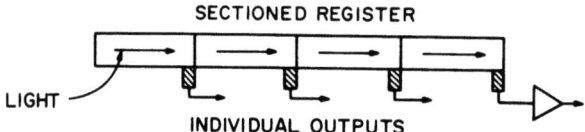

Fig. 3 Charge-coupled device with sectioned channel to increase integration-to-readout duty cycle.

bias charge. This is no longer true when the information falling onto the device can change during the readout period. Furthermore, the ratio of bias charge to actual signal amplitude may become inacceptably high in long devices. The device would then have to be protected from incident light by some mechanical or electronic shutter synchronized with the readout period of the device (Fig. 2). However, this results in an effective loss of integration time and correspondingly in a reduced sensitivity of the sensor. Furthermore, a mechanical shutter would present additional system complexity and would definitely be contrary to the main goals that are being pursued with the introduction of solid-state sensors.

In principle a long transfer channel could be subdivided into several shorter sections each with its own separate output (Fig. 3) [3]. This approach is considerably more complicated to implement, since decoders or address registers are required to switch the readout pulses from section to section, and the video signals from the different outputs have to be multiplexed. The latter is particularly problematic since it has to occur without producing discontinuities in the display due to any switching transients, or nonuniformities due to gain differences of the many individual preamplifiers.

2.2 Separate Integration Sites

A truly practical device results from the organization depicted in Fig. 4. The individual photosensors are located at the side of the readout register from which they are isolated by a transfer gate [3]. When this gate is opened, all the charges integrated in the photosensors, which can be implemented with depleted MOS capacitors or with reverse-biased diodes, transfer in parallel into the readout register. This charge transfer register has typically one complete transfer cell opposite to each photosensor, and during this lateral transfer the electrode receiving the charge packet is biased to a high potential. After the gate is closed, the photosensors immediately start to integrate a new line of information, while the charge packets are read out along the transfer register, which is shielded from incident light. Video information can be read out from this device almost continuously, except for the short time intervals reserved for the lateral transfer, and the sensor element can integrate charge with a 100% duty cycle.

Limitations with respect to the number of resolution elements that can be realized in a single device with this organization arise from two different reasons. Firstly, of course, the number is limited by the maximum feasible overall device length divided by the cell dimension. Secondly, transfer inefficiency also imposes a limit on the maximum length of the readout register. The inefficiency product increases linearly with the distance from the output diode. Inefficiency products $n\epsilon$ of more than 0.4 will lead to a loss of more than a factor of 2 in the modulation transfer function of the highest resolvable spatial frequencies near the Nyquist limit. In a 3-phase surface channel CCD of medium quality the inefficiency ϵ will be about 10^{-4} when the device is operated with some bias charge. In this example more than about 1000 transfer elements would result in a noticeable signal degradation towards one end. In principle it is conceivable to read out both halves of the transfer register of the device from the middle towards both ends in order to reduce the number of transfers by a factor of 2. However, it is then difficult to reconstruct the video signal on a continuous display such as a cathode ray tube without producing a discontinuity at the butt joint in the middle.

In such a device with a single readout register the minimum spatial period may be limited by the minimum feasible transfer cell length rather than by the minimum geometry of the sensor elements. This is normally the case when all transfer electrodes are formed in a single level of metallization, where a complete transfer cell may require a minimum of four to six individual features. In this case, more than one sensor cell could be attached to each readout element, and the various cells would be read out in subsequent scans in an interlaced format. The subject of interlacing will be discussed in a separate lecture.

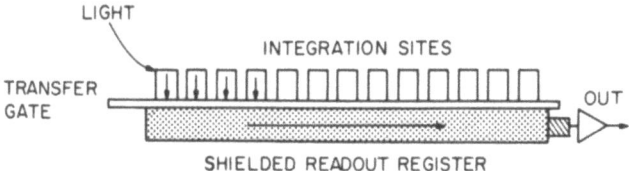

Fig. 4   Charge-coupled readout register with separate integration sites.

2.3 Bilinear Arrangement with Interdigitated Sensors

However, a bilinear arrangement has proven to be more practical to increase the spatial density of the resolution elements and to reduce the required number of transfers in the readout register. Two structures corresponding to Fig. 4 are integrated into a single device by interdigitating the two sets of photosensors (Fig. 5). The strings of charge packets from the two registers are combined at the output with an alternate merge that conserves the proper phase relation. In registers that are driven by clocks using an even number of phases this presents no problem [4]. When 3-phase registers are used, the clock pulses in the two registers have either to be offset properly or the packets emerging from the two registers are read in pairs into an auxiliary register that runs at twice the clock frequency and thus performs the multiplexing function (Fig. 6) [5]. Organizations shown in Figs. 4-6 have led to practical devices, and representatives will be discussed in later lectures.

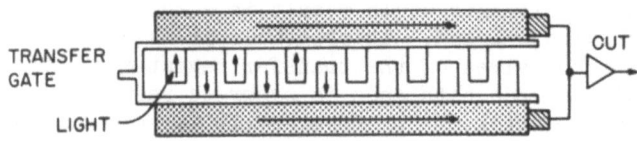

Fig. 5   Bilinear arrangement of transfer channels with an even number of clock phases.

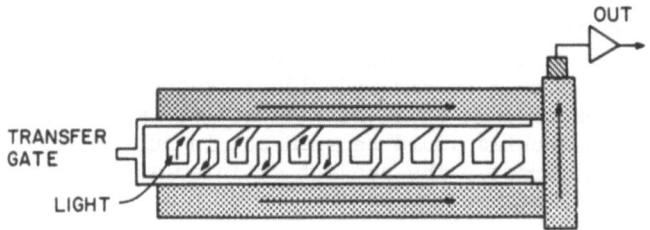

Fig. 6   Bilinear arrangement for the case of 3-phase readout registers.

3. AREA IMAGE SENSORS

3.1 Line-Addressed Devices

An area image sensor can be formed in a straightforward manner by assembling simple linear sensors with illuminated charge transfer registers in a two-dimensional array. This array is then provided with an addressing circuit that switches the clock pulses to a certain line when its readout is desired (Fig. 7). The charge packets from this line are then transferred to a common output diode bus or, preferably, into a common readout register running in the vertical direction, which is shielded from incident light. The latter organization was first demonstrated in a bucket-brigade structure with 16×16 elements [6] and subsequently in devices with 32 rows of 44 elements each [7]. Optical smearing as discussed for the single transfer channel takes place during the readout of each line. However, a reasonable integration-to-readout duty cycle is guaranteed for each line, since in a device with M×N elements the readout of a single line of M elements may last at the most a fraction $1/N$ of the total frame time.

In addition to the array of transfer channels such a device requires the circuitry to address each line in turn. This involves a decoding tree or a shift register. An individual line is addressed for readout by switching the transfer electrodes from their static integration potential into the pulsing mode required for charge transfer. Therefore, electronic switches are required to switch the electrode sets of each line between the proper quiescent or pulsing bus bars.

A further complication arises from the fact that the information from each line has to travel through a different number of elements in the vertical readout register. This has to be taken into consideration in the timing of the start of the readout of each line. The maximum number of elements through which a charge

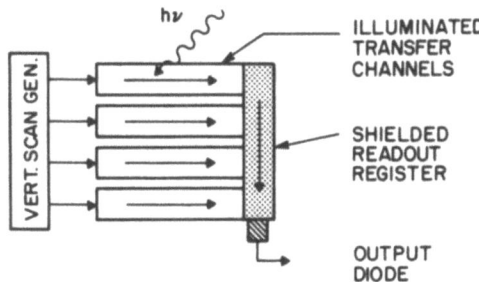

Fig. 7 Line-addressed charge transfer area image sensor.

packet has to travel in an M×N array is M + N, and all transfers take place at the readout rate of the video signal.

During readout of different lines the clock pulses on the chip are applied to different electrodes, and thus the capacitive pickup at the detection node may vary and result in fixed pattern noise.

3.2 Frame Transfer Devices

In the preferable frame transfer approach [3], which avoids some of the shortcomings of the previous device, the illuminated charge transfer channels run in the vertical direction (Fig. 8). They lie side by side and are operated with common electrodes running across all registers to form the imaging area of the image sensor. All registers are extended for the same number of elements as in the imaging area to form a storage area of equal capacity, which, however, is shielded from incident light. During the normal vertical retrace period of the television display system the charge pattern accumulated in the imaging area is shifted quickly and in unison downwards into the storage area. The imaging area is then returned to the integrating mode, and the accumulation of the subsequent frame is started. During the horizontal retrace periods the whole pattern of charge packets in the storage area is moved downwards one line at a time so as to shift the bottom row of charge packets into a horizontal readout register. This line of charge is then transferred serially to the output of the device where the video signal is formed. When the whole pattern of video information has been read out from the storage area in this manner, the next vertical frame transfer is initiated and the overall cycle repeats.

In practice the disadvantage of the need for an additional storage area is offset by the simplicity of the electrode structure and thus the feasibility of more compact transfer cells. Furthermore, in the frame transfer device, the vertically 2:1 interlaced video format can be obtained in a simple way [8], which will be discussed in a later lecture, and which effectively reduces by a factor of two the number of transfer elements required in the vertical direction. The number of transfer cells required to produce a display with M×N resolution elements is then a total of M×N vertical cells in the imaging and storage area, and M horizontal transfer elements in the readout register. The largest number of transfers experienced by any charge packet is again N + M, but N/2 of the vertical transfers occur at the much lower line step rate.

The light which is incident continuously onto the imaging area will cause some optical smearing during the frame transfer.

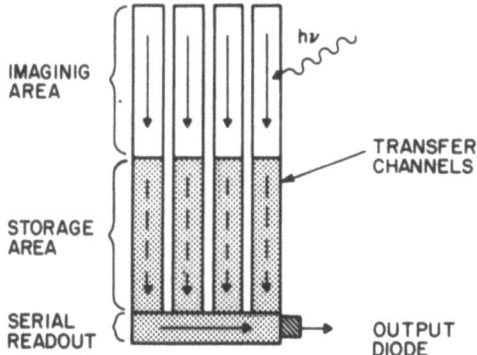

Fig. 8 Frame transfer organization of charge-coupled area image sensor.

To minimize this effect the frame transfer must be reasonably fast. Normally it will be executed at a transfer rate which is about four or eight times lower than the video readout rate, and the frame transfer time T_t will then correspond to a few line times. The carriers generated by the incident light during this time are distributed among the $N/2$ lines transferred from the imaging area, and the contribution to an individual charge packet will be on the average a fraction $2T_t/T_f N$ of the mean signal packet, where T_f is the frame time of the device. For normal lighting conditions this superposed smeared image does not cause any problems. However, a local optical overload, two orders of magnitude or more above the saturation of the device, will superpose a white vertical line onto the image [9].

The frame transfer device requires three sets of clocks for the imaging area, the storage area, and the readout register, respectively. Only the horizontal register has to operate at the high video readout rate. During all times when actual video information is read out, the pulse pattern applied to the device is exactly the same, and no fixed switching noise pattern should be observable. So far the frame transfer approach, which can be implemented with a single level of metallization [10,11], is the one which has been used most extensively.

3.3 Interline Transfer Devices

Optical smearing during readout can be avoided if all the charge transfer registers are shielded from incident light. Just as in the case of the linear image sensors discussed in Section 2.2, separate integration sites are then placed to the

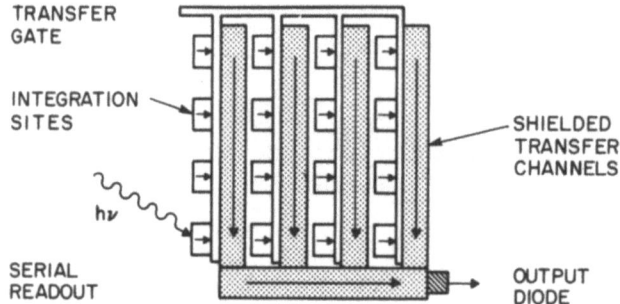

Fig. 9   Interline-transfer charge-coupled area image sensor.

side of the readout registers. In principle, horizontal or vertical readout registers could be used, but the readout logic is much simpler if the transfer channels run in the vertical direction. This leads to the organization of the interline transfer device (Fig. 9) [12]. This device can be understood as a frame transfer device in which the integration sites are interleaved between the vertical registers of the storage area. Frame transfer now takes place in a single lateral step.

The operational simplicity of this structure has to be paid for by a more complex cell design. In addition to the transfer electrodes which run horizontally across the array, an additional set of electrodes is required to form a transfer gate which isolates the photosensors from the vertical transfer channels, or a transparent electrode is required to form special MOS capacitors for the photosensor elements. Furthermore, light shields have to be placed onto all the storage and the readout registers. Such a device can therefore no longer be implemented with a single level of electrodes.

The necessary light shields on the transfer channels also reduce the effective light sensitivity of the device and present serious problems when a device with backside illumination is preferred.

3.4   Backside Illumination

Illuminating a charge transfer image sensor from the electrode side may result in low light sensitivity and poor uniformity of the spectral response. In devices with a single level of an opaque metallization, only the small fraction of the incident light falling on the narrow gaps between the transfer electrodes can produce minority carriers in the silicon. In devices using a single level of selectively doped polysilicon to form the transfer electrodes, the strongest loss of light occurs in the blue part of the spectrum. This loss is particularly serious in large arrays, where the polysilicon has to be about 0.5 µm thick in order to obtain the required conductivity in the transfer electrodes. The situation is even worse in devices using several levels of overlapping polysilicon electrodes [13]. In addition to the absorption losses, further spectral nonuniformities are produced by interference effects due to the partial reflection of the incident light at the various $Si-SiO_2$ interfaces. The curve of the spectral responsivity can then show several maxima and minima, the position of which is a sensitive function of the thickness of the various dielectric layers.

Illuminating the solid-state sensor from the backside, or substrate side, similar to the mode in which silicon diode array targets are normally operated, avoids these problems and can bring the quantum efficiency of the sensor close to unity. The transfer electrodes on the device can then be selected with respect to their electrical properties such as conductivity rather than for their optical characteristics. However, since most of the visible light is absorbed within the first few microns of silicon, backside illumination requires that the device be thinned. The thickness of the substrate has to be considerably less than the diffusion length of the minority carriers to get a high degree of charge collection, and less than the dimensions of the resolution cells, in order not to lose spatial resolution due to the lateral diffusion of carriers [14]. In addition, the backside surface should be provided with an antireflection coating to increase optical transmission and should be kept in accumulation to minimize carrier recombination in surface states. The solution of all these problems in conjunction with a processing sequence which yields high transfer efficiency and low dark current requires a sophisticated technology. Nevertheless, the first experimental results look quite encouraging [15].

Of the area image sensors discussed in the previous sections the best contender for backside illumination is the frame transfer device. In an interline transfer device, the required shielding of the transfer channels presents serious problems. The shields have to be carefully aligned with the transfer channels formed on the front surface, and the thickness of the substrate would have to be reduced so much that it can be depleted almost completely through its whole thickness so that no photogenerated minority carriers can diffuse laterally into the transfer channels.

While speculating about devices which can be built with future improved technology one might consider a device which combines the advantages of a **frame transfer** device with those of the interline transfer structure. In this hypothetical device (Fig. 10) contiguous sensor elements cover the whole backside surface of the device. The substrate is thick enough so that the photogenerated minority carriers can be collected near the back surface without diffusing to the front side. The frame transfer takes place in parallel through the substrate to the front surface upon proper biasing of an electrode at the backside. For this step the substrate has to be suitably structured, e.g., with a grid of a channel stopping diffusion, to maintain the individuality of the different charge packets. The front surface of the device is occupied by an array of charge transfer channels, comparable to the storage area of a frame transfer device, and by a single serial readout register. Readout of the transferred frame occurs in a line-by-line manner as in the frame transfer approach. This device would have the operational simplicity of an interline transfer device, the simple electrode configuration of a frame transfer device, and it would require no special light shields.

In practice, the choice of a particular readout organization will not only depend on operational simplicity or on required performance, but to a large extent on the available technology. In the recent past the frame transfer approach and interline transfer devices have been actively studied. No obvious winner has yet been declared, and the trade-offs between the two organizations will be discussed in a separate lecture.

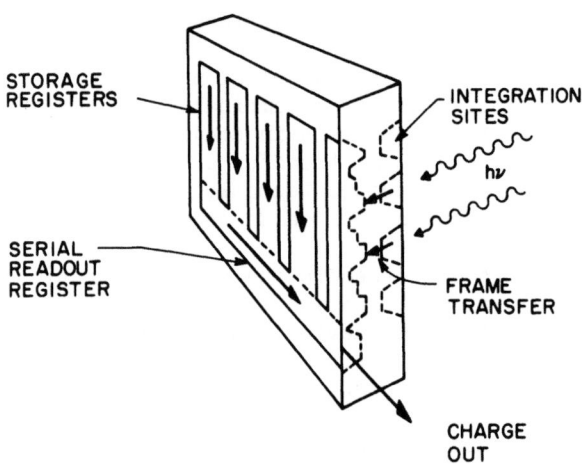

Fig. 10 Hypothetical device with integration and readout on different sides of the silicon substrate.

REFERENCES

[1] Weimer, P. K., "Image Sensors for Solid State Cameras," Advances in Electronics and Electron Physics 37, 182-262 (1975).

[2] Tompsett, M. F., Amelio, G. F., Smith, G. E., "Charge Coupled 8-Bit Shift Register," Appl. Physics Lett. 17, 111-115 (1970).

[3] Tompsett, M. F., Amelio, G. F., Bertram, W. J., Buckley, R. R., McNamara, W. J., Mikkelsen, J. C., and Sealer, D. A., "Charge Coupled Imaging Devices: Experimental Results," IEEE Trans. on Electron Devices ED-18, 992-996 (1971).

[4] Tompsett, M. F., Bertram, W. J., Sealer, D. A., and Séquin, C. H., "Charge-Coupling Improves Its Image, Challenging Video Camera Tubes," Electronics 46, No. 2, 162-168 (1973).

[5] Kim, C-K., and Dyck, R. H., "Low Light Level Imaging with Buried Channel Charge Coupled Devices," Proc. IEEE 61, 1146-1147 (1973).

[6] Weimer, P. K., Kovac, M. G., Shallcross, F. V., and Pike, W. S., "Self-Scanned Image Sensors Based on Charge Transfer by the Bucket Brigade Method," IEEE Trans. on Electron Devices ED-18, 996-1003 (1971).

[7] Kovac, M. G., Pike, W. S., Shallcross, F. V., and Weimer, P. K., "Solid-State Imaging Emerges from Charge Transport," Electronics 45, No. 5, 72-77 (1972).

[8] Séquin, C. H., "Interlacing in Charge Coupled Imaging Devices," IEEE Trans. on Electron Devices ED-20, 535-541 (1973).

[9] Séquin, C. H., Shankoff, T. A., and Sealer, D. A., "Measurements on a Charge-Coupled Area Image Sensor with Blooming Suppression," IEEE Trans. on Electron Devices ED-21, 331-341 (1974).

[10] Séquin, C. H., Sealer, D. A., Bertram, W. J., Tompsett, M. F., Buckley, R. R., Shankoff, T. A., and McNamara, W. J., "A Charge Coupled Area Image Sensor and Frame Store," IEEE Trans. on Electron Devices ED-20, 244-252 (1973).

[11] Weimer, P. K., Pike, W. S., Kovac, M. G., and Shallcross, F. V., "The Design and Operation of Charge Coupled Image Sensors," ISSCC, Philadelphia, Digest of Tech. Papers,

132-133 (1973).

[12] Amelio, G. F., "Physics and Applications of Charge Coupled Devices," IEEE INTERCON, New York, Digest, Vol. 6, paper 1/3 (1973).

[13] Séquin, C. H., Morris, F. J., Shankoff, T. A., Tompsett, M. F., and Zimany, E. J., "Charge-Coupled Area Image Sensor Using Three Levels of Polysilicon," IEEE Trans. on Electron Devices ED-21, 712-720 (1974).

[14] Crowell, M. H., and Labuda, E. F., "The Silicon Diode Array Camera Tube," Bell Syst. Tech. Jour. 48, 1481-1528 (1969).

[15] Hartsell, G. A., and Kmetz, A. R., "Design and Performance of a Three Phase Double Level Metal 160X100 Element CCD Image," IEDM, Washington, D. C., Tech. Digest, 59-62 (1974).

CHARGE TRANSPORT WITHOUT TRAPS

Marvin H. White

Westinghouse Electric Corporation,
Advanced Technology Laboratory,
Baltimore, Maryland U.S.A.

ABSTRACT. The free charge transfer characteristics of charge-coupled devices (CCD's) are determined by incomplete transfer due to (1) insufficient transfer time and (2) trapping in fast interface states and bulk traps. This chapter will concentrate on the former, although a phenomenological treatment of trapping effects will be introduced in the section on the charge-control method of analysis. In this chapter the three basic mechanisms of charge motion, namely (1) thermal diffusion, (2) self-induced drift, and (3) fringe field drift will be discussed with respect to free charge transfer and transfer inefficiency.

1. THE CHARGE TRANSFER EQUATION

In this section the nonlinear, partial differential equation which describes charge transfer in a charge-coupled device (CCD) will be derived from first principles. Through the development of the charge transport equation we will understand the various electric fields which determine the charge transfer characteristics of a CCD. Following the development of the transport equation we will discuss boundary conditions and several limiting transfer mechanisms. Finally, we will introduce the method of charge-control analysis to formulate a closed-form solution for the transfer inefficiency.

1.1 Injected Charge Density and Influence on Surface Potential

In order to understand charge-coupled device operation we begin by writing the relationship between gate voltage V_G and surface potential,[1]

$$V_G = V_{FB} - \frac{Q_s}{C_o} + \phi_s \tag{1}$$

in which V_{FB} is the flat-band voltage, Q_s the semiconductor charge density, C_o the oxide capacitance and ϕ_s the surface potential. Figure 1 illustrates the injection of charge density ρ beneath a metal electrode. The semiconductor charge density may be expressed in terms of the injected charge density by the equation,

$$Q_s = q\, N_D\, x_d + \rho \tag{2}$$

where N_D is the density of ionized donors in the space charge region and x_d the space charge region width. Through considerations analogous to p-n junction theory, the space charge width is related to the surface potential by the expression,

$$x_d = \left[\frac{2K_s\, \epsilon_o\, \phi_s}{q\, N_d}\right]^{1/2} \tag{3}$$

where $K_s\, \epsilon_o$ is the dielectric constant of the semiconductor. Combining equations (1) to (3) we obtain,

$$2q\, K_s \epsilon_o\, N_D\, \phi_s = \left[C_o\,(V_G - V_{FB} - \phi_s) + \rho\right]^2 \tag{4}$$

where $C_o = K_o \epsilon_o / x_o$ and $\phi_s(y)$ is a function of distance from the edge of the gate electrode. Equation (4) is the basic relationship between surface potential, material, and circuit parameters after charge transfer into the well has taken place.

1.2 Drift Field Formulation (Self-Induced Drift)

There are several types of drift fields and we shall first consider the field-driven, diffusion like phenomena[2] by differentiation of equation (4)

$$\frac{-\partial \phi_s}{\partial y} = +E_d(y) = \frac{-1}{C_o + C_s}\frac{\partial \rho}{\partial y} \tag{5}$$

where the space-charge capacitance

$$C_s = K_s\, \epsilon_o / x_d \tag{6}$$

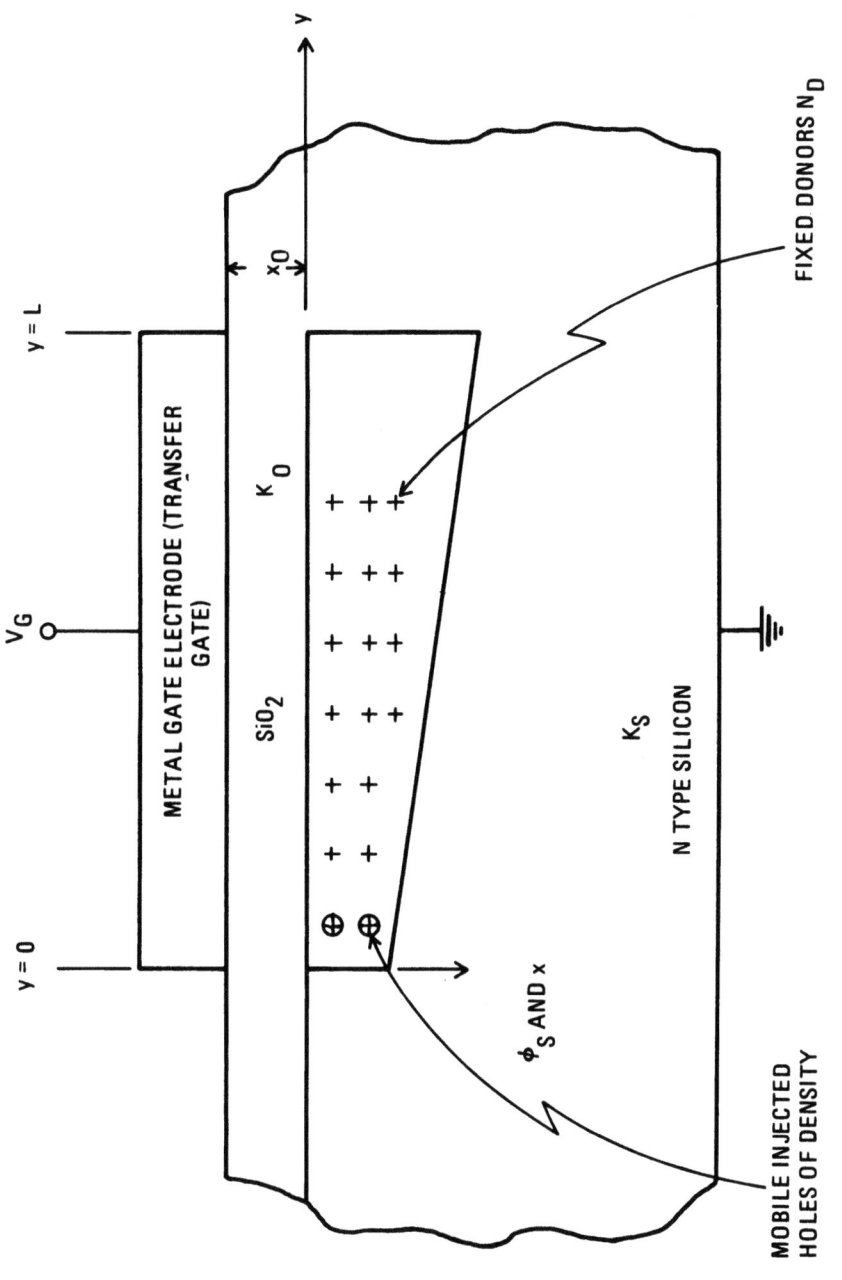

Figure 1. Injection of Charge Density Beneath Metal Gate Electrode

depends on the surface potential through x_d [i.e. see equation (3)]. In general $C_s \ll C_o$ such that the electric field (E_d) is dependent on the gradient of the injected carrier density ρ.

There are other contributions to the drift field which result from a net retarding field due to the electrostatic repulsion of the carriers.[3] The electrostatic repulsion of the carriers is reduced due to the small attractive field of the image force.

The repulsive field is given as

$$E_r(y) = \frac{2 x_o}{(K_o + K_s)\epsilon_o} \frac{\partial \rho}{\partial y} \tag{7}$$

1.3 Dynamics of Charge Transfer

The dynamics of charge transfer can be formulated through the current density equation,

$$j = q \mu p E_{SD}(y) - q D \frac{\partial p}{\partial y} \tag{8}$$

where p is the excess carrier density (holes/cm^3) and E_{SD} is the total drift field beneath the gate electrode. The current is obtained as follows,

$$i = W \int_0^{x_d} j\, dx$$

$$= W \bar{\mu} \left[E_{SD}(y) \rho - \frac{kT}{q} \frac{\partial \rho}{\partial y} \right] \tag{9}$$

where $\rho = q \int_0^{x_d} p\, dx$ and $\bar{\mu} = \frac{q}{kT} \bar{D} = \int_0^{x_d} \mu p\, dx \Big/ \int_0^{x_d} p\, dx$. The continuity equation for charge transfer is,

$$\frac{\partial j}{\partial y} + q \frac{\partial p}{\partial t} = G - R \tag{10}$$

where G and R are the generation and recombination rates respectively.

Combining equations (9) and (10) yields,

$$\frac{\partial \rho}{\partial t} = -\frac{\partial}{\partial y}(\bar{\mu}\rho E) + \int_0^{X_d}(G-R)\,dx \tag{11}$$

where ρ is the surface charge density in coulombs/m^2, $\bar{\mu}$ the effective surface mobility (m^2/volt-sec), y the direction of charge transport, and E the total tangential surface electric field (volts/m). If we assume under the gate electrode the generation is zero and neglect charge trapping (recombination), then equation (11) reduces to:

$$\boxed{\frac{\partial \rho}{\partial t} = -\frac{\partial}{\partial y}(\bar{\mu}\rho E)} \tag{12}$$

The surface tangential field is composed of the following terms:

$$E = E_s + E_{th} + E_F \tag{13}$$

where surface electric field is given as

$$E_s = E_r + E_d = -S\frac{\partial \rho}{\partial y} \tag{14}$$

with
$$S = \frac{1}{C_{eff}} = \frac{1}{C_0 + C_s} - \frac{2X_0}{(K_0 + K_s)\epsilon_0} \tag{15}$$

The thermal field E_{th} is given by,

$$E_{th} = -\frac{kT}{q}\frac{1}{\rho}\frac{\partial \rho}{\partial y} \tag{16}$$

The fringe field E_F is determined by the surface electrode and geometry to be discussed in section 1.5.

1.4 Boundary Conditions

Figure 2[4] shows the configuration of the potential well under the gate electrode during charge transfer. The configuration of the well determines the boundary conditions which must be applied to equation (12) to obtain a proper solution for the charge transfer characteristics for the CCD structure. The two assumptions are

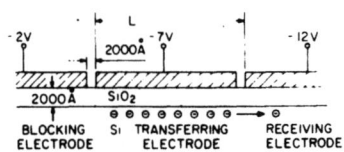

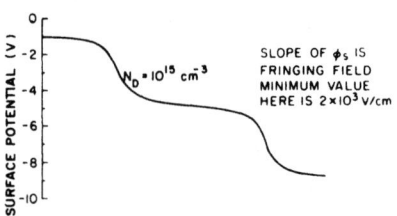

Figure 2. Cross Sectional View of Electrode Geometry for the Unit Cell Used in Charge Transport Analysis. $L = 4 \mu m$ and $N_D = 10^{15}$ cm^{-3}. [After J.E. Carnes, W.F. Kosonocky, and E.G. Ramberg, "Free Charge Transfer in Charge-Coupled Devices," *IEEE Trans. Electron Devices ED-19*, p. 798, (1972)] (Ref. 4)

a. The current flow is in only one direction, that of increasing y, and there is no current flow across the boundary at y = 0. This is expressed mathematically as,

$$i\big|_{y=0} = 0$$

The current density is given by equation (9) as:

$$i = W \bar{\mu} \rho E \tag{17}$$

where $E = -(S + kT/q\rho) \dfrac{\partial \rho}{\partial y} + E_F$

Evaluating (17) at y = 0 yields,

$$\left.\frac{\partial \rho}{\partial y}\right|_{y=0} = \left[\frac{E_{FR}}{(S + kT/q\rho)}\right]_{y=0} \quad t \geq 0 \quad (18)$$

b. There is no charge buildup at the right boundary (y = L), so that this boundary acts as an infinite sink for charge. This results in the boundary condition,

$$\rho(L, t) = 0 \text{ for all } t \quad (19)$$

In addition to the two boundary conditions above, an initial condition is assumed as follows:

c. The initial charge distribution is uniform with magnitude,

$$\rho(y, 0) = \rho_0 \quad y < L \quad (20)$$

The inclusion of a nonuniform fringe field for which the first partial of the field is nonzero requires that a charge conservation scheme be applied to the computed values of ρ. The effects of the nonuniform field are two-fold. When the first partial is positive, it aids the transfer of charge from the storage gate. When it is negative it retards charge transfer, and tends to accumulate charge. The inclusion of charge conservation ensures that a negative first partial does not act as a charge source. It is implemented in the model by restricting the value of ρ at any point to a quantity of charge equal to the total charge stored, minus the charge stored at all preceding points, minus the charge transferred from the point in question.

1.5 Fringe Field Analysis

In order to obtain an accurate solution to the charge transfer problem the electrical fringe field must be analyzed for the unit cell. The lateral fringe field along a semiconductor-oxide interface in the near region of charged plates can be calculated from the two-dimensional Poisson-Boltzmann equation. The solution to this equation can be obtained numerically for the region as delineated by the physical structure of the charge-coupled device. In the semiconductor the Poisson equation becomes

$$\nabla_x^2 \phi = - \frac{q}{K_s \epsilon_0} N_D \quad (21)$$

where N_D is the ionized donor concentration in a region near the surface of the semiconductor depleted of mobile charge carriers. In the oxide the descriptive equation becomes

$$\nabla_x^2 \phi = \frac{Q_{ss}}{K_o \epsilon_o} \delta (X - X_s) \tag{22}$$

where X_s is the interface defining vector for which the oxide charge is nonzero. With suitable boundary conditions these equations, Poisson's equation for the semiconductor and oxide respectively, can be solved for the potential, ϕ, in the region of interest Ω.

The solution to the Poisson equation in a region Ω can be obtained by a number of satisfactory numerical techniques.[5] An efficient system for the condition of simple boundary conditions is an explicit finite difference method employing successive overrelaxation. This method has been applied to the region Ω, which has been segmented into m x n rectangular cells with sides h and K_j. With this notation the potential at any node becomes:

$$\phi_{i,j} = M^2 \left\{ \frac{\phi_{i+1,j} + \phi_{i-1,j}}{2h^2} + \frac{K_{j-1}\phi_{i,j+1} + K_j\phi_{i,j-1}}{K_j K_{j-1}(K_j + K_{j-1})} + \frac{\delta_{i,j}}{2\epsilon_{i,j}} \right\} \tag{23}$$

where

$$M^2 = \frac{h^2 K_j K_{j-1}}{h^2 + K_j K_{j-1}}$$

The solution is approached more rapidly by using a linear combination of potentials at succeeding relaxation points. Thus, the potential at iteration n becomes:

$$\phi_{i,j}^{(n)} = (1 - \alpha) \phi_{i,j}^{(n-1)} + \alpha \phi_{i,j}^{(n)} \tag{24}$$

where α is the relaxation factor determining the rate of converence. For a rectangular region Ω the optimum relaxation factor can be analytically determined as:

$$W = 1 + \left\{ \frac{\tau_L}{\left[1 + (1 - \tau_L)^{1/2} \right]^2} \right\} \tag{25}$$

$$\tau_L = \cos^2 \left[\frac{\pi (p^2 + 1/2\ q^2)^{1/2}}{p\ q} \right] \tag{25}$$

For the nonrectangular region defined by the stepped gates and varying depletion layer of the device under consideration, W can be approximated by equation (25) with pq equal to the number of mesh points in the solution.6

The region Ω is divided into approximately 50 by 50 cells. The nodes are spaced equally in the direction parallel to the semiconductor surface and quadratically in a direction normal to the surface. The maximum mean error in potential is decreased to less than 10^{-2} after less than 10^2 iterations. The surface field is then calculated from the known surface potentials. Figure 3 illustrates the tangential fringing field[7] $E_y = E_F$ versus distance along the gate electrode. A linear approximation to $1/E_F$ is also indicated by the dotted line. This approximation is useful in the evaluation of a single carrier transit time

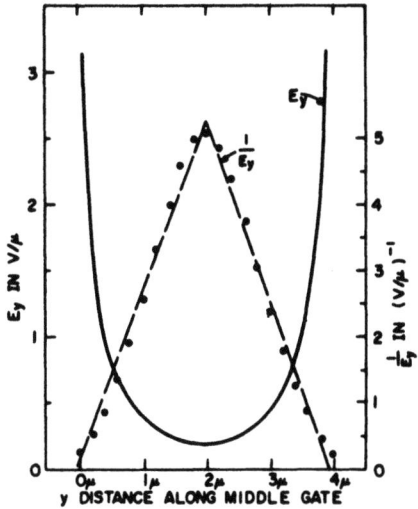

75-0440-VA-3

Figure 3. Silicon Surface Fringing Field $E_y = E_F$ versus Distance Along the Transfer Gate (See figure 2). A Linear Approximation to $1/E_F$ is Indicated by the Dotted Line. [After J.E. Carnes, W.F. Kosonocky, and E.G. Ramberg, "Drift-Aiding Fringing Fields in Charge-Coupled Devices," IEEE Trans. Devices SC-6, p 326 (1971)] (Ref. 7)

$$\tau_f = \frac{1}{\mu} \int_0^L \frac{dy}{E_F(y)} \cong \frac{L}{2\bar{\mu} E_F^{min}} \tag{26}$$

Figure 4 illustrates an approximation for E_F^{min} and the single transit line τ_f as a function of substrate doping density N_D. The minimum value of fringe field may be written in the form,[7]

$$E_F^{min} = \frac{2\pi}{3} \frac{VK_s\epsilon_0}{L^2 C_0} \left[\frac{5 X_d/L}{1 + 5X_d L}\right]^4 \tag{27}$$

where V is the potential difference between adjacent electrodes. In the time τ_f the spatial charge distribution becomes stationary and decays thereafter with a time constant of approximately $1/3\ \tau_f$. A bulk channel CCD, operating under low charge conditions (i.e., $\rho \ll N_D d$, where d is the depth of the channel) has an effective capacitance which is the series combination of the oxide capacitance C_0 and the channel capacitance $C_{ch} \simeq K_s\epsilon_0/d$. For ion-implanted, so-called buried channels, $d \simeq 0.5\ \mu m$ and the

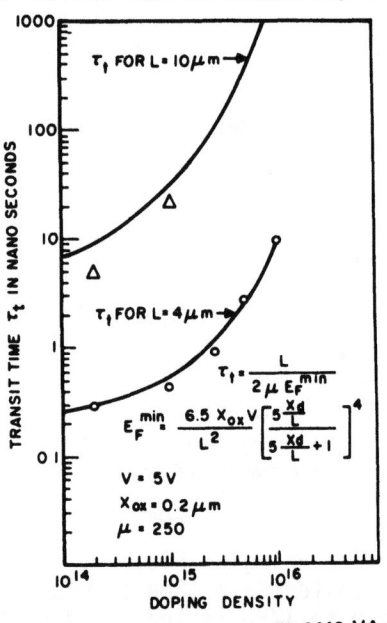

Figure 4. Single Carrier Transit Time τ_f Based on Fringe-Field Drift vs Doping Density of the Substrate [After Carnes, Ref. 7]

effective capacitance is not significantly different from C_0; however, for deep bulk channel CCD's (i.e., peristaltic CCD's) the $d \simeq 5 \ \mu m$ and the effective capacitance is significantly reduced which results in a large increase in $E_F{}^{min}$.

$$E_F{}^{min} \text{ (Bulk Channel)} \cong \left[1 + \left(\frac{K_0 d}{K_s X_0}\right)\right] E_F \text{ min (Surface Channel)} \quad (28)$$

1.6 Computer Simulation of Charge Transport

If we use the charge transfer structure of figure 2 and the boundary conditions discussed in section 1.4, then the charge profiles may be determined for various times as shown in figure 5. The fringe fields dominate the charge transfer process after 800 psec, in this example, and the effects of self-induced drift are negligible. We notice the charge profile tends to move to the right (i.e., in the direction of the receiving or storage electrode); however, instead of a continuous drift to the right, the charge profile becomes stationary after 1400 psec and decays exponentially afterwards. The reason for this stationary process is the large accelerating fringe field near the boundary between the transfer and receiving electrodes. We should also notice the

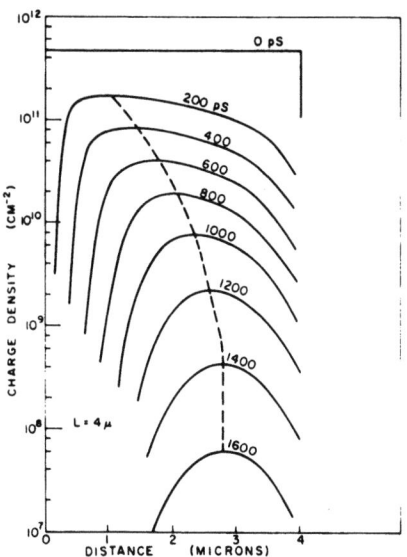

Figure 5. Charge Profiles for Various Times and for the Problem of Figure 2. [After Carnes, Ref. 4]

retarding and accelerating fringe fields tend to contain the charge packet near the center of the transfer gate. In the early part of the charge transfer the self-induced drift may dominate for large charge densities (i.e., $\rho/q > 10^{10}$ cm^{-2}); however, within a nanosecond, the effects of thermal diffusion and fringe fields determine the final charge transfer. Thus, approximately 99 percent of the charge may be transferred by self-induced drift, but to achieve 99.99 percent transfer efficiency (i.e., transfer inefficiency $\epsilon = 1 \times 10^{-4}$), the thermal diffusion and fringe fields are important. Fringe fields enable 99.99 percent transfer efficiencies (in the absence of trapping effects) to be obtained in several nanoseconds. The absence of fringe fields means the final transfer process is done by thermal diffusion with typical times of several hundred nanoseconds.

2. CHARGE-CONTROL ANALYSIS OF CHARGE TRANSPORT AND TRAPPING[8]

In this section we will concentrate on the development of a closed-form solution to charge transport in CCD's with the aid of the charge-control method of analysis. We will develop the transfer inefficiency as a function of device parameters and a phenomenological trapping time constant τ_s.

2.1 Introduction

The classical approach of charge-control theory is to integrate the continuity equation [see equation (10)] over the length L of the discharging potential well or transfer gate. Thus, we obtain

$$\frac{dN_{TOT}}{dt} = \frac{-N_{TOT}}{\tau_s} - i(L,t) + i(o,t) \qquad (29)$$

where

$$N_{TOT} = \int_0^L \rho(y,t)\, dy \qquad (30)$$

Thus, the transient behavior of the total charge under the transfer gate is determined by boundary conditions to the current and the model used to represent charge trapping as indicated phenomenologically by the storage time constant τ_s. We should also note there is no restriction to the particular form of the charge distribution $\rho(y,t)$ under the transfer gate. The boundary conditions to be employed are

$$i(o,t) = 0 \longrightarrow \left.\frac{\partial \rho}{\partial y}\right|_{y=0} = 0$$

$$\rho(L,t) = 0 \tag{31}$$

where fringe fields have been momentarily neglected (later to be considered as part of an effective diffusion constant). The current remains finite at the right boundary since

$$i(L,t) = \lim_{y \to L} V_s \rho(y,t) \tag{32}$$

where V_s, the maximum saturation limited velocity of carriers, $\to \infty$ as $y \to L$. If we combine the boundary conditions of equation (31) with equations (17) and (29), then we have

$$\frac{dN_{TOT}}{dt} = \frac{-N_{TOT}}{\tau_s} + \left(\bar{D} + \frac{\bar{\mu}\rho}{C_{eff}}\right)\left.\frac{\partial \rho}{\partial y}\right|_{y=L} \left[1 + \left.\bar{\mu}E_F \rho\right|_{y=L}\right] \tag{33}$$

in the absence of fringe fields. The concentration dependent diffusion term $\rho \frac{\partial \rho}{\partial y}$ must be evaluated at the boundary $y = L$. A first order approximation sets the charge gradient induced drift current at $y = L$, at any instant of time, equal to its spatial average under the transfer gate:

$$\left.\rho(y,t)\frac{\partial \rho(y,t)}{\partial y}\right|_{y=L} = \frac{1}{L}\int_0^L \rho(y,t)\frac{\partial \rho(y,t)}{\partial y}dy = -\frac{[\rho(o,t)]^2}{2L} \tag{34}$$

2.2 Solution of the Charge-Control Model

The postulate of instantaneous charge redistribution or stationary charge profiles may be used to seek a separation of variables solution in the absence of fringe fields. We note the charge profiles of figure 5 and the stationary form of the charge density within a short time interval. Let us use the substitution

$$\rho(y,t) = \phi(y)f(t) \tag{35}$$

in equations (33) and (34) to obtain[8]

$$\frac{dN_{TOT}}{dt} = -\frac{N_{TOT}}{\tau_s} + \frac{\bar{D}\phi'(L)N_{TOT}}{N_{TOT}(0)} - \frac{\bar{\mu}\phi^2(0)N_{TOT}^2}{2L\,N_{TOT}^2(0)\,C_{eff}} \qquad (36)$$

where

$$N_{TOT}(0) = \int_0^L \phi(y)\,dy \qquad (37)$$

Equation (36) is identical in form to the rate equation for the bimolecular recombination processes in photoconductivity. Since this equation is an ordinary differential equation, we can determine the solution as,

$$F(t, \tau_s) = \frac{N_{TOT}(t)}{N_{TOT}(0)} = \frac{\exp(-Kt/t_{tr})}{1 + \frac{q_{av}L^2\phi^2(0)}{C_{eff}^2 N_{TOT}^2(0)\,K}\left[1-\exp(-Kt/t_{tr})\right]} \qquad (38)$$

where

$$t_{tr} = \frac{L^2}{\bar{\mu}\,U_0} \qquad U_0 = 1 \text{ volt}$$

$$K = \frac{t_{tr}}{\tau_s} - \frac{\bar{D}\,L^2\phi'(L)}{\bar{\mu}\,N_{TOT}(0)} \qquad (39)$$

and $\frac{q_{av}}{C_{eff}}$ is the average potential well level to be transferred.

If trapping effects cannot be neglected, then we write

$$1-\epsilon = -\frac{W}{N_{TOT}(0)}\int_0^t i(L,t)\,dt = -\int_0^t \frac{d}{dt}\frac{N_{TOT}(t)}{N_{TOT}(0)}\,dt - \frac{1}{\tau_s}\int_0^t \frac{N_{TOT}(t)}{N_{TOT}(0)}\,dt$$

$$\epsilon = F(t,\tau_s) + \frac{1}{\tau_s}\int_0^t F(t,\tau_s)\,dt \qquad (40)$$

which is the transfer inefficiency. For no trapping effects we let $\tau_s \to \infty$.

2.3 Stationary Charge Distribution[8] $\phi(y) = \cos\frac{\pi y}{2L}$

If we assume a stationary charge distribution of the above type, then substitution of $\phi(y)$ into equations (38) and (40) yields,

$$\epsilon = \frac{\exp(-Kt/t_{tr})}{1+\frac{\pi^2}{8K}\left(\frac{q_{av}}{C_{eff}}\right)\left[1-\exp(-Kt/t_{tr})\right]} + \frac{8t_{tr}\left\{1+\frac{\pi^2}{8K}\left(\frac{q_{av}}{C_{eff}}\right)\left[1-\exp(-Kt/t_{tr})\right]\right\}}{\pi^2\left(\frac{q_{av}}{C_{eff}}\right)\tau_s} \qquad (41)$$

for the charge transfer inefficiency. The value of K is determined from equation (39) and $\phi(y)$ as

$$K = t_{tr}\left(\frac{1}{\tau_{th}} + \frac{1}{\tau_s}\right) \qquad (42)$$

where $\tau_{th} = 4L^2/\pi^2\bar{D}$ is the thermal time constant of charge decay. Figure 6 illustrates a comparison of the transfer inefficiencies as calculated by the charge-control method [i.e. equation (41)] versus the numerical solution of equation (12) with the Crank-Nicholson technique. The comparison is performed for negligible trapping ($\tau_s \to \infty$) and no fringe field ($E_F = 0$) with the assumed charge distribution of this section.

2.4 Effects of Electrical Fringe Field[8]

In this section we will estimate a fringe field enhanced diffusion constant from equations (33) and (34) as follows:

$$\bar{D}\phi'(y) + \bar{\mu}\phi(y)E_F = \left[\bar{D} + \bar{\mu}E_F\frac{\phi(y)}{\phi'(y)}\right]\phi'(y)$$

$$= \bar{D}_{eff}\,\phi'(y) \qquad (43)$$

where $\bar{D}_{eff}$ may be evaluated near the middle of the transfer gate $y = L/2$ for the distribution of section 2.3.

$$\bar{D}_{eff} = \bar{D} + \frac{2\bar{\mu}LE_F}{\pi} \qquad (44)$$

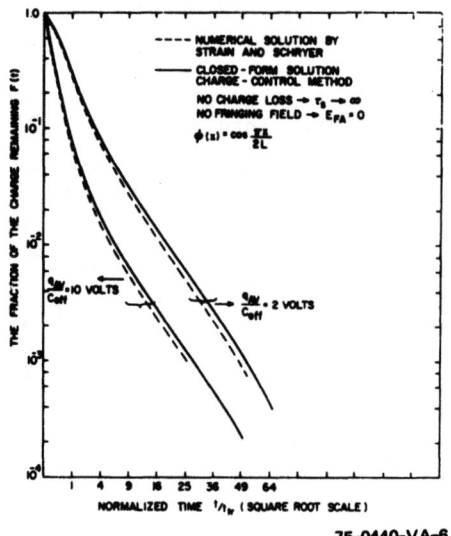

Figure 6. The Fraction of the Charge Remaining Under the Transfer Gate (i.e. ϵ the transfer inefficiency) as a Function of Normalized Time t/t_{tr}. A Comparison of Two Methods to Illustrate the Accuracy of the Charge-Control Technique. [After H.S. Lee and L.G. Heller, IEEE Trans. Electron Devices ED-19, p. 1270 (1972)] (Ref. 8)

which gives a modified K factor

$$K = \frac{\pi}{2} L E_F + t_{tr}\left(\frac{1}{\tau_{th}} + \frac{1}{\tau_s}\right) \tag{45}$$

to be used in conjunction with equations (41) and (27) for the calculation of ϵ under the influence of electrical fringe field. Figure 7 illustrates the effect of fringe field on ϵ and the improvement in charge transfer is apparent.

2.5 Influence of Charge Trapping[8]

A phenomenological approach has been taken in this chapter in which the trapping effects have been lumped into a single time constant τ_s. There are, in general, simultaneous capture and emission processes which occur during charge transfer. Based on

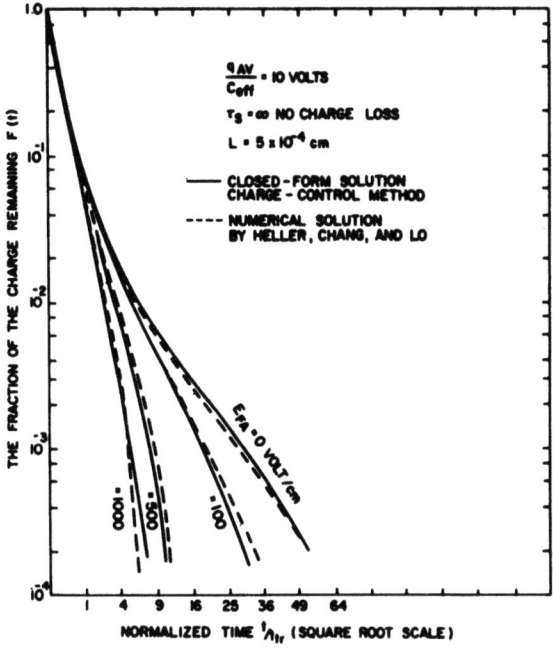

Figure 7. The Effect of Fringing Field E_F on the Fraction of Charge Remaining Under the Transfer Gate (i.e. є the Transfer Inefficiency) [After Lee and Heller, Ref. 8]

the simplified model described, we have assumed τ_s to be independent of the amount of trapped charge; however, we are aware of the use of a background charge or "fat zero" to increase the effective time constant τ_s. In order to understand this mechanism, we must write the coupled differential equations for a single-level trapping state,

$$\frac{dN_{TOT}}{dt} = -\frac{dN_S}{dt} + i(L,t) \tag{46}$$

$$\frac{dN}{dt} = \frac{N_{TOT}}{\tau_c} - \frac{N}{\tau_e} \equiv \frac{N_{TOT}}{\tau_s} \tag{47}$$

where the capture and emission time constants are given as

$$\tau_c = \frac{1}{\sigma V_t / X_i \left[N_s - \frac{N}{LW} \right]} \tag{48}$$

$$\tau_e = \frac{1}{\sigma V_t N_c} \exp\left[E_s/kT\right] \tag{49}$$

where N_s and E_s are the density and energy level of the trapping center, σ the capture cross section, V_t the thermal velocity, X_i the thickness of the inversion layer, and N_c the density of conduction band states (assuming electron transport). If we examine equation (47) we have

$$\frac{1}{\tau_s} = \frac{\sigma V_t}{X_i}\left[N_c - \frac{N}{LW}\right] - \frac{N}{\tau_e N_{TOT}} \tag{50}$$

and the time constant τ_s increases as the number of trapped charges N increases. Figure 8 illustrates the effect of τ_s on the tranfer inefficiency and we see when $\tau_s \longrightarrow t_{tr}$. the carrier transit time, the charge trapping effects degrade severely the transfer process. On the other hand, the fringe fields accelerate the charge transfer process and counteract the effects of trapping. Finally, if the density of trapping centers is high, then a so-called "fat zero" may be inserted to increase the trapping time constant.

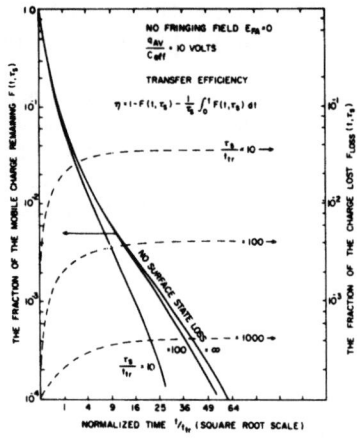

Figure 8. The Effect of the Trapping Time Constant on the Fraction of Charge Remaining Under the Transfer Gate (ϵ the transfer inefficiency) [After Lee and Heller, Ref. 8] The Total Transfer Inefficiency is the Sum of the Solid and Dotted Curves for a Given τ_s/t_{tr}.

REFERENCES

1. White, M.H., MOS Transistors, *Semiconductors and Semimetals*, Vol. 7, Applications and Devices Part A, New York, 1971.

2. Engeler, W.E., Tiemann, J.J., and Baertsch, R.D., Surface Charge Transport in Silicon, *Appl. Phys. Lett.*, 17,469, 1970.

3. Strain, R.J., and Schryer, N.L., A Nonlinear Diffusion Analysis of Charge-Coupled Device Transfer, *B.S.T.J.*, 50, 1721, 1971.

4. Carnes, J.E., Kosonocky, W.F., and Ramberg, E.G., Free Charge Transfer in Charge-Coupled Devices, *IEEE Trans. Electron Devices*, ED-19, 798, 1972.

5. Young, D., Iterative Methods for Solving Partial Difference Equations of Elliptic Type, *Trans. American Math. Society*, 76, 1954.

6. Young, D., *Survey of Numerical Analysis*, McGraw-Hill, 1963.

7. Carnes, J.E., Kosonocky, W.F., and Ramberg, E.G., Drift Aiding Fringing Fields in Charge-Coupled Devices, *IEEE J. Solid-State Circuits*, SC-6, 322, 1971.

8. Lee, H.S., and Heller, L.G., Charge-Coupled Method of Charge-Coupled Device Transfer Analysis, *IEEE Trans. Electron Devices*, ED-19, 1270, 1972.

CHARGE TRANSPORT WITH TRAPS

G. F. Amelio, R. H. Dyck

Research and Development Laboratory
Fairchild Camera and Instrument Corporation
4001 Miranda Avenue, Palo Alto, Ca. 94304

ABSTRACT. In a real charge-coupled device there exists some finite density of allowed levels within the forbidden gap which can interact with the free carriers which comprise the signal. The filling and emptying of such traps distorts the signal in a way which depends on the trap energy and density. In addition, the signal-to-noise ratio is reduced. As a consequence, any reasonably rigorous consideration of charge transport must include treatment of traps. This paper will consider the effect of traps on charge transport and derive quantitative expressions for their effect on transfer efficiency and the signal-to-noise spectral density.

1. INTRODUCTION

A surface channel CCD (SCCD) is of limited use without a means for coping with surface state trapping effects. This is even true for devices made by the best processing techniques which give $<10^{10}$ effective surface states/cm^2. A buried channel CCD (BCCD) can, on the other hand, have such a low trap density that charge packets of the order of 10 electrons in size can be transported efficiently.

2. SCCD TRANSPORT WITH TRAPS

2.1 Model: 2ϕ, overlapping gates.
Barriers may be formed by implantation, stepped-

oxides, or by externally applied potentials.

2.2   Example:

Clock swing = 10 V.
Barrier height = 5 V.

$$n_{SAT} = \left(\frac{C_o}{A}\right)\frac{\Delta V}{q} = \frac{3.8 \times 8.85 \times 10^{-14} F/cm (5V)}{(10^{-5} cm) \quad 1.6 \times 10^{-19}}$$

$$\cong 1 \times 10^{12}/cm^2$$

2.3   Fat Zero

Definition: an injected background signal charge level. The use of a fat zero is appropriate to digital shift registers, analog shift registers and to imagers. In the last case, bias light may be used to provide a fat zero.

Except for edge effects, if $n_{FZ} \stackrel{\sim}{=} n_T$ the traps will not have a chance to empty for more than half an element time and will thus be kept nominally deactivated.

Edge effects can be large even where $n_{FZ} \stackrel{\sim}{=} n_T$ due to the fact the smaller charge packets are smaller in physical dimension.

2.4   Treatment by Mohsen et al (1973)
        (2∅, p-channel)

Trap Kinetics

The change in the filled-trap population is given by the difference between the capture and the emission rates:

$$\frac{dn_{ss}}{dt} = K_1 (N_{ss} - n_{ss})p - K_2 n_{ss} \exp(-E/KT) \quad (1)$$

where

$K_1 = \sigma_h V_{th}/d$

$K_2 = \sigma_h V_{th} N_v$

and d is the average thickness of the inversion layer at the interface.

When charge-coupled devices are operated with the circulating background charge, interface states having an emission time constant larger than the cycle time remain almost completely filled all the time. These states capture carriers every cycle and do not get a chance to reemit an appreciable fraction of the captured carriers during the cycle time. Interface states with an emission time constant much less than the cycle time will be emptying and filling every cycle. Hence, the interface states that make a substantial contribution to the incomplete transfer will be those with a time constant of the order of the clock cycle period and will lie within an energy range of the order of the thermal voltage.

Trap occupation in steady state and in transition

In steady state, the trap occupation can be obtained from (1) and is given by

$$n_{ss} = \frac{N_{ss}}{1 + \frac{K_2 \exp(-E/KT)}{K_1 p}} \tag{2}$$

The interface states are in equilibrium with the mobile carriers. Their occupation is described by the same quasi-Fermi level as the mobile carriers.

$$E_1 = KT \ln \frac{K_2}{K_1 p} = KT \ln \frac{N_v d}{p} \tag{3}$$

Following a change in the mobile carrier concentration, the trap occupation changes to the new steady-state value corresponding to the new mobile carrier concentration p_1 with an effective time constant given by

$$\tau_{eff} = \frac{1}{K_1 p_1 + K_2 \exp(-E/KT)} \tag{5}$$

If the effective time constant of the interface states τ_{eff} is smaller than the time constant τ measuring the variation of the mobile carrier density, then the trap occupation reaches steady state very early in the transition. That is, if $\tau > \tau_{eff}$, then

$$n_{ss}(t) = N_{ss} \left[1 + \frac{K_2 \exp(-E/KT)}{K_1 p(t)} \right] \tag{6}$$

Thus, the quasi-Fermi levels of the traps follows the quasi-Fermi level of the mobile carriers

$$E_f(t) = KT \ln \frac{N_v d}{p(t)} \tag{7}$$

On the other hand, if $\tau < \tau_{eff}$, then the trap occupation fails to follow the variation of the mobile carrier. If we let $K_1 p(t) \gg K_2 \exp(-E/KT)$, then this occurs when the mobile carrier density falls to a level such that

$$K_1 \tau p(t) < 1. \qquad (8)$$

For charge transfer from under a gate, we can define two regimes. First, when $K_1 p(t)\tau > 1$, the mobile charge is in effective equilibrium with the trapped charge. The total number of trapped carriers, p_{tr} is given by

$$p_{tr}(t) = N_{ss} \left[E_g - KT \ln \frac{K_2}{K_1 p(t)} \right] \qquad (9)$$

Second, when $K_1 p(t)\tau < 1$, the mobile charge is no longer in equilibrium with the trapped charge. If we let t_4 be the time emission mechanism becomes dominant, then for $t > t_4$ the trap occupation is given by

$$n_{ss}(t) = \frac{N_{ss}}{1 + \frac{K_2 \exp(-E/KT)}{K_1 p(t_4)}}$$

$$\cdot \exp\left[-(t-t_4) K_2 \exp(-E/KT) \right] \qquad (10)$$

and the interface states start to empty with a time constant that increases exponentially with the trap energy. The total number of trapped carriers is given by

$$p_{tr}(t) = N_{ss} \left[E_g - KT \ln K_2(t-t_4) - \frac{KT}{(t-t_4)K_1 p(t_4)} \right]$$
$$t > t_4 \qquad (11)$$

So in this case, the interface states above $E_f = KT \ln K_2(t-t_4)$ are almost full and those below it are nearly empty. The last terms in (10) and (11) show the dependence of the interface state occupation on the mobile carrier density.

3. BCCD TRANSPORT WITH TRAPS

3.1 Differences from the SCCD case:

(1) Traps are localized at one or a few energy levels, and
(2) Trapping effects are of a much smaller

magnitude.

4. Large Signal BCCD Analysis of Mohsen & Tompsett (1974)

- Model: 3∅ n-channel
- $\tau_e \gg \tau_t$
- At low f, $q_t \simeq q N_t V_s$
- At intermediate f's, and no $n_{F\cancel{B}}$, q_t

depends on number of zero-level bits preceding.

Real devices always have a finite background current level, the dark current or thermally generated current. However, in any given device and operating conditions, this ranges from $n_{F\cancel{B}}$ (OUT) down to zero.

At 25°C and with good processing $j_D = 10$ nA/cm². Then

$$\frac{n_{F\cancel{B}} (OUT)}{A cm^{-2}} = (10^{-8} A/cm^2) \frac{t_I}{q} = 6 \times 10^{10} t_I (sec^{-1})$$

Analogous to the edge effect in the SCCD, here a charge packet changes size in 3 coordinates resulting in edge effects over the entire surface area of the charge packet. In a BCCD with $X_J \sim 1$ μm the vertical dimension of a charge packet can range from ~ 0.5 μm on the high end to ~ 0.1 μm on the low end. This low end is determined by the $\frac{kT}{q}$ isopotential relative to the empty channel potential. Thus there is $\sim 5X$ change in size. In a BCCD in general the lateral fringe field effects are greater than in SCCD's. Thus the $\frac{kT}{q}$ isopotential width can be as narrow as 1 μm with a 10 μm wide channel. This leads to large discrepancies between simple models and actual situations. A total volume range of 500:1 is typical.

4.1 Example:

Assume a string of 1% fat zeroes so that all traps in at least 90% of the well volume become empty. Consider the first signal packet at near the maximum signal level.

Since the carrier density in this packet is approximately N_A,

$$\varepsilon \stackrel{\sim}{=} \frac{N_T}{N_A}$$

4.2 Experimental Results

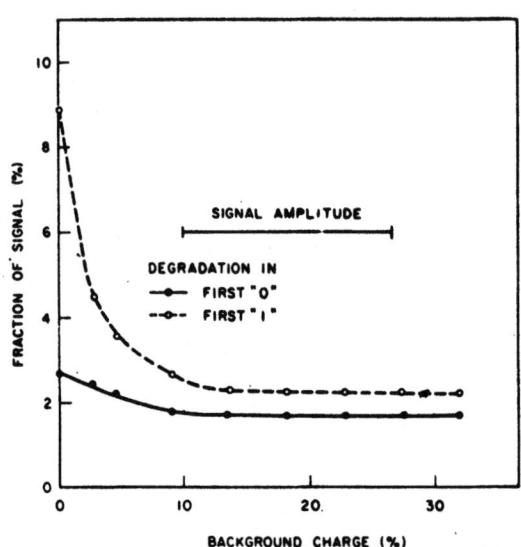

Fig. 1 Charge deficit of first "one" after 4962 "zeros" and charge excess in first "zero" following a group of "ones" in a buried channel device operated at 1-MHz clock frequency as a function of background charge. (Ref: Mohsen & Tompsett 1974)

5. Small-Signal Buried Channel Case

For signals of $\stackrel{<}{\sim} 10^{-3}$ of saturation the localized charge density becomes $<< N_A$ and the possibility of reduced transfer efficiency needs to be considered. In L^3 TV applications it is desired to have $\varepsilon < 10^{-3}$ at signal levels of $\sim 10^{-5}$ of saturation. Cooling is essential to have dark signal low enough.

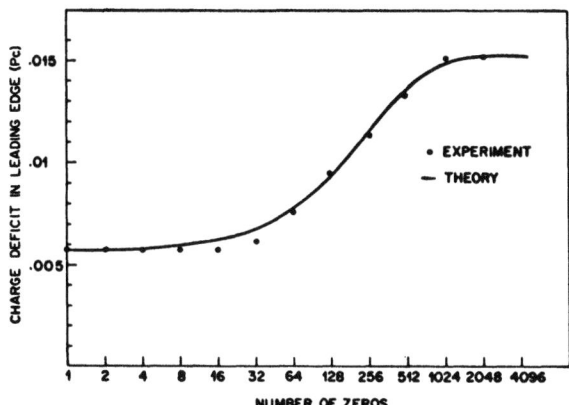

Fig. 2 Charge deficit in the first "one" in picocoulombs for a buried channel device operated at 1-MHz clock frequency as a function of the number of "zeros" between the group of "ones". The solid curve is calculated from (5). (Ref: Mohsen & Tompsett 1974)

5.1 J. M. Early's treatment (unpublished)

Model
Assumes no significant neutralization of the depletion region by the signal carriers.

The trapping probability is position dependent around the potential minimum.

For a given set of device design parameters a worst case trap energy is deduced.

Poisson statistics are used to accurately average the individual trapping events.

$T = -50°C$

Results
At $N_T = 10^{11}/cm^3$, 15 KHz operation, and the worst case trap energy of close to 0.36 eV, the transfer efficiency varies with free charge level as shown in Figure 3.

6. Experimental Results

A 256 photoelement linear imaging device with a two-phase, oxide-gap, overlapping gate electrode, CCD transfer register was selected for the experiments. The device was cooled to obtain an average dark charge per pixel of 4 electrons. A light spot was focused on a photoelement remote from the on-chip amplifier; the photosignal was then transferred to the amplifier at a clock rate of 15.7 KHz. This signal was compared to that for the light spot focused on an element close to the amplifier. It was observed that a 15 $\pm$ 3 electron signal underwent 238 transfers with a loss of 1 $\pm$ 5 electrons. This result confirms the theoretical analysis that there is useful cumulative transfer efficiency after 500 transfers to detect photoelectronic signals at approximately the 10 electron level.

7. Experimental Comparison of SCCD and BCCD Linear Imagers.

See Fig. 4. It shows that with $\sim$ 20% bias light signal the BCCD is performing better after $\sim$ 200 stages of transfer ($\sim$ 600 transfers) than is the SCCD after only about 50 stages of transfer.

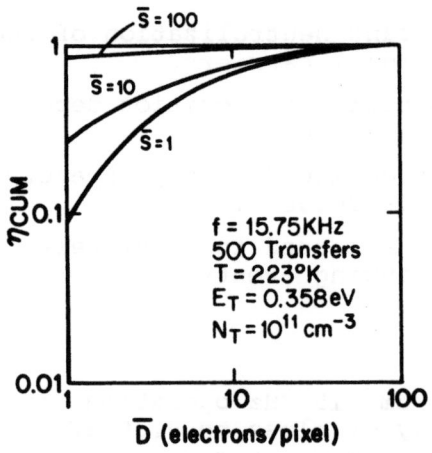

Fig. 3 Cumulative transfer efficiency vs. average dark charge per pixel for several values of average signal charge per pixel

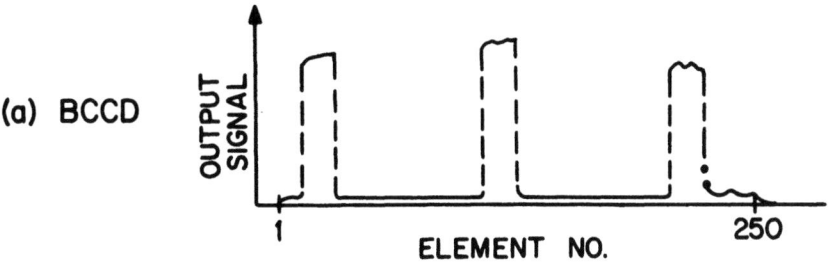

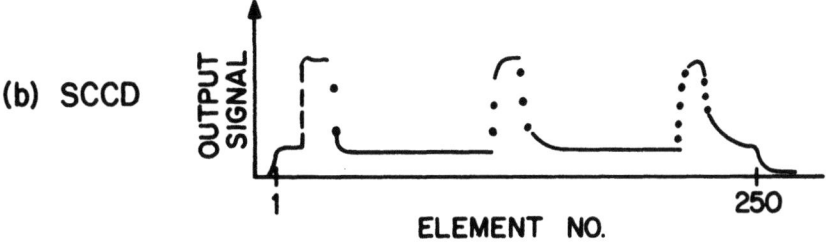

Fig. 4 A comparison of the image response of otherwise identical SCCD and BCCD 250-element linear imagers. The image is 3 white bars on a dark background. The highlight level is near saturation.

8. REFERENCES

1. M. F. Tompsett, "The Quantitative Effects of Interface States on the Performance of Charge-Coupled Devices," IEEE Trans. on Electron Devices ED-20 45-55 (Jan. '73)

2. A. M. Mohsen, et. al, "The Influence of Interface States on Incomplete Charge Transfer in Overlapping Gate Charge-Coupled Devices," IEEE J.S.S.C., SC-8, 125-138 (Apr. '73)

3. A. M. Mohsen and M. F. Tompsett, "The Effects of Bulk Traps on the Performance of Bulk Channel Charge-Coupled Devices," IEEE Trans. on Electron Devices, ED-21, 701-712 (Nov. '74).

IMAGE SENSORS USING SURFACE CHANNEL CHARGE-COUPLED DEVICES

C. H. Séquin

Bell Laboratories
Murray Hill, New Jersey 07974, U. S. A.

ABSTRACT. Several typical surface channel charge-coupled image sensors are described to illustrate the historical development and future potential of these devices. The special properties of these devices are discussed to clarify the advantages and limitations of SCCD's for solid-state imaging.

1. INTRODUCTION

In the first few years after the invention of charge-coupled devices the development of charge-coupled image sensors was mainly determined by the available technologies, and the main goal was to demonstrate the feasibility and potential of such devices. More recently, with many different technologies and a host of possible electrode structures available, the further development of these devices is mainly governed by performance and yield considerations. These devices will now have to prove that they can lead to implementations of all-solid-state cameras with a compactness or performance which cannot be achieved with more conventional devices, or, conversely, that they make possible the construction of such cameras at a lower price in those applications which can use lower resolution devices.

Some results achieved to date with surface channel CCD's and the promises for the near future will be reviewed.

2. FIRST APPROACHES USING SINGLE-LEVEL METALLIZATION

2.1 A 500-Element Linear Sensor

With the goal to demonstrate the potential of CCD's for high resolution linear image sensors, a 500-element device [1,2] was implemented at an early stage of development. It used a single illuminated transfer channel which was alternately kept in the integration mode and subsequently pulsed for serial readout. The light entered the silicon bulk through the gaps between the transfer electrodes. Single-level tungsten or aluminum metallizations were employed and crossings with the bus lines, which are topologically necessary in a 3-phase device, were formed by diffused crossunders. To obtain a high spatial density of sensor elements without jeopardizing yield because of small contact

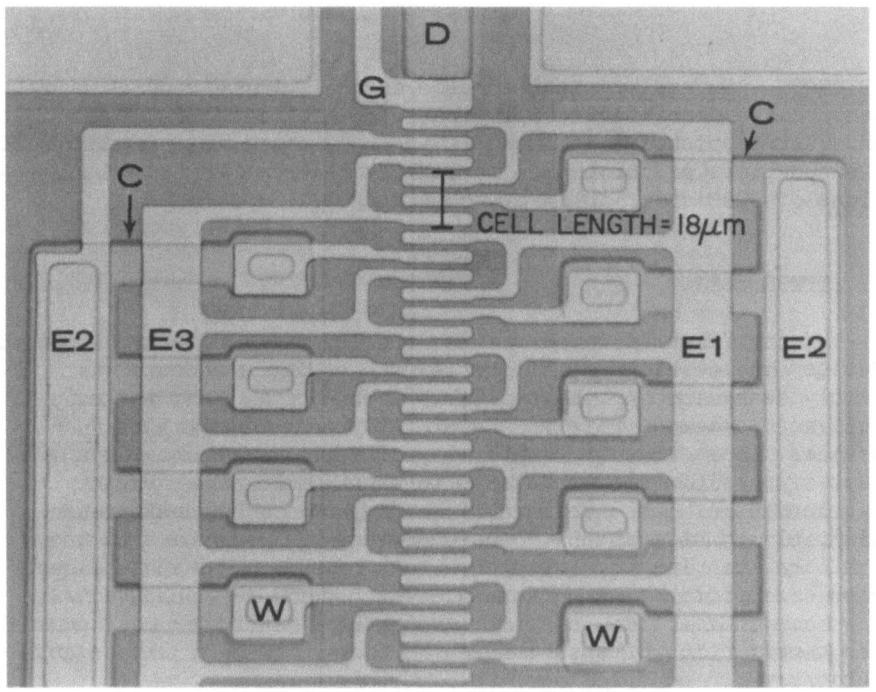

Fig. 1 One end of the linear 500-element 3-phase CCD. Shown are the three sets of transfer electrodes (E1, E2, E3), the input diode (D), and the input gate (G). Electrodes E2 are alternately contacted to diffused crossunders (C) through contact windows (W) lying on either side of the transfer channel. (Ref. 2)

windows, the diffused crossunders for Phase 2 were placed alternately to either side of the transfer channel. The other two phases formed a continuous metallization pattern. This resulted in the electrode structure shown in Fig. 1 with a cell length of only 18 µm. The whole device was symmetrical, having input/output diodes at either end, so that it could also be used as a delay line.

Typically the devices were fabricated on 20-40 ohm-cm p-type silicon. An n-type phosphorus diffusion (5×10^{14} cm^{-2}) provided input and output diodes as well as the crossunders. A lighter p-type boron diffusion (2×10^{14} cm^{-2}) defined the boundary for the transfer channel. After the removal of all the masking oxides used for the diffusions, a dry HCl gate oxide, ranging in thickness from 125 nm to 300 nm for different models, was thermally grown and annealed. Typical values for oxide charge and interface state density were 5×10^{10} cm^{-2} and 1×10^{10} cm^{-2}, respectively. After the contact windows were opened, a single level of a tungsten or aluminum metallization was deposited and defined. This latter step was the most critical process in the fabrication of these devices since the electrodes were nominally only 3 µm long and placed at 6 µm centers. Typically the final electrodes were about 3.5 µm long and thus separated by 2.5 µm gaps. However, the chemical etching especially of the tungsten electrodes was developed to high perfection, and the final electrode pattern often looked as clean as the photomask itself.

Computer modeling of the free carrier motion in such 3-phase electrode structures indicated very rapid and efficient charge transfer [3]. But in the best devices the transfer efficiency was limited by the interaction of the charge packets with interface states, which was not included in the computer modeling. The "edge effect" in this interaction with the interface states [4], which cannot be compensated by the use of a bias charge, was particularly strong in these devices because of the low substrate doping and the rather small potential wells. Transfer inefficiency in the clock range between 100 kHz and a few MHz was typically above 5×10^{-4}. On the other hand, high speed operation was demonstrated up to clock rates of 17 MHz.

However, most of these devices had transfer inefficiencies in the range between 10^{-2} and 10^{-3}. These devices seemed to work in the incomplete charge transfer mode. A potential barrier in the region of the bare gaps was produced by the fixed charge in the gate oxide or by a particular charging condition of the unprotected outer surface and retained a certain amount of residual charge in each transfer. Since the shape of this barrier is modulated by fringe fields and thus depends on the amount of signal charge present, low transfer efficiency resulted, even in the presence of a substantial artificially introduced bias charge

Fig. 2 Image obtained in line scan mode with a linear 500-element charge-coupled sensor which integrates the charge in the transfer channel.

on the order of 30%. In most devices the shape of these barriers and thus the transfer efficiency could be strongly affected by breathing onto the device and thereby changing the charge distribution on the outer surface. In most cases this process improved the performance of the device since the potential in the gaps would approach some kind of a time average of the potentials on the adjacent electrodes, and would thus fall in a range where the spurious barriers underneath can be suppressed by the fringe fields produced by sufficiently high clock voltage amplitudes.

In spite of these difficulties, the best devices produced very good images when used as linear sensors which were mechanically scanned across an original. In order to reduce the effects of optical smearing the integration to readout duty cycle was typically set at about 3:1. An image produced in this mode is shown in Fig. 2.

2.2 A 106×128-Element Area Sensor

The same technology was also used to implement a frame transfer device with 106 vertical transfer registers having a total of 128 elements each [5]. The device was provided with a serial input register in addition to the normal serial output register so that the device could also be used as a serial-parallel-serial delay line or frame store. Figure 3 shows the schematic layout of the device. The transfer channels were about

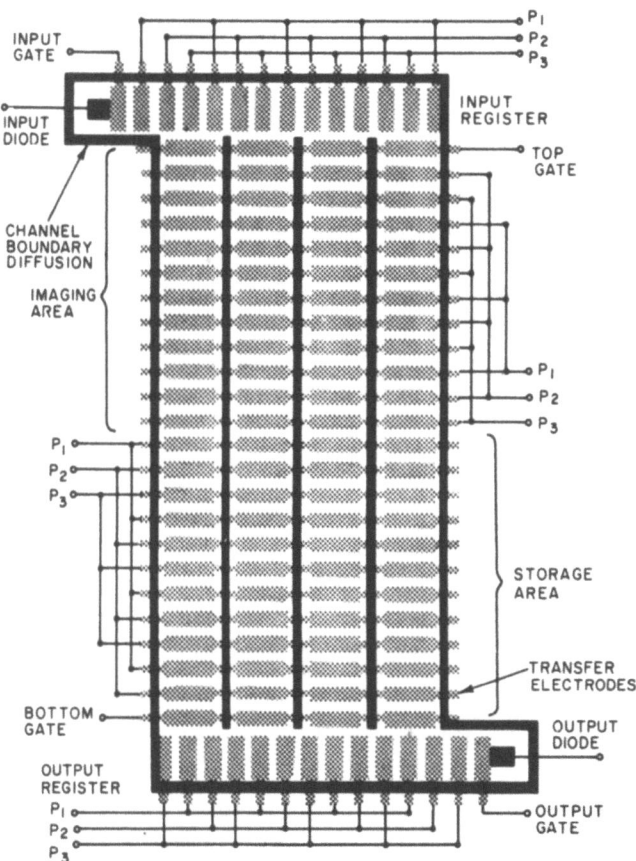

Fig. 3 A schematic diagram of a 3-phase charge-coupled image sensing array organized for frame transfer mode of operation. The upper serial register enables it to be used as a frame store. (Ref. 5)

22 μm wide at a spatial pitch of 30 μm. Each element measured 33 μm in the vertical direction and consisted of three 9 μm electrodes separated by 2 μm gaps. The electrodes were necked down between the transfer channels (Fig. 4) to reduce the total length of 2 μm gaps, to increase light sensitivity and to provide larger islands for better adhesion of the photoresist during the etching of the metallization pattern. In these devices a thinner gate oxide of about 130 to 140 nm was normally employed. Together with the larger electrodes and the narrower gaps, the edge effect was thereby reduced, and the transfer inefficiency produced by the interaction with interface states was on average below 10^{-3} and for the best devices in the low 10^{-4} when driving pulse amplitudes above 10 V and a 20% bias charge were employed. This resulted in typical inefficiency products of less than 0.3 for both the horizontal and vertical directions, respectively, and thus in adequate modulation transfer functions.

The major challenge was to obtain a defect-free device. The intricate metallization pattern is a major source of fatal defects. A short across the 2 μm gaps, the total length of which amounts to about 1 m in a single device, makes the device inoperable. The appearance in the display of some other basic defects is illustrated in Fig. 5. Blocked transfer channels result in vertical black bars [5]. Localized breakdowns or spots of extreme dark current generation produce white spots or white vertical lines, depending on their strength and position. In addition there is the more or less uniform dark current background

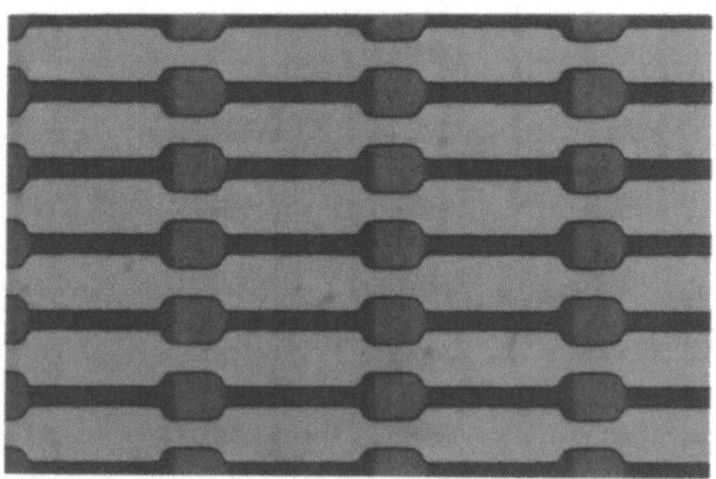

Fig. 4   Close-up on actual electrode geometry in imaging area of the 106x128-element charge-coupled area image sensor.

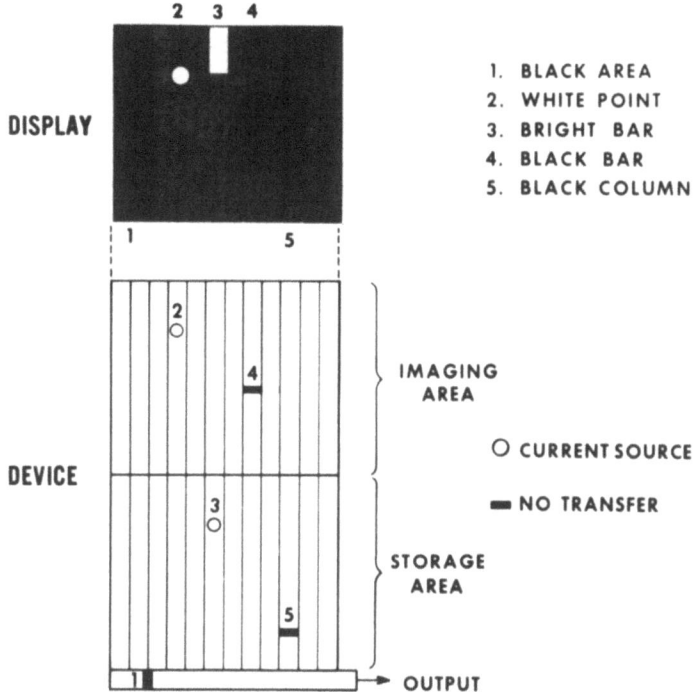

Fig. 5 Some defects observed in charge-coupled area image sensors and their appearance in the display for the case of frame transfer mode of operation. (Ref. 5)

superposed on the video signal. In the better devices this background ranged from 5 to 10 nA cm^{-2} and represented a fraction of less than 1% of a full well when the device was integrating for 1/60 s.

These devices rendered images with enough resolution to reproduce a human face (Fig. 6). They had good uniformity and were free of the striation problems of earlier x-y-addressed sensors. They demonstrated that charge transfer devices are indeed suitable to make solid-state image sensors.

2.3 Blooming

In live demonstrations in actual video cameras these devices showed an objectionable reponse to optical overloads. Saturated white areas in the display, corresponding to points of intense

(a)

(b)

Fig. 6   Image obtained with 106x128-element area image sensor (a) in the frame transfer mode and (b) using the full array at an integration to readout duty cycle of 3:1. (Ref. 5)

illumination, spread out into white bars along the vertical axis. This "channel blooming" [6] is produced by excess charges overflowing into neighboring elements along the vertical transfer channels in the imaging area. This effect was especially strong since originally these devices were operated in a rather simple manner. The electrodes were switched between only two potentials V_R and V_P, and the same potentials were used during integration to form the potential wells and barriers. In this mode, V_R has to be high enough to keep the interface depleted in order to obtain good transfer efficiency, but during integration this provides inadequate isolation. The channel blooming can be suppressed if somewhat more complicated drivers are used, which permit the interface under the nonintegrating electrodes to be biased into accumulation during integration. Excess carriers will then be spilled into the substrate, where they will diffuse randomly in all directions and thereby produce a more circular blooming pattern.

To achieve even stronger blooming protection, a well-defined drain has to be provided for the excess carriers. In a frame transfer device such overflow drains can be provided in the form of reverse biased diodes placed between the vertical transfer channels in the imaging area [6]. If the amount of charge that can be collected is properly limited, these drains do not have to extend into the storage area. The reverse biased overflow diodes not only collect the excess carriers overflowing directly from saturated potential wells, but they also sustain a depleted area along the $Si-SiO_2$ interface, which can sink carriers diffusing through the bulk and thereby reduce the magnitude of circular blooming.

These overflow drains must be separated from the actual transfer channels by a potential barrier. The desired interface potential profiles underneath an integrating electrode biased at V_P and an isolating electrode biased at V_R are illustrated in Fig. 7a. Carriers are collected in the potential wells of initial depth ψ_P. When the potential well is filled to the barrier potential ψ_T, excess carriers will flow laterally into the overflow drain. For proper operation, the electrodes that isolate the resolution elements in the vertical direction have to produce an interface potential ψ_R which is lower than ψ_T, so that excess carriers will flow into the overflow drain before they spill along the transfer channel.

The overflow barrier potential ψ_T can be established in several ways, either by a light channel-stopping implant (Fig. 7b), by a thick oxide region (Fig. 7c), or by a special threshold electrode (Fig. 7d). Whereas a special electrode would permit electrical adjustment of the barrier height, the implanted barrier is a simpler approach practically. The barrier height is chosen in accordance with the anticipated operating potentials of the device, its oxide thickness, and substrate doping.

The previously described area image sensor was later modified and was provided in two-thirds of its imaging area with overflow drains formed in the same step as the input/output diodes and with barriers formed by a light p-type boron implant (2×10^{12} cm^{-2}).

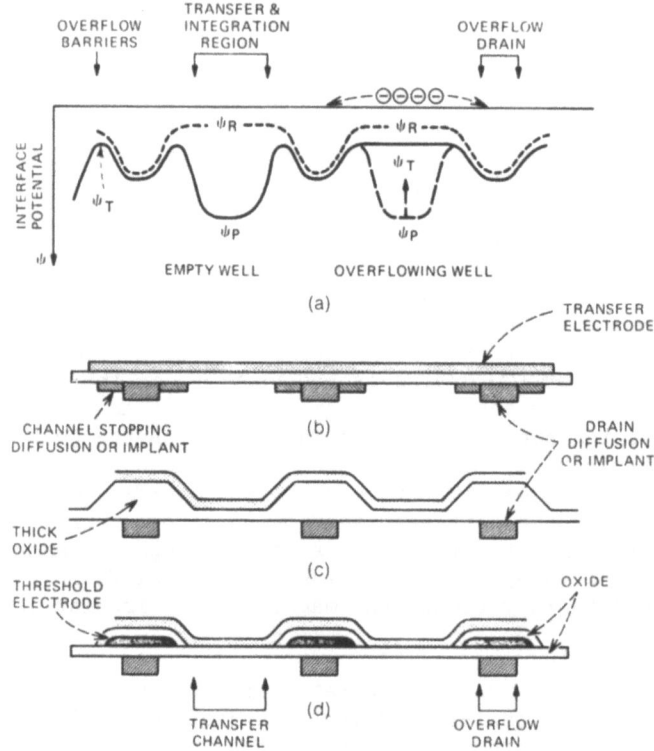

Fig. 7 Cross section through imaging area of a frame transfer image sensor. (a) Desired potential profile at the Si-SiO$_2$ interface, underneath isolating electrodes (---) and underneath integrating electrodes (—). Realization using (b) channel-stopping diffusion or implant, (c) a thick-thin oxide structure, (d) a special threshold electrode. (Ref. 6)

These testers demonstrated the strong blooming protection which can be obtained by such an approach. Figures 8a and 8b show results obtained with a HeCd laser beam falling onto the unprotected and protected areas, respectively, when the beam power is varied from saturation to five orders of magnitude above. While in the unprotected area the saturated region can spread several millimeters and eventually overload the whole device, the size of the saturated spot in the protected area stays within reasonable limits, and generally starts to spread due to internal reflections and lens flare in the optical system before blooming effects become significant. However, at overloads of a factor 100 or 1000 the effect of frame transfer smearing become objectionable.

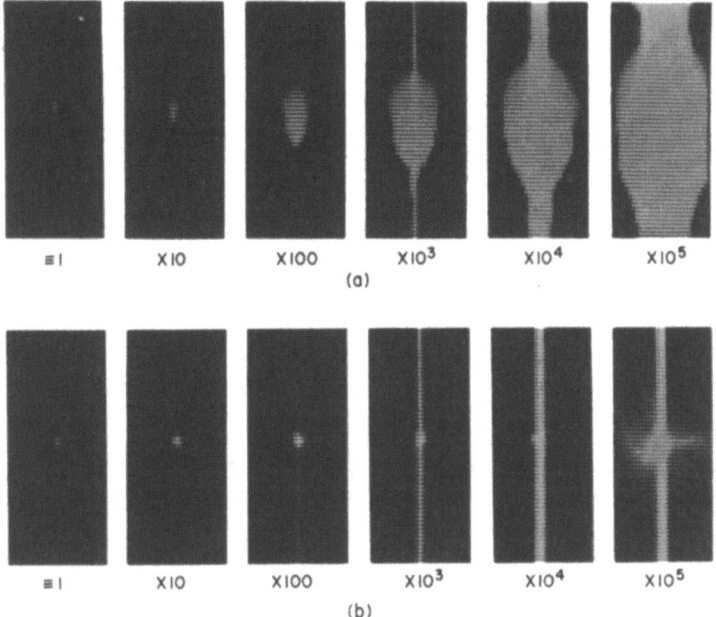

Fig. 8   Pattern produced by a beam of blue light (441.6 nm) of increasing intensity in (a) the unprotected and (b) the protected area of a 106×128 area image sensor with overflow drains and ion-implanted overflow barriers. (Ref. 6)

During frame transfer, the continuously incident laser beam will write a noticeable white line into the moving charge frame. This limits the usefulness of frame transfer devices in applications where strong overloads have to be expected. An optical shutter in front of the device, properly synchronized with the frame transfer, or a special structure of the overflow drains which would permit variation of the sensitivity of the device itself may circumvent these problems. Both approaches, however, are either cumbersome or difficult and have so far not been demonstrated.

3. THE 3-LEVEL POLYSILICON APPROACH

3.1 Fabrication Process

The problems in fabrication and operation of devices with narrow gaps between the transfer electrodes can be overcome when a multilevel metallization with overlapping electrodes is employed. For devices with the illumination incident from the circuit side, the electrodes have to be as transparent as possible. The transmission of a polysilicon layer of about 0.5 μm thickness is adequate for many applications. Other transparent electrode materials exist and have been demonstrated in small imaging devices [7], but the technology of polysilicon electrodes is at a much higher state of development. Devices with 2, 3, or 4 phases formed with two levels of polysilicon could thus readily be built. But in this section, devices which use a 3-phase electrode structure implemented with three levels of polysilicon [8] will be discussed. This approach has the advantage that each set of electrodes can be formed in a separate level. The photolithographical features F in each level can thus be large, while the overall cell length is still only 3F. Furthermore, mask defects leading to intralevel shorts will no longer jeopardize the operation of the device and will only cause a visible perturbation in the image if they occur in the first or second electrode level above an active transfer channel. The probability of fatal shorts between different phases is strongly reduced because it relies on an insulator of high integrity rather than on a perfect photolithographical step. Thermal oxides on the polysilicon electrodes can be grown with pinhole densities as low as 1 per cm^2. This 3-level polysilicon approach is therefore very suitable for the fabrication of large area devices.

The fabrication of such a device starts with the diffusion of the source and drain features and the implantation of the channel stopping areas. A new gate oxide is grown, typically to a thickness of about 1500 angstroms. The first level of polysilicon is deposited, doped with phosphorus, and steam oxidized. This oxide is used as an etch mask to define the first set of electrodes. The remaining masking oxide and the exposed gate oxide are now removed and a new gate oxide is thermally grown. This process partly oxidizes the electrodes and forms an insulator of high integrity between electrode levels. These processes are repeated to form the second and third polysilicon electrode levels. Removal of the exposed gate oxide prior to each oxidation step gives equal gate oxide thicknesses under all three sets of polysilicon electrodes (Fig. 9a).

In a later approach the first gate oxide is made only 100 nm thick and is covered with 100 nm of Si_3N_4 (Fig. 9b). This eliminates the need to etch the oxides back to the substrate in order to obtain equal gate oxide thicknesses for all three electrode systems, and it also reduces the probability of electrode-to-substrate shorts. Storage times of several minutes, observed on test capacitors, make this approach look very promising to obtain a low dark current background.

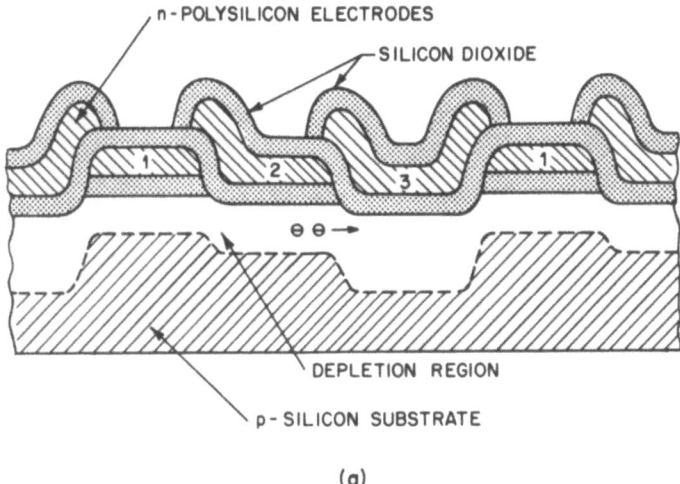

(a)

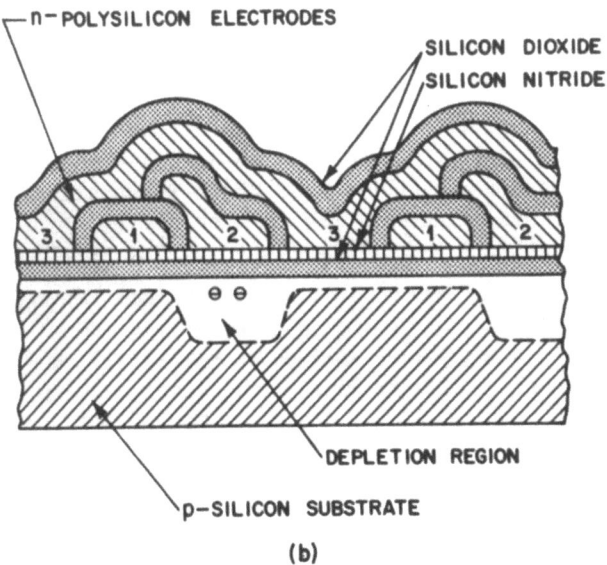

(b)

Fig. 9 Cross-sectional view of the transfer electrode structures in 3-phase 3-level polysilicon charge-coupled devices, (a) approach with normal gate oxide, and (b) with extra layer of Si_3N_4 and with a continuous third level electrode.

To reduce dark current and the number of white video defects, some gettering steps are necessary. After the definition of all three electrode levels, the electrode side of the wafer is covered with 1000 nm of SiO_2, and the backside is stripped to the substrate. The gettering process typically consists of a 30-min phosphorus diffusion at 1000°C.

Contact windows to the polysilicon bus bars and to the substrate are etched. A level of true metal forms high conductivity bus bars, interlevel contacts, and bonding pads. This approach has the advantage that the metal level contains only large features at the periphery of the device.

3.1 A 1600-Element Linear Sensor

With the above-described technology a linear image sensor was built which had 1600 sensor elements attached to a single readout channel [9,10]. This approach was chosen for the following reasons. Firstly, delay lines with 256 and 512 elements fabricated with the 3-level polysilicon approach [8] had already demonstrated very low transfer inefficiencies on the order of a few time 10^{-5}, which promised to give adequate inefficiency products in transfer channels with several thousand transfers. Secondly, the 3-level approach allows the fabrication of short transfer cells (15 μm) without the need for reducing the photolithographical features below 5 μm. For this purpose the third level of polysilicon is made continuous (Fig. 9b). The spatial period of the sensor elements in the device is then limited by the dimension of the integrating potential wells and the defining channel stopping diffusion itself. In this device the spatial period was made 16 μm. Thirdly, a single readout channel avoids all problems with the multiplexing of the outputs of the two 3-phase readout channels. A schematic layout of the device is shown in Fig. 10. The sensor elements are formed under a level 2 electrode, and the transfer gate between the sensors and the readout register is formed in level 1. The charges from the sensors are transferred into the transfer electrodes formed in levels 2 and 3. The final metallization which forms the bus lines and bonding pads also covers the readout registers and defines a slit over the sensor elements.

The lowest observed transfer inefficiency in the 40 μm wide channel at 300 kHz and using a 30% bias charge was 1×10^{-5}. This performance is illustrated by Fig. 11 which represents an electrically injected pulse sequence 0101111 after 4800 transfers. With such devices high resolution images can be obtained in a slow scan mode which are able to resolve even the small print in a footnote on an 8-1/2×11 inch printed page (Fig. 12). However, the alignment of the 5 μm transfer electrodes over the whole

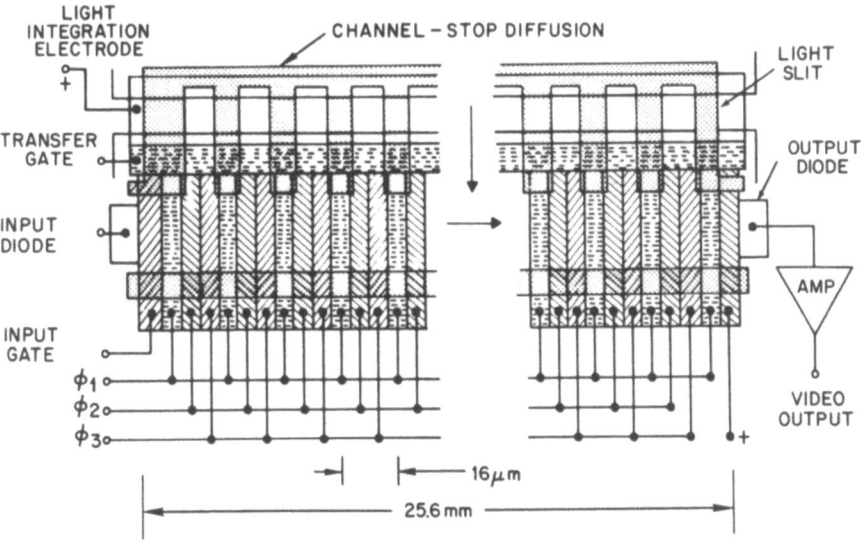

Fig. 10 Schematic diagram of the organization of the 1600-element linear charge-coupled image sensor.

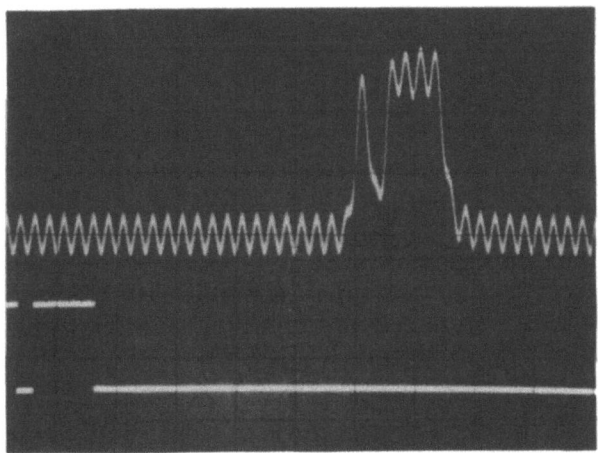

Fig. 11 Upper trace shows the signal output after 4800 transfers in the readout channel of the 1600-element linear image sensor when the input signal shown in the lower trace is electrically injected. (Ref. 10)

Fig. 12 Image obtained with the linear 1600-element charge-coupled image sensor in a line scan mode. (Ref. 16)

25 mm length of the device often caused difficulties and resulted in much higher transfer inefficiencies. The question thus remains open, whether the bilinear readout organization with larger transfer cells is a preferable approach.

3.3 A 220x256-Element Area Sensor

The 3-level polysilicon electrode structure, promising high functional yield even for very large devices, was also adopted to build a frame transfer device to match the video format of the present-day PICTUREPHONE® system. The imaging area of the new device [11] has been provided with 128 rows of elements. The photogenerated charge is integrated alternately underneath different sets of electrodes to match the 2:1 interlaced format of PICTUREPHONE. In this mode, assuming a Kell factor of 75 percent, the limiting vertical resolution is 96 line pairs.

The number of elements in the horizontal direction has been chosen to yield equal resolution density in both dimensions. Since the PICTUREPHONE screen has an aspect ratio of 11:10, the corresponding number of line pairs in the horizontal direction is 106, which necessitates at least 212 elements per line. The element readout rate has been chosen 256 times higher than the line rate, to allow a simple binary countdown from a master clock. Video information is transmitted during 105.5 µs out of the full 125 µs line time. This corresponds to 216 picture elements. The device has thus been provided with 220 elements horizontally, a number that meets the resolution requirements as well as the goal to keep the driving logic simple.

Fig. 13   Frame transfer area image sensor with 220×256 elements mounted in a 24-pin dual-in-line package. Background has 0.05 inch divisions. (Ref. 11)

The horizontal center-to-center spacing of the vertical transfer channels is 30 μm. These channels are separated by 5 μm stripes of a channel stopping diffusion. This leads to an electrode length of 10 μm in the readout register, a dimension which is in accordance with reasonable design rules and with the need to read out charge packets at an element rate of 2 MHz. The aspect ratio of the screen and the 2:1 interlaced format then dictate a vertical cell length of 48 μm in the imaging area and thus an active electrode length of 16 μm. The storage area is not subject to such restriction, and the electrodes are only 10 μm long. The mutual overlap of adjacent electrodes is 2 μm. Enough space is provided in the imaging area so that at a later time overflow drains could be introduced. This would reduce the channel width in this area, and thus match its charge handling capability to that of the storage area. The device is mounted in a 24-pin dual-in-line package (Fig. 13).

This device has been demonstrated in a self-contained black and white camera operating from rechargeable batteries. The camera, including the batteries, measures 13 cm by 8 cm by 7 cm and accommodates any C-mount lens. No special steps to miniaturize the camera have been taken. The driving logic is built with low power TTL dual-in-line packages and consumes about 0.5 W. The pulse drivers for the nine independent electrode systems are built with three or four discrete transistors each and consume a

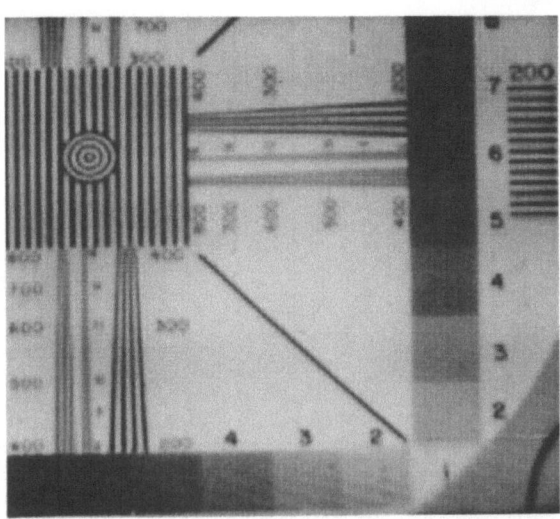

Fig. 14   Image obtained with 220×256-element area image sensor operated in a self-contained battery-operated camera in the 2:1 interlaced PICTUREPHONE® format. (Ref. 11)

total of about 1 W. The video signal is taken off the chip directly from the output diode and is fed to a preamplifier built from discrete components. After the signal has been properly blanked and the synchronization pulses added, it is band limited, and the clock frequency is removed with a notch filter. A final driver stage puts the composite video signal onto a 75 ohm coaxial line. Figure 14 shows results displayed on a monitor operating in the PICTUREPHONE format.

3.4 A 475×496-Element Area Sensor

A scaled-up version of the above device has recently been built [12] to match the American broadcast television format. The imaging area has been designed equivalent in size to the scanned area on a 1 inch vidicon, i.e., 12.8×9.6 mm. The device has 248 rows of sensor cells measuring 39 µm in the vertical direction. This number exceeds the requirement for the standard format by a few lines. In the horizontal direction at least 450 cells are required for a black and white picture. The readout rate of the serial register was thus chosen to be equal to 8.95 MHz, or 2.5 times the color subcarrier of 3.58 MHz. At this rate 470 elements can be scanned during the active line time of 52.5 µs. The horizontal dimension of 12.8 mm is thus subdivided into 475 cells each 27 µm long. On a substrate material with a boron doping of 5×10^{14} cm^{-3}, the 9 µm long electrodes will still give adequate transfer efficiency at the required 9 MHz clock rate. To conserve silicon area, the storage area is designed somewhat more compact than the imaging area with cells which were only 30 µm long. The overall chip dimensions including the bus lines and bonding pads at the periphery are 16×20 mm thus covering a silicon area of more than 3 cm^2. Only two devices can be fabricated on a 2 inch wafer. The photolithographical masks were fabricated on a electron beam exposure system (EBES) [13].

A substantial fraction of the first experimental device lots has been operational even though no perfect defect-free devices have been obtained so far. It is probably fair to state that with any other than the 3-level polysilicon approach none of these devices with equivalent defects in the electrode levels would have been functional.

To demonstrate the device, a small camera (Fig. 15) operating in the normal 525-line TV format has been built. In principle the electronic circuitry was similar to that of the camera described in the preceding section. However, the clock drivers which must operate at the higher speeds and drive the larger capacitances of the bigger chip have to meet much tougher requirements. The image sensor had a short dummy register near the end of the serial

Fig. 15 Television camera for 525-line TV format and corresponding 475x496-element area image sensor.

output register and the video signal was taken from a differential floating gate amplifier. An image obtained with this device is shown in Fig. 16. Although a substantial development effort is still required before a picture of commercial broadcast television

Fig. 16 Image obtained with 475x496-element area image sensor in standard 525-line TV format.

quality can be obtained, this charge-coupled image sensor demonstrates the basic feasibility of such devices.

4. THE SINGLE-LEVEL POLYSILICON APPROACH WITH SELECTIVELY DOPED ELECTRODES

One of the disadvantages of the 3-level polysilicon approach is the long processing sequence. Although devices will still be operational even in the presence of large mask defects, the channel stopping and electrode patterns have to be almost perfect in order to produce a flawless device of high uniformity. The 3-level approach might then become too expensive in a competitive environment. Furthermore, the spectral response of such a device can be rather nonuniform and strongly process dependent because of interferences of the incident light produced by partial reflections at the various interfaces, as illustrated by Fig. 17 [11].

For front-surface illuminated devices it is thus desirable to employ as simple an electrode structure as possible. The use of a single level of high resistivity polysilicon, selectively doped to form individual transfer electrodes, has proven to be a practical approach [14]. This approach has been used with several bulk channel image sensors and has led to the first area image sensor to match the line format of standard 525-line broadcast television [14].

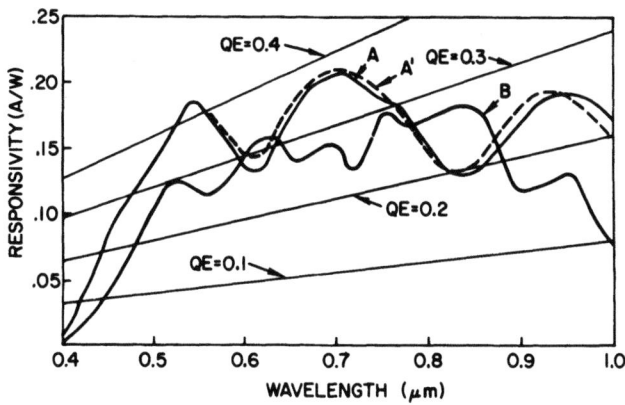

Fig. 17   Example of spectral response curves of two area image sensors with three levels of polysilicon. Two curves were taken on the same device A in the center and A' near the edge. (Ref. 11)

4.1 A 320×512-Element Area Sensor

A frame transfer device with 320×512 elements has been implemented [14] with the single-level polysilicon approach. The imaging area, measuring 9.7×7.3 mm^2, corresponds to that of a 2/3 inch vidicon. The 320 elements in the horizontal direction are sufficient for the requirements of a color channel in the American broadcast TV format. A 3-phase clocking arrangement was chosen to achieve maximum density of the resolution cells. The n$^+$-type transfer electrodes were diffused into a layer of p-type polysilicon to obtain low leakage current between the different electrode systems. The size of each sensor cell is 30×30 μm. Aluminum was used to form the bus lines on both sides of the array and the bonding pads.

To improve horizontal transfer efficiency and thus horizontal resolution, a bias current of 300 nA is injected into the serial output register, which operates at 6.1 MHz. For high vertical transfer efficiency, a bias current of 30 nA has to be injected by uniformly illuminating the array, e.g., with two light emitting diodes placed inside the housing of the TV camera. The peak-to-peak signal current is 250 nA, while the dark current in the imaging area is 4 nA (5 nA cm^{-2}). This results in a maximum peak signal to rms noise ratio of 50 dB. The device has an

Fig. 18   Image obtained with 320×512-element area image sensor in standard interlaced 525-line TV format, displaying a solid-state camera which is equipped with such a sensor. (Courtesy R. L. Rodgers, III, RCA Corporation, Lancaster, Pa.)

output circuit consisting of a resettable floating diffusion and a single amplifying MOSFET, which can be operated as a source follower or as an inverter. An image obtained with this device [14] is shown in Fig. 18, it displays a self-contained, 525-line TV camera which is equipped with such a sensor.

5. DISCUSSION

Frame transfer surface channel devices, because of their relatively simple electrode structure and processing sequence, were the first charge-coupled area image sensors to produce acceptable pictures. Currently, the position of these devices is being challenged by bulk channel approaches and by the interline transfer readout organization. The better transfer efficiency and lower transfer noise of bulk channel devices will give superior performance in low light level applications. Nevertheless, the simpler processing leading more readily to lower dark current and the somewhat higher signal handling capability of surface channel devices will assure that devices of both types will coexist for the near future. Furthermore, "channel blooming" can be suppressed in a relatively simple way by biasing the nonintegrating electrodes into accumulation. This mode is not possible in a bulk channel device because the surface potential under these electrodes will be clamped by charge injection from the channel stopping areas while the bulk transfer channel remains depleted. The introduction of the necessary overflow drains and barriers will then even further complicate the structure of a bulk channel image sensor.

The slower roll-off of the MTF and the more compact inplementation possible with the interline transfer organization may be preferable in some military or scientific applications. On the other hand, once a viable thindown technique for charge-coupled area image sensors has been developed, devices with backside illumination will have a higher quantum efficiency and a more uniform spectral response. The frame transfer approach, being more suitable for backside illumination, will thus retain a significant advantage.

Constituting yet another source of competition, medium sized arrays of, say, 200x200 elements for normal room lighting conditions are now also being fabricated with the x-y-addressed charge injection devices, producing images of equal quality [15]. The low dark current performance demonstrated in these devices, the absence of any problems with transfer efficiency or the uniform injection of background charge, and the possibility of multiple nondestructive readout make the charge injection device the real competitor for surface channel charge-coupled image sensors.

REFERENCES

[1] Bertram, W. J., Sealer, D. A., Séquin, C. H., Tompsett, M. F., and Buckley, R. R., "Recent Advances in Charge Coupled Imaging Devices," IEEE INTERCON, New York, Digest, 291-293 (1972).

[2] Séquin, C. H., "Experimental Investigation of a Linear 500-Element 3-Phase Charge-Coupled Device," Bell Syst. Tech. Jour. 53, 581-610 (1974).

[3] Amelio, G. F., "Computer Modelling of Charge Coupled Device Characteristics," Bell Syst. Tech. Jour. 51, 705-730 (1972).

[4] Tompsett, M. F., "The Quantitative Effects of Interface States on the Performance of Charge Coupled Devices," IEEE Trans. on Electron Devices, ED-20, 45-55 (1973).

[5] Séquin, C. H., Sealer, D. A., Bertram, W. J., Tompsett, M. F., Buckley, R. R., Shankoff, T. A., and McNamara, W. J., "A Charge Coupled Area Image Sensor and Frame Store," IEEE Trans. on Electron Devices ED-20, 244-252 (1973).

[6] Séquin, C. H., Shankoff, T. A., and Sealer, D. A., "Measurements on a Charge-Coupled Area Image Sensor with Blooming Suppression," IEEE Trans. on Electron Devices ED-21, 331-341 (1974).

[7] Brown, D. M., Ghezzo, M., and Garfinkel, M., "Transparent Metal Oxide Electrode CID Imager Array," ISSCC, Philadelphia, Digest of Tech. Papers, 34-35 (1975).

[8] Bertram, W. J., Mohsen, A. M., Morris, F. J., Sealer, D. A., Sequin, C. H., and Tompsett, M. F., "A Three Level Metallization Three-Phase CCD," IEEE Trans. on Electron Devices ED-21, 758-767 (1974).

[9] Séquin, C. H., Sealer, D. A., Bertram, W. J., Buckley, R. R., Morris, F. J., Shankoff, T. A., and Tompsett, M. F., "Charge Coupled Image Sensing Devices Using Three Levels of Polysilicon," ISSCC, Philadelphia, Digest of Tech. Papers, 24-25 (1974).

[10] Sealer, D. A., Séquin, C. H., and Tompsett, M. F., "High Resolution Charge Coupled Image Sensors," IEEE INTERCON, New York, Digest, Session 2, paper 2/1 (1974).

[11] Séquin, C. H., Morris, F. J., Shankoff, T. A., Tompsett, M. F., and Zimany, E. J., "Charge-Coupled Area Image Sensor

Using Three Levels of Polysilicon," IEEE Trans. on Electron Devices ED-21, 712-720 (1974).

[12] Séquin, C. H., Zimany, E. J., Tompsett, M. F., and Fuls, E. N., to be published.

[13] Herriott, D. R., Collier, R. J., Alles, D. S., and Stafford, J. W., "EBES, A Practical Electron Lithographic System," IEDM, Washington, D. C., Tech. Digest, 21-26 (1974).

[14] Rodgers, R. L., "A 512×320 Element Silicon Imaging Device," ISSCC, Philadelphia, Digest of Tech. Papers, 188-189 (1975).

[15] Michon, G. J., and Burke, H. K., "Recent Developments in CID Imaging," Symp. on CCD Tech. for Scientific Imaging Applications, Pasadena (1975).

[16] Séquin, C. H., and Tompsett, M. F., "Charge Transfer Devices," Suppl. #8 to "Advances in Electronics and Electron Physics," Academic Press, Inc., New York (1975).

BURIED CHANNEL CCD's

G. F. Amelio, R. H. Dyck
Research and Development Laboratory
Fairchild Camera and Instrument Corporation
4001 Miranda Avenue, Palo Alto, Ca. 94304

ABSTRACT. A buried channel CCD is one in which the signal carriers are transported within the bulk of the semiconductor rather than along the Si-SiO2 interface. This is accomplished by adding, adjacent to the interface, a region of opposite conductivity type than the bulk. Devices fabricated in this way exhibit better transfer efficiency and lower noise than surface channel CCD's. This paper outlines the theory of buried channel CCD's and gives examples of performance from actual devices.

1. INTRODUCTION

- First published report: Walden et al, BSTJ Brief, Sept. '72.
- Major efforts on BCCD development began at Fairchild in 1971.
- The first CCD product was a BCCD 500-element linear imaging device introduced by Fairchild in early 1973.

2. THE DEVICE STRUCTURE

Fig. 1 shows the simplest BCCD structure in cross-section. The only difference from the SCCD is the n-layer. The n-type layer thickness can range from < 1 µm to greater than 5 µm. Thin n-type layers have the advantage of giving high channel capacitance. Thick layers have the advantage of giving higher speed CCD operation.

The n-type layer is typically implanted such that the peak concentration is below the silicon surface. Compared to having the peak concentration at the surface and with concentration per unit area invariant, this increases the amount of charge that can be handled without surface interaction.

3. COMPARISON OF THE BASIC PHYSICAL PROPERTIES OF A BCCD, TO THOSE OF AN SCCD

3.1 Higher mobility of carriers. Typically:

Type	(cm^2/V sec)
SCCD	500
BCCD	1000

3.2 Higher surface state contribution to dark current,

$$j_D = \frac{qn_i}{2} \left(s + \frac{W_D}{\tau_o}\right)$$

because, where charge is stored in a SCCD, $S \ll S_o$.

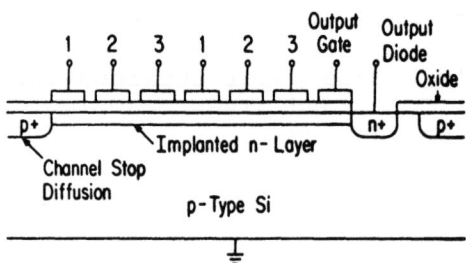

Fig. 1 Cross-section of a 3-phase n-channel BCCD. An analog shift register would have an input gate between the left hand channel stop and the first $\emptyset_1$ gate.

4. VERTICAL POTENTIAL PROFILE

4.1 ZERO CHARGE CONDITION

This treatment was presented by Kim, et al in the NEREM '72 Record.

The first requirement for successful operation

of a buried-channel CCD is that for zero signal the implanted layer should be entirely void of majority carriers associated with the implanted donors. This is easily accomplished by applying a sufficiently large positive voltage to the output diode at the end of the CCD channel, thereby raising the quasi-Fermi level for electrons in the implanted layer. When the implanted layer is completely depleted, an energy well is formed away from the surface and the potential of this well can be modulated by the gate voltage applied. If the implanted donor profile is approximated by a box distribution having doping N_D and depth t, the channel potential ϕ_{max} is related to the gate voltage V_G by

$$V_G - V_{FB} + V_{imp} = \phi_c + \sqrt{V_{ox}\, \phi_c} \qquad (1)$$

$$\phi_{max} = \left(1 + \frac{N_A}{N_D}\right) \phi_c \qquad (2)$$

where

$$V_{imp} = \frac{qN_D t^2}{2\varepsilon_s} \left(1 + \frac{2\varepsilon_s}{\varepsilon_{ox}} \frac{d}{t}\right)$$

$$V_{ox} = \frac{2qN_A t^2}{\varepsilon_s} \left(1 + \frac{\varepsilon_s}{\varepsilon_{ox}} \frac{d}{t}\right)^2$$

In these equations, V_{FB} is the flat-band voltage for the Metal-Oxide-N-layer structure, N_A is the doping in the substrate, q is the magnitude of the electronic charge, and ε_s and ε_{ox} are dielectric sonstants of silicon and silicon dioxide respectively. The appropriate equations for a surface-channel CCD can be obtained from these equations by putting $N_D = 0$ and $t = 0$. It is to be noted that equation (1) is in the same form as the corresponding equation for the surface-channel CCD except the term V_{imp} which is due to the implanted ions. Therefore, clocking requirements for the buried-channel CCD are similar to those for the surface-channel devices, but with a suitable voltage offset. (See Fig. 2).

4.2   FINITE CHARGE CONDITION

Treatment of Barbe (1975).

The profile of potential versus distance into silicon for a buried-channel CCD element is shown in Fig. 3. As for the surface-channel CCD, the basic

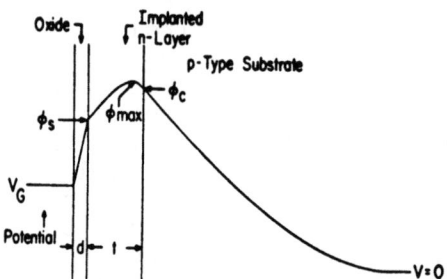

Fig. 2   Formation of a potential well in a Buried-N-channel CCD.

electrostatic design equation relating the minimum electron potential in the channel, ϕ_{min} to the doping density of the substrate, N_A; doping density of the channel, N_D; oxide thickness d; thickness of the donor layer, t; and gate voltage V_G is calculated from Kirchhoff's equation

$$\phi_s + V_G - V_{FB} + V_{ox} = \phi_j + \phi_c. \quad (3)$$

Since $V_{ox} = 1/C_{ox} \times$ (mobile charge density + fixed charge density, it follows that

$$V_{ox} = \frac{eN_D(t - \Delta t - N/N_D)}{\varepsilon_{ox}/d} \quad (4)$$

Using the depletion approximation gives

$$\phi_s = \frac{eN_D(t - \Delta t - N/N_D)^2}{2\varepsilon_s} \quad (5)$$

$$\phi_c = \frac{eN_A L^2}{2\varepsilon_s} \quad (6)$$

$$\phi_j = \frac{eN_D(\Delta t)^2}{2\varepsilon_s} \quad (7)$$

and

$$\phi_{min} = \phi_c + \phi_j. \quad (8)$$

Combining these six equations gives

$$V_G - V_{FB} + V_I - eN\left(\frac{d}{\varepsilon_{ox}} + \frac{t}{\varepsilon_s} - \frac{N/N_D}{2\varepsilon_s}\right)$$

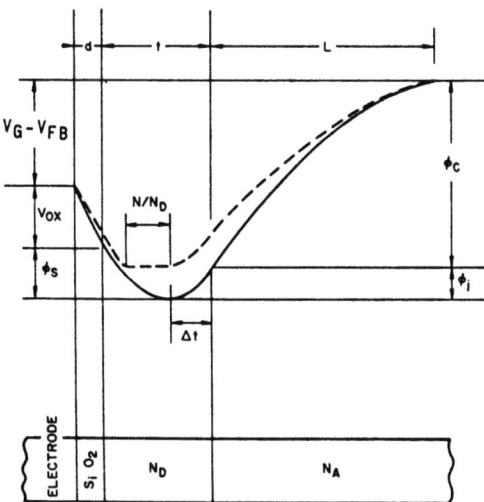

Fig. 3 Energy-band diagram for a buried-channel MOS capacitor showing the potential minimum in the semiconductor

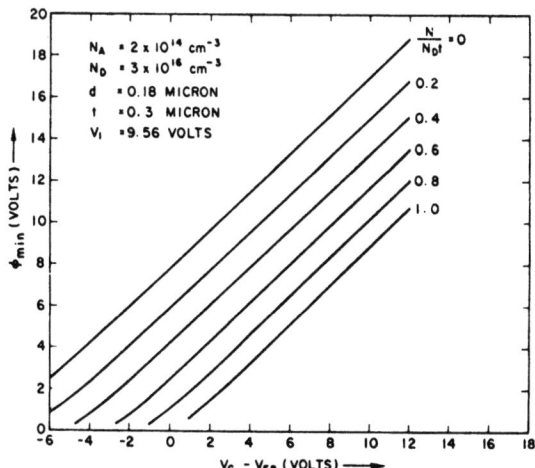

Fig. 4 Minimum potential versus effective gate voltage, with density of carriers in the potential well, N, as a parameter. Curves are calculated from equation (11).

$$= \left[2e\varepsilon_s \left(\frac{N_D N_A}{N_D + N_A}\right) \phi_{min}\right]^{1/2} \left(\frac{d}{\varepsilon_{ox}} + \frac{t}{\varepsilon_s} - \frac{N/N_D}{\varepsilon_s}\right) \quad (9)$$

$$+ \frac{N_D}{N_D + N_A} \phi_{min}$$

where

$$V_I \equiv eN_D t \left(\frac{d}{\varepsilon_{ox}} + \frac{t}{2\varepsilon_s}\right) \quad (10)$$

and V_I is the voltage that must be applied across the p-n junction to deplete the channel. The quadratic formula gives the solution:

$$\phi_{min} = \left\{-\left(\frac{e\varepsilon_s N_A (N_D+N_A)}{2 \; N_D}\right)^{1/2} \left(\frac{d}{\varepsilon_{ox}} + \frac{t}{\varepsilon_s} - \frac{N/N_D}{\varepsilon_s}\right)\right.$$

$$+ \left[\frac{e\varepsilon_s N_A (N_D + N_A)}{2 \; N_D} \left(\frac{d}{\varepsilon_{ox}} + \frac{t}{\varepsilon_s} - \frac{N/N_D}{\varepsilon_s}\right)\right.$$

$$+ \frac{N_D + N_A}{N_D} \left(V_G - V_{FB} + V_I - eN\right.$$

$$\left.\left.\left.\left(\frac{d}{\varepsilon_{ox}} + \frac{t}{\varepsilon_s} - \frac{N/N_D}{2\varepsilon_s}\right)\right)\right]^{1/2}\right\}^2 \quad (11)$$

Fig. 4 shows ϕ_{min} versus $V_G - V_{FB}$, with $N/N_D t$ as a parameter for a shallow buried-channel device. For a voltage ($V_G - V_{FB}$) swing from -2 V to $+5$ V, a full well corresponds to $N/N_D t \approx 0.75$, which gives $N \approx 6.75 \times 10^{11}$ cm^{-2}. Assuming that the area of the well is 10^{-6} cm^2, the number of electrons in the full well is approximately 6.75×10^5.

5. LONGITUDINAL POTENTIAL PROFILE

Because fringe-field effects are so large in BCCD's, it is possible to design devices whose charge transfer speed is virtually entirely determined by drift rather than by diffusion. Hence, it is desirable to know the shape of the longitudinal profile down to field strengths of the order of 25 V/cm, below which the diffusion term dominates over the drift term.

This critical field strength of 25 V/cm is derived approximately by Amelio (1972). The boundary condition in the derivation is appropriate for the removal of the last few electrons from a 10μm-long well at approximately room temperature.

For deep structures ($X_J \sim 5$μm) a 2-dimensional analysis has been reported by El-Sissi and Cobbold (1975).

- ○ The method is capable of handling Gaussian doping profiles and stepped oxides.
- ○ The paper only treats the case of zero signal.
- ○ It uses Fourier analysis.
- ○ Example:

  N_D (bulk) = 1 x 10^{14}/cm^3

  N_A (peak) = 4.8 x 10^{15}/cm^4

  X_0 = 0.10 μm

  X_J = 4.9 μm

  Gate Length = 10 μm

  3 ∅ gate voltages at -5, -10 and -15 volts.

- ○ Results: Average channel potential = -25.8V.

The channel length for which E < 25 V/cm is $\sim$ 0.05 μm. Therefore, drift dominates by a large factor.

For shallow buried channels ($X_J \lesssim 1$ μm), solutions to the potential profile problem lie between the deep-channel case and the surface channel case.

6. PERFORMANCE EXAMPLES

6.1   Wide dynamic range was demonstrated in the CCD101 by Kim and Dyck (1973). The device is a linear array scanned by a pair of 250-stage, 3-phase BCCD registers. A dynamic range of >1000:1 was demonstrated at a 1 MHz output bit frequency. (See Fig. 5).

6.2   Excellent charge transfer efficiency over a wide range of signal level was demonstrated in the CCD201 by Dyck and Jack (1974). The device is a 100 X 100 element array with 100 50-stage vertical registers and a 100-stage horizontal register, all of the BCCD type. No image degradation due to charge transfer inefficiency is detected over a signal range of 1000:1

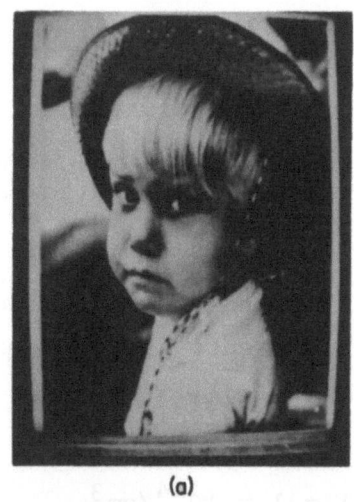

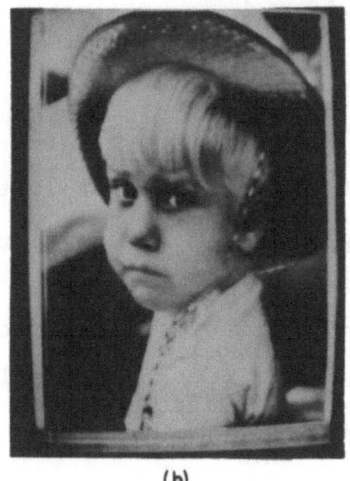 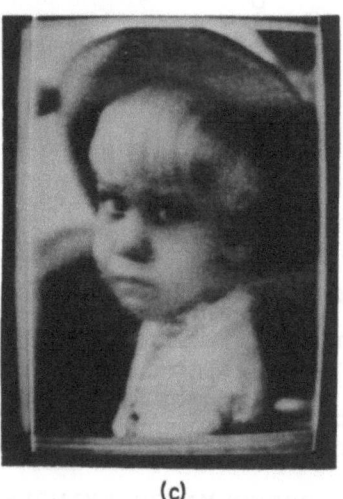

Fig. 5 Pictures taken with a 500-element linear imaging device at three different illumination levels. The highlight signal level in the region around the boy's shoulder in (a) is approximately 90 percent of saturation. The light levels in (b) and (c) are reduced by 100 x and 1000 x from (a). The operating frequency was 1 MHz.

at low temperature and at a 500 KHz output bit frequency. Fig. 6 shows a series of images. Fig. 7 shows light spot linearity results.

6.3   VERY HIGH FREQUENCY OPERATION HAS BEEN DEMONSTRATED BY ESSER (1973).

The device is a 4-phase BCCD with a 4.5 μm junction depth and 7.5 μm gates. Charge transfer efficiencies > 0.999 at >100 MHz were observed.

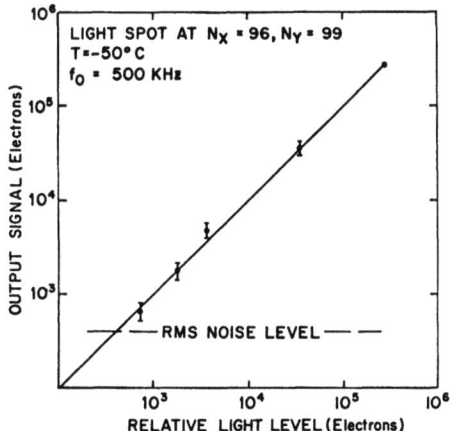

Fig. 7   Relative output signal versus relative light level in electrons

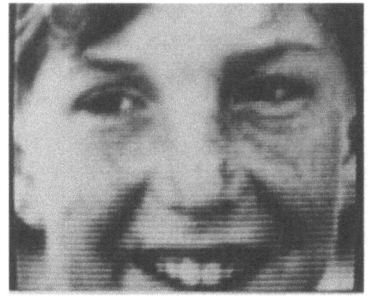
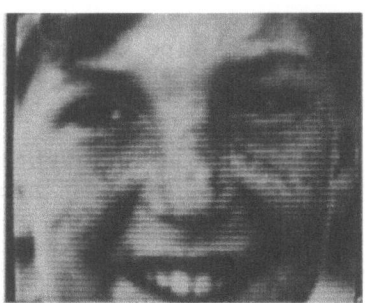

(a) Highlight Signal Level
≃360,000 Electrons/Pixel
1/30 sec. Exposure

(b) Highlight Signal Level
≃36,000 Electrons/Pixel
1/30 sec. Exposure

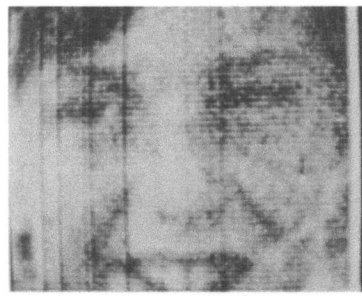
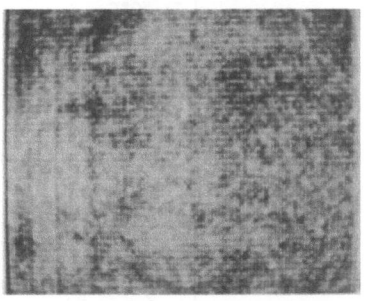

(c) Highlight Signal Level
≃3600 Electrons/Pixel
1/30 sec. Exposure

(d) Highlight Signal Level
≃360 Electrons/Pixel
1/30 sec. Exposure

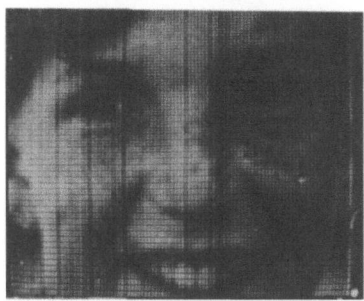
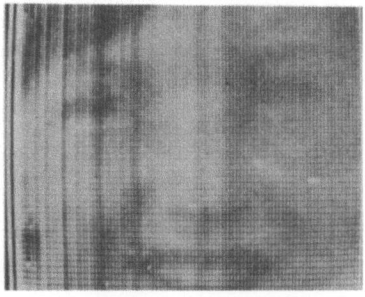

(e) Highlight Signal Level
≃3600 Electrons/Pixel
5 sec. Exposure

(f) Highlight Signal Level
≃360 Electrons/Pixel
10 sec. Exposure

Fig. 6 Low light level images taken with the CCD201 at -40°C. The illumination level was successfully attenuated by a factor of ten in Figs. (a) to (d). Figs. (e) and (f) were made at the same illumination levels as Figs. (c) and (d), respectively.

7. REFERENCES

1. R.H. Walden et. al, "The Buried Channel Charge-Coupled Device", <u>Bell Sys. Tech. J.</u> <u>51</u> 1635-40 (Sept. '72)

2. C. K. Kim et. al, "Buried Channel Charge-Coupled Devices," <u>NEREM 72 Record</u>, 161-164 (1972)

3. D. F. Barbe, "Imaging Devices Using the Charge-Coupled Concept," <u>Proc. IEEE</u> <u>63</u> 28-67 (Jan. '75)

4. G. F. Amelio, "Computer Modeling of Charge-Coupled Device Characteristics", <u>Bell Sys., Tech. J.</u> <u>51</u> 705-730 (Mar. '73)

5. H. El-Sissi and R. S. Cobbold, "Potentials and Fields in Buried Channel CCD's: Two Dimensional Analysis and Design Study", <u>IEEE Trans. on El. Devices</u>, <u>ED-22</u> 77-90 (Mar.'75)

6. C-K. Kim and R. H. Dyck, "Low Light Level Imaging with Buried Channel Charge-Coupled Devices", <u>Proc. IEEE</u> <u>61</u> 1146-7 (Aug. '73)

7. R. H. Dyck and M. D. Jack, "Low Light Level Performance of a Charge-Coupled Area Imaging Device", <u>Proceedings of International Conf. on the Tech. and Appl. of Charge-Coupled Devices</u>, Edinburgh, Sept. 25-27, 1974

8. L. J. M. Esser "The Peristaltic Charge Coupled Device," <u>Proc. of the CCD Applications Conf.</u>, San Diego, Sept. 19-20, 1973, pp 269-77. Also, L. J. M. Esser et. al, "The Peristaltic Charge-Coupled Device", <u>Tech. Dig.</u>, 1973 El. Devices Meeting, pp. 17-20

PERISTALTIC CHARGE COUPLED DEVICES:
WHAT IS SPECIAL ABOUT THE PERISTALTIC MECHANISM*

L.J.M. Esser

Philips Research Laboratories,
Eindhoven, The Netherlands

ABSTRACT. In the present paper we discuss how the advent of the Charge Transfer Devices (CTDs) may bring about a revolution in the field of signal-processing and imaging. The following main types of CTD are discussed: the Bucket Brigade Device (BBD), the surface Charge Coupled Device (surface CCD), the buried channel Charge Coupled Device (buried channel CCD) and the Peristaltic Charge Coupled Device (PCCD). With the aid of simple but quite accurate CTD models a detailed description is given of their operation (basic charge transfer process, transfer efficiency, transfer speed and charge handling capacity) as well as of the main technological aspects of their specific structure.

It is shown, both theoretically and experimentally, how limitations in transfer efficiency, transfer speed and charge handling capacity can be largely overcome by the twin-channel mode of operation on which the PCCD concept is based.

It may be expected that advanced types of CTD can be used for signal processing up to the GHz region.

---

* Also presented at the Semiconductor Device Course on Semiconductor Technology for Electronic Devices, 8-19 May, 1975. "Ettore Majorana" Centre for Scientific Culture
International School of Physics for Industry
Erice-Sicily-Italy, under the title:
Charge Transfer Devices, a Revolution in Signal Processing and Imaging.

ABSTRACT
1. INTRODUCTION
    1.1 Basic Operation
    1.2 Historical Review
2. THE CHARGE TRANSFER
    2.1 The main types of Charge Transfer Devices
        2.1.1 The MOS Bucket Brigade Device
        2.1.2 The Surface Charge Coupled Device
        2.1.3 The Bulk Charge Coupled Devices
            2.1.3.1 The Buried Channel Charge Coupled Device
            2.1.3.2 The uniformly doped Peristaltic Charge Coupled Device
            2.1.3.3 The profiled Peristaltic Charge Coupled Device
    2.2 The Transfer Mechanism
    2.3 Transfer Rate
        2.3.1 Self-induced Field Transfer
            2.3.1.1 Bucket Brigade Mode
            2.3.1.2 Charge Coupled Mode
            2.3.1.3 Surface Charge Coupled Device
                a. Surface Mode
                b. Self-induced Transfer Rate
                c. Thermal Diffusion Transfer Rate
            2.3.1.4 Bulk Charge Coupled Devices
                2.3.1.4.1 Buried Channel Charge Coupled Device and uniformly doped Peristaltic Charge Coupled Device
                    a. bulk mode operation
                    b. surface and bulk mode (PCCD twin-channel mode)
                2.3.1.4.2 Profiled Peristaltic Charge Coupled Device
                2.3.1.4.3 Self-induced Transfer Rate
        2.3.2 Externally-induced Fied Transfer
            2.3.2.1 The externally-induced transfer rate
3. CHARGE HANDLING CAPACITY
    3.1 Surface Charge Transfer Devices
    3.2 Uniformly doped Bulk Charge Coupled Device
        a. bulk mode
        b. surface+bulk mode (PCCD twin-channel mode)
    3.3 Profiled Peristaltic Charge Coupled Device
4. IMPROVEMENT OF THE TRANSFER EFFICIENCY
5. POTENTIAL DISTRIBUTION
    5.1 Self-induced Field-type Charge Transfer Devices
        5.1.1 Surface Charge Coupled Device
        5.1.2 Bulk Charge Coupled Devices
            5.1.2.1 Uniformly doped Bulk Charge Coupled Device
                a. bulk mode
                b. surface+bulk mode (PCCD twin-channel mode)
            5.1.2.2 Profiled Peristaltic Charge Coupled Device

5.2 Externally-induced Field-type Charge Transfer Devices
6. EXPERIMENTAL RESULTS
　　6.1 Bucket Brigade Devices
　　6.2 Surface Charge Coupled Devices
　　6.3 Bulk Charge Coupled Devices
7. TECHNOLOGICAL ASPECTS
　　7.1 Gate structures
　　7.2 Channel definition
　　7.3 Bulk layers
　　7.4 Leakage current
8. INPUT AND OUTPUT CIRCUITS
　　8.1 Input Circuits
　　9.2 Output Circuits
9. APPLICATIONS
　　9.1 Time Conversion
　　　　9.1.1 <u>Analogue time delay</u>
　　　　9.1.2 <u>Television standards conversion</u>
　　　　9.1.3 <u>Time error correction</u>
　　9.2 CTD image sensors
　　　　9.2.1 <u>CTD line sensors</u>
　　　　9.2.2 <u>CTD area sensors</u>
　　9.3 Memories
　　9.4 Filters
10. REFERENCES
11. FIGURES

1. INTRODUCTION

The Charge Transfer Device (CTD) is a new type of silicon integrated circuit invented at Philips Research Laboratories [1]. Since that time several fundamental improvements of the device have been realized [7-11]. Basically the CTD is a dynamic analogue shift register; analogue signal charges are stored and transferred, in clocked shift register mode, under a periodic electrode structure. These electrodes form an array of control and storage gates (capacitors). CTDs can be used in signal processing applications such as electronically variable delay lines, programmable and reprogrammable filters. They can also be constructed for operation as self-scanning photo-sensor arrays and as memories.

1.1 Basic operation

The basic operation of a CTD will be demonstrated on a mechanical model consisting of a flexible water hose covered with many valves (fig. 1). The valves have a periodic structure of three sets. In this model 3 valves form a basic cell which is repeated many times. In the hose are many water packets, which may differ in volume and so represent a signal. It is the aim to transfer the packets to the right completely so that they do not interfere. At least one set of valves has to be kept closed to insure the separation of the packets. At the same time another set is opened to contain the packets. Another set of valves is necessary to gain a one step transport of the water packets. A minimum of three valves per cell is necessary. This is called a 3-phase system. Fig. 1a shows the situation in which valves 1 are closed and valves 3 are fully open. To obtain a transfer valves 2 will be closed. By that they take over the separation function of valves 1. After that the valves 1 can be fully opened. The packets are now below valves 1 and 3. Valves 3 will be closing next (fig. 1b). In comparison with fig. 1a the packets are now transferred one step. It is obvious that the model can be operated with more than 3 valves per cell.

A similar process takes place in a CTD. The instantaneous value of an electrical signal at the input is transformed into an analogue amount of charge in the first cell (fig. 2). According to the rhythm of an applied clock signal this charge quantity will be transferred together with all foregoing charge packets to their next cell. Then a new charge sample is taken at the input and so the cycle is continued.

A solid state analogy of the waterhose type, the Peristaltic Charge Coupled Device, is shown in fig. 3. An N-type twin-layer (N_1 and N_2) with a thickness of some microns, is placed on a negatively biased (isolating) P-type substrate. At its top the N-

layer is isolated from the gates by an oxide layer. In this N-layer the transfer of packets of electrons takes place. The gates 1 to 3 included have the function of the valves in fig. 1. Separation of the charge packets is achieved by a negative voltage applied to the gate set concerned (no. 3 in the situation shown). Because of this the negatively charged electrons from different packets cannot interfere. Storage of the packets is accomplished by a positive voltage applied to the gate set concerned (no. 2). At least one gate set more is necessary to gain a transfer. When this set (no. 1) is connected to a negative voltage to maintain the separation of the charge packets, next set no. 3 connected to a positive voltage and no. 2 to 0 volts potential, then all electron packets are transferred one step, etc.

Fig. 3 shows the situation in which a charge sample has just been taken at the input. The transformation of the input signal into an analogue charge packet happens as follows. At the moment that the first two gates near the input are at a positive potential there is a number of electrons below these gates which is connected to the input. If now the potential at gate 1 is changed to a negative value then a wedge (depletion) is driven between the charge quantity below the second gate and the input. The quantity of the split charge corresponds with the potential of the input signal. A varying input signal gives a charge packet varying from instant to instant, which is successively transferred to the output.

In general for every Charge Transfer Device it is of importance that:
1. the charge packets are completely transferred, so that the packets do not interfere.
2. the charge will be transferred in a very short time, so that high bandwidth signals, as in radar and television, can be processed.
3. a high dynamic range in charge quantities can be covered, resulting in a high signal to noise ratio.

The time alloted for transfer of the charges decreases with a decreasing time period of the clock drivers. Below a certain period of time the transfer becomes increasingly incomplete. This transfer inefficiency (ϵ) means interference of the signal charges. The shortest transfer time for an efficient transport realized up till now is 1.4 nsec for a gate length of about 7.5 µm. This corresponds with a transfer rate of 0.72 GHz or a clock rate of 180 MHz for a 4-phase system. The transfer efficiency is already non ideal for relative long clock periods. (transfer efficiency $\eta = 1-\epsilon$). Per transfer 10^{-5} to 10^{-4} of the initial charge stays behind. The error at the output depends on the total number of transfers of each charge packet in the device.

1.2 Historical Review

The first concept of an analogue shift register was given in 1943 by Schlesinger [3]. The circuit is shown in fig. 4. If the potential ϕ_1 is positive with respect to ϕ_2 then all the odd-numbered capacitors are charged to the inverted potentials of the preceeding capacitors $(V_{C2n+1} = -V_{C2n})$. If ϕ_2 is made positive with respect to ϕ_1 then all the even-numbered capacitors are charged to the inverted potential of their preceding odd-numbered capacitor $(V_{C2n+2} = -V_{C2n+1})$. So every second capacitor contains the analogue information. After the described period (T) the information voltages at the even-numbered capacitors have been transferred two steps $(V_{C2n}(t+T) = V_{C2n}(t))$. Basically this Schlesinger register is a two phase voltage transfer register. The electron tubes cannot be replaced by transistors because of the fact that the current flow in a transistor can be reversed and in a tube one has only a one-directional flow. So the circuit is not suited for solid state integration.

Another concept of a voltage transfer register was given in 1952 by Janssen [4]. It is built up with capacitors, unity gain amplifiers with high input impedance and low output impedance and switches (fig. 5). If the even-numbered switches are opened then the odd-numbered ones are closed. In this situation the information voltages of the even-numbered capacitors are transferred to their succeeding capacitors $(V_{C2n+1} = V_{C2n})$. The odd-numbered switches are opened and the even-numbered ones are closed. Now the signal voltages are transferred one step again $(V_{C2n+2} = V_{C2n+1})$ and so on. As in Schlesinger's register after every clock period (T) the signal voltages at the even-numbered capacitors have been transferred two steps. $(V_{C2n}(t-T) = V_{C2n-2}(t))$. It is obvious that Schlesinger's register can be seen as a special case of Janssen's register.

A bipolar circuit operating according to Janssen's proposal was given in 1965 by Hannan et al [5].

An integrated MOS-circuit based on Janssen's scheme (fig. 6) was designed in 1969 by Mao et al [6]. The unity gain inverting amplifier is built up by the two vertically drawn transistors and in between the switches by the horizontally outlined transistors. The even-ordered switches are connected with the ϕ_2-clock line and the odd ones with the ϕ_1-clock line. Each storage cell contains only 6 transistors and two capacitors. A disadvantage of the circuit is that the signal transfer efficiency is basically dependent on the variation of the properties (e.g. variation in threshold voltages) of the vertically outlined transistors.

An analogue shift register whose signal transfer efficiency is basically independent of the variation of the properties of its elements was proposed by Sangster [1,2] as the Bucket Brigade Device (BBD). This was the first and basic Charge Transfer Device. As is seen from fig. 7 it is designed with bipolar transistors.

Every storage cell is built up with two transistors and two capacitors. If V_1 is a positive voltage with respect to V_2 then the signal charges are transferred to the succeeding even-numbered capacitors (see dashed line in fig. 7). The transfer stops when the emitter voltage of the conducting transistor approaches the base voltage. If the V_2 potential is made positive with respect to V_1, then the odd-numbered capacitors will receive the signal charge of the preceeding capacitors. In the following period V_1 is applied to the positive potential and the respective signal charges will be transferred from the odd to the even-numbered capacitors. When the transfer stops the potential at the odd capacitors will be equal to the corresponding potential in the forgoing period. Hence the last received signal charge is "totally" transferred. So variations in properties of the elements, like variation in the size of capacitors and transistors have no influence. The base current gives rise to an attenuation of the signal but can be regenerated. The charge lost, due to the base current, is eliminated when the bipolar transistors are replaced by MOS FET's (fig. 8). This scheme was proposed by Sangster [8] and realized in 1968 as an integrated circuit. Fig. 9 shows the very simple IC design. If V_2 is positive with respect to V_1 then the odd-numbered capacitors are discharged via their succeeding even-numbered transistors (fig. 8). The transfer stops if the sources of these transistors reach their threshold voltages. During the period that V_1 is positive with respect to V_2, the odd-numbered capacitors receive the succeeding signal charge packet. If V_2 becomes positive with respect to V_1 these capacitors are discharged again to their respective threshold voltages. Thus the signal charge packet is basically totally transferred independent of the mutual variation in threshold voltages of the transistors or mutual variation in the value of the capacitors or other properties of the elements. So it may be concluded that the MOS-BBD is very suited for integration.

A disadvantage of the circuit is the feedback of the drain to the source. By this feedback the threshold voltage is modulated depending on the charge quantity. This limits the transfer efficiency to 0.999 per transfer. An improvement to 0.9999 is obtained by having such a low number of donors for the N-regions, that the transfer process stops when all the charges are transferred before the threshold voltage is reached. Thus the feedback effect is eliminated. The idea of transferring all the charge together with the three phase CTD was proposed in 1969 and published in 1970 by Boyle and Smith [9]. The principle of this so called Surface Charge Coupled Device (SCCD) is shown in fig. 10. It is basically a three phase system, which has been explained before. The inversion charge packets are transferred at the Si-SiO_2 interface. They are stored and transferred by moving potential wells generated by the periodically altering potentials applied to the gates. The surface states interact with the charges by capture and emission. Captured charges of one packet might be emitted later on and be added to another packet. This would strongly limit the transfer efficiency. However, if the charge packets always exceed a certain number (bias charge or fat zero) the surface states are kept filled, reducing their contribution to the transfer inefficiency considerably. Efficiencies of 0.9999 have been reported for bias charge operated SCCD's.

Complete elimination of the influence of the surface states on the transfer efficiency will be obtained if the storage and transfer of the charges takes place in a bulk-layer away from the surface. This has led to a type of bulk CTD viz. the buried channel CCD (fig. 11). This device was proposed by Boyle, Smith, Early et al in 1972 [7,10]. Transfer and storage inside an implanted layer (bulk mode) prevents interference of the charge packets with the surface states. Only interaction with bulk traps is left. Presently their concentration can be better controlled than that of the surface states which leads to better transfer efficiencies. Furthermore for a thick layer device high drift fields, directly induced by the externally applied gate potentials, exist deep in the bulk. This results in a high transfer speed. However the charge handling capacity is one order of magnitude lower than that for the BBD and the SCCD, caused by the larger distance of the charge to the gates. A thin layer device has a high charge handling capacity due to the high storage capacitance, but a low speed by the strongly reduced drift fields.

Independently and simultaneously with the buried channel CCD another type of bulk layer CTD, viz. the Peristaltic Charge Coupled Device (PCCD) (fig. 12) was invented and partly published by Esser in 1972 [11]. The concept of the PCCD was found in a search for CTDs which fundamentally have a complete transfer of all the charges at a very high speed, while maintaining the high charge handling capacity of the BBD and SCCD. The first concept is shown in fig. 12. The epitaxial N-layer is relatively low doped (5×10^{14} cm^{-3}). The main part of the full well charge packet (80-100%) is stored at the Si-SiO$_2$ interface, in the high oxide capacitance. This contributes to the high charge handling capacity. Only a small part of the full well packet is stored in the bulk layer, mainly owing to the low dope level. The charges are transferred by two channels (Twin-channel mode). The first and main part of the charge is mainly transferred by a high-conducting channel at the surface and especially the last fraction of the charge via a weak-conducting channel deep in the bulk layer where high drift fields exist. The result is fundamentally a complete transfer at a high speed combined with a high charge handling capacity. The pinch-off point of the surface channel moves during transfer from $x = L$ to $x = 0$ and then only a bulk channel is left. The surface channel can be seen as a moving source of charges which are driven into the bulk. As in the SCCD, the transfer efficiency is limited by the surface states. However high speed and high charge handling capacity are obtained. To eliminate the effect of the surface states on the transfer behaviour one can operate the device as a bulk channel CCD and make use only of bulk transfer. Then the charge handling capacity is considerably reduced. However, maintenance of the high charge handling capacity together with the elimination of the surface states effect on the transfer behaviour, has been obtained in the profiled PCCD (figs. 3, 13). This device has been introduced by Esser [15,16] in 1973. Now the main part of the charge (80-100%) is stored in a high capacitance, almost at the $Si-SiO_2$ interface. It is stored in a highly doped (10^{17} cm^{-3}) and very thin layer (0.2 μm) on top of the relatively thick low doped layer. This device maintains the advantage of the uniformly doped PCCD and eliminates the interaction between the charges and surface states. The main part of the charge is mainly transferred by a high-conducting channel in the surface-layer and especially the last fraction of the charge via a weak-conducting channel deep in the bulk-layer (Twin-channel mode). Here the charges are subjected to strong drift fields. An improvement of the transfer efficiency to 0.99999 has been found, coupled with a high charge handling capacity (1.2×10^{12} electrons cm^{-2}) and very high speed of operation (> 180 MHz).

2. THE CHARGE TRANSFER

2.1 The Main Types of Charge Transfer Devices

A CTD-classification is given by:
1. Bucket Brigade Devices (BBDs)
2. Surface Charge Coupled Devices (SCCDs)
3. Bulk Charge Coupled Devices
   where bulk CCDs can be subdivided into:
3a. Buried channel Charge Coupled Devices
3b. Peristaltic Charge Coupled Devices (PCCDs).

As can be seen from the introduction, in CTDs a row of periodic potential wells is present. Their depth can be varied and moved by the potentials applied to a periodic gate structure. The thermal equilibrium of the semiconductor is disturbed if a deep potential well is induced by such a gate structure. As a result electron-hole pairs are generated so that after some time the potential wells are filled and the thermal equilibrium is restored. We assume the transfer time of the charge packets from input to output so short that the number of thermally generated charges is neglible.
As the bipolar BBD and the MOS BBD operate in a similar way [18] we will restrict ourselves to MOS CTDs.

2.1.1 MOS Bucket Brigade Device

Sangster reported on the MOS BBD in 1970 [8]. A cross-section of this device is shown in fig. 9. It is made up from an array of silicon diodes coated by a dielectric layer (SiO_2) on top of which conductive gates are deposited. Each gate overlaps an N-islands and the P-region between two adjacent islands. The capacitance between the island and the substrate can be much smaller than that between the gate and the island. Hence the capacitance of the PN-junction may be neglected. In that case the potential of an N-island follows the variations in potential of the overlapping gate. The potential of an N-island (also referred to as bucket) depends also on the electronic charge quantity, thus:

$$\delta V = - \frac{\delta Q_r}{C} \qquad (2.1)$$

where δV is the change in potential caused by an increment δQ_r in the electronic charge density and C the capacitance per unit area between island and gate. This is shown in fig. 14 where ϕ_s, the surface potential at the $Si-SiO_2$ interface is plotted for one stage of the device. The N-islands constitute potential wells for electrons. Equal and large quantities of electrons are absent from both buckets and this is represented in fig. 14 by a solid line. The right-hand gate is at a higher positive potential than the left-hand one, giving rise to an electron transport from left to right which is indicated by a dashed line. Ultimately; a stable situation is reached in which there is still a potential difference between the two wells (chain-dotted line). A large number of electrons remains in the left-hand bucket because the potential barrier between the buckets stops the electron flow.

The MOS-BBD fits the description of a CTD, where an array of potential wells with an externally variable depth is present, making it possible to introduce, transfer and read out charge packets (see fig. 9). However, only the charge carriers in excess of a fixed reference number are transferred. This reference number is determined by the potential depth of the well and the potential of the barrier region and hence by the voltages on the gates. The signal charge in the left-hand bucket is given by:

$$Q_r = C(V_s - V_{Ref.}) \qquad (2.2)$$

where V_s is the actual potential on the left-hand bucket and V_{Ref} the potential on this bucket when it contains the reference number of electrons. This mode of operation will be referred to as the "reference charge mode" or "Bucket Brigade Mode". (section 2.3.1.1).

With a limited number of implanted donors for the N-islands and a sufficiently large clock voltage, no reference charge will be left. In this case V_{Ref} is the surface potential of the N-island region when no electrons are present. That situation is shown in fig. 15, where V_{Ref} is at a higher potential than the barrier surface potential. This mode of operation will be referred to as a "complete" transfer mode or Charge Coupled Mode. (section 2.3.1.2)

Both modes of operation are very similar to the surface mode of operation of the SCCD, as will be shown.

It should be noted that the described BBD is a 2-phase device (2 clock lines) seen from users point of view. As has been shown in the introduction, in fact 3 is the minimum number of phases. However, from physical point of view the BBD is a 4-phase device, 4 clock phase lines are combined to 2 pairs. This is possible due to the incorporated potential difference between the barrier and the N-region, which assures unidirectional charge transport.

2.1.2 The Surface Charge Coupled Device

A cross-section of the SCCD introduced by Boyle and Smith in 1970 [9] is shown in fig. 10.

If a positive voltage is applied to the metal gate of a MOS structure on P-type silicon, a depletion of holes is caused in the surface region, thus creating a potential well for electrons at the Si-SiO$_2$ interface. If two or more MOS structures are a sufficiently small distance apart (smaller than 3 µm), the potential diagram is obtained as shown in fig. 16. Each of the three flat portions of this diagram refers to its corresponding electrode. The charge is stored below the second gate and is transferred to the right by setting the right-hand gate at its highest potential. The dashed line relates to a transient situation, in which electrons flow to the right due to the electric field generated by the electron concentration gradient until all the electrons have transferred to the deepest well. In the final situation, which is indicated by the chain-dotted line, the entire signal charge is underneath the right-hand electrode and equals

$$Q_r = C(V_s - V_{Ref}) \tag{2.3}$$

Here V_s is the surface potential under the right-hand electrode in the presence of the signal charge and V_{Ref} the surface potential with no electrons present (solid line). C approaches the oxide capacitance. The device is in accordance with the description of the CTD given in the introduction. One stage of such a 3-phase SCCD consists of three gates. The charge transfer behaviour is described by the surface mode of operation (section 2.3.1.3).

2.1.3 The bulk Charge Coupled Devices

2.1.3.1 The Buried Channel Charge Coupled Device

In 1972, Walden, Early et al [7,10] published a paper on the Buried Channel Charge Coupled Device. The main objective was to eliminate the interaction between the carriers and the surface states, which improves the transfer efficiency. For that reason the minimum potential, situated at the surface for the previous device, is displaced to the bulk and the charges are transferred in the bulk layer. Fig. 11 shows a cross section of such a device. It consists of an N-type bulk layer implanted in a P-type substrate and covered by thermally grown silicon dioxide. Conductive gates are deposited on top of the device in a manner similar to the SCCD, however preferably by an overlapping gate structure to minimize the gaps.

Fig. 11c shows the potential for a cross-section perpendicular to the $Si-SiO_2$ interface. The solid curve indicates full depletion of the N-type layer by reverse biasing the PN-junction and by the application of a gate voltage which is negative with respect to the potential well in the N-layer. When a less positive voltage is applied to the N-layer, it will not be fully depleted which is indicated by the dashed curve. The electron concentration is equal to the concentration of ionized donors in the flat part of this curve. The signal charge is formed by the electronic charge quantity.

The depth of the potential well is a function of the voltage between the gate and the substrate. The variation of the potential minimum with the distance in the direction of charge flow for three gate voltage levels resembles the graph shown in fig. 16. Hence the charge transfer mechanism in the bulk layer is in good agreement with the description of a CTD given in the introduction. The transfer process will be described by the bulk mode operation (section 2.3.1.4).

2.1.3.2 The uniformly doped Peristaltic Charge Coupled Device

In 1972, Esser [11] published a paper on the Peristaltic Charge Coupled Device. The main objective was to obtain efficient and high speed transfer together with a high charge handling. For that reason a low doped and relatively thick (~ 5 μm) N-type epitaxial layer was used as bulk layer [11,15]. The potential diagram of fig. 12c shows that the full well charge packet is mainly stored at the surface. The portion stored in the bulk layer is small, owing to the low doped bulk layer. This results in a small effective distance of the charge packet to its gate, that is a high charge handling capacity. The last vestige of charge is transferred deep in the bulk layer at a depth d (fig.

12b) - the depth of the pinch off level below the 0 gate - thus at a high speed. Small charge quantities are stored and transferred in the bulk layer only, as can be seen from the potential diagrams. In that case there is no interference with the surface states.

For larger charge quantities the charge transfer of this device is described by the surface + bulk mode or twin-channel mode (section 2.3.1.4). For small quantities it is described by the bulk mode (section 2.3.1.4).

2.1.3.3 The profiled Peristaltic Charge Coupled Device

The uniformly doped PCCD is limited in efficiency by the interaction of the charge carriers with the surface states. However its twin channel operation combines the advantages of charge transfer at the surface (high charge handling capacity) and charge transfer deep in the bulk (high speed). In order to maintain these advantages and to eliminate the interference of the charges with the surface states, the profiled PCCD was proposed in 1973 by Esser [15,16]. This device is obtained by implanting a very thin and relatively high doped N-type surface layer on top of the relatively thick and lowly doped N-type layer of the uniformly doped PCCD (fig. 13). Most of the charge is stored in the highly doped and thin surface layer, very near to the Si-SiO$_2$ interface. This results in a high charge handling capacity while the interaction with the surface states is eliminated. This profiled PCCD or twin-layer PCCD is superior to all other CTDs in terms of the combination of high charge handling capacity, very high speed transfer and high transfer efficiency. The charge transfer is described by the surface + bulk mode or twin-channel mode (section 2.3.1.4).

2.2 The transfer Mechanism

The mechanisms of charge transfer are:
- self-induced drift[*]
- diffusion
- externally-induced drift (fringing field drift)[*]

The first portion of a relatively large charge packet is transferred by self-induced fields. At the edge of the transferring electrode there are strong electric fields which reduce the local electron concentration to a very low value (channel

---

[*] The terms self-induced field and fringing field have been introduced by Kosonocky et al [12,65]. The term fringing field is questionable. For that reason we will use a term which is complementary to self-induced field, viz. externally induced field [11].

pinch-off) within a short time after applying the transfer pulse to the gate. This results in a charge-carrier gradient which generates self-induced fields in the transfer direction. These fields are the main factor determining the charge transfer as long as large charge concentrations are present.

The self-induced fields can be calculated [11,70] from the formula:

$$E = \frac{1}{C} \frac{\partial Q}{\partial x} \qquad (2.4)$$

where Q is the mobile charge carrier concentration per unit area multiplied by the electronic charge, x is the charge transfer position and C the capacitance per unit area between the charge at x and the overlying gate electrode. C is inversely proportional to the distance between charge packet and gate.

When the remaining mobile charge under the transferring gate has decreased to a sufficiently low value, $\partial Q/\partial x$ and hence E will be negligible. The remainder of the transfer process will then be governed either by diffusion or by externally induced fields, depending on the geometry of the device. These fields are generated directly by the external potentials applied to the gate electrodes. Since the total amount of charge is proportional to $\partial Q/\partial x$ [12], the residual charge needed to allow the self-induced drift to prevail over diffusion or externally induced drift becomes smaller when the distance between the charge packet and the gate is increased. This is the case with bulk CTDs. At the same time the transfer speed is increased during the period in which the self-induced field is dominant [11].

Depending on the ratio L/d - where L is the gate length and d the effective distance of the last charge fraction to its gate - we distinguish:
(a) <u>self-induced field transfer</u> (SF type) ($L \gg d$). For this type the charge transfer is determined by drift fields, caused by variations in carrier density, and by thermal diffusion.
(b) <u>externally-induced field transfer</u> (EF type). For this type L is of the same order of magnitude as d. Drift fields directly induced by external voltages on the electrodes become dominant for the charge transfer.

From the viewpoint of mode of operation, Charge Transfer Devices can be divided into three classes.(This will be shown in section 2.3 and 3). These classes with their qualities and their main representatives are summarized in the following table. (see next page)

By <u>surface mode</u> we understand charge transfer at the surface or near the surface, so that the capacitance (C) between the charge and its gate approximates the oxide capacitance (C_{ox}). The <u>bulk mode</u> is defined as the charge transfer in a bulk layer at a considerable distance from the surface, so that $C \ll C_{ox}$. By the <u>twin-channel mode</u> is meant an appropriate combination of the surface mode and bulk mode.

mode of operation	Qualities (H=high,L=low)		representative types
	charge handling	transfer speed	
1. surface mode	H	L	BBD; surface CCD; shallow-layer buried channel CCD
2. bulk mode	L	H	deep buried-channel CCD; bulk-operated uniformly doped PCCD
3. surface + bulk mode (twin channel mode)	H	H	uniformly doped layer PCCD; twin layer (profiled) PCCD

2.3 Transfer Rate

Using a one-dimensional model, the charge transfer equation (2.7) may be derived with the aid of the relationship

$$J = - Q\mu E - D\frac{\partial Q}{\partial x} \tag{2.5}$$

where

$$D = \mu \frac{k\Theta}{q}$$

and the divergence equation for current

$$\frac{\partial Q}{\partial t} = - \frac{\partial J}{\partial x} \tag{2.6}$$

$$\frac{\partial Q}{\partial t} = \frac{\partial \left(D_{eff} \frac{\partial Q}{\partial x}\right)}{\partial x} \tag{2.7}$$

where J current density for electrons
  μ mobility of the free charge
  Q free charge concentration per area
  x coordinate of the transfer direction
  E drift field in transfer direction
  D thermal diffusion coefficient
  t time
  k Boltzmann's constant
  Θ absolute temperature
  q magnitude of electronic charge

For the remaining charge concentration per unit area we define

$$Q_r(t) = \frac{1}{L} \int_0^L Q(x,t) dx \qquad (2.8)$$

$x = 0$ to $x = L$ is the area where the charge is stored initially and for the initial charge we define

$$Q_i = Q_r(0) \qquad (2.9)$$

2.3.1 Self-Induced Field Transfer

This section deals with the main types of CTDs designed with relatively large gates (L) in comparison to the distance (d) of the last charge fraction - being transferred - to its gate.

$$L \gg d$$

In this case we may neglect externally induced fields.

2.3.1.1 Bucket Brigade Mode

The surface potential along one stage of a MOS-BBD is shown in fig. 14. The N-regions are heavily doped, hence the variation in potential within these buckes will be negligible. The potential drop between the buckets is sustained by the barrier region L_B. The chain dotted curve shows the potential when the left bucket contains the reference number of electrons. The dashed line represents the situation immediately after initiation of transfer.

The positively biased bucket, which receives the transferring charge, generates a very strong field in the adjacent part of the barrier region. In this field-strength region, the electron concentration will be very low resulting in a channel pinch-off. Over the rest of the barrier channel the field strength will be sufficiently strong to justify neglect of thermal diffusion currents with respect to self-induced drift currents, obtaining for the current

$$J = -Q\mu E = Q(x,t)\mu \frac{\partial \phi_s}{\partial x} \qquad (2.10)$$

Since the gate on top of the oxide causes an equipotential plane, the surface potential along the barrier can be related to the charge by

$$\frac{Q(x,t)}{C_{ox}} = \phi_s(x,t) - V_{Ref} \qquad (2.11)$$

where C_{ox} is the oxide capacitance per unit area. Combination of (2.10) and (2.11) results in

$$J = \mu C_{ox}(\phi_s(x,t) - V_{Ref})\frac{\partial \phi_s}{\partial x} \quad (2.12)$$

The current in the barrier region is almost independent of x ($L_C \gg L_B$). Hence by integration one obtains

$$J = \frac{C_{ox}}{2L_B}(V_s(t) - V_{Ref})^2 \quad (2.13)$$

Integration of equation (2.6) over the area L_c combined with (2.8) yields

$$J = -L_C \frac{dQ_r(t)}{dt} \quad (2.14)$$

Using (2.2), (2.13) and (2.14) leads to

$$\frac{dQ_r(t)}{dt} = -\frac{\mu}{2L_B L_C C_{ox}} Q_r^2(t) \quad (2.15)$$

Integration combined with (2.9) yields

$$\frac{Q_r(t)}{Q_i} = \left(1 + \frac{t}{T_B}\right)^{-1} \quad (2.16)$$

where

$$T_B = \frac{2C_{ox} L_B L_C}{\mu Q_i} \quad (2.17)$$

$Q_r(t)$ is the signal charge per unit area still to be transferred at time t, Q_i is the total signal charge per unit area to be transferred at $t = 0$, or the charge at $t = 0$ present in the emitting bucket in excess of the reference charge. Figure 17 shows Q_r versus time for two initial signal charge quantities Q_i. A large part of the charge is transferred in a relatively short time. After that the charge transfer speed decreases sharply. Although the last charge quantity decreases relatively slowly, the charge difference δQ_r falls off much faster. This effect has been used by Sangster to improve the efficiency of the BBD as will be shown. It is related to the definition of charge transfer inefficiency. When the gate voltages vary with a clock frequency f_c, the allotted time for each transfer equals

$$t = \frac{1}{2f_c} \quad (2.18)$$

Hence

$$\frac{Q_r}{Q_i} = \frac{2f_c T_B}{1 + 2f_c T_B} \quad (2.19)$$

The most logical definition of the transfer inefficiency is

$$\varepsilon = \frac{Q_r}{Q_i} \qquad (2.20)$$

that is the fractional charge remaining. We will define it as the transfer inefficiency for zero bias charge operation, the "zero bias transfer inefficiency" (ε). Suppose that after $t = 1/2f_c$ the variance charge δQ_r becomes negligible. In that case we would have an ideal CTD. For, independent of the initial charge, the transfer process is stopped at a (by the transfer time) modified reference level. This means that just before receiving a packet and just after transfer of the same packet the bucket has the same modified reference level. In other words all the signal charge received will be transferred completely. For this reason Sangster [66,67] defined the transfer inefficiency as

$$\varepsilon_B = \frac{\delta Q_r}{\delta Q_i} \qquad (2.21)$$

A similar definition has been given by Berglund et al [17]. We will refer to this as the transfer inefficiency for bias charge operation, the "bias charge transfer inefficiency". Equation (2.21) represents a more appropriate definition of transfer inefficiency for self-induced field type CTDs.

Using equations (2.16), (2.17) and (2.20), equation (2.21) results in

$$\varepsilon_B = \left(\frac{Q_r}{Q_i}\right)^2 = \varepsilon^2 \qquad (2.22)$$

For relatively low clock frequencies where

$$\frac{1}{2f_c} \gg T_B$$

equation (2.19) and (2.20) yields

$$\varepsilon \simeq 2f_c T_B = \frac{4f_c C_{ox} L_B L_C}{\mu Q_i} \qquad (2.23)$$

Substitution in (2.22) leads to

$$\varepsilon_B \simeq \frac{16 f_c^2 C_{ox}^2 L_B^2 L_C^2}{\mu^2 Q_i^2} \qquad (2.24)$$

Since ε_B depends quadratically on the frequency and the signal-charge handling capacity is reduced by the bias charge, the operation of the device is limited at high frequencies, as can be deduced from eq. (2.24).

For simplicity, the most important parasitic capacitances, i.e. the capacitance between the gate and the preceding bucket and the PN junction capacitance of the bucket, have not been included in eq. (2.24). In [18] these parasitics have been taken into account indeed.

Equation (2.24) predicts ε_B to increase in proportion with the square of the operating frequency. However, measurements [19] have shown that ε_B is constant at lower frequencies. This low-frequency ε_B increases with decreasing doping level of the P-type substrate. This can be accounted for in terms of feedback through channel length modulation [17]. Thornber [20] showed the effect to be frequency-dependent. This frequency dependence has not been confirmed by experiments.

A plausible alternative explanation of the low-frequency transfer inefficiency is that it is caused by feedback through modulation of the barrier height. This has been confirmed by Sangster [18] in a special experiment. Fig. 18 shows two adjacent buckets. The boundary of the depletion region at the end of the charge transfer is indicated by a dashed line. The barrier height (V_{Ref}) varies under the influence of the fringing field produced by the potential of the receiving bucket. This feedback effect becomes weaker as L_B is larger and the depletion layers are narrower (i.e. a heavier doping of the substrate). Besides, this effect is independent of frequency. A change in V_{Ref} by about 10 mV is sufficient to explain the low-frequency behaviour of the transfer inefficiency.

2.3.1.2 Charge Coupled Mode

As is shown in section 2.1.1, for relatively low doped N-regions the buckets can be emptied completely. So we have a zero reference charge level. The potential diagram is shown in fig. 15. Now the bucket region (L_c) is no longer an equipotential plane. As can be seen from section 2.3.1.3, for this mode of operation of the 2-phase MOS device can be derived for $L_B \ll L_c$

$$\frac{Q_r}{Q_i} = \left(1 + \frac{t}{T_c}\right)^{-1} \tag{2.25}$$

with

$$T_c = \frac{2L_c^2 C_{ox}}{\pi \mu Q_i} \tag{2.26}$$

These equations are very similar to those for the BBD mode viz. (2.16) and (2.17). However, a lower limit in frequency is set in comparison with the BBD mode. The limit in frequency is decreased by $\pi L_B/L_c$ as is seen from equations (2.17) and (2.26).

2.3.1.2 Surface Charge Coupled Device

a. Surface Mode

Fig. 16 shows the surface potential diagram of the three-phase SCCD. The solid curve shows the situation that the electron concentration at the surface is negligible. The dashed line shows the potential modification due to the electron concentration shortly after the initiation of transfer. Self-induced drift and thermal diffusion currents are responsible for the charge transfer. At $x = L$ (channel pinch-off point) the charge concentration is supposed to be zero as a result of the positive biased third gate. The drift field in the transport direction can be approximated by

$$E = \frac{1}{C_{ox}} \frac{\partial Q}{\partial x} \tag{2.27}$$

The capacitance of the charge with respect to the substrate has been neglected. Combination of the equations (2.27), (2.5), (2.6) results in equation (2.7), with effective diffusion coefficient

$$D_{eff} = \mu \left(\frac{Q}{C_{ox}} + \frac{k\Theta}{q} \right) \tag{2.28}$$

Numerical solution of the charge transfer (eq. (2.7), (2.8) and (2.28)) with $Q_i = 5 \times 10^{11}$ q Coul/cm^2 is shown in fig. 19 curve 1. It shows the plot of the fractional charge versus time. After a few times T, the fractional charge remaining is a few percent. During this time the transfer is governed by self-induced drift. After that time, when $Q/C_{ox} \leq k\Theta/q$, the transfer process is dominated by thermal diffusion.

b. Self-induced Transfer Rate

As long as the electron concentration is sufficiently large, thermal diffusion may be neglected. Then equation (2.28) becomes

$$D_{eff} = \frac{\mu Q(x,t)}{C_{ox}} \tag{2.29}$$

Substitution in (2.7) leads to

$$\frac{\partial Q(x,t)}{\partial t} = \frac{\mu}{2C_{ox}} \frac{\partial^2 Q^2(x,t)}{\partial x^2} \tag{2.30}$$

As shown by Carnes et al [12] a short time after the initiation of transfer, the electron concentration gets a shape which does not change in time. Hence, separation of variables can be applied and equation (2.30) can be integrated to yield

$$\frac{Q_r(t)}{Q_i} = \left(1 + \frac{t}{T_c}\right)^{-1} \tag{2.31}$$

where

$$T_c = \frac{2L^2 C_{ox}}{\pi\mu Q_i} \tag{2.32}$$

is the self-induced transfer time constant for the surface mode of operation. Since these equations are very similar to the corresponding equations for the BBD, the SCCD and the BBD will have similar high frequency limitations. Similar expressions are found for the zero bias transfer inefficiency (ϵ) and the bias charge transfer inefficiency (ϵ_B).

c. Thermal Diffusion Transfer Rate

For small charge fractions, when $Q/C_{ox} \ll k\Theta/q$, self-induced field transfer may be neglected. Then equation (2.28) becomes

$$D_{eff} = \mu \frac{k\Theta}{q} \tag{2.33}$$

Substitution in (2.7) yields

$$\frac{\partial Q(x,t)}{\partial t} = \mu \frac{k\Theta}{q} \frac{\partial^2 Q(x,t)}{\partial x^2} \tag{2.34}$$

Solution of this equation [21] yields

$$\frac{Q_r}{Q_i} = \exp\left|-\frac{t}{T_d}\right| \tag{2.35}$$

where

$$T_d = \frac{4}{\pi^2} \frac{qL^2}{\mu k\Theta} \tag{2.36}$$

This diffusion time constant is responsible for the relatively slow transfer of the last charge fraction. Eq. (2.35) represents the transfer inefficiency of small charge fractions, valid for all self-induced field CTDs.

2.3.1.4 Bulk Charge Coupled Devices

2.3.1.4.1 Buried channel Charge Coupled Device and uniformly doped Peristaltic Charge Coupled Device

a. Bulk mode

This mode of operation is valid for the buried channel CCD and also the PCCD in case of small charge quantities (fig. 11b). For the sake of simplicity we assume a very high-ohmic substrate, so we can neglect the charge-substrate capacitance as well as the depletion region in the N-layer, caused by the negatively biased P-substrate. From the one-dimensional Poisson equation we derive for the potential difference between the channel and its gate

$$V = qN_2 d' \left(\frac{d_1}{\varepsilon_1} + \frac{d'}{2\varepsilon_2} \right) \qquad (2.37)$$

and for the self-induced drift field

$$E = \left(\frac{d_1}{\varepsilon_1} + \frac{d'}{\varepsilon_2} \right) \frac{\partial Q}{\partial x} \qquad (2.38)$$

with

$$Q = qN_2(d-d') \qquad (2.39)$$

where d_1 thickness of the oxide layer
N_2 dope level of the N-layer
ε_1 dielectric constant of the silicon oxide
ε_2 dielectric constant of the silicon.

Equations (2.5) to (2.7), and (2.38) result in

$$D_{eff} = \mu \left| Q \left(\frac{d_1}{\varepsilon_1} + \frac{d'}{\varepsilon_2} \right) + \frac{k\Theta}{q} \right| \qquad (2.40)$$

At the depletion depth d' (figs. 11b, 12b, 13b) we assume a discontinuity of zero concentration to dope level concentration of the mobile charge. The term d' is place and time dependent. When, initially, the bulk is completely filled then d' varies from zero to d during transfer. One may expect an effective value (d_{eff}) which lies between zero and d. Calculations show [11] that d_{eff} is almost equal to d. (fig. 19, curve 2, $N_2 = 10^{15}$ cm^{-3}, $d = 5$ μm, $Q_i = 5 \times 10^{11}$ q Coul/cm^2). This is due to the fact that the last fraction of the charge, which determines the transfer time, is almost at a distance d. This means that in principle the transfer behaviour in the bulk mode of operation is almost the same as in the surface mode of operation with an increased oxide thickness of

$$t_{ox} = Bd_1 \qquad (2.41)$$

where

$$B = \left(1 + \frac{d\varepsilon_1}{\varepsilon_2 d_1}\right) \tag{2.42}$$

It is the ratio of the oxide capacitance (surface mode) and the bulk capacitance for the last charge fraction to its gate ($d' \simeq d$) and thus the gain factor of the self-induced drift field (eq. (2.27), (2.38)). So, for bulk transfer, the self-induced drift prevails over diffusion to a B times smaller charge value than for surface transfer. This decreases the self-induced transfer time constant by a factor of B [11]. If $d = 5$ μm and $d_1 = 0.1$ μm then $B \simeq 16$.

It may be concluded that for a relatively high-ohmic substrate the transfer rate is not dependent on dope level and dope profile of the N-layer, but depends only on the depth d where the last charge fraction is transferred.

The degree of doping of the P-substrate and the N-layer does not significantly alter the principle of operation of the device (fig. 20). It can be shown that B is the ratio of the oxide capacitance to the parallel capacitance of the charge bunch at pinch off (d) to its electrode and substrate [11].

$$B = \frac{(d_2-d)(1+\frac{N_2}{N_a})(1+\frac{d\varepsilon_1}{\varepsilon_2 d_1})}{d_2 + (d_2-d)\frac{N_2}{N_a} + \frac{\varepsilon_2}{\varepsilon_1}d_1} \tag{2.43}$$

with N_a the substrate dope level. A disadvantage of bulk mode operation is that the charge handling capacity is reduced in the order of B, compared to surface mode operation.

b. Surface and bulk mode operation

This is the basic PCCD twin-channel mode of operation for the uniformly doped bulk CCD. In the following considerations, for convenience, we will again assume a very high-ohmic substrate. Fig. 12b shows the situation when a charge packet is transferred. At $x = x_1$ the accumulation charge is zero. During transfer x_1 moves from $x = L$ to $x = 0$. For the self-induced drift fields one finds

$$E = \frac{d_1}{\varepsilon_1}\frac{\partial Q}{\partial x} \quad, 0 < x < x_1 \tag{2.44}$$

and

$$E = \left[\frac{d_1}{\varepsilon_1} + \frac{d'}{\varepsilon_2}\right]\frac{\partial Q}{\partial x} \quad, x_1 < x < L \tag{2.45}$$

These equations yield the corresponding effective diffusion coefficients given by eq. (2.28) and (2.40). The bulk- and surface mobility are assumed to be equal, so that Q and $\partial Q/\partial x$ are continuous at the moving boundary x_1.

A result of the numerical solution of the charge transfer equations is shown in fig. 19 (graph 3) for $N_2 = 2 \times 10^{14}$ cm^{-3} and the same amount of initial charge as for graph 1 and 2. When the surface channel dominates the transfer process, curve 3 approximately follows curve 1 of the surface mode. After a few time constants, when the surface charge has disappeared, ($x_1 \approx 0$) curve 3 follows almost curve 2 (bulk transfer). Curve 3 is almost identical to curve 2, when curve 2 is shifted to the right by an amount equal to the time needed for the transfer of the surface accumulation charge (in this case 80%). Numerical calculations show that this still holds good when the difference in surface- and bulk mobility is taken into account.

The time needed for the transfer of the first 80-95% of the charge is negligible compared to the transfer time of the last fraction. This fraction is transferred at a depth d. Thus, from the point of view of the transfer rate the device behaves almost as if all the charge is transferred at the pinch-off level (d).

The same gain factor B, given by the equation (2.42) and (2.43) for the respective cases, is found. The charge handling capacity is greatly increased by the high storage capacitance for the main part of the charge packet at the surface, combined with the relatively low dope level of the bulk layer. It demonstrates the advantages of having a twin channel, one highly conducting channel at the surface (moving source) and a weak channel deep in the bulk layer.

With the present state of technology the transfer efficiency is determined by the surface states for this mode of operation. The following device, the profiled PCCD, however, has the advantages of the surface- and bulk mode operation, that is the high charge handling capacity and high transfer rate and furthermore it eliminates the influence of the surface states.

2.3.1.4.2 <u>Profiled Peristaltic Charge Coupled Device</u>

As has been shown, when the last fraction of the charge is transferred at a depth d - from the point of view of the charge transfer rate - the device behaves as if all the charge is transferred at that depth. Consequently, basically the transfer inefficiency (ϵ) is decreased by the corresponding factor B (eq. (2.42) and fig. 19) in comparison with the surface mode of operation. By storing the former accumulation part, very near its electrode, in a very thin relatively highly doped N_1 layer on top of a substantially thick and relatively low-doped N_2 layer, (fig. 13b) then all the charge is confined to the bulk and the influence of the surface states is eliminated. This type of bulk CCD is known as the profiled or twin layer PCCD [15, 16, 22, 71]. It combines high speed transfer and high charge handling capacity while interaction of the charge with the surface states is avoided.

For both regions $0 < x < x_1$ and $x_1 < x < L$ (fig. 24) equations (2.38) and (2.40) are valid. With

$$Q = qN_1(d_3-d') + qN_2 d, \qquad 0 < x < x_1 \qquad (2.46)$$

$$Q = qN_2(d-d'), \qquad x_1 < x < L \qquad (2.47)$$

in this case the influence of the substrate is also neglected for the sake of simplicity, because the degree of doping of the substrate has no significant influence on the transfer rate and charge handling capacity.

At the moving boundary x_1, Q and $\partial Q/\partial x$ are continuous (fig. 13b). The graph, from numerical calculations of the appropriate transfer equations, as one would expect, is found to lie between graph 3 and 2 of fig. 19, for the same amount of initial charge, where was taken
$N_1 = 1.4 \times 10^{16}$ cm^{-3}, $d_3 = 0.3$ μm
$N_2 = 2 \times 10^{14}$ cm^{-3}, $d = 5$ μm and
$Q_i/q = 5 \times 10^{11}$ cm^{-2}, $d_1 = 0.1$ μm.

2.3.1.4.3 Self-induced Transfer Rate

As long as self-induced drift prevails over the thermal diffusion transfer one can neglect the diffusion. For surface mode operation (2.3.1.3) this is the case as long as

$$\frac{Q}{C_{ox}} > \frac{k\Theta}{q} \qquad (2.48)$$

which leads to the time constant given by equation (2.32). Equation (2.31) gives the high-frequency limitation when the charge transfer is dominated by self-induced fields. For a bulk CCD, when the charge packet has its pinch-off level at depth d (fig. 11b, 12b and 13b) equation (2.40) can be approximated by

$$D_{eff} \simeq \mu \left[BQ\frac{d_1}{\epsilon_1} + \frac{k\Theta}{q} \right] \qquad (2.49)$$

Hence, for the self-induced transfer time constant for a bulk CCD one may write [11]

$$T_B \simeq \frac{1}{B} \frac{2L^2 \epsilon_1}{\pi \mu d_1 Q_i} \qquad (2.50)$$

This is an improvement of about B times in comparison to the surface CCD. For $d = 5$ μm and $d_1 = 0.1$ μm it is found $B \simeq 16$. Thus, for the same transfer efficiency, self-induced transfer bulk CCDs will operate to a B-times higher frequency than the SCCD [11].

For the ratio of the thermal diffusion time constant and the self-induced time constant for a bulk CCD we find (eqs. (2.36) and (2.50)).

$$\frac{T_d}{T_B} = B \frac{2}{\pi} Q_i \frac{d_1}{\varepsilon_1} \frac{q}{k\Theta} \qquad (2.51)$$

with $Q_i \dfrac{d_1}{\varepsilon_1} = 2.5$ Volts ($Q_i/q \simeq 5 \times 10^{11}$ electrons/cm^2)

$d_1 = 0.1$ μm
$d = 5$ μm

substituted in (2.51) yields

$$\frac{T_d}{T_B} \simeq 10^3 \qquad (2.52)$$

and for surface transfer

$$\frac{T_d}{T_B} \simeq 65 \qquad (2.53)$$

2.3.2 Externally-induced Field Transfer

Now L and d are of the same order of magnitude. The drift fields induced by externally applied potentials become very important, especially for the transfer of the last charge fraction. When there is no mobile charge in the device the potentials very near to the transferring electrode will be determined by that electrode so that only a small drift field may be expected. Deeper in the bulk the potential becomes more dependent on the potentials of the adjacent electrodes so that larger drift fields will exist there [10,11,12,24,25,26]. As a function of x (transfer direction) at $x \simeq L/2$ a minimum is found for the drift field, below the transferring gate. The drift field as a function of the depth y, at $x = L/2$, is given in fig. 21 [25,26]. A maximum of $E \simeq V/2L$ is found at a depth of $y \simeq L/2$ (saddle point). If the last fraction is transferred at a depth around $L/2$ we will have a very short transfer time leading to high transfer efficiencies (complete transfer) at very high clock rates. This situation will occur in all EF-type CTDs.

In the twin-layer PCCD the main part of the charge is very near to its electrode, what results in a high charge handling capacity, and will mainly be transferred by self-induced fields. As is seen from fig. 19 (curve 3) the self-induced transfer speed is very high for charge values larger than about 3×10^{10} electrons per cm^{-2}. This last charge fraction is driven deeper into the layer where high, externally-induced drift fields exist, leading to high efficient and high speed transfer. Hence all the charge will be transferred at a high speed. This again shows the advantages of the PCCD twin-channel operation.

2.3.2.1 The externally-induced transfer rate

If the externally-induced drift is dominant the continuity equation becomes

$$\frac{\partial Q}{\partial t} = \mu \frac{\partial (QE)}{\partial x} \tag{2.54}$$

If we assume a constant field strength E for $0 < x < L$, zero charge concentration at $x = L$ and zero gradient in charge at $x = 0$ then (2.54) results in [27]

$$\frac{Q_r}{Q_i} = \exp\left[-\frac{t}{T_e}\right] \tag{2.55}$$

$$T_e = \frac{4}{\pi^2} \frac{L}{\mu E} \tag{2.56}$$

for $E = \frac{V}{2L}$, $V = 5$ Volt (fig. 21)

$$\frac{T_d}{T_e} = \frac{qV}{2k\Theta} = 10^2 \tag{2.57}$$

An improvement of two orders of magnitude for the EF-time constant, in comparison to the diffusion time constant for the SF-type. Equations (2.52), (2.53) and (2.57) show that for high speed and high charge handling self-induced fields have to be exploited (see last paragraph before this section). The inverse transfer time derived from eq. (2.55)

$$t^{-1} = \frac{-1}{T_e \log\left(\frac{Q_r}{Q_i}\right)} \tag{2.58}$$

can be seen as a minimum upper bound in frequency. This means a transfer rate of 1.8 GHz for $L = 7.5$ μm, $\varepsilon = Q_r/Q_i = 10^{-4}$ and $V = 5$ volts. This result depends very strongly on the boundary conditions at $x = 0$ and $x = L$. Practical results will be different and will be more favourable. The transfer gates induce a quasi travelling potential wave in the bulk. If we consider the charge transfer at a depth of about $L/2$ as "surf riding" caused by a travelling wave and assume an effective drift field under the full gate length of

$$E = \frac{V}{2L} \tag{2.59}$$

(which will be higher in practice [68]) then the maximum upper bound is found from

$$t^{-1} = \frac{\mu E}{L} \approx 0.4 T_e^{-1} \tag{2.60}$$

where $V = 5$ volts, half the peak to peak clock voltage, $\mu = 1300$ cm^2/Vs. This means transfer rates of 6.5 GHz, and a clock frequency of 1.6 GHz for a 4-phase gate structure. Note that the clock rate is also the sampling rate of the signal at the input of the device.

Equation (2.55) has been drawn in fig. 19 graph 4. It shows the tremendous increase in transfer speed, minimally obtained for the externally-induced field type, in comparison to the self-induced field type. The curves of fig. 19 show that for an EF type of bulk CCD, and particularly the profiled PCCD, the self-induced transfer will prevail over the externally-induced transfer until the charge is reduced to about 3×10^{10} electrons cm^{-2}. Then the transfer speed is governed by the externally-induced transfer.

<u>Conclusion from this section</u>: If the last charge fraction is transferred at a depth d, the bulk device almost behaves as if all charge is transferred at that depth d. This fortunately only holds true from the point of view of speed and not from the point of view of the charge handling capacity.

3. CHARGE HANDLING CAPACITY

The charge handling capacity of a CTD is the maximum amount of charge which can be handled for a given clock voltage. The signal range is given by the charge handling capacity minus the bias charge (fat zero), which may be necessary for a good operation of the device (see section 2.3 and 4).

All externally-induced field-type bulk CCDs can be operated with a reasonable transfer efficiency without bias charge (real zero mode).

The most simplified model for the determination of the charge handling capacity will be obtained by the following considerations:
1. A one-dimensional model gives a tremendous simplification of the problem for the externally-induced field-type CTD and it is a realistic model for the self-induced field-type CTD.
2. Use of trapezoidally shaped clock pulses causes a quasi sta-

tionary transfer. This effects that the voltage necessary for storing a charge packet equals the voltage to obtain a one-directional transfer.
3. If the last charge fractions are transferred in the bulk, comparison between the uniformly doped and twin-layer device is made such that these last fractions are transferred at the same depth (d), thus for the same transfer speed.

Point 2 will be clear from fig. 22, which shows the situation where the gate potentials V_2 and V_1 are just sufficient to store the charge packet under the second gate $(V_2 > V_1)$. To obtain a transfer the third gate is applied to the potential V_2 (fig. 23) and the potential of the second gate is varied relatively slowly from potential V_2 to potential V_1 (The last charge fraction is transferred almost at depth d). Thus V_2-V_1 is the minimum peak to peak value of the clock voltage for storing and transferring the charge packet shown. If one uses a staircase shape as shown in fig. 24 then the peak to peak value of the clock voltage is about $2(V_2-V_1)$.

3.1. Surface CTDs

To surface CTDs belong the MOS BBD and the SCCD. Since the charge packets in these devices are transferred and stored at the surface in the oxide capacitance, the relation between the stored charge and its minimum peak-to-peak clock voltage is given by

$$V_{cl} = V_2 - V_1 = \frac{Q_i}{C_{ox}} \tag{3.1}$$

In this situation, the surface potentials below the two gates are just equal (fig. 25). The charge handling capacity does not vary with the substrate dope level. Equation (3.1) is plotted in fig. 28 curve 1. The surface CTDs have the highest charge handling capacity.

3.2 Uniformly doped Bulk Charge Coupled Device

Very shallow layer devices where

$$d_2 \leqslant \frac{\varepsilon_1}{\varepsilon_2} d_1 \tag{3.2}$$

have the charges transferred and stored almost at the surface in the oxide capacitance. Thus their charge handling capacity is qualitatively well approximated by the surface charge handling given by curve 1 fig. 28 (eq. (3.1)).

a. bulk mode

In order to store the charge in a thick layer device (fig. 26), the potential of any point in the charge packet has to be equal to or somewhat higher than the potential below the neighbouring gates. In other words the depletion region of these gates has to be equal to or somewhat below the depth d of the charge packet (fig. 22,26). Thus the minimum clock voltage (V_{cl}) can be calculated from the difference in depletion voltage of the two gates.

$$V_{cl} = V_2 - V_1 = Q_i \left(\frac{d_1}{\epsilon_1} + \frac{z}{\epsilon_2} \right) \qquad (3.3)$$

where $Q_i = q N_d (d-d')$, $C_{ox} = \frac{\epsilon_1}{d_1}$

and $z = \frac{d+d'}{2} = d - \frac{Q_i}{2qN_2}$.

z is the depth of the centre of gravity [15] of the charge packet below the Si-SiO$_2$ interface. Equation (3.3) has been plotted in fig. 28 curve 2 (d = 5 µm, N_2 = 10^{15} cm^{-3}) [16]. At the intersection with the dot dash line the bulk is completely filled (d'=0). This intersection shifts proportionally with the dope level N_2. The dot dash line can be seen as a qualitative approximation of the charge handling capacity for bulk mode operation. The corresponding equation is given by

$$V_{cl} = Q_i \left(\frac{d_1}{\epsilon_1} + \frac{d}{2\epsilon_2} \right) \qquad (3.4)$$

where $Q_i = qN_2 d$, $\left(z = \frac{d}{2} \right)$

b. Surface+bulk mode (PCCD twin-channel mode)

If we add more charge, when the bulk is filled (d'=0), this charge will accumulate at the surface. The following relationship is found between the minimum clock voltage and the charge $Q_i = (Q_a + Q_b)$ to be handled (fig. 27).

$$V_{cl} = Q_b \left(\frac{d_1}{\epsilon_1} + \frac{d}{2\epsilon_2} \right) + Q_a \frac{d_1}{\epsilon_1} \qquad (3.5)$$

where $Q_b = q N_2 d$ is the bulk charge and Q_a is the accumulation charge. Equation (3.5) is shown in fig. 28 by the part of curve 2 above the dash dot line. This part of the curve is parallel with curve 1. Equation (3.5) can be written as

$$V_{cl} = Q_i \left(\frac{d_1}{\epsilon_1} + \frac{z}{\epsilon_2} \right) \qquad (3.6)$$

where

$$z = \frac{Q_b}{Q_a + Q_b} \frac{d}{2} = \frac{qN_2 d}{Q_i} \frac{d}{2}$$

d/z is a measure for the charge handling and is proportional to Q_i. It can be seen that the clock voltage is proportional to the centre of gravity (z). In its turn, the centre of gravity is proportional to the dope level of the bulk layer. Thus the clock voltage can be considerably reduced for the same amount of charge by taking a lower doped bulk layer, e.g. one with a dope level of 2×10^{14} cm^{-3}. This is shown by curve 3 of fig. 28.

3.3 Profiled Peristaltic Charge Coupled Device

It is obvious that curve 3 is also valid for a profiled PCCD with a very shallow N_1 layer ($d_3 \leqslant \varepsilon_2/\varepsilon_1 \, d_1$) [16]. It shows the tremendous increase in charge handling capacity of the profiled PCCD compared with that of the uniformly doped bulk CCD for sole bulk transfer (curve 2, lower part). Note that for both curves 2 and 3 the last charge fractions in the corresponding devices are transferred at the same depth ($\simeq d$). Thus curves 2 and 3 compare in charge handling capacity of the uniformly doped bulk CCD with the twin-layer PCCD for the same transfer speed.

The depth of the centre of gravity for a completely filled twin-layer PCCD is given by [15]

$$z = \frac{(N_1 - N_2)d_3^2 + N_2 d^2}{2[(N_1 - N_2)d_3 + N_2 d]} \qquad (3.7)$$

For $d_3 \to 0$ and $(N_1 - N_2)d_3 = \dfrac{Q_a}{q}$ is constant

$$z = \frac{Q_b}{Q_a + Q_b} \frac{d}{2} \qquad (3.8)$$

where $Q_b = qN_2 d$

From the sections 2 and 3 it can be concluded:
1. The transfer speed is determined by the depth d (pinch off level) at which the last charge fraction is transferred for both the SF type and EF type of CTD. The transfer rate increases with depth d. (For the EF type upto $d \simeq L/2$).
2. The charge handling capacity depends on the depth (z) of the centre of gravity of the charge packet.
3. For the PCCD (twin-channel operation) d and z can be chosen more or less independent of each other, so high-speed transfer and high charge handling capacity for sole bulk transfer (high

efficient transfer) can be combined. As has been verified experimentally in the profiled or twin-layer PCCD.

4. IMPROVEMENT OF THE TRANSFER EFFICIENCY

The transfer inefficiency ε has been dealt with in section 2, where it is defined as

$$\varepsilon = \frac{Q_r(t)}{Q_i} \qquad (2.20)$$

The transfer efficiency is defined by

$$\eta = 1 - \varepsilon \qquad (4.1)$$

An improvement in transfer efficiency for the basic transfer process has been found by bias charge operation. A bias charge gives a smaller time constant for the transfer process of SF-type CTDs. As will be shown there is another reason for using bias charge operation for some types of CTDs. For lower clock frequencies the transfer inefficiency is no longer determined by the basic transfer process, but comes to be dominated by an almost frequency independent term. In CCD this effect is attributed to the interaction of the carriers with the surface states [28, 29] and trapping centres in the bulk [30,31]. These traps are situated at the Si-SiO$_2$ interface and in the bulk, respectively. Both types of trap are at an energy level situated between the valence and conduction band edges of the semiconcuctor. The capture and emission rate of these traps [32] are described by:

$$\frac{dN_t^-}{dt} = \frac{N_t^o}{\tau_c} - \frac{N_t^-}{\tau_e}$$

$$\tau_c = (\sigma_n \bar{v} n)^{-1}$$

$$\tau_e = \left(\sigma_n \bar{v} n_i \exp\frac{E_t - E_i}{k\Theta}\right)^{-1} \qquad (4.2)$$

Here N_t^- = number of traps occupied by an electron,
N_t^o = number of non-occupied traps
E_i = intrinsic Fermi level,
E_t = energy level of the trap,
$\bar{v}$ = average thermal velocity of electrons,
σ_n = trap capture cross section for electrons,
n = concentration of electrons.
τ_c = capture time constant
τ_e = emission time constant

In the presence of a charge packet, n is large so that τ_c is small and the traps will capture electrons until N_t^o has become small (eq. (4.2)). n decreases during the transfer of the charge packet and hence the rate of electron emission exceeds the capture rate. If the trap lies sufficiently far from the conduction band, then the emission time will be longer than the time allowed for charge transfer. Thus the captured electrons will be injected into following charge packets, so that ε decreases. The spread in emission time constants of the surface states is large because they are distributed throughout the band gap [33]. The contribution of the surface states to the transfer inefficiency is almost frequency-independent. Bulk centres are not found to be distributed throughout the band gap [118], so that a frequency dependence of the bulk transfer inefficiency might be expected. For the SCCD and uniformly doped PCCD (in the surface mode and surface plus bulk mode, respectively) a bias charge (fat zero) is used to reduce the influence of the surface states on the transfer efficiency. Each passing charge packet causes the electron concentration at the surface n to be high once every clock period, thus keeping most of the surface states filled. In other words they do not contribute to the transfer inefficiency. The area occupied by the charge packet varies with its magnitude [34,35]. The surface states at the boundary of the charge regions still contribute to the charge transfer inefficiency. Kosonocky et al have given an analysis of this so called "edge effect" [34,39] supplemented by experimental results. For sole bulk transfer in a bulk charge-coupled device, the electron concentration at the surface is always low enough to keep the surface states empty. This prevents the surface states from contributing to ε.

In bulk CCDs one expects bulk trapping centres to increase the charge transfer inefficiency. As with the surface CCD, one would like to eliminate the influence of these centres. However, in a uniformly doped layer bulk CCD the volume occupied by a bulk charge packet is proportional to its magnitude. When the trapping centres are homogeneously distributed, the number of traps occupied in the presence of a packet is proportional to the magnitude of the packet. Obviously a bias charge will not be useful.

In the profiled PCCD most of the volume occupied by a charge packet is taken by a small fraction of the total charge. A fat zero packet of this magnitude will eliminate the major part of the effect of bulk traps and an improved transfer will be found [16,22].

5. POTENTIAL DISTRIBUTION

5.1 Self-induced field-type Charge Transfer Devices

For the SF type CCD $(L \gg d)$ the potential distribution below one gate is negligibly influenced by the potentials of the neighbouring gates. So a one dimensional analysis is appropriate for this case. A simplified yet realistic approach is the problem for bulk CCDs is found by assuming a discontinuity of zero mobile charge outside the packet to dope level concentration at the boundary of the depletion region of the charge packet and or the same assumption at the boundary of the substrate depletion region.

The potential distribution will be found by using Poisson's equation

$$\frac{d^2 V}{dy^2} = -\frac{\rho}{\epsilon} \qquad (5.1)$$

with the conditions of zero electric field at the boundary of the charge packet (for the bulk layer CCDs) and at the boundary of the substrate depletion region, and together with the defined potentials at the gate and substrate.

5.1.1 Surface Charge Coupled Device

Solving equation (5.1) with the appropriate boundary conditions for the situation in fig. 29 yields

$$V_g = Q\frac{d_1}{\epsilon_1} + qN_a d_a \left(\frac{d_1}{\epsilon_1} + \frac{d_a}{2\epsilon_2}\right) \qquad (5.2)$$

where V_g can be seen as the actual gate potential corrected for the flat band potential. For the surface potential

$$V_s = qN_a \frac{d_a^2}{2\epsilon_2} \qquad (5.3)$$

is found by elimination of d_a from equations (5.2) and (5.3)

$$V_s = V_g - Q\frac{d_1}{\epsilon_1} - qN_a \epsilon_2 \frac{d_1^2}{\epsilon_1^2} \left[\left(1 + \frac{2\left(V_g - Q\frac{d_1}{\epsilon_1}\right)}{qN_a \epsilon_2 \left(\frac{d_1}{\epsilon_1}\right)^2} \right)^{\frac{1}{2}} - 1 \right] \qquad (5.4)$$

From equations (5.2) and (5.3) can also be derived

$$V_g = V_s + Q\frac{d_1}{\epsilon_1} + \frac{d_1}{\epsilon_1}\left(2V_s qN_a \epsilon_2\right)^{\frac{1}{2}} \qquad (5.5)$$

For the case that N_a is small and or

$$d_a \gg \epsilon_2 \frac{d_1}{\epsilon_1} \qquad (5.6)$$

equation (5.5) reduces to

$$V_s = V_g - Q\frac{d_1}{\varepsilon_1} \tag{5.7}$$

This expression is valid if the capacitance of the charge to the substrate can be neglected (equation (5.6)). The reference potential V_{Ref} is found by setting $Q = 0$ in equations (5.4) and (5.7).

5.1.2 Bulk Charge Coupled Devices

5.1.2.1 Uniformly doped bulk Charge Coupled Device

a. Bulk mode

In the area where mobile charge is present $\rho = 0$ and for the remaining part of the semiconductor $\rho = qN_2$ or $\rho = -qN_a$. Together with Poisson's equation and the mentioned boundary conditions, where $V_{substrate} = 0V$, this leads to the following equations (fig. 26)

$$V(d') - V_2 = qN_2 d' \left(\frac{d_1}{\varepsilon_1} + \frac{d'}{2\varepsilon_2}\right) \tag{5.8}$$

$$V(d') = qN_2 \frac{(d_2-d)^2}{2\varepsilon_2} \left(1 + \frac{N_2}{N_a}\right) \tag{5.9}$$

$$d = d' + \frac{Q}{qN_2} \tag{5.10}$$

V_2 is the gate potential corrected for the flat band voltage. These equations show the relations between the potential of the charge packet $V(d')$, the charge quantity Q and the position of the charge between d' and d. The extremum of the potential at d_m will be found by setting $Q = 0$.

b. Surface+bulk mode (PCCD twin-channel mode)

The equations for this situation (fig. 27) are found from equations (5.8)-(5.10) by setting $d' = 0$ (5.11)
and correcting V_2 by

$$\Delta V_2 = -Q_a \frac{d_1}{\varepsilon_1} \tag{5.12}$$

where Q_a is the accumulation charge at the surface.

5.1.2.2 Profiled Peristaltic Charge Coupled Device

The same method, as is used in 5.1.2.1 a, can be applied for the potential distribution in the profiled PCCD. If the thickness of the top layer is negligible (fig. 13c)

$$d_3 < \frac{\varepsilon_2}{\varepsilon_1} d_1 \tag{5.13}$$

the same equations are found with V_2 shifted by

$$\Delta V_2 = qI\frac{d_1}{\varepsilon_1} \tag{5.14}$$

(I is the implantation level in cm^{-2}) if there is no mobile charge in the surface layer. (For the situation of fig. 13c is $\Delta V_2 \simeq 10$ V. Compare fig. 13c with fig. 12c) In the case of mobile charge (Q_a) in the surface layer, V_2 is corrected by

$$\Delta V_2 = (qI-Q_a)\frac{d_1}{\varepsilon_1} \tag{5.15}$$

with

$$d' \simeq 0 \tag{5.16}$$

and Q is the mobile charge in the low doped thick layer.
In [22] this will be treated more accurately where the thickness (d_3) of the top layer is taken into account.

5.2 Externally-induced field-type CTDs

For the EF-type bulk CCD ($L \simeq d$) the potential is influenced by the potentials of the neighbouring gates. Thus a two-dimensional analysis is necessary for a good understanding of these CTD types. McKenna and Schryer [24] and Collet and Vliegenthart [25] have given a two-dimensional analysis of the potential distributions in the bulk CCDs without charge packets. Both have made use of Fourier series to solve the problem. Moreover in [25] calculations on the charge distribution in the uniformly doped bulk CCD and the profiled PCCD are presented. A different analysis is given by Hanneman et al [26], based on a proposal of Chi-Shin Wang [36]. The analysis has led to an accurate and simple approximation and has been extended to a model with an oxide layer (equation (5.17)).

$$\phi(x,y) = \frac{1}{\pi} \sum_{i} \Delta V_i (\pi - \arctan\frac{y+d_1}{x-iL}) \tag{5.17}$$

The gates are speced between $x = iL$ and $x = (i+1)L$. ΔV_i is the potential jump of these gates at $x = iL$ and $y = -d_1$. In the region of interest $0 < y < 2L$ and $-2L < x < 2L$ for a four phase device, equation (5.17) is sufficiently accurate for $i = -6$ to $+6$. The outer gates for $x < -6L$ and $x > 6L$, which extend to infinity, are at the average gate potential (zero in this case).

For one potential jump from 0 to ΔV at $x = 0$ and $y = -d_1$ the exact solution is

$$\phi(x,y) = \frac{\Delta V}{\pi} \sum_{n=0}^{\infty} \left(\frac{\varepsilon_1-\varepsilon_2}{\varepsilon_1+\varepsilon_2}\right)^n \left(\pi - \arctan\frac{(2n+1)d_1+y}{x}\right) \qquad (5.18)$$

for

$y > 0$

The one term approximation ($n=0$) gives an accuracy within 1.5%. The presented method is also applicable for non periodic and limited gate structures. Figures 30 and 31 show the equipotentials for a uniformly doped bulk CCD and the profiled PCCD, respectively. The equipotentials in the profiled PCCD are more concentrated below one gate, so it may also be expected that a charge packet will be more concentrated below one gate, as has been shown in [25]. Furthermore, the profiled PCCD shows a considerably deeper potential well in correspondence with the higher charge handling capacity.

6. EXPERIMENTAL RESULTS

The transfer efficiencies η mentioned below give the efficiency per transfer. The efficiency form one stage to the other is found by

$$\eta_{stage} = 1 - p(1-\eta) \qquad (6.1)$$

where p is the number of phases of the device.

6.1 Bucket Brigade Devices

For bipolar BBDs transfer efficiencies of $\eta = 0.999$ up to clock frequencies of 30 MHz [19,37] have been measured. Base current loss gives an attenuation of the signal but has not been taken into account, because charge amplifiers can correct for this signal loss in a long delay line [1,2,37]. With p-channel MOS BBDs transfer efficiencies of $\eta = 0.999$ have been achieved up to 3 MHz clock frequencies [19,37]. With the MOS Tetrode BBD [42] transfer efficiencies of $\eta = 0.9999$ have been obtained.

6.2 Surface Charge Coupled Devices

Typical performances obtained for SCCDs are given in [38,39,40, 57]. The best results reported are $\eta = 0.99995$ at 1 MHz clock frequency, and $\eta = 0.9998$ at 10 MHz clock frequency, both results for bias charge operation (30% of full well). These transfer efficiencies correspond to a density of fast interface states of 3×10^{10} (cm^2 eV)$^{-1}$ [38,39] and a few 10^9 (cm^2 eV)$^{-1}$ [40,57].

For an interesting report on "Experimental Characterization of Transfer Efficiency in CCDs", the reader is referred to Brodersen et al [41].

6.3 Bulk Charge Coupled Devices

The first experimental results [10,11,15] were obtained on self induced field-type bulk CCDs with uniformly doped layers. These devices were provided with a one level metallization. It was found that the gate spacings together with the dope level of the bulk layer have a considerable influence on the transfer efficiency. (See also [44]). These gaps between the gates give rise to potential minima (wells) in the bulk layer. Mobile charge will fill the wells, but the amount of charge depends on the potentials of the spaces under the neighbouring CCD plates. This means that the wells can exchange charge with any signal that might be present, leading to a deterioration of the transfer efficiency.

The first operating bulk CCD was reported in [11,15]. It was operated in the twin-channel mode (surface+bulk mode). This PCCD showed a limited efficiency of $\eta = 0.99$ due to 10 µm gaps between the electrodes. These devices were made by J.G. van Lierop. In following experiments, gate spacings are eliminated by applying overlapping gate structures. The data of dope levels and geometry for some of the present devices are shown in the table. (see next page)

Device nr. 3 shows no difference in ε for real zero or fat zero operation. At 100 kHz its worst response was found for a "one" after a very long range of real "zeros" (no bias charge). Then all bulk states are empty and will be filled by the first charge packet ("one") ($\tau_c \ll \tau_e$ eq. (4.2)). $\varepsilon_{01} = 3 \times 10^{-4}$ was measured. If we assume a uniform distribution for the charges and bulk centres then ε_{01} corresponds with an effective bulk centre concentration of about 2×10^{11} cm^{-3} [15,57]. Because of the much larger emission time constant of trapped electrons compared to the capture time constant, the transfer inefficiency of a "zero" after a large number of "ones" was always smaller namely $\varepsilon_{01} < 5 \times 10^{-5}$.

Device 4 shows for zero bias charge operation a transfer efficiency of 0.9998 and 0.9999 for a bias charge operation of

table

U = uniformly doped layer, P = profiled PCCD

device nr	1[45]	2[46]	3[15,16,47]		4[16]		5[48]		6[49,22]	
Bulk CCD	U	U	U	P	U	P	U	P	U	P
surface layer thickness, μm	–	–	–	0.5	–	0.5	–	–	–	0.2
" dope level, cm^{-2}	–	–	–	5x10^{11}	–	5x10^{11}	–	–	–	2x10^{12}
bulk layer thickness, μm	1	0.7	4.5	4.5	4.5	4.5	1.8	1.8	4.5	4.5
" dope level, cm^{-3}	10^{16}	~2x10^{16}	7x10^{14}	3x10^{14}	7x10^{14}	3x10^{14}	3x10^{15}	3x10^{15}	7x10^{14}	7x10^{14}
substrate dope level, cm^{-3}		10^{15}	5x10^{14}	5x10^{14}	5x10^{14}	5x10^{14}	5x10^{14}	5x10^{14}	5x10^{14}	5x10^{14}
oxide thickness, μm			0.15	0.15	0.1	0.1	0.1	0.1	0.1	0.1
number of phases	4	2	4	4	4	4	4	4	4	4
gate length per phase, μm	7.5	9(3+6)	7.5	7.5	7.5	7.5	7.5	7.5	7.5	7.5
transfer efficiency	0.99998	0.9998	0.99995	0.99995	0.9999	0.9999	0.9998	0.9998	0.99998	0.99998
charge handling capacity, electrons cm^{-2}	4x10^{11}		1.5x10^{11}	1.5x10^{11}	3x10^{11}	3x10^{11}	3x10^{11}	3x10^{11}	1.2x10^{12}	1.2x10^{12}
clock frequency, MHz	5	10	135	135	135	135	135	135	135*	135*
" voltage swing, volt	10	10	10	10	10	10	10	10	10	10

A charge handling of about 7x10^{11} may be possible for device 2. However this figure has not been quoted in [46].
The devices 3 to 5 included have been made by M.J.J. Theunissen. These PCCDs were the first CTDs which could be operated at clock rates in excess of 100 MHz. The measured results on devices 3 to 6 included have been obtained by L. Heldens and B. Visser.

---

*Recently this device has been operated at a clock frequency of 180 MHz [122].

1% or higher (see section 4). The charge handling is not higher than 3×10^{11} electrons cm^{-2} as a result of the relatively low dope level of the surface layer.

Device 5 contains an implanted 1.8 µm layer. It demonstrates that $d_2/L \simeq 0.25$ (see fig. 21) is sufficient for high speed operation. Recently a profiled PCCD [22,49] (device 6) has been made by H.L. Peek. This device has a much higher dope level for the surface layer. Consequently a charge handling capacity of 1.2×10^{12} electrons cm^{-2} together with a transfer efficiency ($\eta=1-\varepsilon_{01}$) of 0.99998 after 100 zeros is found. ε_{01} is always found to be smaller than 5×10^{-5}, and ε_{10} to be smaller than 2×10^{-5}. The best efficiency found for these devices $\eta = 1 - \varepsilon_{01} = 0.99999$ for a series of ones after 100 zeros ($Q_i = 0$, zero bias) (fig. 32). No degradation in efficiency for smaller charge packets is observed. Difference in efficiency is found for operation with and without bias charge.

These experimental results show that the PCCD combines high charge handling capacity (approaching that of a surface CTD) and high transfer speed (> 135 MHz clock frequency) and a high transfer efficiency (0.99999).

7. TECHNOLOGICAL ASPECTS OF CTDs

7.1 Gate structures

Fig. 33 shows some gate structures which are in use for CTDs. The single-level metallization [37,9,51] (fig. 33a) is the simplest one and the easiest to fabricate. It is extremely well suited for the MOS BBD. This device does not require a very accurate definition of the metal pattern. Since the bucket islands conduct well, the gaps between the gates are not critical [37].

The surface CCD in its original form [9,51] - three phase structure, one level of metallization - demands very accurate definition of the gate pattern since the distance between the gates should not be greater than 3 µm. The potential on top of the oxide between the gates is not defined. The instabilities associated with the exposed channel oxide in the gaps, together with the potential barrier (SCCD) or potential well (bulk layer) associated with the gaps, has led to the development of more reliable sealed gate structures.

The single layer doped polysilicon [52,53] gate structure (fig. 33b) has partially doped regions which form the gates. They are conductively connected by the undoped high resistivity spaces in between, which passivates the inter-electrode regions.

A double level metallization of Al-Al$_2$O$_3$-Al [54,55] is another possiblity for a sealed CTD gate structure (fig. 33c). This structure has low-resistivity conductors and therefore has advantage over all structures where long polysilicon gates are used.

The RC time delay in long polysilicon gates may set too low a limit on operating frequency. Here C is mainly effected by the overlapping area of the double-level gate structure.

A more common double-level gate structure is polysilicon-aluminium [38,39,56,16,46,48] (fig. 33d). It is a self-aligned overlapping gate structure. The isolation between the gates is formed by thermally grown SiO_2 and can have the same thickness as the channel oxide layer. This structure is successfully used for the PCCD.

A three-level polysilicon gate structure is used for three-phase CCDs [57]. The advantages of the construction are that the spaces between the gates at one gate level are maximized for a given gate length and the shorts between the gates on the same polysilicon level are not necessarily catastrophic.

7.2 Channel definition

Some methods by which the width of the surface CTD channel can be defined are shown in fig. 34. For N-channel surface CTDs one can use thick oxide and or P^+ stop diffusions (fig. 34a). For P-channels thick oxide may be sufficient if low resistivity N-type substrates are used. Fig. 34b shows a method where gates are used for the definition of the channel. These gates can also serve as transfer gate, for a transfer perpendicular to the CTD line. This can be used, for instance in image sensors and SPS structures (fig. 43b). It is obvious that the method can be applied also to bulk-layer devices.

Fig. 35 presents some more methods for bulk-layer CTDs. Channel definition can be obtained by masking during implantation (fig. 35a). If one starts with epitaxial layers or implanted layers (N-type) P^+ channel isolation diffusions, which contact the substrate, may be used. Fig. 35c shows a field effect channel isolation, as in use with the PCCD [16,48]. It is a shallow P^+ diffusion, which is contacted for definition of its potential. A contact with the area outside the channel is advantageous. By biasing this contact positive, the leakage current from this area is drained and is prevented from penetrating into the channel area.

7.3 Bulk layers

Bulk CCDs can be fabricated with implanted and epitaxial layers. Some remarks on both types of layers:
- High ohmic layers on low ohmic substrates can be fabricated more accurately by epitaxial growth than by implantation.
- Ion implantation seems to be more appropriate for low ohmic shallow layers than epitaxial growth.

Dope profiled layers for the profiled PCCD (fig. 35c) can be

made by:
1. thick epitaxial layer + implanted shallow surface layer
2. both the thick and surface layer epitaxially grown
3. both layers by implantation.

By using two donors - as predeposition - with a different diffusion rate one can obtain the desired dope profile.

Some remarks on surface and bulk CTDs:
- Surface CTDs and, in particular, the MOS BBD offers simpler processing.
- Bulk CCDs operated with zero bias charge already achieve relatively good transfer efficiencies, in contrast with surface CCDs.
- A bulk CCD, in particular the profiled PCCD, has a higher dynamic signal range.
- Bulk CCDs have very low noise operation [58]
- Bulk CCDs have larger leakage currents because of the larger depletion regions and of the larger thermal generation by the empty surface states [59].
(5 nA/cm^2 has been obtained for the PCCD).
- Bulk CCDs, in particular PCCDs, have very high frequency operation.

7.4 Leakage current

All CTDs are operated under non-thermal equilibrium conditions. Under these circumstances net generation of electron-hole pairs occurs. These leakage currents fill the potential wells of a CTD and may therefore obliterate the signal charges. This limits the storage time. A leakage current of 1 nA/cm^2 allows a storage time of about 1 sec for about 2% of full-well charge contribution (average value). The leakage current can be measured as output current during CTD operation. In this way one obtains the average leakage current. The distribution of the leakage current can be measured by periodically stopping the clock for a fixed collection time of the leakage charge. The leakage current profile is then read out [38]. By this method the magnitude and localization of current spikes, which are sensitive to the applied gate voltage, can be determined. The control of the leakage current is very important for CTD image sensors. Average leakage currents of 5 nA/cm^2 have been found for the PCCD. However, a typical value is 20 nA/cm^2.

8. INPUT AND OUTPUT CIRCUITS

All the circuits discussed in this section can be used for all types of CTDs.

8.1 Input circuits

Charges can be generated in the CTD by optical input and by electrical input. Fig. 36 shows some ways by which an electrical input can generate charge packets in a CTD. Fig. 36a shows a method where the input voltage is transformed into an equivalent amount of charge. The first potential well is filled to the input voltage, if a positive potential is applied to the first gate. This potential, in general, equals or exceeds the highest potential of the clock generator. This assures a good conducting channel between the input source diffusion and the first potential well. If the first gate is connected to the most negative potential of the clock generator then the charge packet becomes isolated from the source input and can be transferred through the CTD. The amount of charge is given by

$$Q = C(V_{in} - V_H) \qquad (8.1)$$

where C is the effective capacitance of the charge to its gate and substrate, and V_H is the potential of the well for zero mobile charge. For the uniformly doped bulk CCD (bulk mode) non linearities of 8% from full-well charge have been observed in 0.8 part of the signal range. The profiled PCCD showed non-linearities of 0.3% from full-well charge in 0.8 part of the signal range, due to the input capacitance being mainly determined by the surface layer [83]. Both for the described input method.

Fig. 36b shows another sampling method [60,61]. A potential well is formed by the first and second gate. The input signal is applied between these gates. The source voltage is made more negative than the barrier below the first gate, or the barrier is raised above the source voltage. This overfills the well below the second gate. By raising the source voltage the excess charge returns to the source diffusion (solid arrow). The charge quantity is given by

$$Q = CV_{in} \qquad (8.2)$$

where

$$C^{-1} = \frac{d_1}{\varepsilon_1} + \frac{z}{\varepsilon_2} \qquad \text{see} \quad (3.6)$$

if the oxide of both input gates has the same thickness and if both gates have equal channel pinch off voltages. C is the effective capacitance of the charge quantity to its gate. For surface CTDs C is equal to C_{ox}.

It has been predicted by Kosonocky that this sampling method is the least noisy way of introducing the input into the first potential well [60,62,117]. Fig. 36c shows another input scheme. The input signal V_{in} is converted into a current I_{in}. The channel formed below the first gate is pinched off by the draining area

below the second gate. The input source and first gate act as a
MOS transistor in the saturation mode. This forms a current source
that fills the first well as long as the source is switched on
(Δt). The charge quantity is given by

$$Q = I_{in} \Delta t \tag{8.3}$$

The current I_{in} is not proportional with V_{in}. So this method is less
attractive, for a linear input circuit is desirable.

8.2 Output circuits

The simplest output circuit is shown in fig. 37a. Instead of the
charge packet it senses the current spike in an RC-circuit. This
output circuit is extremely well suited for very wide band signal
detection. Fig. 37b presents a sense-reset output scheme or floating diffusion amplifier (FDA) [63]. It is the simplest output
circuit with an on-chip amplifier and has been introduced by
Sangster [37]. The floating diffusion is reset every clock period
to a reference voltage V_R. This can be done by a succeeding CTD
(reference charge mode) or a transistor. After resetting when the
reset transistor is switched off, a following charge packet arrives at the floating drain. This charge gives a proportional
voltage change, which is sensed by the MOS transistor (inverter or
source follower). It is a charge sensing circuit.

Similar to the FDA is the floating gate amplifier (fig. 37c),
proposed by Early [64]. The DC gate determines the DC-level of
the floating gate, e.g. at the average clock voltage level. The
charge to be sensed is brought under the floating gate by appropriate potentials on the adjacent electrodes. This charge produces
electrostatically a potential change at the floating gate and
this is detected by the floating-gate transistor. It is obvious
that both the FDA and FGA can be used as a tapping technique for
transversal filters.

A distributed floating-gate amplifier (DFGA) [64,69] is illustrated in fig. 37d. If a charge packet is sensed it induces
a current in the floating gate transistor. This current can be
integrated during a part of the clock period and stored in a CTD
capacitor. In this way a larger copy of the original charge quantity is obtained. The signal amplitude grows in proportion to the
integration time Δt and the amplifier noise amplitude in proportion to $(\Delta t)^{\frac{1}{2}}$. So the signal-to-noise power ratio of the DFGA increases with the integration time Δt. The DFGA makes it possible
to extend the integration time over a much longer time than the
clock period. This leads to an even larger improvement in the
signal-to-noise ratio. This is achieved by shifting the amplified
charge in the lower CTD synchronously with the original charge to
the following stages which have the following FGA in between.

Here another amplified copy of the original charge quantity is taken and added to the corresponding preceeding copy. For N stages the DFGA signal-to-noise ratio is proportionally improved with the effective integration time $N \times \Delta t$. The DFGA is thus useful for detection of small charge quantities with a minimum increase of the original signal-to-noise ratio.

The DFGA is used as an output stage for CTD image sensors. It is stated by Wen et al [69] that with the DFGA the low-light-level imaging capacity of bulk CCDs can be fully exploited. 10 to 20 signal electrons can be detected.

Kubo et al published an integrated sample holder output circuit based on the FDA [119]. This sample holder circuit has a reduced Nyquist noise compared to the conventional FDA.

9. APPLICATIONS

CTDs can store and transfer charge packets. These charges form a sampled analogue representation of the processed information. Charge packets can be generated by light or by an electrical input. The transfer of a charge packet gives in its gate capacitance an induced charge transport. Therefore the charge packet can be sensed non-destructively by its gate.

The basic functions that CTDs can perform are:
1. storing
2. transferring
3. electrical generation
4. generation by light

and 5. non-destructive and destructive sensing
of charges.

Proper combinations of these functions give a large field of applications [120], including:
- delay lines
- image sensors
- memories
- filters

Bulk CCDs and in particular PCCDs offer new possibilities especially in those fields where signals with a large bandwidth have to be processed, such as in radar, television, communication and instrumentation.

CTDs perform their basic functions in an extremely simple way. They handle analogue signals. Their manufacturing process is simple and compatible with standard MOS manufacturing technology, what permits integration with other MOS circuitry. It is by this means that CTDs can perform complex system functions in applications and will be competitive with many present digital or other analogue solutions.

Some CTD applications are:
Time conversion

- TV-standard conversion.
- Dynamic time error correction in video recording systems.

Image sensors
line sensor facsimile,
area sensor for all solid state TV-camera.

Memories
computer, TV-line and TV-frame memories.

Filters
with programmable and reprogrammable response, e.g. for correlation techniques such as in radar (matched filters, pulse compression), bandpass filters, adaptive equalization, etc.

9.1 Time Conversion

9.1.1 Analogue time delay

The simplest application of CTDs is analogue time delay. The maximum delay time is given by the leakage current in the device. Delay times up to 1 sec, with 10% average leakage contribution to the charge packet, are feasible. Fig. 38a shows the common serial CTD. The delay time per stage is given by the clock period (T). Thus a delay of NT sec. for an N-stage device. For a p-phase device, with a transfer inefficiency per step ε, the fractional loss per stage is $p\varepsilon$. Thus for an input signal $s(t)$ the "relative" output signal becomes [107]

$$s_0(t) = s(t-NT) + \frac{Np\varepsilon}{1 - Np\varepsilon} s(t - [N+1]T) \qquad (9.1)$$

It consists of the desired signal and an echo with a modified amplitude and an extra delay of one clock period, that is one sampling period. If the echo amplitude becomes too large - e.g. caused by inefficient transfer, too many stages or too high clock rate - one can use a phase multiplexed CTD (fig. 38b). A number of K CTD lines operate in parallel, each having F stages ($K \times F = N$). The periodically (T) sampled input signal is sequentially clocked into the lines. Thus the parallel driven lines have a clock period of KT sec, during which the input circuits are sequentially switched on and off. At the output the same function is performed in the opposite sequence. For $K > p$ one has many input and output clocking (switching) lines. In that case a series-parallel-series (SPS) structure is more attractive. Two vertical CTD registers perform the input and output functions. In every period of KT secs, the input register contains K samples, which are then shifted in parallel into the horizontal lines (demultiplexing). The output register performs the same function in the opposite sequence (multiplexing). The clock periode of the two vertical registers is T and that of the parallel lines is KT. Fig. 43b shows an SPS structure in use as a digital memory.

The "relative" signal response of the circuit of fig. 38b is given by

$$s_0(t) = s(t-NT) + \frac{Fp\varepsilon}{1 - Fp\varepsilon} s(t - [N+1]T) \qquad (9.2)$$

where

$$N = KF$$

For the signal response of the SPS structure is found

$$s_0(t) = s(t-NT) + \frac{(F+K)p\varepsilon}{1 - (F+K)p\varepsilon} s(t-[N+1]T) \qquad (9.3)$$

The two possible techniques have some disadvantages:
1. The leakage current and ε in the parallel lines have to be identical, otherwise one gets a fixed pattern with period kT at the output.
2. The clock frequency of the parallel lines ($f = 1/KT$) is in the signal band.

For the first reason the first system needs in- and output circuits with identical characteristics. The SPS structure has one in- and output circuit and, together with a simpler layout, is therefore preferable

For N parallel registers with one stage, one has in fact N sample and hold circuits where charge transfer techniques are utilized. Inefficient transfer is no problem. This structure has been proposed by Tiemann et al for structures with very long delays [72,73,74]. Time delay can be achieved by sampling charge on a capacitor and reading that charge later at the desired delay time [75]. An important drawback is the non-uniform leakage distribution which gives rise to a periodic noise pattern. This sets a lower limit to the maximum delay time than the serial CTD does, because the leakage current in semiconductors is very inhomogeneously distributed [38]. Every charge packet in a serial CTD passes the same leakage current sources and the noise contribution is drastically lowered. These drawbacks argue for serial CTDs with high transfer efficiency and high-speed transfer. With such devices there will be less reason to use multiplexed CTD structures.

9.1.2 Television standards conversion

TV standards conversion is for instance important for conversion of the European standard 625 lines/picture and 25 pictures/sec into the U.S. TV standard 525 lines/picture and 30 pictures/sec. The standards conversion is also of great potential importance for the conversion of the European TV standard into the proposed videophone standard 313 lines/picture and 25 pictures/sec. An analogue circuit by which this function can be performed is shown in fig. 39 [76]. It consists of two lines with e.g. 320 stages. The original TV standard line - which will be limted in bandwidth to 2 MHz - is read into the CTD register at a clockrate of 5 MHz. When this register contains one line and the other CTD line has been shifted out for videophone presentation, both switches are changed over. A new line is shifted into the other CTD register and the stored line is shifted out at a clock rate of 313/625 x 5 MHz. So the time required for shifting the signal out takes 625/313 more than for reading the signal into the shift register. This means that every second line of the original signal is used. The ratio of 625 to 313 lines makes a very simple system possible.

It is obvious that, for conversion from 313 to 625 lines, every line of the 313 system has to be repeated [77].

9.1.3 Time error correction

In video cassette recorders (VCR) and video long play records (VLP) the video information may be stored as a TV standard signal. Elongation of the tape or excentricity of the hole in the record and other mechanical inaccuracies are the reasons that the recorded TV lines appear asynchronously at the recorder or player output (fig. 40a). This may result in a ragged picture with colour fringing. In order to correct the time error ($\Delta\tau_i$) in the signal, one uses a flywheel circuit to obtain a periodic line sync pulse. The time error between the sync pulse of the tape signal and the flywheel sync pulse is measured ($\Delta\tau_i$) and transformed into a voltage Δ_i as shown by

$$\Delta_i = c_1(\tau - \Delta\tau_i) \tag{9.4}$$

The clock frequency of the CTD delay line becomes

$$f_i = \frac{c_2}{\Delta_i} \tag{9.5}$$

This clock frequency gives a delay time for the video signal

$$\tau_i = \frac{N}{f_i} \tag{9.6}$$

By choosing

$$\frac{c_2}{c_1} = N \tag{9.7}$$

the result is

$$\tau = \tau_i + \Delta\tau_i \tag{9.8}$$

where $\tau > \Delta\tau_i$ has to be fulfilled.

Every TV line signal gets the same delay τ compared to the "original" signal. Thus a time-error-corrected signal (synchronous signal) at the output of the CTD line (fig. 40b) is obtained. This method is adequate if the variation of $\Delta\tau_i$ is small during one line.

Professional qualities have been obtained with experimental time error correcting circuits provided with a PCCD register [78].

9.2 CTD image sensors

The CTD has introduced a revolutionary possibility of designing self-scanned solid-state image sensors. The idea of using the CTD as a selfscanned image sensor has originally been proposed by Teer and Sangster [79]. In this application the basic functions 1,2 and 3 are combined. The generation of the charges by the light input can be obtained

1. in the shift register itself, or
2. in separated photosensing areas, so that the light integrating and shifting functions are independently performed. Furthermore, one may have
1. top side illumination, through tranparent electrodes, or via gaps between the gates,
2. back-side illumination, where the photosensitive area is thinned (10-30 μm) and covered only with an oxide layer [84].

Some advantages of the self-scanning function over the X-Y scanned arrays are

1. that all the generated charge packets are detected at the same output amplifier with a relatively low output capacitance, hence no fixed pattern noise due to variation in output amplifiers;
2. the frequency of the clock noise at the output is outside the video band and can easily be filtered out.

9.2.1 CTD line sensor

A CTD line with one or more transparent gates per stage or one level of metallization (gaps between the gates) can be used as a line imager. When the clocks are stopped and fixed at the appropriate voltages, then one has a CTD line with one potential well per stage. Projection of a line image generates charges (electron-hole pairs). The electrons are collected in their respective wells and the holes are drained by the substrate and the channel-stop regions. The number of electrons is linear with the intensity of the light at the corresponding stage and the integration time. To prevent smearing of the image, the total time required to transfer the detected image from the CTD line sensor should be very short compared to the integration time. This can be done by high-speed transfer of the charges into another line which is shielded from light. This line is used as an analogue buffer memory, from which the image is read out at video clock rate. Another possibility for prevention of smearing is obtained by separating the light-integration and transfer functions. Fig. 41 shows line sensors organized in that way. After integration of the light pattern in the photosensing elements, the corresponding charges are transferred - on command of a shared gate - into the corresponding CTD stage. Subsequently the image information is shifted down the CTD register to the output. The shift registers are shielded from light. This method of organization was first proposed by Heyns and Sangster [80]. Other devices which are or may be used as line sensor are given in references [16,39,48,54, 81]. The largest line sensor is reported in [82].

9.2.2 CTD area sensors

Three basic types of organization for CTD area image sensors are shown in fig. 42. Fig. 42a shows the interline transfer system [85,86,87]. It can be visualized as consisting of a parallel array of the line sensor shown in fig. 41a. For purposes of interlacing there are two photosensing elements per stage. Assume that the shift registers have 2-phase clock lines. When the clocks of the parallel registers are stopped in phase 1 and the shared gates of the sensors are opened, then only the photo elements numbered 1 are discharged into their corresponding phase-1 bucket of the shift register. At the same time the charges of the photo elements numbered 2 are prevented from entering the phase-2 bucket. Now the first field charges are shifted down the CTD lines and transferred into the output register one horizontal line at a time. The horizontal line is then transferred to the output stage at normal video clock rate. After that the next horizontal line is shifted in. When the first field has been read out the vertical registers are stopped in phase 2 and the charges of the field-2 photoelements are shifted into their corresponding phase-2 bucket of the shift register and shifted out in the same way as the previous field 1. Interlaced pictures are thus obtained.

Fig. 42b shows the frame transfer system [121]. The imaging area is formed by the upper part. The optical image is integrated directly in the upper vertical shift registers. So one may use top-side or back-side illumination. After integration during one field time the detected image is transferred down to the storage area, which is shielded from light. This is done at a relatively high speed, during the vertical blanking time, to prevent the detected image from smearing. After that the horizontal lines are shifted one by one into the output register and read out at video clock rate.

Vertical interlacing [88] can be obtained by shifting the potential wells half the pitch length of a stage from one field to the other.

Fig. 42c shows the <u>line transfer</u> system [79]. Weimer, one of the early workers in the field of CTD imaging, has done experiments on the CTD line transfer system and xy-addressed image sensors [89].

The optical information is directly applied to the horizontal CTD lines. The detected image is read out line by line at video clock rate into the output register and transferred at the same clock rate to the output stage. This has the advantage that smearing effects are negligible due to the relatively short line-transfer time compared to the picture duration. Interlacing can be performed by reading all the odd and all the even numbered field lines in succession.

The first bulk CCD imager was reported by Walsh [52]. The largest area-image sensor was published by Rodgers [90] and Vogl et al [95].

9.3 Memories

CTDs are analogue solid state devices. Analogue and thus also digital information can be handled because CTDs are operated in non thermal equilibrium. As a result the information will disappear for long storage times by thermally generated charges. Another non ideal property of a CTD is its limited transfer efficiency which also leads to a loss of information. Both properties make periodical refreshing of the information necessary for long storage times and for large numbers of transfer. The most reliable and the simplest methods for regeneration (refreshing) of information are found for binary information [63,92,93,94]. This demands two-level decisions and relatively large errors in the analogue representation of the binary "1" and "0" are acceptable.

There are various system organizations for CTD memories [94, 96-101]. The most interesting organizations are shown in fig. 43. Fig. 43a,b shows two types with one single large storage loop. The bit flow in the first type follows a serpentine route. The bits are refreshed at every turn. A two-phase CTD is necessary if one wants to avoid multilayer crossovers for the transfer gates. The second system is an SPS structure with only one regenerator. It needs two clock drivers operated at different frequencies with the ratio K/F. The advantage of the SPS structure is that it may be constructed in such a way that it can be used as a frame transfer image sensor, as a line transfer image sensor, as a delay line or as a digital memory.

The lowest frequency at which the system can be run is determined by the leakage current and the delay time of a charge packet from one to another refreshing stage. Thus the serpentine memory can be driven at a much lower sampling frequency. This may be of interest for some applications.

Shorter access times can be obtained by using systems with a larger number of parallel-operated and smaller storage loops. Fig. 43c shows a one-dimensional extension of single storage loops. This demands extra circuitry for the purposes of addressing and controlling the individual storage loops.

In a similar way two dimensional extensions of single storage loops can be obtained. The most parallel system is that in which one CTD stage is used per xy-address. Actually it is a CTD version of a refreshable memory with one transistor per bit proposed by Engeler et al [94]. It consists of x-addressed MOS storage gates, y-addressed MOS transfer gates and x-addressed diffused bit lines with an output amplifier and regenerator per bit line. The bit lines are used for sensing and regeneration or rewritting the information. The transfer gates are used to control the barrier between the storage areas and the bit lines. When a y-addressed MOS transfer gate is switched on, the barriers between the x-addressed storage areas and bit lines disappear. All stored information at (x_i, y) for i = 1 to N, is detected via the corresponding bit lines (N in parallel). Rewriting is performed in the opposite way, like sampling of a charge packet at a CTD input circuit.

The power dissipation per bit of the clock driver for square wave operation [91] is given by

$$P_{bit} = C_{bit} V^2 f_c \qquad (9.9)$$

where C_{bit} is the capacitance per bit,
V_{bit} the clock voltage swing and
f_c the clock frequency.

A practical value for C_{bit} is 0.1 to 0.3 pF. This results in a clock power dissipation of 10^{-4} to 3×10^{-4} watt per bit for a clock voltage swing of 10 volts at a frequency of 10 MHz. The power dissipation for the charge transfer in the CTD line per bit is one to two orders of magnitude lower [91].

9.4 Filters

Transversal CTD filters with fixed and electronically variable weights are possibly the most interesting and effective devices for processing analogue signals [37,50,72,74,102-116]. Fig. 44 shows a block diagram of a transversal filter. The input signal samples (S_i) are shifted in the delay line. At each stage the samples are sensed and multiplied be a weighting factor (h_i). The outputs of the multipliers are summed in the filter output circuit. The response of the filter at $t = mT$ is [107]

$$S_{out}(mT) = \sum_{i=1}^{n} h_i S_i[(m-i)T] \qquad (9.10)$$

The filter impulse response is determined by the weighting factors h_i. Thus the filter performs the sampled data convolution or correlation between the input signal and the filter impulse response.

A very simple circuit performing the weighting and summing function has been proposed by Sangster [1,104,37] and is shown in fig. 45a. The ϕ_1 gates are split up each into two parts. The charges are consequently proportionally divided into two parts. When the charges arrive they induce a displacement current of an equal amount of charge. These displacement currents can be applied to an integrator at the output. This is done for both parts of the gates. The displacement charge (Q_{ia}) of the upper part of the gate equals

$$Q_{ia} = \frac{(W/2) - a_i}{W} Q_i \qquad (9.11)$$

and that of the lower part of the gate

$$Q_{ib} = \frac{(W/2) + a_i}{W} Q_i \qquad (9.12)$$

Subtraction of both charge quantities at the output results in

$$\Delta Q_i = -\frac{2a_i}{W} Q_i \qquad (9.13)$$

Thus the weighting coefficient equals

$$h_i = -\frac{2a_i}{W} \qquad (9.14)$$

and can be given any value between + 1 and - 1. Summation is simply performed by connecting all the (+) parts and all (-) parts of the gates so that the circuit fulfils Kirchoffs' law $\Sigma I = 0$.

The split electrode weighting and summing is a very effective way of constructing transversal filters with fixed weights. In general the filter response is determined only by one mask. CTD transversal filters using split electrode weighting have been implemented for spread-spectrum communication applications, radar pulse compression, low-pass filters, band-pass filters, broadband 90° phase shifters for single sideband systems and chirp-z transform filters for spectral analysis [74,107,112]. Buss has reported a 500-stage CTD band pass filter and 500-stage chirp-z transform filter [107].

Another method for weighting and summing proposed by Buss et al, where the floating diffusion [103] or the floating gate technique (chapter 8.2) may be employed is shown in fig. 45b.

The voltage change of the floating gate, induced by a charge packet, is detected by a source follower. This source follower is loaded in its source by a transistor. The conductance of the transistor load is adjustable by its gate potential. The current change passing the two transistors is a product of the adjustable conductance and the charge-induced voltage change of the floating gate. Positive and negative weighting coefficients can be performed by splitting the current-summing lines into two parts and by taking their difference, similar to the split electrode weighting.

Another method of adjustable tap weights have been announced by White et al [114,111]. The floating gate with source follower tapping is used. Two transistor loads are used in parallel as modulated conductances, one is connected to a positive and the other to a negative summation line. The conductance modulation is performed by an MNOS structure. In this way a transversal filter with variable analogue tap weights is obtained. The weighting factors can be stored for a very long time by the MNOS system.

Another approach with variable tap weighting has been published by Tiemann et al [72,115,116] (fig. 45c). Their starting point was to achieve transversal filters with variable tap weights for correlation techniques in such a way that individual chips provided with many stages can be connected together to form a filter with hundreds of stages. The charge-sloshing principle [113] was introduced to eliminate the influence of the limited transfer efficiency in CTDs. The charge sloshing stages are shown in fig. 45c, every stage being indicated by the letters A, B and C. The charge quantities are stored under the C-gate. Every clock period the charge is transferred to and from the A-gate if the corresponding binary reference shift register stage contains a "1". If this stage contains a "-1" then the charge quantity in the sloshing stage is transferred to and from the B-gate. The charge quantity is detected by the displacement of the same amount of charge at the corresponding summing line A or B. By subtracting the summed results of the A- and B-lines one obtains the summed products of the analogue charges in the sloshing stages with the corresponding binary contents of the reference shift register. The advantage of the charge sloshing is that after transfer the remaining charge is later on added to the original charge quantity. In this way 2×10^5 transfers can be achieved in practice, what is equivalent to a transfer efficiency of a serial CTD of 0.999999 [113]. Therefore a good transfer efficiency is not required for the charge sloshing system.

The operation of the filter shown in the block diagram of fig. 44 can also be obtained by shifting the weighting factors and keeping the signal samples at the same place. Every clock period the oldest signal sample is rewritten by a new one. The result obtained at the output is exactly the same as for the original operation. This method is followed for the filter of fig.

45c. It is a filter with an analogue input, binary weighting and analogue output. The analogue input signal is periodically sampled and stored in the form of an equivalent amount of charge in the sloshing stages. This is performed by shifting a "1" down the scan shift register. This "1" activates every clock period the input circuit of the corresponding charge sloshing stage where the old sample is rewritten by a new one. The weighting factors are shifted down the reference shift register. Every clock period the output gives the summed products of the weighting factors with their corresponding signal samples. A filter of any length can be made by simply connecting the input pins of one chip with the output pins of the previous chip.

It should be noted that for the described tapping methods it is desirable to have a much higher capacitance between the charge and its gate than between the charge and the substrate and the other gates. This is important in order to detect a linear copy (displacement charge) of the original charge quantity. This requirement can be fulfilled for all self-induced field-type CTDs. For the externally-induced field-type CTDs the requirement is fulfilled for the twin-channel operated devices (PCCDs).

ACKNOWLEDGEMENT

Special thanks are due to M.G. Collet and A.C. Vliegenthart for valuable contributions to the present paper.

10. REFERENCES

1. F.L.J. Sangster, US patent 3546490, filed Oct. 25, 1966, granted Dec. 8, 1970.
2. F.L.J. Sangster and K. Teer, "Bucket-brigade Electronics", IEEE J. Solid-State Circuits, 1969, SC-4, pp. 131-136.
3. K. Schlesinger, US patent 2403955, filed May 11, 1943, granted July 16, 1946.
4. J.M.L. Janssen, "Discontinuous Low-Frequency Delay Line with Continuously Variable Delay", Nature, pp. 148-149; Jan. 26, 1952.
5. W.J. Hannan et al, "Automatic Correction of Timing Errors in Magnetic Tape Recorders". IEEE Transact. MIL, pp. 246-254; July-Oct. 1965.
6. R.A. Mao, K.R. Keller and R.W. Ahrons, Integrated MOS Analog Delay Line", IEEE Intl. Solid State Circuits Conf., Philadelphia, pp. 164-165, Feb. 1969.
7. C.K. Kim, J.M. Early, G.F. Amelio, "Buried-Channel Charge-Coupled Devices", NEREM, Boston, pp. 161-164, Oct. 1972.
8. F.L.J. Sangster, "Integrated MOS and Bipolar Analog Delay Lines using Bucket-Brigade Capacitor Storage". IEEE Inter-

national Solid-State Circuits Conf., Philadelphia, pp. 74, 75, 185, Feb. 1970.
9. W.S. Boyle and G.E. Smith, "Charge-coupled Semiconductor Devices", Bell Syst. Techn. J., 1970, 49, pp. 587-593.
10. R.H. Walden, R.H. Krambeck, R.J. Strain, J. McKenna, N.L. Schryer and G.E. Smith, "The Buried Channel Charge Coupled Device", Bell Syst. Techn. J., pp. 1635-1640, Sept. 1972.
11. L.J.M. Esser, "Peristaltic Charge Coupled Device: A new type of Charge Transfer Device", El. Letters, pp. 620-621, Dec. 1972.
12. J.E. Carnes, W.F. Kosonocky and E.G. Ramberg, "Free Charge Transfer in Charge-Coupled Devices", IEEE Trans. Electron Devices, Vol. ED-19, No. 6, pp. 798-808, June 1972.
13. R.H. Krambeck, R.H. Walden and K.A. Pickar "Implanted Barrier Two-Phase Charge Coupled Device", Appl. Phys. Lett., Vol. 19, no. 12, pp. 520-522, Dec. 15, 1971.
14. C.N. Berglund, R.J. Powell, E.H. Nicollian, J.T. Clemens, "Two-Phase Stepped Oxide CCD Shift Register Using Undercut Isolation", Appl. Phys. Lett., Vol. 20, no. 11, pp. 413-414, June 1, 1972.
15. L.J.M. Esser, "The Peristaltic Charge Coupled Device", Charge Coupled Device Applications Conference, San Diego, pp. 269-277, Sept. 1973.
16. L.J.M. Esser, "The Peristaltic Charge Coupled Device for High-Speed Charge Transfer", IEEE Intl. Solid-State Circuits Conf., Philadelphia, pp. 28, 29, 219, Feb. 1974.
17. C.N. Berglund, H.J. Boll, "Performance Limitations of the IGFET Bucket Brigade Shift Register", IEEE Trans. Electron Dev., ED-19, no. 7, pp. 852-860, July 1972.
18. F.L.J. Sangster, "A Review of Bucket-Brigade Charge Transfer Devices", Paper presented at the Semiconductor Device Course on Charge Transfer Devices, Catholic University of Leuven, Heverlee, Belgium, Sept. 30-Oct. 3, 1974.
19. L. Boonstra, F.L.J. Sangster, "Progress on Bucket Brigade Charge Transfer Devices", IEEE Intl. Solid-State Circuits Conf., Philadelphia, pp. 140-141, 228, Feb. 1972.
20. K.K. Thornber, "Incomplete Charge Transfer in IGFET Bucket Brigade Shift Register", IEEE Trans. Electron Dev., ED-18, no. 10, pp. 941-950, Oct. 1971.
21. C.K. Kim, "Carrier Transport in Charge-Coupled Devices", IEEE Intl. Solid-State Circuits Conf., Philadelphia, pp. 158-159, Feb. 1971.
22. L.J.M. Esser, A.C. Vliegenthart, L.G.M. Heldens and B. Visser, "The PCCD an efficient analogue shift register for low noise and wide band signal processing", to be published in Philips Res. Rep..
23. H. El-Sissi, R.S.C. Cobbold, "One-Dimensional Study of Buried-Channel Charge-Coupled Devices", IEEE Trans. Electron Devices, Vol. ED-21, No. 7, pp. 437-447, July 1974.

24. J. McKenna and N.L. Schrijer, "The Potential in a Charge Coupled Device with no Mobile Minority Carriers and zero Plate Separation", The Bell System Technical Journal, Vol. 52, No. 5, pp. 669-696, May-June 1973.
25. M.G. Collet, A.C. Vliegenthart, "Calculations on Potential and Charge Distributions in the Peristaltic Charge-Coupled Device", Philips Research Reports, Vol. 29, No. 2, pp. 25-44, Febr. 1974.
26. H.W. Hanneman and L.J.M. Esser, "Field & Potential Distributions in Charge Transfer Devices", Philips Research Reports, Vol. 30, no. 1, pp. 56-72, Feb. 1975.
27. G.F. Amelio, "Computer Modelling of Charge-Coupled Device Characteristics", The Bell System Technical Journal, Vol. 51, No. 3, pp. 705-730, March 1972.
28. C.N. Berglund and R.J. Strain, "Fabrication and Performance Considerations of Charge-Transfer Dynamic Shift Registers", The Bell System Technical Journal, Vol. 51, No. 3, pp. 655-703, March 1972.
29. J.E. Carnes and W.F. Kosonocky, "Fast-Interface-State Losses in Charge Coupled Devices", Appl. Phys. Letters, Vol. 20, No. 7, pp. 261-263, April 1972.
30. H.J. Stein, "Radiation Effects in Semiconductors", eds., J.W. Corbett and G.D. Watkins, p. 125, Gordon and Breach, Science Publishers Ltd., London (1971).
31. C.T. Sah, L. Forbes, L.J. Rosier, A.F. Tash Jr., A.B. Tole, "Thermal and Optical Emission and Capture Rates and Cross Sections of Electrons and Holes at Imperfection Centers in Semiconductors from Photo and Dark Junction Current and Capacitance Experiments", Solid-State Electronics, Vol. 13, pp. 759-788, June 1970.
32. A.S. Grove, "Physics and Technology of Semiconductor Devices", pp. 129-134, John Wiley and Sons, Inc., New York (1967).
33. W. Fahrner and A. Goetzberger, "Energy Dependence of Electrical Properties of Interface States in Si-SiO_2 Interfaces", Applied Physics Letters, Vol. 17, No. 1, pp. 16-18, 1 July 1970.
34. W.F. Kosonocky and J.E. Carnes, "Polysilicon-Aluminium Gate CCD", Charge Coupled Device Applications Conference, San Diego, pp. 217-227, Sept. 1973.
35. M.F. Tompsett, "The Quantitative Effects of Interface States on the Performance of Charge-Coupled Devices", IEEE Trans. on Elec. Dev. ED-20, No. 1, pp. 45-55, Jan. 1973.
36. C.S. Wang, private communication, Sept. 1973.
37. F.L.J. Sangster, "The bucket brigade delay line, a shift register for analogue signals", Philips Technical Review, Vol. 31, No. 4, pp. 97-110, 1970.
38. W.F. Kosonocky and J.E. Carnes, "Two-Phase Charge Coupled Devices with Overlapping, Polysilicon and Aluminium Gates", RCA Review 34, No. 1, pp. 164-202, March 1973.

39. W.F. Kosonocky and J.E. Carnes, "Design and Performance of Two-Phase Charge-Coupled Devices with Overlapping Polysilicon and Aluminium Gates", 1973 International Electron Devices Meeting Technical Digest, pp. 123-125, Washington, D.C., Dec. 3-5, 1973.
40. C.H. Sequin, D.A. Sealer, F.J. Morris, R.R. Buckley, W.J. Bertram and M.F. Tompsett, "Charge-Coupled Image-Sensing Devices Using Three Levels of Polysilicon", 1974 Intl. Solid-State Circuit Conference Digest of Technical Papers, pp. 24, 25, 218, Philadelphia, Feb. 13-15, 1974.
41. R.W. Broderson, D.D. Buss and A.F. Tash Jr., "Experimental Characterization of Transfer Efficiency in Charge Coupled Devices", IEEE Trans. on Elec. Dev. ED-22, No. 2, pp. 40-46, Feb. 1975.
42. F.L.J. Sangster, "Integrated bucket-brigade delay line using MOS tetrodes", Philips Techn. Rev., Vol. 31, No. 7/8/9, p. 266, 1970.
43. L.J.M. Esser, "The Peristaltic Charge Coupled Device", IP & IEE Meeting on Charge Transfer Devices, paper no. 3, 11 June 1973. See also Nature, Vol. 244, pp. 11-12, July 6, 1973.
44. I. Takemoto, H. Sumami, S. Ohba, M. Aoki, M. Kubo, "Bulk Charge Transfer Device", Japan, Intl. Conf. on Solid State Devices, Tokyo, Aug. 1973.
45. R.W. Broderson and D.R. Collins, private communication, Sept. 1974.
46. I. Takemoto, M. Ashikawa, M. Kubo and S. Ohba, "A Charge Transfer Device with Self-Aligned Electrodes", Intl. Electron Devices Meeting, Washington, pp. 473-476, Dec. 1973.
47. L.J.M. Esser, M.G. Collet and J.G. van Santen, "The Peristaltic Charge Coupled Device", Intl. Electron Device Meeting, Washington, pp. 17-20, Dec. 1973.
48. M.J.J. Theunissen and L.J.M. Esser, "PCCD Technology and Performance", CCD 74, Intl. Conf. on Technology and Appl. of CCDs, Edinburgh, pp. 106-111, Sept. 1974.
49. H.L. Peek, "Twin Layer PCCD Performance for Different Doping Levels of the Surface Layer", to be published in the joint issue of IEEE J. Solid-State Circuits and Trans. Electron Devices, Feb. 1976.
50. C.M. Puckette, W.J. Butler and D.A. Smith, "Bucket-Brigade Transversal Filters", IEEE Trans. on Communications, Vol. COM-22, No. 7, pp. 926-934, July 1974.
51. G.F. Amelio, M.F. Tompsett and G.E. Smith, "Experimental Verification of the Charge Coupled Device Concept", Bell System Tech. J., Briefs 49, No. 4, p. 593, Apr. 1970.
52. L. Walsh and R.H. Dyke, "A New Charge-Coupled Area Imaging Device", CCD Applications Conference Proceedings, San Diego, Ca., pp. 21-22, Sept. 18-20, 1973.
53. C.K. Kim and E.H. Snow, "P-channel Charge-Coupled Device

with Resistive Gate Structure", Appl. Phys. Lett., 20, 514, 1972.
54. D.R. Collins, W.C. Rhines, J.B. Barton, S.R. Shortes, R.W. Brodersen and A.F. Tash, "Electrical Characteristics of Long CCD Shift Registers Fabricated Using Al-Al$_2$O$_3$-Al Double Level Metallization", 1973 Intl. Electron Devices Meeting, Technical Digest, p. 29, Washington, D.C., Dec. 3-5, 1973.
55. D.R. Collins, W.C. Rhines, J.B. Barton, S.R. Shortes, R.W. Brodersen and A.F. Tash Jr., "Electrical Characteristics of 500-Bit Al-Al$_2$O$_3$-Al CCD Shift Registers", IEEE Proceedings, Vol. 62, No. 2, 282-284, Feb. 1974.
56. N.A. Patrin, "Performance of Very High Density CCD", IBM Journal of Res. and Dev. Vol. 17, No. 3, pp. 241-248, May 1973.
57. W.J. Bertram, A.M. Mohsen, F.J. Morris, D.A. Sealer, C.H. Séquin and M.F. Tompsett, "A Three-Level Metallization Three-Phase CCD", IEEE Trans. on Electron Dev., ED-21, Dec. 1974.
58. S.B. Campana, "Charge-Coupled Devices for Low Light Level Imaging", CCD Applications Conference Proceedings, San Diego, Ca., pp. 235-246, Sept. 1974.
59. A.F. Tash, R.W. Brodersen, D.D. Buss and R.T. Bate, "Dark Current and Charge Storage Considerations in Charge-Coupled Devices", CCD Applications Conference Proceedings, pp. 179-185, San Diego, Ca., Sept. 1973.
60. J.E. Carnes, W.F. Kosonocky and P.A. Levine, "Measurements of Noise in Charge-Coupled Devices", RCA Review, 34, pp. 553-565, Dec. 1973.
61. M.F. Tompsett and E.J. Zimany, "Use of Charge-Coupled Devices for Delaying Analog Signals", IEEE J. of Solid-State Circuits SC-8, No. 2, p. 151, Apr. 1973.
62. J.E. Carnes and W.F. Kosonocky, "Noise Sources in Charge-Coupled Devices", RCA Review, 33, pp. 327-343, June 1972.
63. W.F. Kosonocky, "Charge-Coupled Digital Circuits", IEEE J. of Solid-State Circuits SC-6, No. 5, pp. 314-322, Oct. 1971.
64. D.D. Wen and P.J. Salsbury, "Analysis and Design of a Single-Stage Floating Gate Amplifier", 1973 IEEE Intl. Solid-State Circuits Conference, Digest of Technical Papers, Philadelphia, pp. 154-155, Feb. 1973.
65. J.E. Carnes, W.F. Kosonocky and E.G. Ramberg, "Drift-Aiding Fringing Fields in Charge-Coupled Devices", IEEE J. Solid-State Circuits, vol. SC-6, pp. 322-326, Oct. 1971.
66. F.L.J. Sangster, unpublished report, 1970.
67. F.L.J. Sangster, "The Bucket Brigade and other Charge Transfer Devices", Proc. of the 5th Conf. (1973 Intl.) on Solid-State Devices, Tokyo, 1973. Suppl. to the J. of the Japan Society of Appl. Physics, Vol. 43, 1974.
68. C.H. Chan and S.G. Chamberlain, "Numerical Methods for the Charge Transfer Analysis of Charge-Coupled Devices", Solid-State Electronics, Vol. 17, pp. 491-499, 1974.

69. D.D. Wen, J.M. Early, C.K. Kim and G.F. Amelio, "A Distributed Floating Gate Amplifier in Charge-Coupled Devices", IEEE Intl. Solid-State Circuits Conf., Philadelphia, Digest of Technical Papers, pp. 24-25, Feb. 1975.
70. W.E. Engeler, J.J. Tiemann and R.D. Baertsch, "Surface Charge Transport in Silicon", Applied Physics Letters, Vol. 17, No. 11, pp. 469-472, Dec. 1970.
71. L.J.M. Esser, "The Bulk Charge Coupled Devices", paper presented at the Semiconductor Device Course on Charge Transfer Devices, Catholic University of Leuven, Heverlee, Belgium, Sept. 30 - Oct. 3, 1974.
72. J.J. Tiemann, R.D. Baertsch and W.E. Engeler, "A Surface Charge Correlator for Signal Processing", CCD Applications Conf., San Diego Ca., pp. 103-109, Sept. 1973.
73. J.J. Tiemann, R.D. Baertsch and W.E. Engeler, "A Surface Charge Correlator", IEEE Intl. Solid-State Circuits Conf., pp. 154-155, Feb. 1974.
74. R.D. Baertsch, W.E. Engeler, H.S. Goldberg C.M. Puckette and J.J. Tiemann, "Two classes of Charge Transfer Devices for Signal Processing", CCD 74, Intl. Conf. on Technology and Applications of CCDs, Edinburgh, pp. 229-236, Sept. 1974.
75. G.P. Weckler, "The Serial Analog Processor", IEEE Intl. Solid-State Circuits Conf., Philadelphia, pp. 142, 143, 226, Feb. 1975.
76. M.C.W. van Buul and L.J. van de Polder, "Standards Conversion of a TV Signal with 625 Lines into a Videophone Signal with 313 Lines", Philips Research Reports, Vol. 28, pp. 377-390, 1973. See also Funkschau 1975, Heft 5, p. 48.
77. M.C.W. van Buul and L.J. van de Polder, "Standards Conversion of a Videophone Signal with 313 Lines into a TV signal with 625 Lines". Philips Research Reports, Vol. 29, pp. 413-428, 1974.
78. A.J.J. Boudewijns, B.H.J. Cornelissen and G.A. van Kempen, private communication, April 1975.
79. F.L.J. Sangster and K. Teer, "Bucket-brigade electronics", IEEE,J. Solid-State Circuits, 1969, SC-4, pp. 131-136, in particular p. 135 application 3.
80. H. Heyns, "New IC Concepts for Linear Optical Scanning - A Bucket Brigade Application", IEEE Eurocon 71, Lausanne Switzerland, pp. C3-6 (1,2), Oct. 1971.
81. A.L. Solomon, "Parallel-Transfer-Register Charge-Coupled Imaging Devices", 1974 IEEE Intercon Technical Papers, Session 2, New York, March 26-28, 1974.
82. D.A. Sealer, C.H. Sequin and M.F. Tompsett, "High Resolution Charge Coupled Image Sensors", 1974 IEEE Intercon Technical Papers, Session 2, New York, March 26-29, 1974.
83. L.J. van de Polder and P.J. Snijder, private communication, Jan. 1975.

84. S.R. Shortes, W.W. Chan, W.C. Rhiner, J.B. Barton and D.R. Collins, "Characterization of Thinned Back-side Illuminated CCD Imager", Appl. Phys. Lett., Vol. 4, pp. 565-567, 1 June 1974.
85. G.F. Amelio, "Physics and Applications of Charge-Coupled Devices", 1973 IEEE Intercon Technical Papers, Session 1, New York, March 26-30, 1973.
86. D.F. Barbe and M.H. White, "A Tradeoff Analysis for CCD Area Imagers: Front-side Illumination Interline Transfer vs. Back-side Illuminated Frame Transfer", CCD Appl. Conference, Proceedings, pp. 13-20, San Diego, Sept. 18-20, 1973.
87. G.F. Amelio, "The Impact of Large Image Sensing Area Arrays", CCD 74, Intl. Conf. on Technology and Applications of CCDs, Edinburg, pp. 133-152, Sept. 1974.
88. C.H. Sequin, "Interlacing in Charge-Coupled Imaging Devices", IEEE Trans. on Electron Dev. ED-20, No. 6, p. 535, June 1973.
89. P.K. Weimer, "Systems and Technologies for Solid-State Image Sensors", RCA review, Vol. 32, No. 2, June 1971.
90. R.L. Rodgers, "Charge-Coupled Imager for 525 Line Television", IEEE Intercon Technical Papers, Session 2, New York, March 26-29, 1974.
91. D.F. Barbe, "Imaging Devices Using the Charge-Coupled Concept", Proc. of the IEEE, Vol. 63, No. 1, Jan. 1975.
92. M.F. Tompsett, "A Simple Charge Regenerator for Use with Charge-Transfer Devices and the Design of Functional Logic Arrays", IEEE Journ. of Solid-State Circuits, SC-7, No. 3, p. 237, June 1972.
93. W.E. Engeler, J.J. Tiemann and R.D. Baertsch, "A Memory System Based on Surface-Charge Transport", 1971 Intl. Solid-State Circuits Conf., Digest of Technical Papers, pp. 164-165, Philadelphia, Feb. 1971.
94. W.E. Engeler, J.J. Tiemann and R.D. Baertsch, "Surface-Charge RAM System", 1972 IEEE Intl. Solid-State Circuits Conference, Digest of Technical Papers, pp. 18-19, Philadelphia Feb. 1972.
95. N.G. Vogl Jr. and B.J. Deliduka, "A Half-Million Pel Bucket-Brigade Optical Scanner", IEEE Intl. Solid-State Circuits Conf. Philadelphia, pp. 30-31, Feb. 1974.
96. R.A. Belt, "CCD Digital Memory for Radar Applications", CCD Applications Conference Proceedings, pp. 63-68, San Diego, Sept. 1973.
97. A. Ibrahim and L. Sellars, "4096-Bit Charge Coupled Device Serial Memory Array", 1973 Intl. Electron Device Meeting, Technical Digest, pp. 141-143, Washington D.C., Dec. 1973.
98. S.D. Rosenbaum and J.T. Caves, "CCD Memory Array with Fast Access by On-Chip Decoding", 1974 IEEE Intl. Solid-State Circuits Conference, Digest of Technical Papers, pp. 210-211, Philadelphia, Feb. 1974.
99. R. Agusta and T.V. Harroun, "Conceptual Design of an Eight-

-Megabyte High Performance Charge-Coupled Storage Device", CCD Applications Conf. Proceedings, pp. 55-62, San Diego, Sept. 1973.
100. D.R. Collins, J.B. Barton, D.D. Buss, A.R. Kmetz and J.E. Schroeder, "CCD Memory Options", 1973 Intl. IEEE Solid-State Circuits Conf., Digest of Technical Papers, pp. 136,137,210, Philadelphia, Feb. 1973.
101. H.A.R. Wegener, "Appraisal of Charge Transfer Technologies for Peripheral Memory Applications", CCD Applications Conf. Proceedings, pp. 43-54, San Diego, Sept. 1973.
102. D.R. Collins, W.H. Bailey, W.M. Gosney and D.D. Buss, "Charge-Coupled Device Analog Matched Filters", Electr. Lett. 8, pp. 328-329, June 29, 1972.
103. D.D. Buss, W.H. Bailey and D.R. Collins, "Matched Filtering Using Tapped Bucket-Brigade Delay Lines", Electr. Lett. 8, No. 4, pp. 106-107, Feb. 1972.
104. F.L.J. Sangster, "MOS Integrated Bucket-Brigade Transversal Filters", IEEE Eurocon 71, Lausanne Switzerland, pp. C3-5 (1,2), Oct. 1971.
105. D.R. Collins, W.H. Bailey, D.D. Buss, "Analog Matched Filter Using Charge-Coupled Devices", NEREM 72, Boston, pp. 165-7, Oct. 1972.
106. D.D. Buss, D.R. Collins, W.H. Bailey and C.R. Reeves, "Transversal Filtering Using Charge-Transfer Devices", IEEE J. of Solid-State Circuits, SC-8, No. 2, pp. 138-151, April 1973.
107. D.D. Buss, W.H. Bailey and A.F. Tasch, "Signal Processing Applications of Charge-Coupled Devices", CCD 74, Int. Conference on the Technology and Applications of Charge-Coupled Devices, Edinburgh, pp. 179-197, Sept. 1974.
108. D.A. Smith, W.J. Butter and C.M. Puckette, "Programmable Bandpass Filter and Tone Generator Using Bucket-Brigade Delay Lines", IEEE Trans. on Communications, Vol. COM-22, No. 7, pp. 921-925, July 1974.
109. W.J. Butler, M.B. Barron and C.M. Puckette, "Practical Considerations for Analog Operation of Bucket-Brigade Circuits", IEEE J. of Solid-State Circuits, SC-8, 2, pp. 157-168, April 1973
110. D.D. Buss, W.H. Bailey and D.R. Collins, "Spread Spectrum Communication Using Charge Transfer Devices", Proc. of 1973 Symposium on Spread Spectrum Communications, San Diego, March 1973.
111. M.H. White, D.R. Lampe and J.L. Fagan, "CCD MNOS Devices for Programmable Analog Signal Processing and Digital Nonvolatile Memory", 1973 Int. Electron Devices Meeting, Tech. Dig., pp. 130-133, Washington, Dec. 1973.
112. L.R. Rabiner, R.W. Schafer and C.M. Rader, "The Chirp Z-Transform Algorithm", IEEE Trans. on Audio and Electroacoustic Au-17, pp. 86-92, June 1969.
113. R.D. Baertsch, W.E. Engeler and J.J. Tiemann, "A New Surface Charge Analog Store", 1973 IEEE Intl. Electron Device Meeting, Technical Digest, pp. 134-137, Washington, Dec. 1973.

114. D.R. Lampe, M.H. White, J.H. Mims, J.L. Fagan, "An Electrically Programmable LSI Transversal Filter for Discrete Analog Signal Processing (DASP)", CCD Applications Conf., San Diego, pp. 111-125, Sept. 1973.
115. J.J. Tiemann, R.D. Baertsch and W.E. Engeler, "A Surface-Charge Correlator", 1974 IEEE Intl. Solid-State Circuits Conf. Digest of Technical Papers, pp. 154-155, Philadelphia, Feb. 1974.
116. J.J. Tiemann, W.E. Engeler, R.D. Baertsch and D.M. Brown, "Intracell Charge-Transfer Structures for Signal Processing", IEEE Trans. Elec. Dev., ED-21, pp. 300-308, May 1974.
117. A.M. Mohsen, M.F. Tompsett and C.H. Séquin, "Noise Measurements in Charge-Coupled Devices", IEEE Trans. Elec. Dev., ED-22, pp. 209-218, May 1975.
118. M.G. Collet, "The Influence of Bulk-Traps on the Charge Transfer Inefficiency of Bulk Charge Coupled Devices", to be published in the joint issue of IEEE J. Solid-State Circuits and Trans. Electron Devices, Feb. 1976.
119. M. Kubo, M. Ashikawa, I. Takemoto and S. Ohba, "A Wideband Low-Noise Analog Delay Line", IEEE Intl. Solid-state Circuits Conf., Philadelphia, pp. 152, 153, 245.
120. J.B.G. Roberts, "The Application of Charge Transfer Devices to Signal Processing", New Directions in Signal Processing in Communication and Control, edited by J.K. Skwirzynski, NATO Adv. Study Inst. Series, pp. 589-599, Noordhoff Intl. Publishing, Leyden 1975.
121. M.F. Tompsett, G.F. Amelio, W.J. Bertram Jr., R.R. Buckley, W.J. McNamara, J.C. Mikkelsen,Jr., D.A. Sealer, "Charge-Coupled imaging devices: Experimental results", IEEE Trans. Electron Devices, Vol. ED-18, pp. 992-996, Nov. 1971.
122. "Philips achieves speed, efficiency in Peristaltic CCD", Electronics, pp. 5E-6E, Sept. 4, 1975.

11. FIGURES

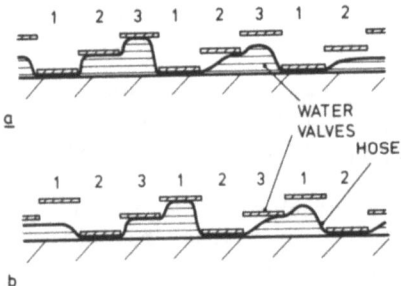

Fig. 1. Peristaltic Transfer Model. A mechanical model of a Charge Transfer Device register. By means of three sets of valves (3-phase system) one can transfer water packets, while keeping them seperated, in a flexible hose pipe. The differential volumes of the packets represent information.

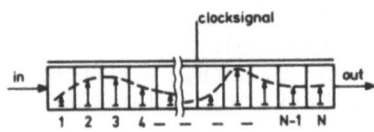

Fig. 2. A block diagram of a Charge Transfer Device (CTD). The input signal is successively transformed into analogue amounts of charge (represented by arrows). These packets are shifted down the register stages to the output in the rhythm of a clock signal.

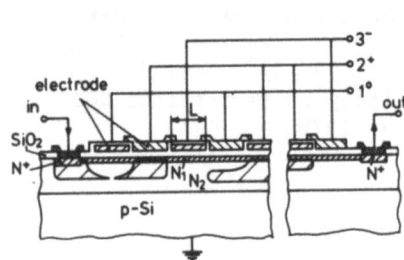

Fig. 3. Cross section of an integrated CTD. It is known as the profiled Peristaltic Charge Coupled Device (PCCD). This CTD combines a very wide frequency range, a large dynamic range and a highly efficient transfer. The very thin (0.2 μm) relatively highly doped N-type surface layer (N_1) is responsible for the high charge handling. The relatively thick (4μm) low doped N-type bulk layer together with the gate length (L = 7.5 μm) effectuates a high transfer efficiency and a very wide frequency range.

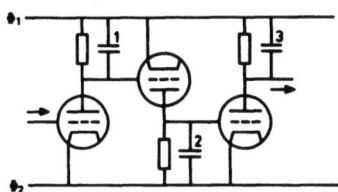

Fig. 4. The first concept of a clocked analogue shift register. The information is stored in the form of a voltage at a capacitor. This information is transferred clockwise from the odd- to the successively even-numbered capacitors, etc. by alternating potentials at the clock lines ϕ_1 and ϕ_2. It is a voltage transfer register.

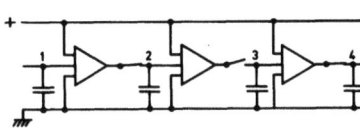

Fig. 5. Sampled analogue delay line. The signal voltage is stored at the even numbered capacitors. When the even-numbered switches are opened and the odd-numbered ones closed, then the signal voltage is transferred, via the unity gain amplifiers, to the successively odd-numbered capacitors. It is a voltage transfer register.

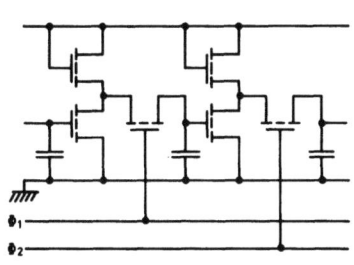

Fig. 6. Integrated sampled analogue delay line according to the concept of fig. 5.

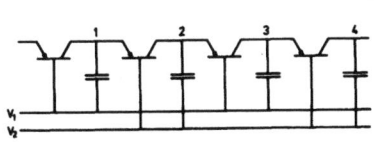

Fig. 7. The bipolar Bucket Brigade Device (BBD). It is a sampled analogue delay line which has the information stored in the form of charge quantities in capacitors. The charges are transferred clockwise from the even-numbered capacitors via transistors to the successively odd-numbered capacitors, etc. The BBD is the first Charge Transfer Device (CTD)

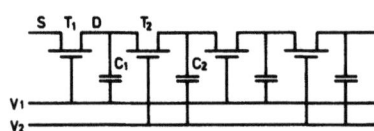

Fig. 8. The MOS Bucket Brigade Device

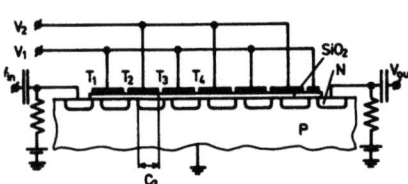

Fig. 9. A cross section of the integrated MOS BBD. The signal charges are stored in MOS capacitors. The charges are transferred by moving potential wells produced by alternating potentials at the gates and the asymmetrically situated N-regions.

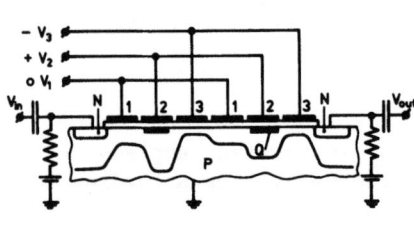

Fig. 10. Cross section of a Surface Charge Coupled Device. The charges (Q) are stored and transferred at the Si-SiO$_2$ interface by moving potential wells induced by the gates. The solid curve shows the depletion depth in the P-type silicon. This depth depends on the positive-biased gates and the amount of charge (electrons).

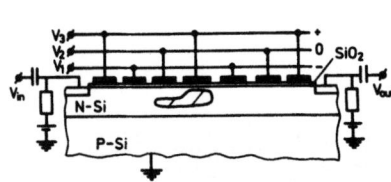

Fig. 11 a. The buried channel Charge Coupled Device. The charges are constrained between the depletion regions produced by the gates and the substrate. The electron packets are transferred in the N-layer. In this way interaction of the charges with the surface states can be avoided. This device combines a high charge handling capacity with a low transfer speed or a low charge handling with a high transfer speed, which depends on the depth d (fig. 11b).

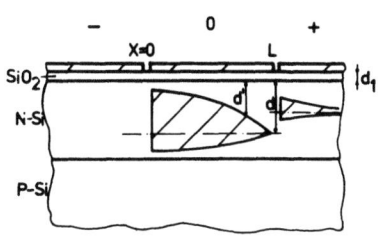

Fig. 11b. <u>Bulk mode</u> of operation. A situation during transfer is shown. At x=L lies the channel pinch-off point. At x=0 the electron concentration gradient equals zero.

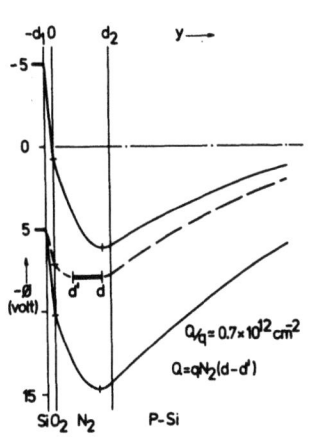

Fig. 11 c. Potential distribution perpendicular to the Si-SiO$_2$ interface for two different gate potentials (-5 and +5 volts). The solid curve shows the situation when no electrons are present. The dashed line shows the situation when there is a charge quantity present. This packet is situated in the horizontal area of the dashed curve in the N-layer where it compensates the donor dope level. (d_1=0.1 µm, d_2=0.7 µm, N_2=2x10^{16}cm^{-3},

$N_a = 5 \times 10^{14}$ cm^{-3})

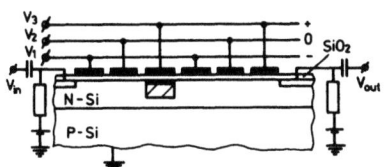

Fig. 12a. The uniformly doped layer <u>Peristaltic Charge Coupled Device (PCCD)</u>. The charges are mainly stored at the Si-SiO$_2$ interface by appropriate potentials at the gates and small fractions are stored in the relatively low-doped bulk layer.

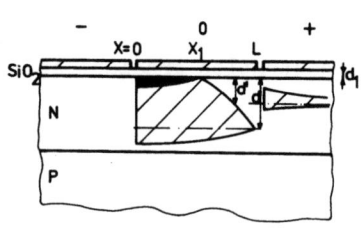

Fig. 12b.
Surface + bulk mode (PCCD twin channel mode)
Charge transfer via a twin-channel for the uniformly doped PCCD. At x=L lies the bulk-channel pinch off point. At $x=x_1$ lies the pinch-off point for the surface channel. During transfer x_1 moves from x=L to x=0. The surface channel together with the low-doped bulk layer effectuates a largely increased charge handling capacity compared to the bulk mode operation (fig. 11b). This holds true if the depths d for both modes are similar. In this case they have similar transfer speeds.

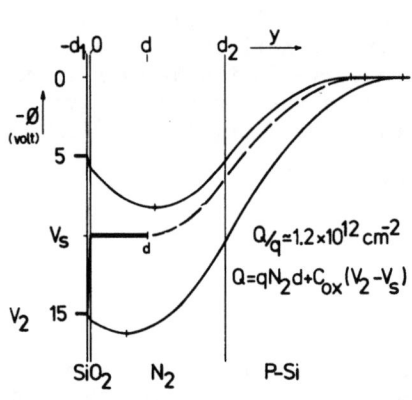

Fig. 12c. The potential distribution perpendicular to the Si-SiO$_2$ interface for two different gate potentials (5 and 15 volts). The solid curves show the situation when there is no mobile charge below the gates. Small charge quantities will be stored in the potential well in the bulk below the +15V gate. The curve in the middle shows the situation when there is a large amount of charge below the +15V gate. The surface charge $Q_a = C_{ox}(V_2-V_s)$ reverses the potential gradient in the oxide.

This device combines high-speed transfer and high charge handling capacity. However, the surface states set a limit to the transfer efficiency.

(d = 0.1 μm, d_2 = 4.5 μm, N_2 = 7×10^{14} cm^{-3}, N_a = 4×10^{14} cm^{-3}).

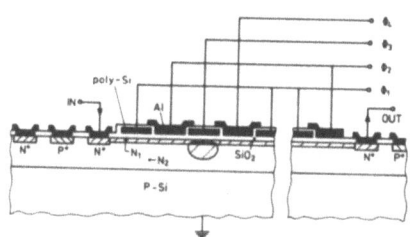

Fig. 13a. Cross section of the profiled PCCD. A charge packet is stored below the third gate. The main part of the charge packet is stored in the highly doped thin surface layer (N_1). This results in a high charge handling capacity and avoids interaction of the charge with the surface states. Only a fraction of the charge packet is stored in the low-doped bulk layer (N_2). Especially the last charge fraction is transferred at a substantial distance from the surface ($\simeq d$), which results in a very high speed transfer. The P^+ regions function as channel isolation. The extreme left N^+ region is a leakage drain.

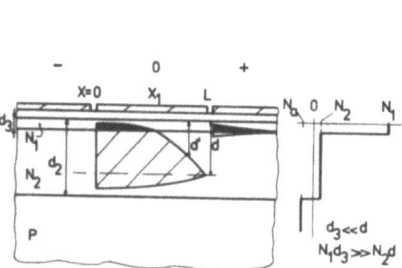

Fig. 13b
"Surface"+bulk mode (PCCD twin channel mode)

A situation during transfer is shown. It demonstrates the twin channel mode of operation similar to that of fig. 12b. However the surface channel is confined to the surface layer (N_1) to avoid interference of the charges with the surface states, while maintaining the high charge handling capacity.
This PCCD mode combines high-speed transfer, high charge handling capacity and high-efficient transfer. Only bulk centers set a limit to the transfer efficiency.

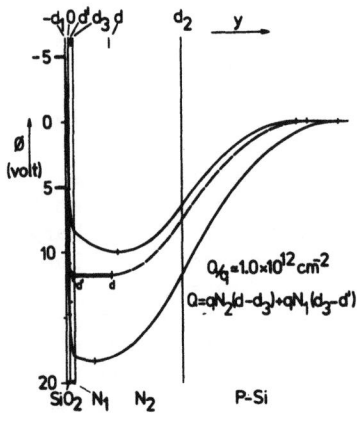

Fig. 13c. The potential versus depth for the situation of two different gate voltages (-5 and +5 volts). The charge is stored below the +5V gate between d' and d.
The main part of the charge is stored between d' and d_3 as a result of the high dope level of the very shallow surface layer (N_1). The solid curves show the potential when no mobile charge is present. Extremely large positive gate voltages are required to shift the potential minimum to the surface. This demonstrates that the surface layer operates also as an effective barrier between the charge and the surface.
(d_1=0.1 µm, d_3=0.2 µm, d_2=4.5 µm, N_1=10^{17}cm^{-3}, N_2=7×10^{14} cm^{-3}, N_a=4×10^{14}cm^{-3}).

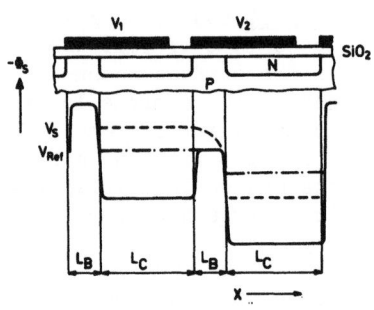

Fig. 14. Surface potential versus distance for one stage of a MOS BBD in case of relatively high-doped buckets. The dashed curve demonstrates a transient situation during the charge transfer. The potential drop falls within the barrier region L_B. The chain-dotted line represents the situation at the end of the transfer. The solid curve shows the potential when the N-Si buckets are missing all their free electrons. When V_2 and V_1 are interchanged then the charges again shift half a stage to the right.

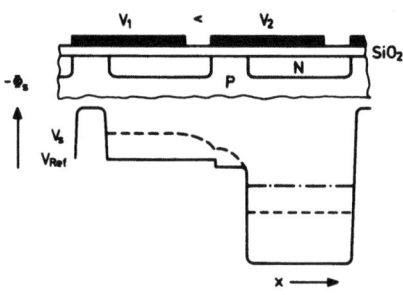

Fig. 15. Surface potential versus distance for one stage of a MOS BBD, where the voltage difference (V_2-V_1) is increased and or the dope level of the N-buckets is lowered. The dashed curve demonstrates a transient situation during charge transfer, the dotted line at the end of transfer. The solid curve shows the situation when the $N\text{-}Si$ buckets miss all their free electrons.

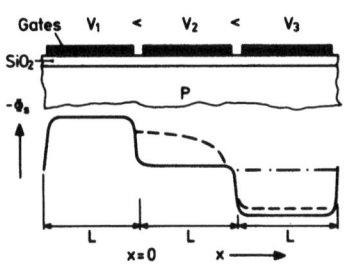

Fig. 16. Surface potential versus distance along one stage of a three-phase surface CCD. The dashed line demonstrates a situation during charge transfer, the chain-dotted line shows the final situation. The solid curve presents the situation when no free electronic charges are available.

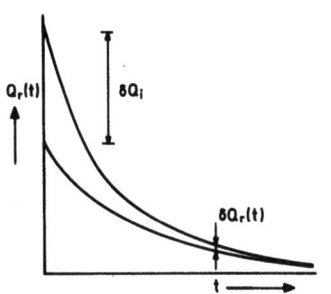

Fig. 17. The amount of signal charge still to be transferred versus time for two different initial charge quantities. The difference between the two curves (δQ_T) decays faster than the signal charge itself.

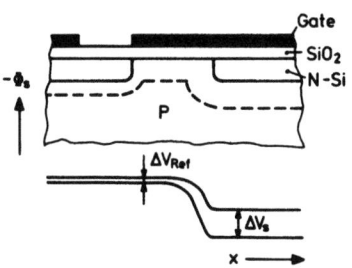

Fig. 18. The feedback in a MOS BBD through modulation of the barrier height by the drain potential.

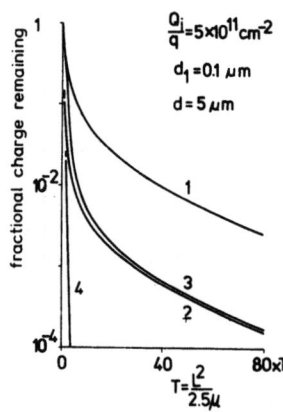

Fig. 19. The fractional charge remaining ($\varepsilon = \dfrac{Q_r}{Q_i}$) versus normalized time.
Curve 1: surface mode of operation
Curve 2: bulk mode of operation
Curve 3: surface+bulk mode, PCCD twin channel operation

Curves 1-3 concern the self-induced field type CTDs.
Curve 4 concerns the externally induced field type CTDs.

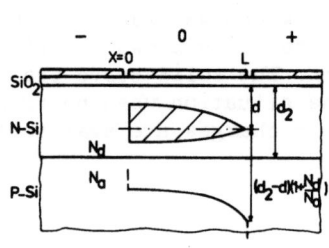

Fig. 20. Bulk mode of operation for the uniformly doped bulk CCD for any degree of dope level for the substrate and the bulk layer.

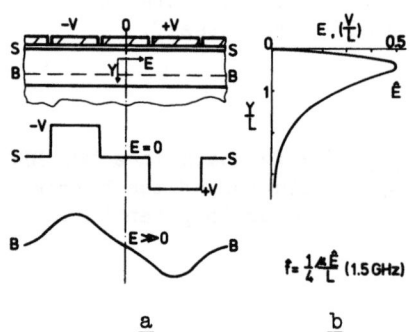

Fig. 21a. The potential distribution in the transfer direction at the surface ($S-S$) and in the bulk ($B-B$) is shown. The smoothed potential in the bulk gives rise to a strong drift field. This is a direct result of the neighbouring gate potentials.
b. shows this externally induced drift field versus depth under the middle of the transferring gate.

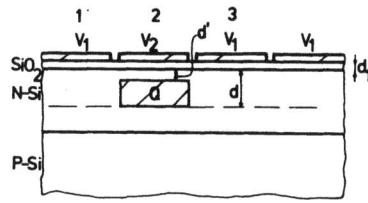

Fig. 22. Model for determination of the charge handling capacity. The voltage $V_{CL}=(V_2-V_1)$ is just sufficient to store the charge quantity Q.

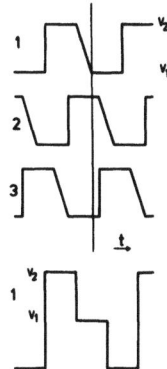

Fig. 23. Clock waveforms for obtaining the maximum charge handling capacity.

Fig. 24. Three level clock waveform, gives the same charge handling capacity as the one in fig. 23 if the voltage swing is increased to about twice its value.

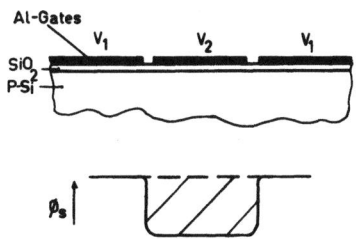

Fig. 25. Surface potential for a surface CCD when the oxide capacitance contains zero charge (solid curve) and when the oxide capacitance contains the maximum charge quantity (dashed curve). see curve 1 of fig. 28.

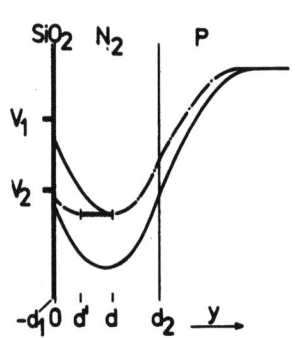

Fig. 26. Potential distribution for the bulk-mode-operated uniformly doped bulk Solid curves for zero mobile charge. Dashed curves for the maximum charge quantity for the clock voltage swing $|V_2-V_1|$. The charge packet is situated between d' and d. See curve 2 of fig. 28.

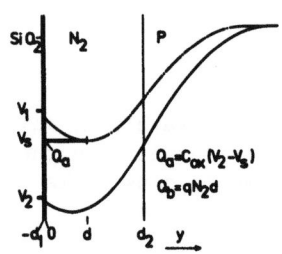

Fig. 27. Potential distribution for the twin-channel-operated uniformly doped PCCD. Solid curves for zero mobile charge, dashed curve for the maximum amount of surface charge (Q_a) + bulk charge (Q_b). See curve 3 of fig. 28.

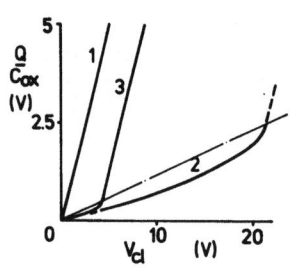

Fig. 28. Charge handling capacity versus clock voltage swing.
Curve 1 for surface mode operated
Curve 2 for bulk mode operated
Curve 3 for the PCCD twin-channel operation.

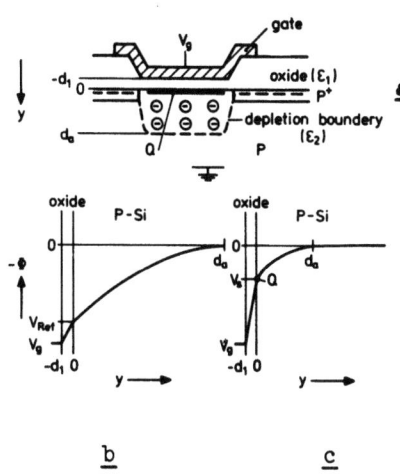

Fig. 29a. MOS capacitor containing the surface charge Q.

b. Potential distribution perpendicular to the surface when the mobile surface charge is zero ($Q=0$).

c. Potential distribution perpendicular to the surface with mobile surface charge Q.

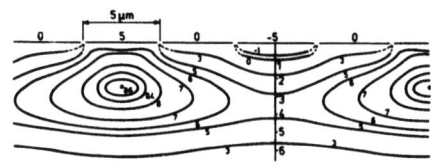

Fig. 30. Equipotential lines in a uniformly doped bulk CCD (EF-type). The potentials are indicated in volts. ($N_2=7.5\times10^{14}$ cm^{-3}, $N_a=2\times10^{14}$ cm^{-3}, $d_1=0.1$ μm, $d_2 = 5$ μm.)

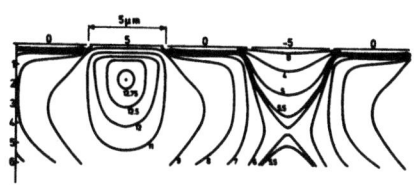

Fig. 31. Equipotential lines in a twin layer PCCD. The potentials are indicated in volts. The surface layer dopant-concentration (N_1) linearly varying from 1.5×10^{6} cm^{-3} to 2×10^{14} cm^{-3} over a depth $d_3=1$ μm. $N_2=N_a=2\times10^{14}$ cm^{-3}, $d_1=0.1$ μm, $d_2=7.7$ μm.

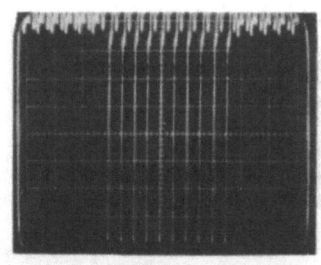

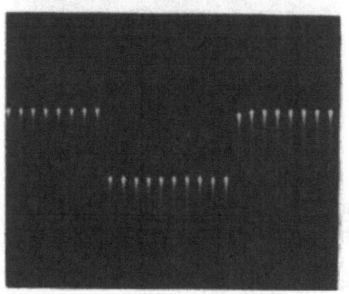

Fig. 32. Pulse response of a profiled PCCD after 512 transfers for zero bias charge operation. A series of ten ones after 90 zeroes.

a. 1 MHZ clock frequency, showing inefficiencies of $\varepsilon_{01} \approx \varepsilon_{10} \approx 10^{-5}$.

b. 135 MHz clock frequency, showing inefficiencies of $\varepsilon_{01} \ll \varepsilon_{10} \approx 10^{-4}$. At this frequency the clock noise is considerably larger than the signal. In applications this clock noise can be filtered out.

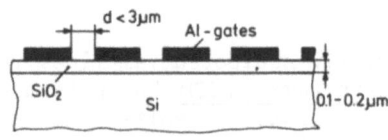

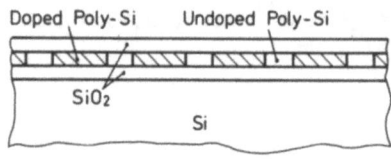

Fig. 33. CTD gate structures

a. single level Al gates

b. single level of doped polysilicon gates

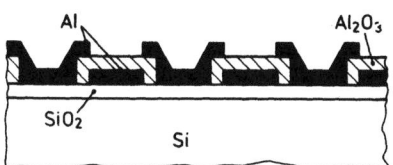

Fig. 33 c. double level of anodized aluminium gates.

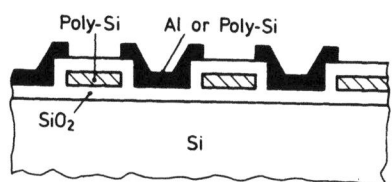

Fig. 33 d. two levels of polysilicon or polysilicon and aluminium gates

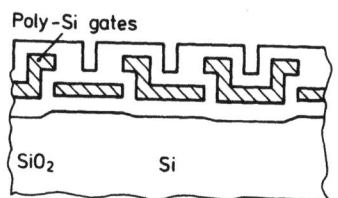

Fig. 33 e. three levels of polysilicon

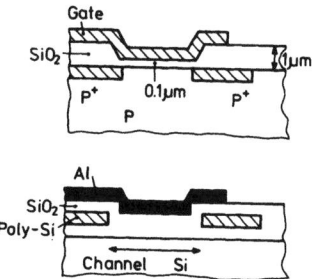

Fig. 34. Channel definition
a. for surface CTDs.
Thick oxide possibly combined with P^+ diffusions.

b. by field effect, by means of polysilicon gates. This method is suitable both for surface and bulk CTDs.

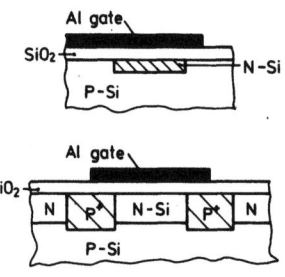

Fig. 35. Channel definition for bulk CTDs.

a. implanted N-channel

b. channel isolation by means of P^+ diffusions through an implanted or epitaxially grown N-layer.

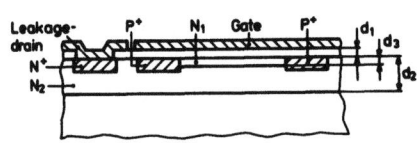

c. field effect channel isolation by means of shallow P^+ diffusions in use with the PCCD.

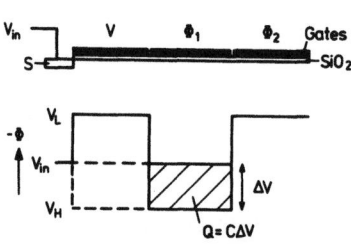

Fig. 36. Three electrical input methods for the introduction of charge quantities into a CTD.

a. The area below the first gate functions as a switch between source (s) and drain area below ϕ_1. The charge quantity depends also on the threshold voltage V_H.

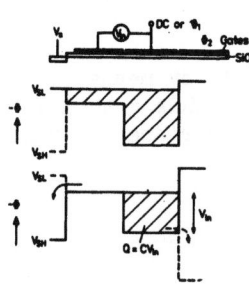

b. Fill and spill method. The charge quantity depends only on V_{in} if the two input gates have equal threshold voltages.

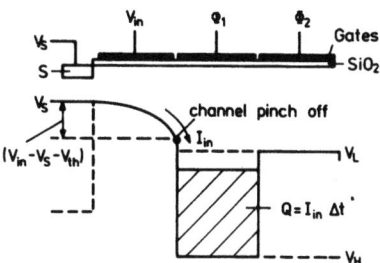

Fig. 36 c. The charge quantity depends on the input current I_{in} and the time Δt that the source is switched on.

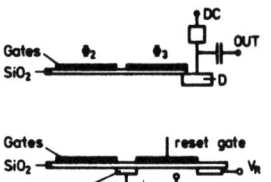

Fig. 37. CTD output circuits

a. current output

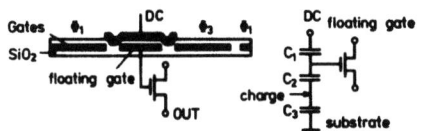

b. floating diffusion amplifier (FDA).

c. floating gate amplifier (FGA)

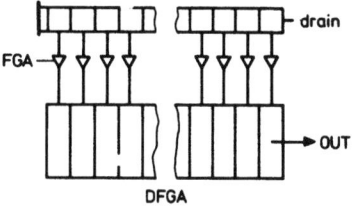

d. distributed floating gate amplifier (DFGA)

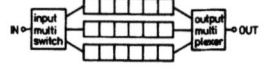

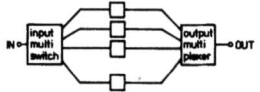

Fig. 38. CTD delay line systems
a. Serial CTD

b. K serial CTDs, operated in parallel.

c. Sequentially operated single-stage CTDs.

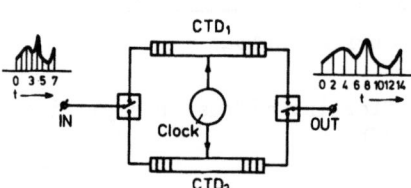

Fig. 39. Block diagram for the TV standards conversion from 625 lines to 313 lines.

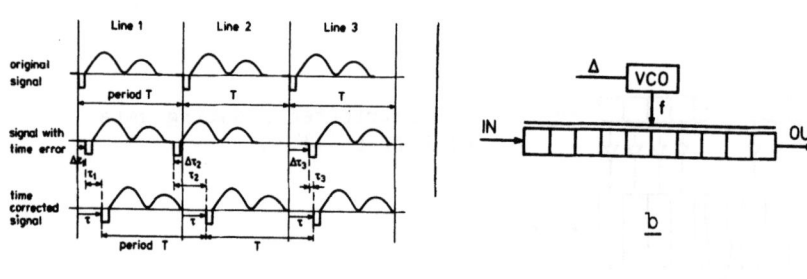

Fig. 40. Block diagram for time error correction of video signals. The delay time of the CTD is controlled by the clock frequency. The delay is set to the value τ_i so that $\tau=\tau_i+\Delta\tau_i$, where $\Delta\tau_i$ is the time error.

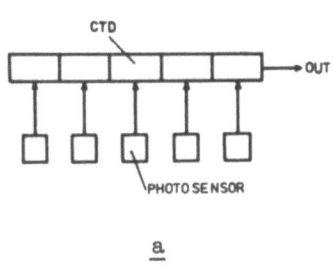 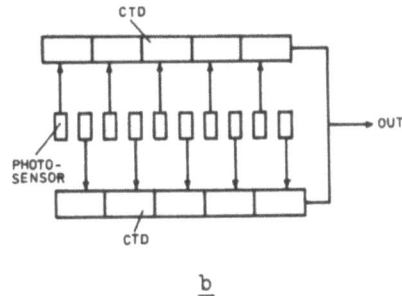

Fig. 41. CTD line sensors

a. one CTD line
b. two CTDs operated in parallel
   to obtain the double amount of photosensors

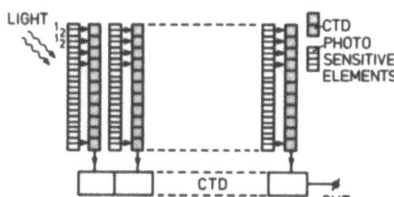

Fig. 42 CTD area image sensors

a. interline transfer imager

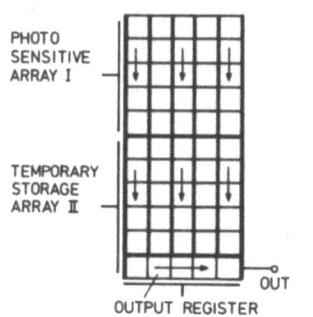

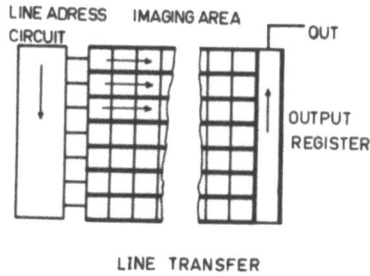

c. line transfer imager

b. frame transfer imager

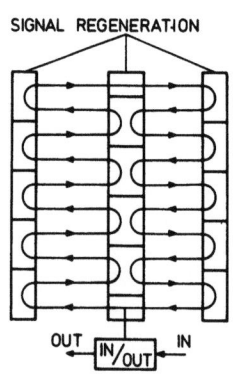

Fig. 43. Memory systems

a. single serpentine storage loop

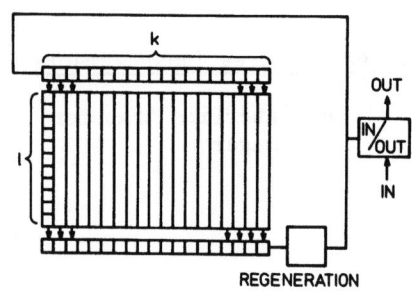

b. series-parallel-series single storage loop with only one regenerator

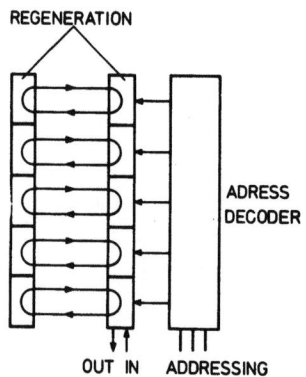

c. one dimensional extension of single storage loops for obtaining smaller loops and shorter access times.

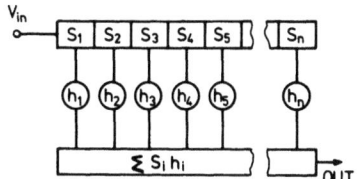

Fig. 44. Diagram of an n-stage transversal CTD filter.

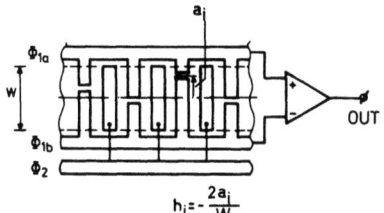

Fig. 45. Weighting methods

<u>a</u>. split electrode weighting and summing

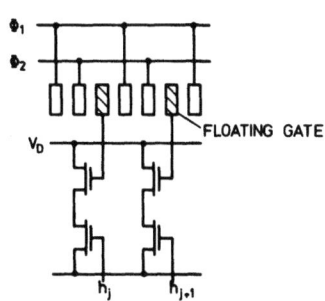

<u>b</u>. Floating gate output. Weighting coefficients are obtained by an electrically adjustable conductance in the source follower lead.

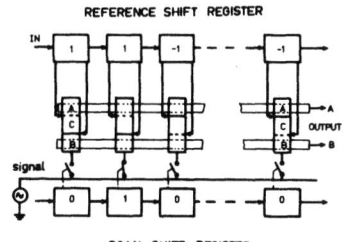

<u>c</u>. Filter with analogue input and digital weighting coefficients. This circuit employs the delay system of fig. 38c.

ELECTRICAL CHARGE INJECTION INTO CCD'S

C. H. Séquin

Bell Laboratories
Murray Hill, New Jersey 07974, U. S. A.

ABSTRACT. Charge-coupled image sensors, when provided with an electrical input, can perform certain signal processing operations such as dark current background subtraction or motion detection. Various methods of electrical charge injection are reviewed and their relative performance with respect to noise and linearity of the input characteristics are discussed.

1. INTRODUCTION

The main source of signal charge in a solid-state image sensor is, of course, the incident light producing hole electron pairs. In principle a charge-coupled image sensor would therefore not require any means for electrical charge injection. However, surface channel CCD's need a certain minimum bias charge to exhibit their best transfer efficiency. This bias charge will keep the interface states in the transfer channel near saturation at most times so that in each signal packet the charge lost by trapping is compensated to first order by the charge left behind from the previous packet. This bias charge need only be a few percent of the maximum signal which can be handled by the device in order to achieve optimum performance. But if this bias charge level is not extremely constant and uniform, it will degrade the superposed video signal by random or pattern noise. These requirements are especially stringent in low light level imaging where the video signal may be smaller than the bias charge level.

The electrical injection of charge is also required in image sensors in which the video signal is recirculated in order to perform signal processing operations such as multiple readouts, dark current cancellation or motion detection. In the following section we will review some of these signal processing concepts before we start the discussion of the various charge injection methods and of their performances with respect to noise and linearity.

2. SIGNAL PROCESSING IN IMAGE SENSORS

Charge-coupled linear or area image sensors can be provided with electrical inputs [1,2]. Because of their analog storage capability these devices can then be used to perform certain video signal processing operations which otherwise would have to be performed in additional devices at greater expense. By recirculating the output signal produced by such a device back into the image sensor itself [3], operations such as multiple readout, motion detection and dark current background subtraction can be performed.

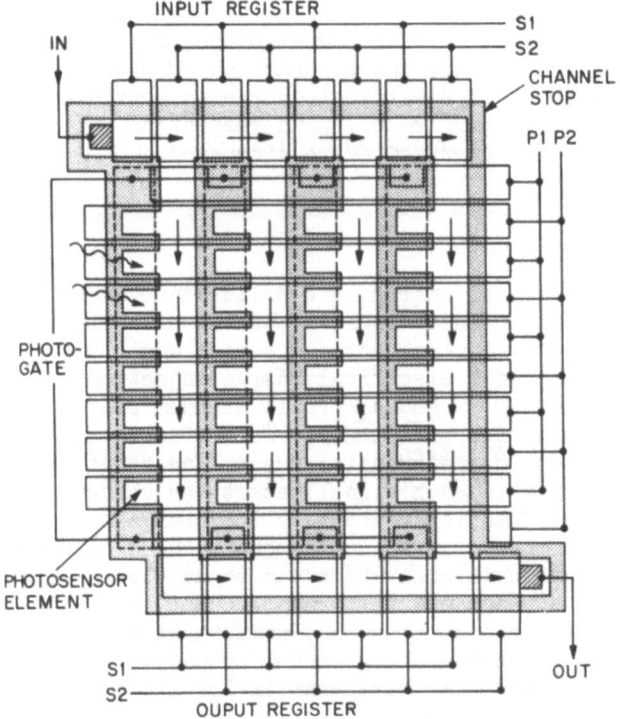

Fig. 1 Interline-transfer charge-coupled area image sensor with a serial input register which permits simple signal processing by recirculation of the video information.

As an example we will discuss the implementation of these operations in an interline transfer device [4] with a suitable input register at the end of the array opposite to the output register (Fig. 1). In the simplest case, the output video signal can be reread back into the device to permit multiple readouts of the same image. During this operation the interleaved integration sites are kept isolated from the transfer channels. However, due to the cumulative effects of dark current and transfer inefficiency, the signal will degrade rapidly. As a variant the feedback path could contain an inverter to reduce the degradation due to dark current. Although four recirculations have been demonstrated in preliminary experiments by Weimer et al. [3], a large number of recirculations of the video signal in purely analog form seems impractical. A possible solution is a quantization of the signal amplitude into several levels and subsequent recirculation and regeneration of the signal coded in this multilevel form.

The described image sensor with an electrical input register can also be used as a simple motion detector. The array is constantly exposed to the optical input, but only every alternate field is fed back into the device in complementary form. After the second lateral frame transfer from the integration sites into the vertical registers the stationary parts of the image will effectively cancel out by adding up to a uniform bias charge level. At the output of the device a video signal is then obtained, which alternates in subsequent fields between the normal image and a difference signal, indicating changes or motion between successive images.

With a slight variation this latter recycling scheme can be used to subtract the fixed pattern noise produced by a spatially nonuniform dark current. An optical shutter in front of the sensor is closed during alternate fields to allow the device to integrate the dark current pattern. During the following field this dark current signal is inverted and recirculated back into the same positions in the vertical registers, while the optical signal plus the dark current signal are being integrated in the adjacent sensor sites. In the lateral frame transfer the two signals are combined and the dark current contributions complement each other to a uniform bias level.

With a suitable modification, these signal processing schemes can also be used in frame transfer structures [5]. In the case of background subtraction, for example, the shutter would have to be closed during two out of three frames, and every third frame transfer would have to be suppressed. A new field of optical information would then be integrated on top of the recirculated inverted background pattern. However, the useful duty cycle of optical input would be reduced to only one-third [3].

In frame transfer structures an optical shutter would also be required for multiple readouts of the video signal to prevent mixing with new optical input.

The described signal processing functions could be performed with other devices, but it is clearly more economical if these operations can be performed within the image sensor itself. The addition of the necessary electrical input represents only a small increase in device complexity. However, subtraction schemes for motion detection or background cancellation will result in a reduced dynamic range, because each potential well must handle both the original and the recycled signal, and suitable performance can only be obtained in a device with good transfer efficiency and with low noise and low distortion in the electrical input/output circuits.

3. REVIEW OF VARIOUS INJECTION METHODS

3.1 Dynamic Current Injection

In almost all electrical injection methods the source of carriers is a special input diode. In many early demonstrations of charge-coupled devices [1,6] the input signal was simply applied to the input diode and referenced with respect to an input gate which was held at a relatively low dc potential (Fig. 2a). In this mode, a certain amount of charge flows across the gate electrode (length L_{IG}) into the potential well under the first transfer electrode (length L_{P2}) as the latter is pulsed to a high potential. The size of the injected charge packet Q_S is dynamically determined by the input diode potential V_{ID}, the channel conductance under the gate, and by the available injection time τ and thus by the clock frequency. Using the assumption that the input gate acts as a MOSFET in saturation, we obtain

$$q_S = \tau\mu \frac{W}{L_G} \frac{C_{ox}}{2} (V_G - V_T - V_{ID})^2 \tag{1}$$

where μ is the carrier mobility, W is the channel width, C_{ox} is the oxide capacitance per unit area and V_T is the threshold voltage of the input gate. Typically, to cover the range of packets from zero charge to a full well at a clock rate of 1 MHz, voltage changes of a few hundred millivolts were found to be necessary.

The roles of the input diode and of the input gate can be reversed, and complementary signals would then be injected for

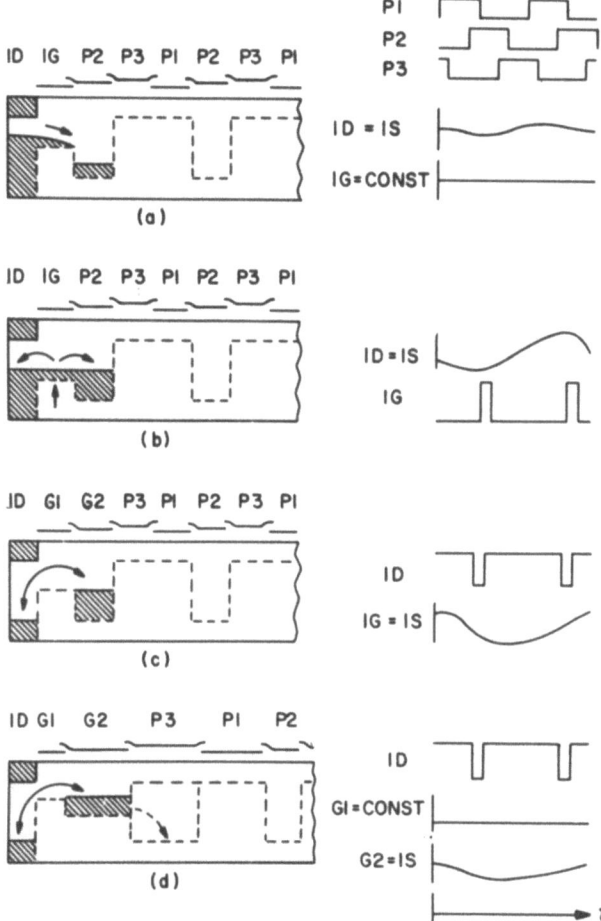

Fig. 2 Various schemes for charge injection into a CCD. The waveforms applied to the electrodes are shown on the right: (a) dynamic current injection, (b) signal voltage applied to the input diode, and input gate pulsed to cut off the input diode, (c) potential equilibration method with a pulsed metering electrode, and (d) improved equilibration method with input signal applied to enlarged metering electrode. (Ref. 18)

the same input signal applied to the input gate. Both versions are typically subject to rather strong nonlinearities and high noise.

3.2 Voltage Inputs

In these methods the input diode is used to define the interface potential in the metering potential well, which is typically formed under the first regular transfer electrode. The input gate is pulsed to perform the sampling function and subsequently to isolate the charge packet in the metering well. In one approach (Fig. 2b) the input signal is applied to the input diode and the input gate is strongly turned on during the time when the first transfer electrode is biased to the full pulse potential (V_R+V_P). But the amount of charge flowing into the metering well will not be a truly linear function of the diode potential because of the depletion capacitance changing with the diode potential. The interface potential under an electrode held at potential V_G is related in a nonlinear manner to the signal charge per unit area Q_S [7]

$$\varphi_s = V_G' + V_B - (2V_G'V_B + V_B^2)^{1/2} \qquad (2)$$

with $V_G' = (V_G-V_{FB}) + Q_S/C_o$ and $V_B = qN_A\varepsilon_S/C_o^2$ where V_{FB} is the flatband voltage; C_o is the oxide capacitance per unit area; ε_S is the dielectric constant of the substrate and N_A its doping density. In practice the nonlinearities of the above equation become significant for substrate dopings higher than 10^{15} cm^{-3}. Furthermore, the gate does not act as an "infinitely sharp knife" in cutting off the charge packet. The charge residing underneath the gate itself is distributed in a poorly controlled manner between the diode and the signal charge packet. The randomness in this process introduces excess noise [8-10] into the charge packets. Furthermore, the distribution depends on the exact shape of the barrier formed under the gate. The variation of the shape of this barrier with changing potentials on the diode and the associated change in the dynamic partitioning of the charge generate additional nonlinearities in the input characteristics.

3.3 Charge Preset Methods

In the charge preset or potential equilibration method a potential well of the required size is formed under a metering electrode which is separated from the input diode by another gate electrode (Fig. 2c). The input diode is pulsed for a short time to a low potential (V_{IDL}) to inject charge across a potential barrier formed under G1 into the potential well under G2. Then the diode is reverse biased, thus draining all excess charge from under G2 until the remaining surface potential under G2 is equal to the barrier potential under G1.

Starting from Eq. (2), it can be shown [11] that the amount of signal charge Q_S retained under G2 is then given by

$$Q_S = (V_{G2}-V_{G1}) \cdot A_{G2} \cdot C_{ox}, \qquad (3)$$

where A_{G2} is the active area of gate 2 and C_{ox} the constant oxide capacitance for all electrodes. After potential equilibration has taken place, the interface potentials under the two gates G1 and G2 are equal. Thus the terms V'_G in Eq. (2) written for the two gates, respectively, must also be equal and it follows that

$$V_{G1} - V_{FB} = V_{G2} - V_{FB} + Q_S/C_{ox}. \qquad (4)$$

In the simplest implementation of the charge preset or potential equilibration method the input signal is applied to G1, and G2 is pulsed like a regular transfer electrode. The input diode is pulsed low when G2 is turned on to V_P, and the signal charge packet should then be a linear function of the potential difference (V_P-V_{G1}).

In this form the described injection method has three disadvantages. Firstly, any noise on the driving pulses for electrode system P2 appears on G2 and thus enters the signal directly [10]. It is therefore preferable to define the injected charge packet by dc voltages. This can be achieved by holding G2 at an intermediate dc potential between V_R and V_P. The charge trapped under G2 during the injection process will empty into the potential well under the next transfer electrode when P3 is pulsed to V_P. In a device designed to be used in this mode, G2 and the following two transfer electrodes have to be made at least twice as large in order to handle the charge that corresponds to a full size packet in the rest of the 3-phase device.

Secondly, in the potential equilibration method using a pulsed metering electrode (G2), the active area A_{G2} changes with varying interface potential due to varying fringing fields. For low potentials on G1, i.e., large wells under G2, the charge packet occupies a larger area, and thus the slope of the curves is somewhat steeper. For very small packets the effective area which holds the signal charge is smaller, which causes a rounding of the input characteristics at the high end of the input range. This component of the nonlinearity can be eliminated by reversing the function of the two input gates with respect to the method described above. Gate G1 is now held at a low reference voltage V_{G1} at some safety margin above V_R, and the input signal is applied to G2 (Fig. 2d). Surface potential equilibration now occurs always at the same interface potential φ_{G1} determined by the fixed reference V_{G1}, and the influence of the fringe fields is thus the same for all values of signal charge. Contrary to the previous two methods the size of the signal charge packet now increases with increasing input signal voltage. The maximum value is reached when the potential of G2 is about midway between V_{G1} and V_P. For larger voltages applied to G2, not all signal charge can be handled by the subsequent electrodes, and excess charge still residing under P3 when this electrode turns off spills forward as well as backward, and thus not the whole charge packet is transferred through the device.

Thirdly, if the two input gates are on different oxide thicknesses, the terms V'_G from Eq. (2) can no longer be equated in the simple manner expressed in Eq. (4), since two different terms V_{B1} and V_{B2} appear in the equations for the two gates G1 and G2. A nonlinear term is then introduced in the relationship between gate potential V_{G1} and signal charge Q_S. However, when the input signal is applied to G2 and G1 is held at a fixed reference value V_{G1}, potential equilibration always occurs at a fixed interface potential and thus for a fixed value of V'_G in Eq. (2) written for G2. Therefore,

$$V_{G2} - V_{FB} + Q_S/C_{ox2} = \text{constant}, \qquad (5)$$

and thus the signal voltage on gate 2 is related in a linear manner to the charge held underneath, regardless of the gate oxide thicknesses involved. Of all the charge preset methods discussed so far this last version shows thus the best performance with respect to the linearity of the input characteristics [12].

3.4 Complete Cancellation of Threshold Variations

A special problem may arise in frame transfer structures, when exactly the same amount of bias charge would have to be injected into every vertical transfer channel during the frame transfer operation. If the threshold voltages under G1 and G2 have different variations across the array, strong pattern noise will result.

Near perfect cancellation of the effects of threshold differences can be obtained if the signal charge is defined as a difference of two subsequent metering processes in the same potential well [13]. The basic scheme is illustrated in Fig. 3. As in

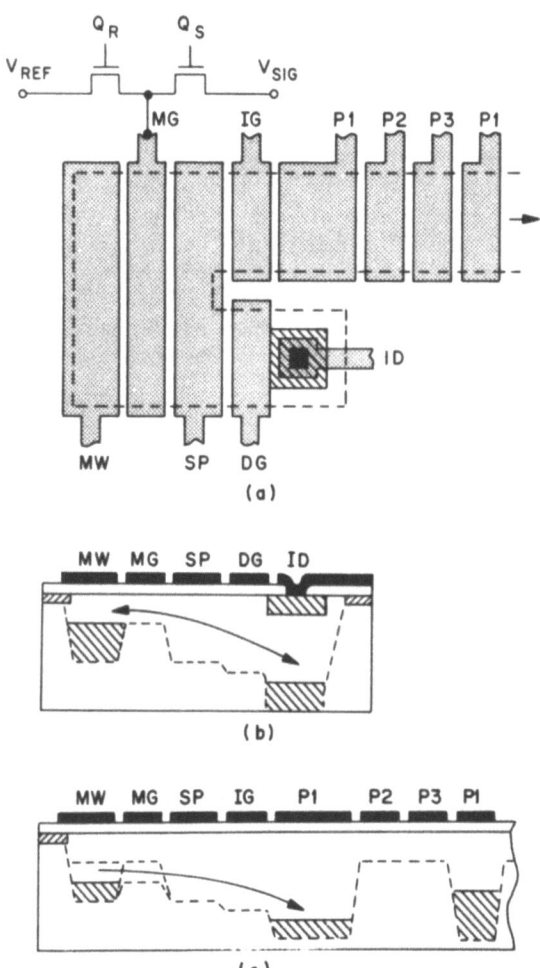

Fig. 3 (a) Input electrode structure for a CCD that is independent of threshold voltage difference. Schematic cross sections showing (b) the precharging of the metering well MW with the potential equilibration method and (c) a subsequent additional discharge from MW constituting the actual signal charge packet. (Ref. 18)

the previously described potential equilibration methods, a metering well (MW) formed under an MOS electrode is filled and subsequently drained to the potential under MG by suitably pulsing the input diode (Fig. 3b). During this operation the input gate IG to the actual transfer channel is closed, and a reference voltage V_{REF} is applied to MG. For the second metering process the drain gate DG is closed, IG is opened and the potential of MG is changed to the higher signal voltage. This will force an additional amount of charge to flow from the metering well into P1 of the transfer channel (Fig. 3c). This signal charge packet is thus given by the capacitance of MW and by the potential difference ($V_{SIG}-V_{REF}$) and does not depend on any threshold voltages. In this method too, as in all previous methods, the functions of the two gates MW and MG can be reversed, and the signal and reference potential could be applied in sequence to MW. A reduction in the size of the metering well would then also push a corresponding amount of charge out of MW into the transfer channel.

4. LINEARITY

While the linearity of the dynamic current injection schemes are unsatisfactory having second harmonic distortions on the order of 20% of the fundamental, the charge preset or potential equilibration methods exhibit much better performance. Figure 4 shows the input characteristics for the three versions discussed in Section 3.3. All measurements were taken on the same device, in which the input gates G1 and G2 were of the same size as the transfer electrodes. Therefore methods B and C (see Fig. 4), in which the reference gate is held at a dc voltage, show only half the voltage range of the original method using a metering well under a pulsed transfer electrode. Nevertheless, it is evident that a linear relationship exists for all three methods over a rather large amplitude range.

Quantitative results of an examination of the generated second and third harmonic distortions are displayed in Fig. 5. The method using the pulsed metering well (on the left) is compared to the "reverse" version (on the right) in which G1 is held at a reference potential. More than a factor of 2 improvement in linearity is observed. This holds true for measurements made on devices with different channel widths and with transfer electrodes of different lengths. In general, the nonlinearities increase with a reduction of the size of the metering well because fringe effects become proportionally more significant [12]. Distortions of more than 40 dB below the fundamental have been observed on wide channel devices, which, however, were not designed particularly with the linearity of the input characteristics in mind. Second and third harmonic distortions 60 dB below the fundamental over a wide range of the input voltage window seem achievable with a proper design of the input stage.

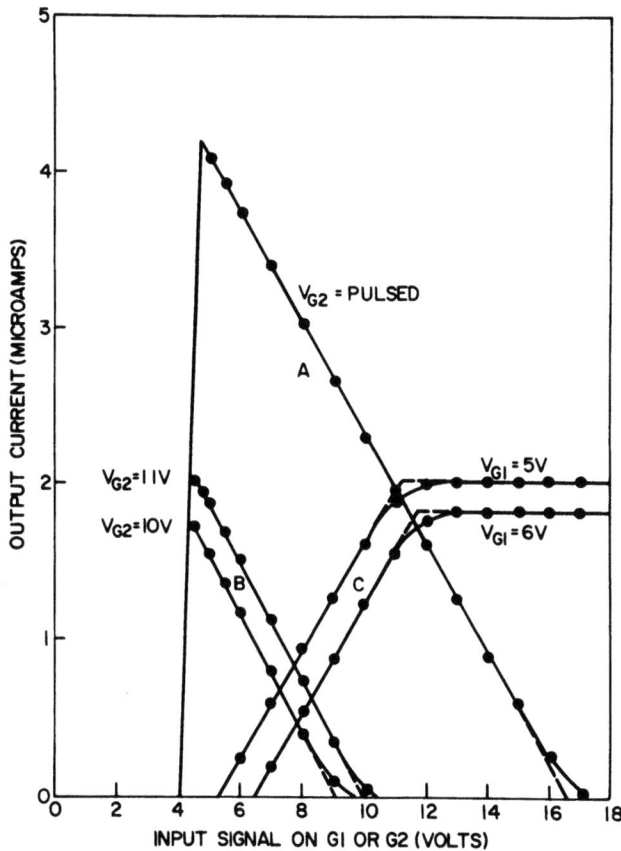

Fig. 4 Input transfer functions for different versions of the potential equilibration method in an SCCD. The input signal is applied to either G1 or G2, and the other gate is either pulsed (a) or held at a reference potential (b,c). (Ref. 12)

If electrical charge injection is simply used to provide a bias charge to obtain the best values of transfer efficiency, the linearity of the process is of no concern. In signal processing applications, however, it has to be given some consideration. In the background subtraction schemes only small signals, which cover not more than a limited part of the full amplitude range, are recirculated in complementary form. The gain of the inverter is simply adjusted for best cancellation, and the nonlinearities due to the electrical charge input should be no problem. In the motion detection mode the recirculated signals can cover the full

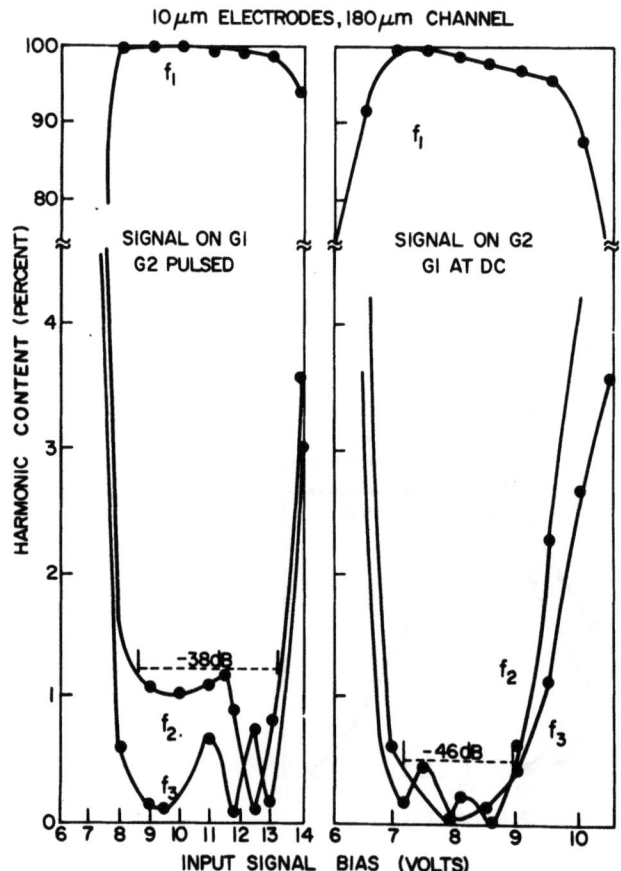

Fig. 5. Fundamental (f_1), second (f_2), and third (f_3), harmonic components in the output signal of an SCCD operated with the two variations of the potential equilibration methods corresponding to Figs. 4(c) and (d), respectively. (Ref. 12)

amplitude range. But the detection of motion will involve a thresholding function, and any small spurious signals which may result from the nonperfect cancellation due to the nonlinearities in the recirculation process will normally be below this threshold. Multiple recirculation to achieve several subsequent readouts of the video signal is most critically dependent on linear input/output characteristics since any distortions may build up. It appears that the low harmonic distortions obtainable with potential equilibration processes will be adequate, and in most practical cases it will probably be less problematic than the required gain constancy of the feedback loop.

5. NOISE

The noise associated with electrical signal injection is more important and has to be considered in all cases, whether only a small amount of bias charge is inserted or whether the whole video signal of the previous frame is being recirculated since this noise will eventually appear on the output video signal. The worst case, of course, would be a small video signal in presence of a relatively large bias charge.

There are several sources of noise associated with electrical injection. A lower limit is given by the thermodynamic fluctuations of the charge retained in the metering potential well. At the beginning of the equilibration process, while excess charge is still drained off through an inverted MOSFET channel, the mean square uncertainty of the charge retained in this well of capacitance C_M is given by

$$\overline{\Delta Q_r^2} = \frac{2}{3} kTC_M. \tag{6}$$

In the final stage, where the discharge process is limited by thermal emission of carriers over the potential barrier, the fluctuations are reduced to

$$\overline{\Delta Q_e^2} = \frac{1}{2} kTC_M. \tag{7}$$

These lower limits can in principle be reached by properly designed input stages operating in the potential equilibration method.

Other injection methods are typically subject to additional sources of noise. Especially poor is the performance of the dynamic current injection since the amount of charge injected is critically affected by the voltage levels on the input diode and on the input gate and also by the exact injection time. In the voltage input method the turn-off of the input gate will introduce additional noise because of the fluctuations in the partitioning of the charge underneath flowing to the input diode or into the metering well, respectively.

In the potential equilibration methods an additional noise contribution of $2kTC_M/3$ or $kTC_m/2$ is introduced if the metering well is formed with a reverse biased diode rather than with a well under an MOS electrode. The second contribution originates when the signal charge is separated from the majority carriers in the diode. For the same reason the noise contributions according to Eqs. (6) or (7) are also doubled if a complete cancellation of threshold effects is attempted by forming the signal charge packet as the difference of two subsequent measuring processes.

Special care has to be taken to shield the electrodes involved in the metering process from the capacitive pickup of clock signals. Clock noise of magnitude $\overline{\Delta V_{cl}^2}$ would introduce noise on the signal on the order of

$$\overline{\Delta Q_{cl}^2} = \overline{\Delta V_{cl}^2} C_m^2 \tag{8}$$

where C_m is the mutual coupling capacitance to the clocked line.

Figure 6 compares experimental results [10] obtained on 3-phase CCD's [14] with overlapping polysilicon electrodes which were operated in various ways. As can be seen, the noise is highest in devices where the metering electrode G2 is connected to the transfer electrodes P2. That the noise on the clock drivers constitutes indeed a significant contribution can be seen from the lower second curve, which was obtained by holding the set of electrodes to which G2 was connected at an intermediate dc value, thus operating the device in a 2-1/2 phase mode. However, there is still a considerable amount of clock noise coupled to the metering well via the mutual overlap capacitance of the transfer electrodes. In devices with two separate input gates, where one of the input gates is biased with a suitably stabilized or filtered dc potential, the electrical injection noise is smallest. However, the measured values are still significantly above the theoretical limit, since no special care was taken in the layout of these devices to shield the input gates from capacitive clock pickup. Better results have been observed on devices with nonoverlapping electrodes [9] and thus with less mutual capacitance.

Since the electrical injection noise is proportional to the metering capacitance C_M it is advisable to keep the metering well as small as possible. This requirement is in contradiction with the design principle for best linearity of the input characteristic. But, as discussed in the previous section, in the case of charge-coupled image sensors the linearity of the electrical

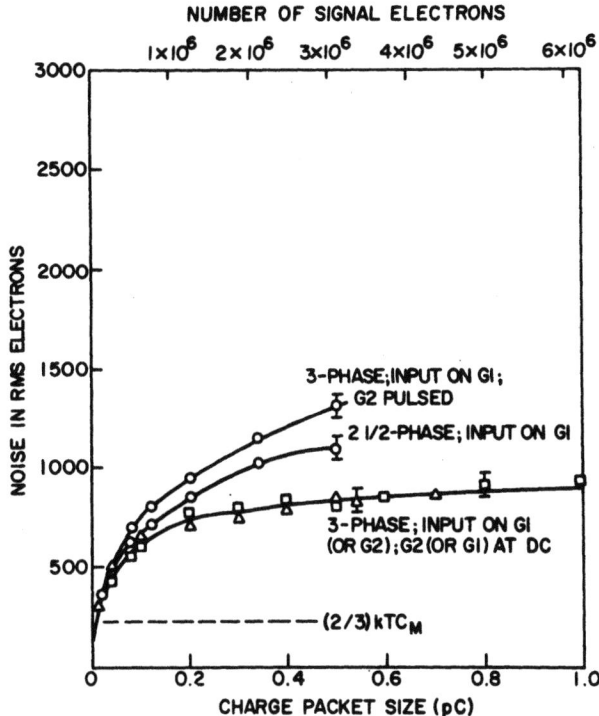

Fig. 6 Electrical injection noise, as a function of charge packet size for various potential equilibration methods, compared to the theoretically expected value of $2kTC_M/3$. (Ref. 10)

input is only of secondary importance and the considerations of low noise are dominant.

6. DISCUSSION

In frame transfer devices the requirements on the bias charge injection are especially severe, since it has to occur at high speed in parallel into all vertical registers. The scheme described in Section 3.4, which should give near perfect cancellation of threshold effects, has so far not been tried in this application. Normal potential equilibration schemes have proven inadequate since they resulted at best in nonuniformities of several percent. To circumvent these problems the following approaches have been taken by different workers. Bulk channel devices, which do not require any bias charge in order to exhibit good transfer efficiency, can be used [15] rather than surface channel devices. However, it appears to be more difficult to develop a bulk channel fabrication process which gives a low dark current performance comparable to that of surface channel devices. The bias charge required in surface channel devices could also be produced optically by uniformly illuminating the array with a small bulb inside the camera [16]. Alternatively a compromise could be taken by using electrical charge injection to produce a few almost completely filled lines at the time when frame transfer is initiated. These lines of charge will be stored at the end of the imaging area near the boundary to the storage area. In the subsequent frame transfer these lines will sweep through the empty storage area, ahead of the actual video charge pattern, and thereby they will refill the interface states in the storage registers which have emptied since the readout of the previous frame. Even though this scheme is not equivalent to the use of bias charge, it can compensate for a substantial amount of the signal degradation experienced in surface channel frame transfer devices.

In interline transfer devices the bias charge could suitably be entered through a serial input register at the top of the device (Fig. 1) at the same rate as the previous video frame is read out from the device. In this approach all bias charge packets would be metered by a single circuit, which could therefore be built with more sophistication, and this could mitigate the uniformity problem considerably.

REFERENCES

[1] Tompsett, M. F., and Zimany, E. J., "Use of Charge Coupled Devices for Delaying Analog Signals," IEEE Jour. of Solid-State Circuits SC-8, 151-157 (1973).

[2] Séquin, C. H., Sealer, D. A., Bertram, W. J., Tompsett, M. F., Buckley, R. R., Shankoff, T. A., and McNamara, W. J., "A Charge Coupled Area Image Sensor and Frame Store," IEEE Trans. on Electron Devices ED-20, 244-252 (1973).

[3] Weimer, P. K., Pike, W. S., Shallcross, F. V., and Kovac, M. G., "Video Processing in Charge-Transfer Image Sensors by Recycling of Signals Through the Sensor," RCA Review 35, 341-354 (1974).

[4] Amelio, G. F., "Physics and Applications of Charge Coupled Devices," IEEE INTERCON, New York, Digest, Vol. 6, paper 1/3 (1973).

[5] Tompsett, M. F., Amelio, G. F., Bertram, W. J., Buckley, R. R., McNamara, W. J., Mikkelsen, J. C., and Sealer, D. A., "Charge Coupled Imaging Devices: Experimental Results," IEEE Trans. on Electron Devices ED-18, 992-996 (1971).

[6] Tompsett, M. F., Amelio, G. F., Smith, G. E., "Charge Coupled 8-Bit Shift Register," Appl. Physics Lett. 17, 111-115 (1970).

[7] Boyle, W. S., and Smith, G. E., "Charge Coupled Semiconductor Devices," Bell Syst. Tech. Jour. 49, 587-593 (1970).

[8] Carnes, J. E., Kosonocky, W. F., and Levine, P. A., "Measurements of Noise in Charge Coupled Devices," RCA Review 34, 553-565 (1973).

[9] Emmons, S. P., and Buss, D. D., "Noise Measurements on the Floating Diffusion Input for Charge-Coupled Devices," Jour. of Appl. Physics 45, 5303-5306 (1974).

[10] Mohsen, A. M., Tompsett, M. F., and Séquin, C. H., "Noise Measurements in Charge Coupled Devices," IEEE Trans. on Electron Devices ED-22, to be published (1975).

[11] Tompsett, M. F., "Surface Potential Equilibration Method of Setting Charge in Charge Coupled Devices," IEEE Trans. on Electron Devices ED-22, June (1975).

[12] Séquin, C. H., and Mohsen, A. M., "Linearity of Electrical Charge Injection into Charge Coupled Devices," IEEE Jour. of Solid-State Circuits SC-10, 81-92 (1975).

[13] Emmons, S. P., Tasch, A. F., and Caywood, J. M., "A Low-Noise CCD Input with Reduced Sensitivity to Threshold Voltage," IEDM, Washington, D. C., Tech. Digest, 233-235 (1974).

[14] Bertram, W. J., Mohsen, A. M., Morris, F. J., Sealer, D. A., Sequin, C. H., and Tompsett, M. F., "A Three Level Metallization Three-Phase CCD," IEEE Trans. on Electron Devices ED-21, 758-767 (1974).

[15] Hartsell, G. A., and Kmetz, A. R., "Design and Performance of a Three Phase Double Level Metal 160×100 Element CCD Image," IEDM, Washington, D. C., Tech. Digest, 59-62 (1974).

[16] Rodgers, R. L., "A 512×320 Element Silicon Imaging Device," ISSCC, Philadelphia, Digest of Tech. Papers, 188-189 (1975).

[17] Séquin, C. H., Morris, F. J., Shankoff, T. A., Tompsett, M. F., and Zimany, E. J., "Charge-Coupled Area Image Sensor Using Three Levels of Polysilicon," IEEE Trans. on Electron Devices ED-21, 712-720 (1974).

[18] Séquin, C. H., and Tompsett, M. F., "Charge Transfer Devices," Suppl. #8 to "Advances in Electronics and Electron Physics," Academic Press, Inc., New York (1975).

Section V

CID ARRAYS

CHARGE-INJECTION DEVICES FOR SOLID STATE IMAGING*

G.J. Michon and H.K. Burke

General Electric Company, Corporate Research and Development
Schenectady, New York U.S.A.

I. INTRODUCTION

The CID is a surface channel device that employs intra-cell charge transfer and charge injection to achieve the solid state image sensing function. Photon generated charge signals are collected and stored in an array of MOS charge storage capacitors. The level of signal charge is detected *in situ* so that excess charge transfer structure is avoided. Charge injection into the underlying semiconductor is used to clear the sensing region of accumulated signal charge and, in some cases, to provide a readout means.

II. BASIC OPERATION

The charge injection approach to solid state image sensing employs MOS capacitor structures to collect and store photon generated charge signals. Charge is injected from the MOS storage (inversion) region into the substrate to clear the storage region and, in some cases, to provide a signal readout means. Charge storage sites can be designed for linear addressing (line imager) or for two-dimensional coincident voltage addressing (area imager).

---

* CID imager development was sponsored in part by the United States Army Electronics Command, Night Vision Laboratory, the Advanced Research Projects Agency (ARPA) and the Air Force Systems Command, United States Air Force.

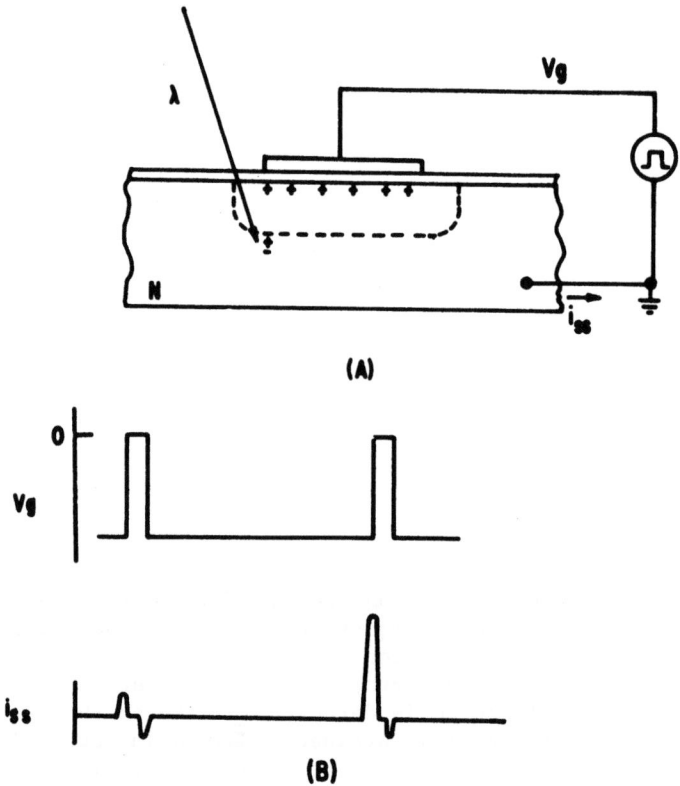

Fig. 1 MOS capacitor operated in the charge injection mode (a) device cross-section, (b) drive voltage and substrate current, without and with injected charge, respectively.

A. Linear Structure

The simplest charge injection device is the MOS capacitor. If voltage is applied to an MOS capacitor, the underlying silicon is depleted of majority carriers and photon generated minority carriers can be collected and stored in a surface inversion region. If voltage is subsequently removed from the capacitor, the stored charge will be injected into the substrate where it can either recombine or be removed by a charge collection structure.

An MOS capacitor operated in a charge injection readout mode is illustrated in Fig. 1. Operation of the device is described for a standard p-channel (N-type substrate) MOS structure. Voltage in excess of the threshold voltage is applied during the charge collection interval, driving the device into deep depletion.

Minority carriers (holes) generated by radiation are collected and stored in a surface inversion layer. The first gate voltage pulse (V_g) of Fig. 1(b) illustrates readout of a capacitor empty of charge. The substrate current signals result from capacitive coupling of the drive voltage to the substrate via the oxide capacitance in series with the silicon depletion capacitance. The second voltage pulse shows injection of stored charge upon voltage turn-off. The residual capacitive current flows upon re-application of gate voltage.

A linear array of MOS capacitors operated in the charge injection mode would be suitable for one-dimensional (line) imaging. A more complex structure is required for area imaging.

B. X-Y Addressable Sensing Site

If two MOS capacitors are used at each sensing site and are coupled together so that stored charge can be transferred from one capacitor to the other, then a two-axis selection method can be used for scanning. The basic approach is to design each capacitor such that it can store the signal charge when voltage is removed from the other capacitor electrode. Injection will then occur only when both electrode voltages are switched off.

Various methods can be used to couple surface charge between adjacent electrodes. Among these are fringing fields, which require a very narrow inter-electrode gap, overlapping but insulated electrodes, or the use of an interconnecting conductive region. The last method is compatible with standard MOS processes and metal patterning capability and has been used for most of the CID imagers fabricated to date.

A very simple topological structure results with this image sensing technique. The MOS capacitors at each sensing site are coupled with a p diffusion. There is no contact made to this diffusion. Array interconnections are readily made using the two conductor level capability of self-registered MOS processes such as Silicon Gate.

Fig. 2(a) shows the cross section of a sensing site with voltages applied to both electrodes. If voltage is removed from either of the electrodes Fig. 2(b) charge stored under that electrode will transfer to the other capacitor through the coupling P region. Charge is injected only from the sensing site that has both electrode voltages switched off simultaneously, Fig. 2(c). This arrangement permits coincident, two-dimensional (X-Y) scanning, in any order. Sequential scanning can be implemented by including MOS shift registers along two edges of the array. Non-sequential ("random") scanning can be mechanized with digital

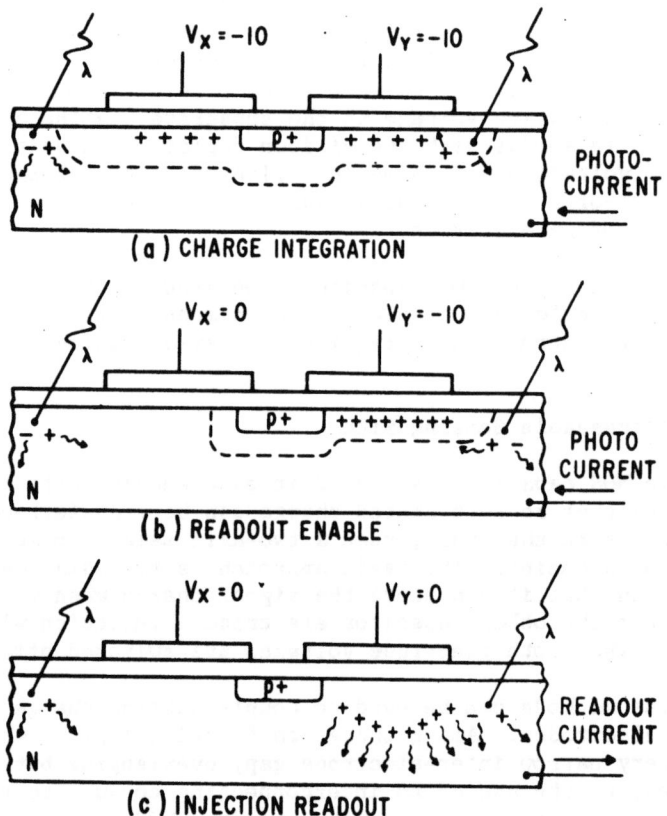

Fig. 2 Cross-Section of X-Y addressable sensing site showing location of stored charge under (a) integration, (b) readout enable, and (c) injection conditions.

decoders for row and column selection.

If array drive voltages are switched below the threshold voltage during half-select operation, charge pumping (1) will cause a signal charge loss. A bias voltage, slightly greater than the threshold voltage, can be added to the drive levels to avoid this loss. A bias charge, similar to the CCD fat zero, will then accumulate at each site.

C. Charge Injection

Charge is removed from CID image sensors by injecting the minority carriers, which have been stored in surface inversion

regions, into the underlying semiconductor. The first devices
(2) were constructed on bulk silicon and relied on recombination
as the primary method of charge removal. In this approach there
is a tradeoff among site density, dark current, and cross talk.
The diffusion length for minority carriers is proportional to the
square root of carrier lifetime, while the thermal charge generation rate in depleted bulk silicon is inversely proportional to
lifetime (3). If the spacing between sensing sites is much less
than the diffusion length, a portion of the charge injected at
one site will be collected by adjacent sites with a resulting
loss of resolution. In addition, the injection pulse width cannot be much shorter than carrier lifetime or else part of the
injected charge will be re-collected and result in image lag.

The solution to these problems has been to fabricate CID
imagers on epitaxial wafers (4). The epitaxial junction, which
underlies the imaging array, acts as a buried collector for the
injected charge. If the thickness of the epitaxial layer is less
than, or comparable to, the spacing between sensing sites, the
injected charge will be collected by the reverse-biased epi junction and injection cross talk is avoided. The rate at which charge
injected at the surface is removed from the epitaxial layer can
be analytically determined. The results of an approximate analysis
are shown in Fig. 3. Diffusion of a quantity of charge, from the
surface to the epitaxial depletion region boundry, has been calculated assuming low level injection and infinite carrier lifetime. Fig. 3(a) shows the charge distribution in the epitaxial
layer at various times after injection, while Fig. 3(b) shows
the fraction of injected charge remaining in the epitaxial layer
as a function of time from injection. Measured injection efficiency (percent of stored charge injected) as a function of
injection pulse width is shown in Fig. 4 for bulk and epitaxial
imagers.

The epitaxial collector also affects imager sensitivity.
Part of the charge generated in the silicon between sensing sites
can be collected by the epi-junction instead of the storage capacitors. This is particularly true for long wavelength radiation
which generates charge further from the imager surface.

III. READOUT METHODS

The signal charge stored in CID imaging arrays can be sensed
by measuring either the displacement current that flows upon
injection or the voltage change induced by charge transfer between
the two storage capacitors that comprise the X-Y addressable
storage site. If charge is injected for readout then the readout process is destructive because the injection operation clears
the sensing sites of signal charge. This readout method has been

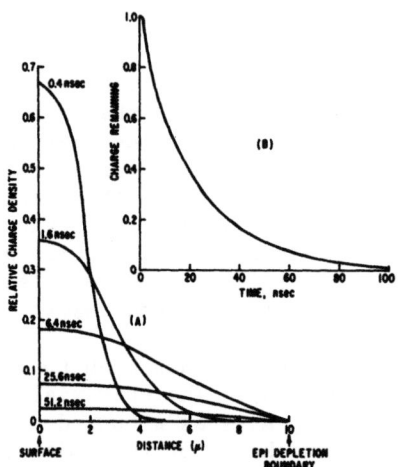

Fig. 3 Plot of the calculated charge-collection characteristics of an epitaxial junction. (a) shows the distribution of the injected charge in the epitaxial layer at various times after injection, while (b) shows the relative amount of charge remaining as a function of time.

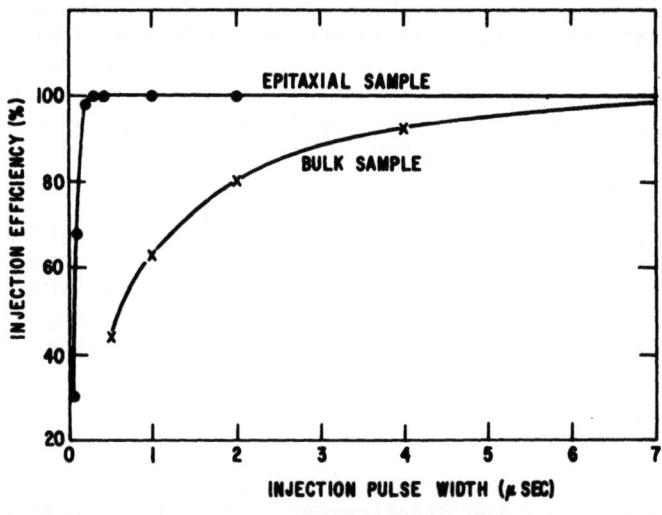

INJECTION RESPONSE

Fig. 4 Measured injection efficiency (% of stored charge injected) of bulk and epitaxial imaging devices.

termed "sequential injection" since, regardless of the scan geometry, the charge storage sites are injected in time sequence.

Readout can also be implemented by measuring the voltage induced by charge transfer between the two storage capacitors that are used at each sensing site in an array (5). In an X-Y addressable array the transfer can be performed on all sensing sites along a row in parallel. Each row can also be cleared of signal charge by performing the injection operation in parallel at all sites in the addressed row. This readout technique has been termed "parallel injection". It is non-destructive in nature because the readout function has been separated from the injection operation. Image charge can be readout and retained or injected dependent upon the array drive voltage conditions.

A. Sequential Injection

The initial charge injection imagers utilized the substrate as a readout port common to all array sensing sites. With this approach, scanning can be implemented by first removing voltage from an array row (X-line) and then pulsing each column (Y-line) in sequence to readout the selected row. All rows can be read out in this manner, with or without interlace, dependent upon the order of row selection.

Video signal waveforms obtained using substrate readout are illustrated in Fig. 5. The raw video signal consists of the substrate charge injected from each sensing site in sequence. This

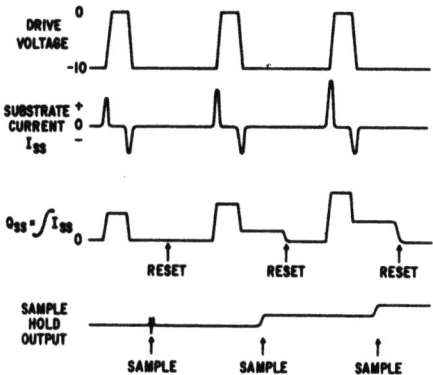

Fig. 5 Video signal waveforms for substrate readout illustrating column drive voltage, substrate current, net injected charge, and sampled video for different levels of injected charge.

signal appears as a displacement current in the presence of drive line interference that results from parasitic capacitive coupling of the drive voltage to the substrate. The drive voltage interference can be made to cancel itself through the use of an integrating readout technique. The first pulse time of Fig. 5 shows the substrate current signal that results from capacitive coupling of the Y-line drive pulse in the absence of signal charge. The substrate current is simply $C(dV_Y/dt)$. If this current signal is integrated, the drive voltage waveform is recovered. The second pulse time shows the substrate current when signal charge is injected upon Y-line drive voltage turn-off. The positive current pulse contains both the parasitic capacitance charge and the signal charge. The negative substrate current pulse that results from the reapplication of Y-line drive voltage contains only the parasitic capacitance charge. The integral of this current waveform results in a net signal proportional to the injected signal charge. This net voltage is sampled to provide the video output voltage.

This signal recovery system results in self-cancellation of the parasitic drive line interference. For complete cancellation it is only necessary that the drive voltage return to its initial state - variations in the magnitude of the drive voltages or individual variations in line capacitances do not affect cancellation.

The signal voltage developed at the output of the image sensor is equal to the net injected charge divided by the total load

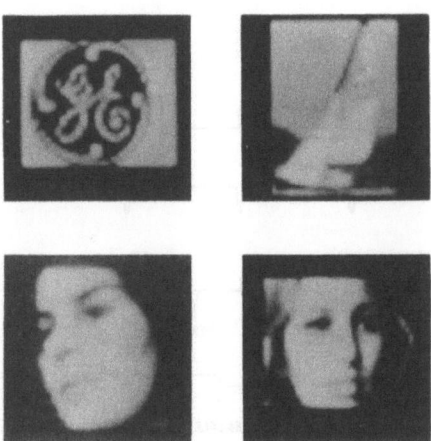

Fig. 6 Images obtained from 32 X 32 self-scanned imagers employing substrate readout.

capacitance. Capacitive loading imposed by the array lines can be minimized by allowing all but the selected row and column conductors to float during each line scan interval. If scanning circuitry is integrated into the imaging array then capacitance between the scanning shift registers and substrate will represent the bulk of the load capacitance. Imagers obtained from early 32 line by 32 element self-scanned imagers using substrate readout are shown in Fig. 6.

The displacement current that flows in the substrate upon charge injection also flows in the driven array line. The capacitive load of on-chip scanning circuitry can be avoided by sensing current in the driven line instead of the substrate.

An array designed for drive line readout which includes integral shift registers is diagrammed in Fig. 7(a). A larger voltage is applied to the row electrodes than to the column electrodes so that photon-generated charge collected at each site is stored under the row electrode thereby minimizing the capacitance of the column lines. The sensing site cross-sections,

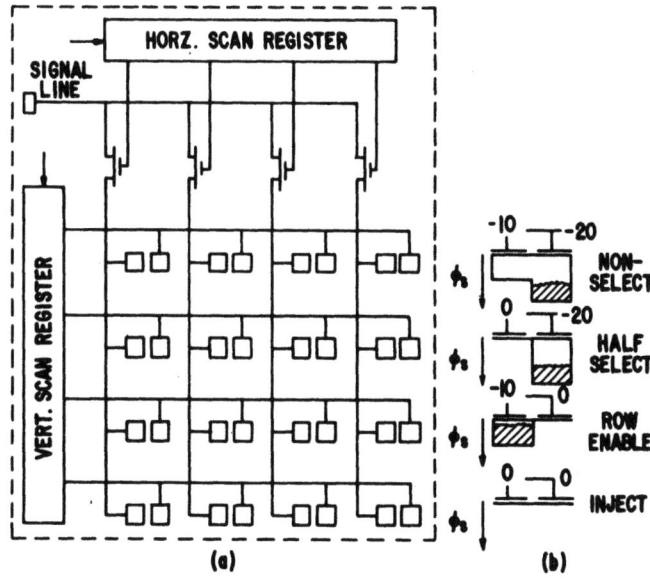

Fig. 7 Diagram illustrating basic X-Y accessing scheme for a CID imager. (a) is a schematic diagram of a 4 X 4 array, while (b) shows the sensing site cross-section showing silicon surface potentials and location of stored charge for various operating conditions.

Fig. 7(b) illustrate the silicon surface potentials and locations of stored charge under various applied voltage conditions.

A line is selected for readout by setting its voltage to zero by means of the vertical scan register. Signal charge at all sites of that line is transferred to the column capacitors, corresponding to the Row Enable condition shown in Fig. 7(b). The charge is then injected by driving each column voltage to zero, in sequence, by means of the horizontal scan register and the signal line. The net injected charge is measured by integrating the displacement current in the signal line, over the injection interval. Charge in the unselected lines remains under the row-connected electrodes during the injection pulse time (column voltage pulse). This corresponds to the half-select condition of Fig. 7(b).

Integration of the net injected charge of each site can begin after column selection but before the column injection pulse. The result is that many sources of pattern noise such as scanner interference or threshold voltage variation across the array are eliminated from the output signal.

The image obtained with a 100 line by 100 element array employing drive line sensing is shown in Fig. 8.

Fig. 8 Image obtained from a 100 X 100 self-scanned imager constructed on an epitaxial substrate and employing drive line sensing.

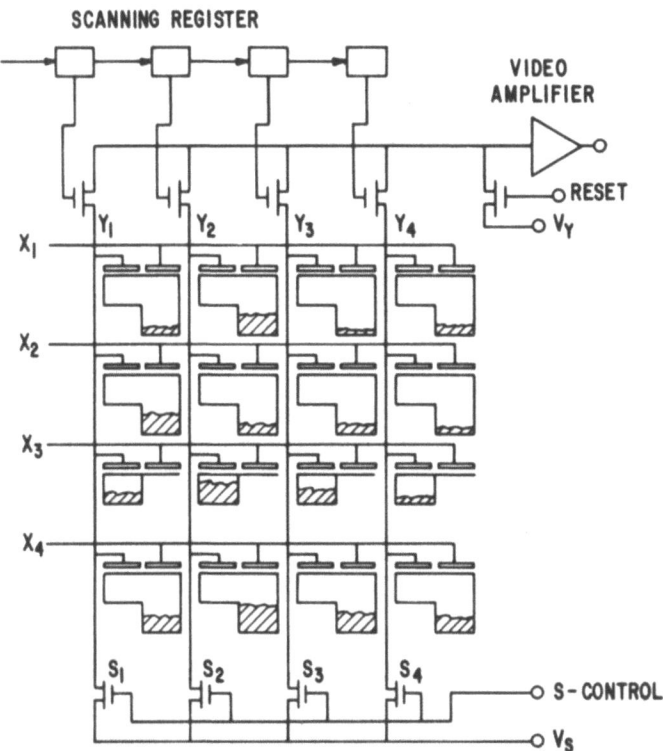

Fig. 9 Schematic diagram of a 4 X 4 CID array designed for parallel-injection readout. Silicon surface potentials and signal charge locations are included.

B. Parallel Injection

The parallel injection technique allows the functions of charge injection and charge detection to be separated. Signal charge levels can be sensed at high speed during a line scan, and during the line retrace time interval all of the charge in the selected line can be injected in parallel. If injection is deferred, nondestructive readout results. The parallel injection technique is well adapted to TV scan formats in that the signal is read out line by line. It is not adapted to random scan applications.

In an array of MOS coupled-capacitor pairs, as is used in the present charge injection imagers, all of the signal charge can be stored under the row-connected electrodes if the row voltages are larger than the column voltages. This condition is illustrated in Fig. 9 for rows X_1, X_2, and X_4. This method of biasing effectively prevents the charge stored under the row-connected electrodes from affecting column voltages. The voltage on all array columns, Y_1 through Y_4 in Fig. 9, can be set to a reference value either by means of a previous column scan readout, or through the use of the column switches, S_1 through S_4.

If the voltage on a row electrode is then switched to zero, signal charge will transfer from the row-connected electrodes to the column-connected electrodes in the selected row of sensing sites. This has been diagrammed in Fig. 9 for row X_3. The voltage on each of the column lines will then be reduced by an amount equal to the signal charge divided by the column capacitance.

The signal can be sensed by sequentially connecting each column line to a video amplifier by the use of a scanning register and MOS switches. The readout operation consists of resetting the video amplifier input to the reference voltage, and then stepping the scanning register to the next column line. After all columns of the array have been scanned, charge can be returned to the row-connected electrodes by reapplying voltage to the previously selected row. This action retains the signal charge for future processing.

Alternately, at the end of readout of the selected row, while the row voltage is maintained at zero volts, the signal charge can be injected from the selected row to the substrate, all sites in parallel, by switching all column voltages to zero simultaneously. This action clears the sensing sites of charge and allows the start of a new signal integration time interval for that row.

Two experiments were performed to identify the precision to which the readout is truly nondestructive readout. First, the charge pattern of an image was generated and stored by momentarily opening a shutter, and then the image was readout continuously at 30 frames per second, until image degradation was noted. At a chip temperature of 200°K, images were readout for three hours (324,000 NDRO operations) with no detectable charge loss. The charge lost during each NDRO operation was, on the average, much less than one carrier per pixel per frame.

The second experiment was performed to insure that charge could be generated and stored at very low light levels under continuous (30 frames per second) NDRO conditions. A series of time exposures was made at successively lower light levels and the time required to reach a given level of signal voltage was measured. The results, Fig. 10, show that the exposure time is inversely proportional to light level with no measurable charge loss for exposure times up to three hours. The lowest light level used was equivalent to about two (2) carriers per

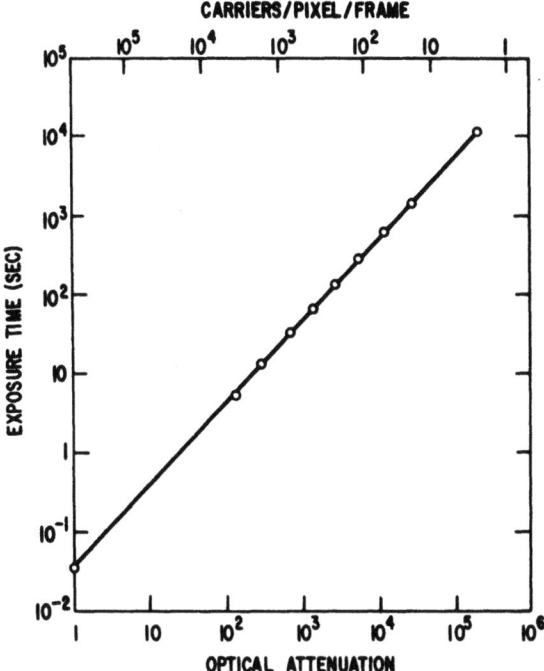

Fig. 10 Plot showing exposure time required to accumulate a given quantity of signal charge as a function of incident light level. Time is plotted as a function of optical attenuation. Measurements were made on a 244 X 248 array, cooled to 200°K and operated at 30 frames per second, NDRO.

pixel per frame in the highlight regions of the image. Here again, the readout loss was not measurable and was much less than one carrier per pixel per frame.

These results demonstrate that the nondestructive charge readout technique of Tiemann et al (6) can be effectively applied to charge injection imagers.

C. On-Chip Amplifier

The first stage of the video amplifier can be integrated into the CID imager chip to allow the signal charge to be sensed with minimum capacitance load. This transistor should be designed for the lowest possible thermal noise. Johnson noise in the transistor channel is proportional to the channel resistance;

therefore, a high g_m is desirable to minimize this noise component.

MOS transistors exhibit a relatively large excess low-frequency noise (1/f noise). It has been shown, both theoretically and experimentally, (7) that the magnitude of this noise component is inversely proportional to the transistor channel area. At a given level of power dissipation, maximum g_m (minimum Johnson noise) occurs with maximum channel width-to-length radio. Low-frequency noise is minimized by using maximum channel area. It is not necessary to optimize on the on-chip amplifier for absolute minimum noise. Correlated double sampling techniques can be used to attenuate noise components that are lower in frequency than the sampling frequency so that 1/f noise need only be lower than Johnson noise at this frequency.

If a switch is used to set the voltage across a capacitor, Johnson noise in the switch resistance results in an uncertainty in the capacitor voltage. The magnitude of this uncertainty (8) is:

$$V_n = (KT/C)^{1/2}$$

where K is Boltzman's constant = 1.38×10^{-23} W-S/°K

and T is the Temperature in degrees Kelvin

Sequential injection CID imagers are not limited by this noise component because it is possible to reference the net injected charge signal to the input capacitor voltage after reset has been completed. The parallel injection technique does not allow complete elimination of KTC noise. The column reset transistors introduce KTC noise that is not rejected.

REFERENCES

1. J.S. Brugler and P.G.A. Jespers, "Charge Pumping In MOS Devices", IEEE Transactions on Electron Devices, ED-16, No. 3 (March 1969).

2. G.J. Michon and H.K. Burke, "Charge Injection Imaging", ISSCC Dig. Tech. Papers, Feb. 1973, pp. 138-139.

3. A.S. Grove, "Physics and Technology of Semiconductor Devices", John Wiley and Sons, 1967 pp. 120-125.

4. G.J. Michon and H.K. Burke, "Operational Characteristics of CID Imager", ISSCC Dig. Tech. Papers, Feb. 1974, pp. 26-27.

5. G.J. Michon, H.K. Burke and D.M. Brown, "Recent Developments In CID Imaging", Proceedings of Symposium on Charge-Coupled Device Technology for Scientific Imaging Applications (JPL), pp. 106-115, (March 6-7, 1975).

6. J.J. Tiemann, W.E. Engeler, R.D. Baertsch, and D.M. Brown, "Intracell Charge-Transfer Structures for Signal Processing", IEEE Transactions On Electron Devices, ED-21, pp. 300 (1974).

7. S. Christensson, L. Lundstrom, and C. Svensson, "Low Frequency Noise in MOS Transistors", Solid-State Electron, VOL 11, pp. 797-812 (1968).

8. J.E. Carnes and W.F. Kosonocky, "Noise Sources in Charge Coupled Devices", RCA Rev. $\underline{33}$, pp. 327-343 (June 1972).

Three-Terminal Charge-Injection Device (*)

P. Jespers and J.M. Millet

Catholic University of Louvain
Microelectronics Lab
Bâtiment Maxwell
1348 Louvain-la-Neuve

Abstract

A new configuration of the Charge-Injection Device (CID) Image Sensor is described in this paper. Readout is performed by means of buried stripes (e.g., P stripes in a N-type substrate) acting as collectors of the injected minority carriers previously stored under the gates. This scheme provides separate paths for the displacement and diffusion currents and consequently improves the signal-to-noise ratio. Furthermore, the new configuration yields simple X-Y selection without the need for extra crosspoint switches. In the present device, a word organized readout is achieved by means of polysilicon stripes perpendicular to the buried P stripes. The silicon area therefore can be used very efficiently.

Introduction

Capacitor photodiode image sensors were first discussed by Arnold et al. (1) (2) and renamed charge-injection devices (CID's) by Michon and Burke (3). CID arrays use pairs of MOS photosensors that can be addressed individually as in a photodiode array. Readout of the integrated charge in each cell occurs sequentially by vertical injection into the bulk of the photon-generated minority carriers previously stored under the gate electrodes when both are simultaneously addressed. In this manner measurable recombination current pulses are produced to compensate exactly the photon-generated charge. The recombination time, although only a small fraction of the light integration time, is often too long to be suitable for

(*) *Copyright 1976 by the Institute of Electrical and Electronics Engineers, Inc. Reprinted, with permission, from IEEE JOURNAL of Solid-State Circuits, February 1976, Vol. SC-11, N°1, pp. 133-139.*

existing timing formats. Michon and Burke therefore introduced
a reverse biased epitaxial junction beneath the surface in order
to substitute the transit time through the epi layer for the
actual recombination time (4). The transit time indeed may be
as small as a few nanoseconds, while the recombination time
amounts to microseconds. The epilayer offers a second advantage
for it reduces the thickness of the volume wherein photon-
generated minority carriers are collected. In this manner
optical crosstalk between adjacent cells due to deep generated
carriers can be reduced drastically.

In the three-terminal CID (3T-CID) cell presented in this
paper which shows evident similarities with the proposal of
Brojdo (7), separate paths are provided for the minority carrier diffusion current (the useful signal) and the "displacement" current (fixed pattern noise signal) produced by the voltage steps applied to the gate electrodes. For the purpose
of detecting useful signals, it is no longer necessary thus
to distinguish the recombination current from the displacement
current by means of sophisticated integration techniques. In
other words, the readout used in the 3T-CID practically eliminates switching transients providing almost noise-free output signals as will be proven later. Furthermore, it will be
shown that three-terminal CID photosensors do not require
additional switching elements for X-Y selection compared to
those normally required in photodiode arrays (5);(6). This
last property greatly simplifies the design of CID arrays resulting in an effecient use of the available silicon elementary photosensor area.

II. Description and Principle of operation.

A microphotograph view of a 3T-CID linear array is shown
in Fig. 1 (a). The basic difference with respect to two-
terminal CID's is the introduction of a buried P stripe (assuming the substrate is N-type) provided with a surface contact
near one end of the stripe. This contact may be obtained by
a local P-type diffusion or by means of orientation dependent
V-groove etching (8). The role of the buried stripe is similar to that of the single large epijunction introduced by
Michon and Burke, namely to collect the injected minority
carriers diffusing through the epi layer. The so-called third
electrode in this paper is the substrate contact which gives
access to the thin N^--epi layer by means of the bulk N^+ silicon
surrounding the buried P stripe.

Seven masks are necessary in order to fabricate a 3T-CID
array. The technology thus is more complex than that of
Michon and Burke. The fabrication involves in fact standard

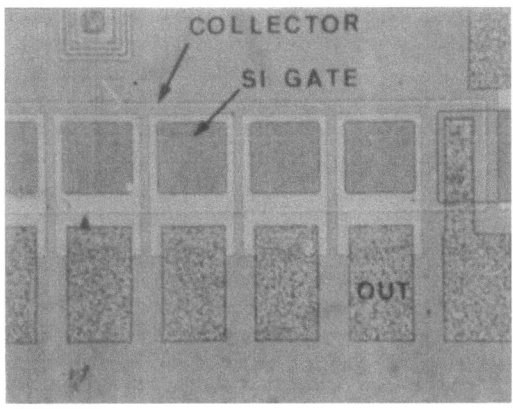

Fig. 1 (a) Microphotograph of a linear 3T-CID array.

buried-layer technology followed by conventional Si-gate technology. No incompatibility exists between these steps as long as the fabrication of the 3T-CID is concerned.

The actual biasing of a typical 3T-CID is illustrated in Fig. 1 (b).

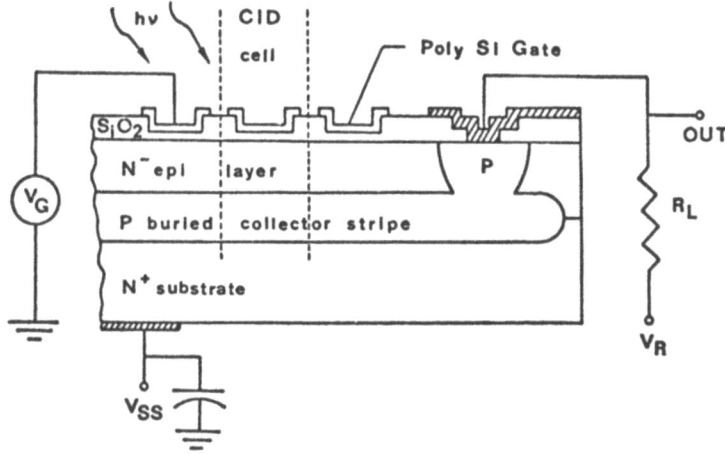

Fig. 1 (b) Cross-sectional view of the array and biasing network.

V_{ss} is a positive voltage applied to the substrate in order to reverse bias the collector terminal, which is grounded through an output resistor R_L. C is a decoupling capacitor. V_G is the applied negative gate voltage, which is periodically switched to zero during short time intervals. Each time the gate voltage is turned off, a corresponding transient capacitive drift current flows trough the silicon gate capacitor, around the P stripe to the grounded N^+ substrate. Meanwhile, the injected minority carriers diffuse from the Si-SiO$_2$ interface through the N^--epi layer, where some of them recombine with majority carriers, and others are collected by the P stripe. This is very similar to the behavior of a bipolar PNP transistor where the grounded N^--epi layer is the base and the buried stripe represents the collector. The only difference with respect to bipolar transistors is in fact the replacement of the diffused emitter by a time-dependent field induced junction.

Since three terminal CID's are so similar to bipolar transistors, the symbolic representation shown in Fig. 1 (c) is proposed which suggests naturally the combination of a bipolar transistor with a P-channel MOS capacitor.

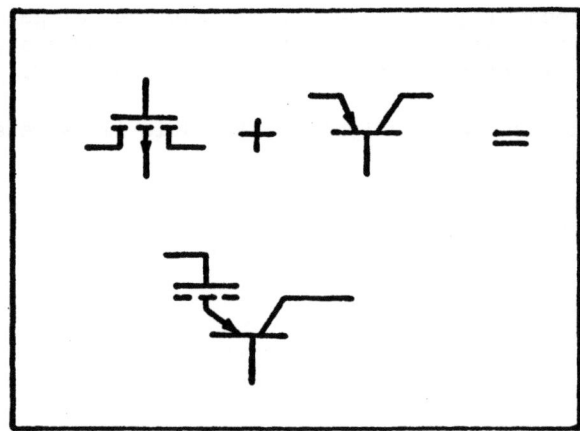

Fig. 1. (c) Proposed symbol.

III. Analysis of the output signal

When the gate potential is switched from a high negative voltage to a lower voltage, readout occurs. The corresponding positive gate voltage step will be called ΔV_G further in the paper. Since the transit time usually is shorter than the rise time t_1 of ΔV_G, injection of minority carriers into the collector is governed by mainly t_1. This time can easily be made shorter than the collector time constant RC_c whereby R_L represents the external collector load resistance and C_c represents the collector junction capacitance including parasitic output capacitance. Assuming that this condition is verified, the charge packet reaching the collector produces a positive voltage step with rise time t_1 whose maximum absolute amplitude is given by :

$$\frac{C_{ox}}{C_c} \cdot \Delta V_G \qquad (1)$$

The rapid increase of collector voltage stops as soon as injection is completed, and is followed by an exponential decay with time constant $R_L C_c$. The output signal thus looks roughly like an exponential pulse with a short rise time equal to t_1. Usually t_1 is of the order of a few tenths of a nanosecond and $R_L C_c$ is chosen around 0.1 to 0.3 µs.

No coupling exists between gate and collector terminals, except for the small voltage drop across the substrate impedance which is due to the capacitive gate-to-substrate current. This voltage drop is coupled to the output port trough the collector junction capacitance.
If we consider for simplicity identical linear rise and fall times t_1 of the gate signal ΔV_G, the interfering displacement current produces successively positive and negative square wave noise signals across the substrate impedance R_{BB}, whose amplitude is given by :

$$C_D \cdot \frac{\Delta V_G}{t_1} \cdot \frac{R_{BB} \cdot C'}{C_c} \qquad (2)$$

where C_D represents an average value of the deep depletion capacitance, while the ratio $(R_{BB} \cdot C')/C_c$ is subject to the following comments :

1) R_{BB}, in fact is formed by the sum of the bulk N^+ substrate, eventually the substrate contact resistance, and mainly by the thin-epi layer comprised between the gate and the buried P region.
2) C' represents the fraction of collector capacitance which is influenced by the voltage drop across the thin-epi layer.

3) C'/C_c thus describes the capacitive divider action due to C' and the remaining part of total collector capacitance C_c.

4) Due to the distributed nature of R_{BB}, as well as C', no easy way exists to determine each of these values, but their product can be measured indirectly as will appear further in this paper.

Assuming that the voltage drop given by (2) appears superimposed on the output signal without distorsion for t_1 is smaller than the time constant formed by the external load resistor R_L and C_c, it is possible to define an "achievable signal-to-noise ratio" by dividing expression (1) by (2). Therefore,

$$\frac{S}{N} = \frac{C_{ox}}{C_D} \cdot \frac{t_1}{R_{BB} \cdot C'} \qquad (3)$$

Typical values of this ratio range from 30 to 50 dB for t_1 between 10 and 100 ns, assuming that the time constant $R_{BB} \cdot C'$ times C_D/C_{ox} is of the order of 0.2 ns.

It is obvious from expression (3) that the collector capacitance C_c that governs C' must be kept as low as possible in order to improve the signal-to-noise ratio. Low doping of the collector thus is a profitable goal regardless of the corresponding increase of series resistance which results. The number of lines in three terminal CID arrays thus should not exceed reasonable limits in order to keep C_c low. This is certainly a major limitation inherent to the voltage mode of operation. The problem may be circumvented, however, if the output signal is stored on the collector capacitance and sampled after resetting of the gate potential. In this manner residual switching noise signals compensate and the final voltage across C_c remains unaffected from switching signals at all.

According to (3), any increase in t_1 would result in an improved S/N ratio. This has been verified experimentally (Fig. 2), but it should be emphasized that the above-mentioned expression ceases to be valid when t_1 becomes of the same order as $R_L C_c$, for the maximum of the output signal is no longer given by (1) but is smaller, thereby decreasing the S/N ratio. In practice, t_1 should not exceed one third of the collector time constant $R_L C_c$.

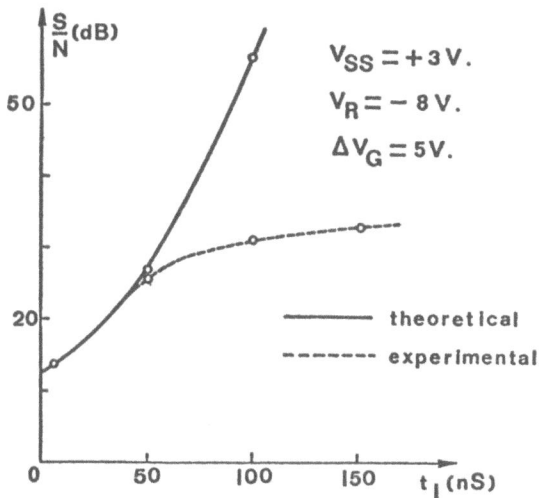

Fig. 2 (a) S/N ratio versus the rise time t_1 of the gate voltage.

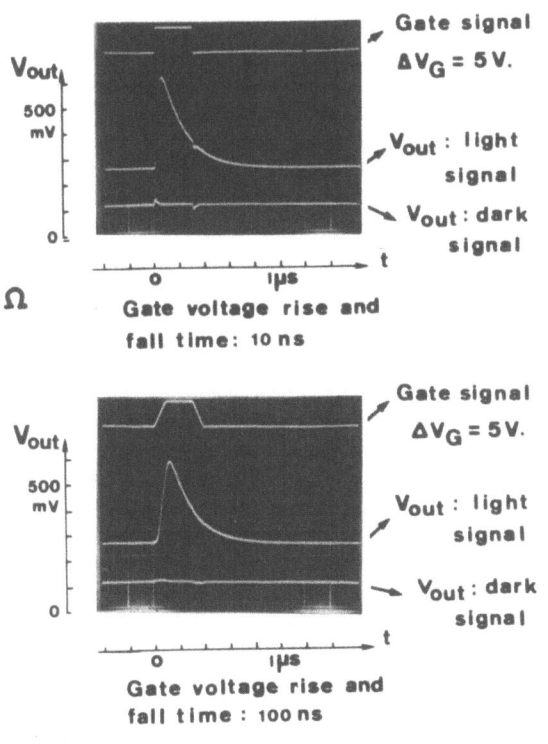

Fig. 2 (b) Photographs showing the output signal with light and without light for two different rise and fall times of the gate voltage.

It may be interesting to point out here that the output signal provides a simple means to estimate the average transit time of minority carriers traveling across the N^--epi layer. Indeed, changing the duration of the gate signal t_2 does not affect the peak amplitude of the output signal, unless t_2 becomes so short that not all minority carriers can be collected by the buried P region anymore. When this happens, the minority carriers remaining in the epi layer return to the surface as soon as V_G becomes negative again and the output signal peak amplitude is consequently reduced. Experimental investigations carried out on samples which are described further on in the paper have shown that this time usually is of the order of 30 ns for a 4 µm-thick epi layer. This is similar to the result found by Michon and Burke [4]

Since it is the life time or diffusion length that governs the recombination in the N^- layer, it seems reasonable to choose the minimum possible thickness for the epi layer. However there is a compromise to achieve between sensitivity (which depends on the penetration depth of light thus on the epi layer thickness) and losses due to recombination during the readout phase. A thin epi layer also results in improved crosstalk characteristics, as will be shown later in this paper. This is particularly relevant in applications such as character recognition where the amount of near infrared light produced by the source shining on the page can be quite large.

One should also keep in mind that reducing the thickness of the epi layer increases the photoleakage current of the buried collectors. A basic difference between this leakage current and the injected minority carrier current is that the last one varies proportionally with the light integration time while the first one is time independant. Therefore the leakage current increases continuously with light intensity, but the size of the charge packets injected in the bulk usually is kept within a given dynamic range due to the proper adjustment of the light integration time. In fact this leakage current does not cause problems in most applications, because it is either too small, or it is easy to perform the separation between the DC leakage current and the pulsed signal. This is less true of course in case the collector load resistor is not connected, because the leakage current then smoothly discharges the collector junction. In such a case, the collector must be sampled immediately after charge injection has been completed and it must be switched back to zero volt immediately after.

IV. Influence of electrical control signals.

The first experimental device used in this paper was a five-gate linear array, shown in Fig. 1(a) and described in Table I. All experiments carried out in this section and in section V were based on a quartz iodine light source shining on the array. No attempt was made to measure the incident energy nor the spectral distribution, for only relative measurements were to be considered. Furthermore, in some experiments an inverted microscope was used in order to isolate a small light spot (70 x 70 µm^2) on the array that was scanned mechanically.

TABLE I

Features of five-Gate Linear Array shown in Fig. 1 (a)

Gates dimensions	150 by 150 µm
Thin-oxide thickness	1300 Å
Thick-oxide thickness	12000 Å
Polysilicon thickness	7000 Å
Final epitaxial layer thickness after boron redistribution	4 µm
Epitaxial layer resistivity	3-5 Ω.cm
Bulk resistivity	3-5 Ω.cm
Buried collector dimensions	230 by 1160 µm
Gate capacitance C_{ox}	6.23 pF
C_{dep} (formula 4)	4 pF
C_c (with reverse biased voltage of 8 V)	50 pF
R_L	3.9 kΩ

Since the collector capacitance decreases with increasing reverse bias, the output peak amplitude consequently increases slightly. Because of the widening of the buried collector depletion region, the part of the neutral N$^-$ region which contributes to light-generated minority carriers thus shrinks progressively. The width of this useful neutral region may become thinner than the light penetration depth especially with long wavelengths, thereby reducing the sensitivity of the photosensor. It could be possible to take advantage of this effect in order to compensate the decrease of collector capacitance resulting from an increasing reverse bias. In this manner the output signal amplitude could be kept more or less constant depending on the epi-layer thickness ,doping level, and on the light spectrum.

Some results are presented in Fig. 3 showing the output signal amplitude plotted versus V_R.

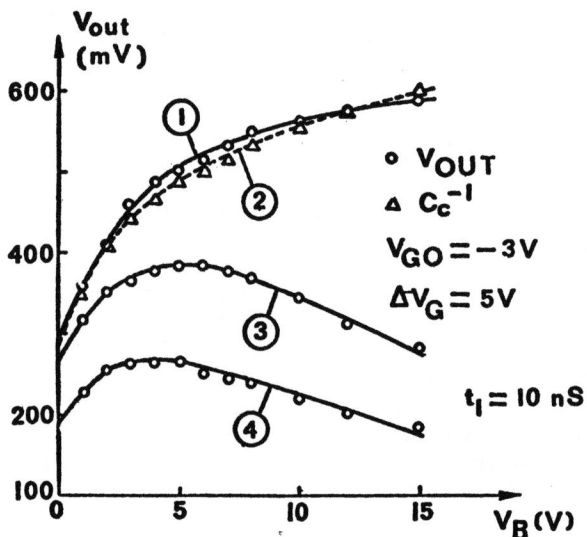

Fig. 3. Peak output signal versus collector voltage. Curve 1 corresponds to light saturated; curves 3 and 4 correspond to no light saturation; and curve 2 represents the reciprocal of C_c in relative scale.

Curve 1 corresponds to light saturation conditions. Curve 2 (dashed line), which represents C_c^{-1} in relative scale, is quite close to curve 1. Curves 3 and 4 give the output signal amplitude for two different nonsaturation illumination levels. It is obvious that the compensation effect cannot take place when the light intensity reaches the saturation level, for the integration time is then always longer than the time needed to create the inversion layer regardless of sensitivity variations. It should be noticed that the curves shown in Fig. 3 were obtained using the bias conditions shown in Fig. 4. which allow collector voltage modifications without influencing the gate-to-substrate dc bias voltage V_{GO}. This of course would not have been true if we had modified V_{ss} in Fig. 1 (b).

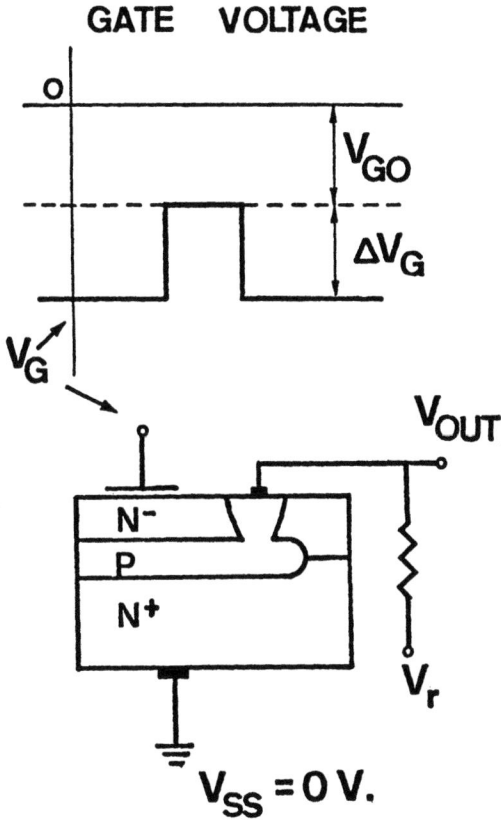

Fig. 4. Biasing network to obtain separate control of V_{GO} and V_R.

It has been verified experimentally that the gate voltage pedestal V_{GO} does not influence the output signal as long as the surface is kept inverted. One must remember indeed that when the gate voltage returns to V_{GO}, the surface inversion layer finds itself necessarily in steady-state conditions. The depletion layer width is then almost independent of the actual gate voltage. In other words, the initial conditions existing before ΔV_G is applied are always the same, regardless of V_{GO}.

The effect of ΔV_G on the output signal amplitude is shown in Fig. 5.

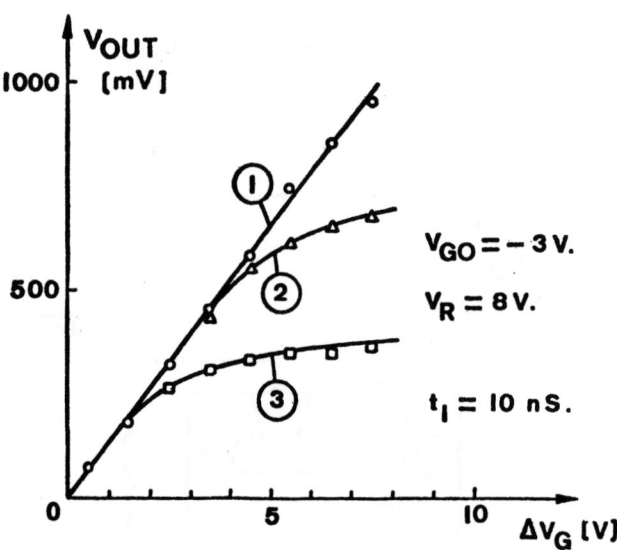

Fig. 5. Peak output signal versus gate voltage step. Curve 1 corresponds to light saturated; curves 2 and 3 corresponds to no light saturation.

Curve 1 corresponds to light saturation conditions. In this case, the output signal amplitude is linearly dependent on ΔV_G according to (1). Curves 2 and 3 represent the output signal amplitude in nonsaturation conditions and for two different illumination levels. For small values of ΔV_G, the output signal is linearly dependent on ΔV_G again according to (1), for the generation rate is sufficient to fill up the potential well corresponding to ΔV_G. When the potential well is deep enough in order not to be completely filled, the only remaining factor that influences the output signal is the collection efficiency which increases slightly with ΔV_G. Indeed, most of the minority carriers are collected within the depletion region (10), and this region increases with ΔV_G according to a square root law. The output signal indeed varies approximatively as $\Delta V_G^{1/2}$, which can be seen in this same figure. From Fig. 5, one may conclude that increasing ΔV_G improves both sensitivity and illumination dynamic range.

If we consider the bias network of Fig. 1(b), we may notice that the choice of V_{ss} in fact is not very critical provided it is large enough to always ensure strong inversion conditions under the gates even when V_G is equal to zero volt (FAT ZERO). A sufficiently high value of V_{ss} means a high reverse bias across the collector junction which substantially reduces the output signal sensitivity to V_{ss}. The upper limit of V_{ss} of course is fixed by the avalanching of the collector junction, or more likely by punchthrough of the thin N$^-$ layer.

Typical output signal values obtained with the linear array described in Table I are as follows. With $V_{GO} = -3$ V, $\Delta V_G = 5$ V, $V_R = -8$ V (bias conditions of Fig. 4), and $t_1 = 10$ ns the peak output amplitude under saturation conditions reaches 580 mV, which is in rather good agreement with the 623 mV predicted from (1). The noise signal amplitude was equal to 80 mV for $t_1 = 10$ ns and the S/N ratio was then given as 18 dB in this case.

The R_{BB}, C' time constant deduced from (2) was equal to 2 ns. Taking $t_1 = 100$ ns, reduced the output signal amplitude to 400 mV and the noise amplitude to 8 mV with a corresponding S/N ratio of 34 dB.

V. Matrix organization of 3T-CID arrays

One of the essential features of 3T-CID's is that no additional elements are required in order to achieve X-Y selection in a matrix organization. Each cell in the array shown in Fig. 6 is addressed directly by selecting a given silicon gate row and by sensing the corresponding collector column. The silicon in the photosensing area therefore may be used very efficiently.

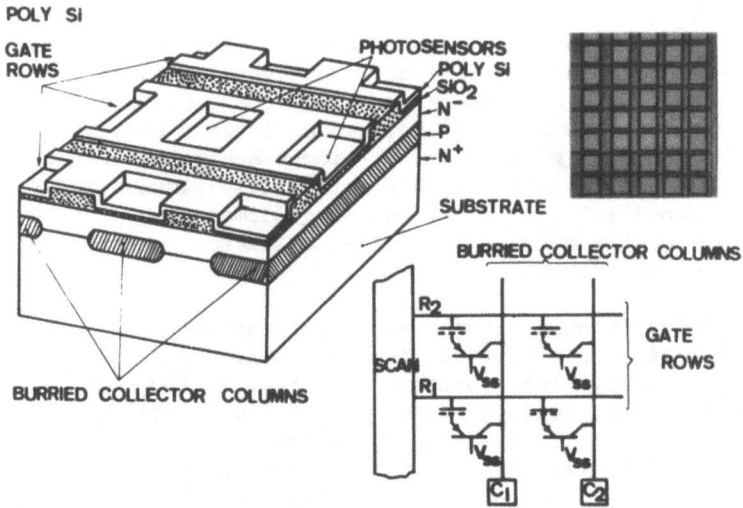

Fig. 6. Matrix organization of a 3-T CID array.

A matrix array shown in Fig. 7 was built for experimental verification.

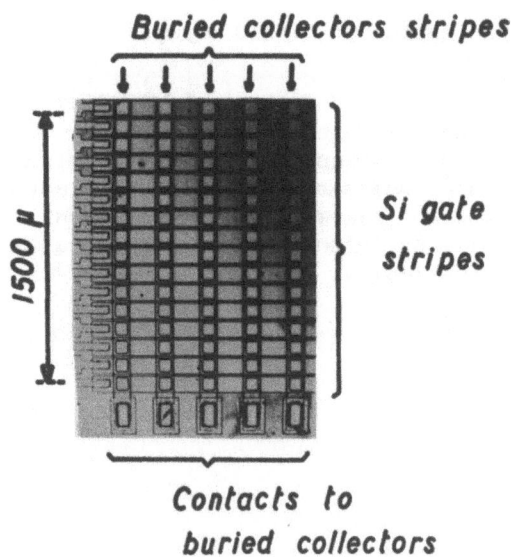

Fig. 7. Microphotograph of an experimental 3-T CID matrix.

Each Si-gate row provided with a metal contact was connected to an external address decoder. The rows were kept at a high negative potential except for the selected one which was driven to zero volt during one period of the clock driver. Scanning of the array occurred by sequential addressing of the decocoder. This approach was chosen for it allows separate adjustments to be made regarding ΔV_G, V_{GD}, and rise and fall times. No attempt was made to integrate the scanner with the array at this stage.

The matrix was used in order to evaluate the influence of minority carrier generation between the cells and to investigate blooming effects. The dimensions of the square photosensors of this array were, respectively, 70 µm in the lower part and 80 µm in the upper part. The matrix had 24 lines. Six columns were widely spaced (150 µm), followed by columns with decreasing space width (from 50 to 30 µm, respectively). The vertical center-to-center spacing was 100 µm. No optical crosstalk or blooming effects were reported in the vertical direction, along the columns. In the horizontal direction, however, some interesting results were obtained which are illustrated in Fig. 8.

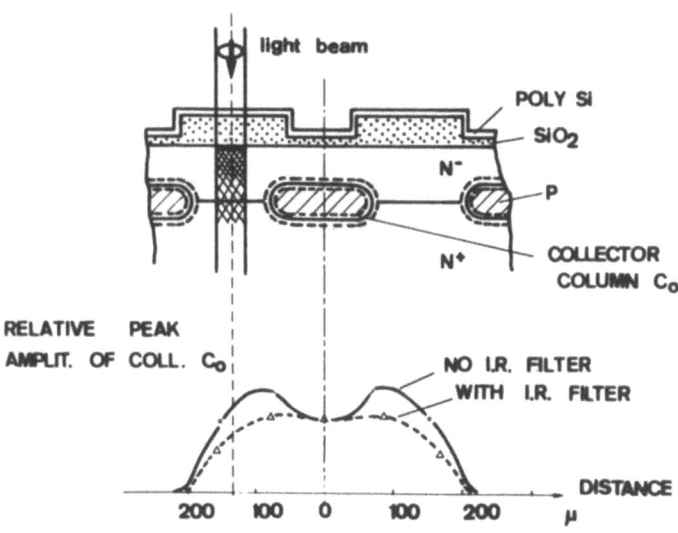

Fig. 8. Experimental curves showing the effect of lateral sensitivity with and without IR filter.

For this experiment we used a 70-by 70-μm size quartz iodine light beam mechanically scanned over the array. The continuous curve shown in Fig. 8 represents the relative peak amplitude of the output signal for a column C_o when the light beam is swept along the Si-gate stripe across column C_o. There are two peaks in the sensitivity curve which do not correspond to direct illumination of the photocell. These peaks are only visible when the distance between neighbouring columns is large.

It is likely that this anomalous lateral sensitivity is due to the transfer of minority carriers from the illuminated site towards the next photocells due to the lateral electrical fringing field induced at the surface by the stepped oxide layer. The reduced sensitivity experienced above the photosensing area could be explained by the fact that the neutral photon generating volume is smaller above a column than aside. Experimental confirmation of this fact was obtained using an IR filter which eliminates the deep penetrating wavelengths of light. This is clearly indicated by the dashed curve represented in Fig. 8.

Another proof of the existence of lateral field aided motion of minority carriers along the Si-gate stripe was given in the following manner. Suppose we illuminate strongly one of the photosensor and saturate it. Excess carriers then become available which fill in the potential wells of the adjacent cells first, then the immediate neighbours, and so on. This blooming effect which was evidenced experimentally with the matrix, shown in figure 7, can be avoided by diffusing a P junction between adjacent columns in order to drain away excess minority carriers resulting from intense local light spots. The same solution is adopted for charge-coupled (CCD) imagers (11),(12).

VI. Concluding remarks

Three-terminal CID's offer some interesting characteristics.
1) They deliver output signals with very little switching noise superimposed without the need for complex circuitry to detect the useful signals. A simple resistance is sufficient for this purpose.
2) They are X-Y addressable.
3) They share with diode arrays the possibility to perform reading out sequences during the light integration time. Consequently, no buffer memory is needed as in certain CCD arrays when very high frame rates are required.

Three-terminal CID arrays, however, suffer from some drawbacks listed below.

1) The peak amplitude is inversely proportional to the collector capacitance C_c in the voltage mode of operation reducing substantially the photosensitivity for large arrays.
2) The relatively high collector output capacitance C_c yields inferior kTC noise performances compared to CCD's, making CID's less suited for very low light level applications.
3) The technology required in order to build 3T-CID devices is more complex than that of diode arrays.

It must be pointed out that by replacing the N^+ substrate by P^+ Si and using As or Sb for the buried-collector stripe, a better control of the epitaxial layer thickness can be achieved.

Considering the list of properties and drawbacks, it is likely that 3T-CID arrays will not compete with CCD arrays for TV imaging applications, but they offer other unique characteristics making them very attractive for applications such as data capture or target search which require poor resolution imagers. Due to the small number of lines, the collector capacitance may be kept small enough to obtain a high sensitivity. Moreover, data capture arrays need very high frame rates for which purpose 3T-CID arrays are ideally suited because no buffer memory is required, and because the output signal is obtained easily almost free from switching noise. This last remark is important because it means that interfacing each column with a very simple individual amplifier is economically feasible, usually one single transistor is sufficient. When the outputs can be processes in parallel, and thus no multiplexing of the output signal within the lines is required, very high frame rates can be achieved which depend more on the shortest light integration times achievable than on limitations inherent in the addressing circuitry. Frame rates of 20 kHz are easily achievable in this manner.

References

(1) E. Arnold, M.H. Crowell, R.D. Geyer, and D.P. Mathur, "New solid state imaging array with reduced switching noise", in Int. Solid-State Circuits Conf., Dig. Tech. Papers, Feb. 1971, pp. 128-129.

(2) ——, "Video signals and switching transients in capacitor-photodiode and capacitor-phototransistor image sensors", IEEE Trans. Electron Devices, vol ED-18, pp. 1003-1010, Nov. 1971.

(3) G.J. Michon and H.K. Burke, "Charge injection imaging", in Int. Solid-State Circuits Conf., Dig. Tech. Papers, Feb. 1973, pp. 138-139.

(4) ——, "Operational characteristics of CID imager" in Int. Solid-State Circuits Conf. Dig. Tech. Papers, Feb. 1974, pp. 26-27.

(5) R.H. Dyck and F P. Weckler, "Integrated arrays of silicon photodetectors for imaging sensing", IEEE Trans. Electron Devices, vol ED-15, pp. 196-201, Apr. 1968.

(6) P.K. Weimer, W.S. Pike, G. Sadasiv, F.V. Shallcross, L. Meray-Horvath, "Multielement self-scanned mosaic sensors", IEEE Spectrum, pp. 52-65, Mar. 1969.

(7) S. Brojdo, U.S. Patent n° 3.676.715, July 1972.

(8) D.B. Lee, "Anisotropic etching of silicon", J. Appl. Phys., vol. 40, pp. 4569-4574, Oct. 1969.

(9) M.J. Declercq, L. Gerzberg, and J.D. Meindl, "Optimization of the hydrazine-water solution for anisotropic etching of silicon in integrated circuits technology", J. Electrochem Soc., vol 122, pp. 545-552, Apr. 1975.

(10) C. Anagnotopoulos and G. Sadasiv, "Collection efficiency and transfer characteristics of CID image sensors," Solid-State Electron., vol 18, pp. 771-776, Sept. 1975.

(11) B.M. Singer and J. Kostelec, "Theory, design, and performance of low-blooming silicon diode array imaging targets," IEEE Trans. Electron Devices, vol ED-21, pp. 84-88, Jan. 1974.

(12) C.H. Sequin, T.A. Shankoff, and D.A. Sealer, "Measurements on a charged-coupled area image sensor with blooming suppression", IEEE Trans. Electron Devices, vol ED-21, pp. 331-341, June 1974.

Section VI

SIGNAL EXTRACTION

DESIGN OF SOLID-STATE IMAGING ARRAYS

Marvin H. White

Westinghouse Electric Corporation,
Advanced Technology Laboratory,
Baltimore, Maryland U.S.A.

ABSTRACT. In this chapter the various factors which influence the design of solid-state imaging arrays will be discussed. Responsivity, spectral response, resolution, noise, streaking, dynamic range, etc, will be treated with the concept of a figure-of-merit based upon the product of S/N and M.T.F. (eff.).

1. SPECTRAL RESPONSE R_λ

A sensor, such as a silicon photodiode, constructed in an intrinsic semiconductor has a spectral response given as,

$$R_\lambda = \frac{e\,\eta(\lambda)}{hc/\lambda} U(\lambda_G - \lambda) \qquad (1)$$

where $\lambda_G = hc/E_G$ is the cut-off wavelength for an intrinsic band gap E_G and $\eta(\lambda)$ is the effective quantum efficiency. $U(\lambda_G - \lambda)$ is a unit step function defined as,

$$U(\lambda_G - \lambda) = 1 \quad \lambda \leq \lambda_G$$

$$= 0 \quad \lambda > \lambda_G$$

Figure 1 illustrates the theoretical and experimental spectral responses for a silicon photodiode. The departure at short wavelengths ($\lambda < 400$ nm) is attributed to surface trapping of the generated minority carriers, while the long wavelength fall-off is due to bulk trapping ($\lambda > 800$ nm). In the wavelength region

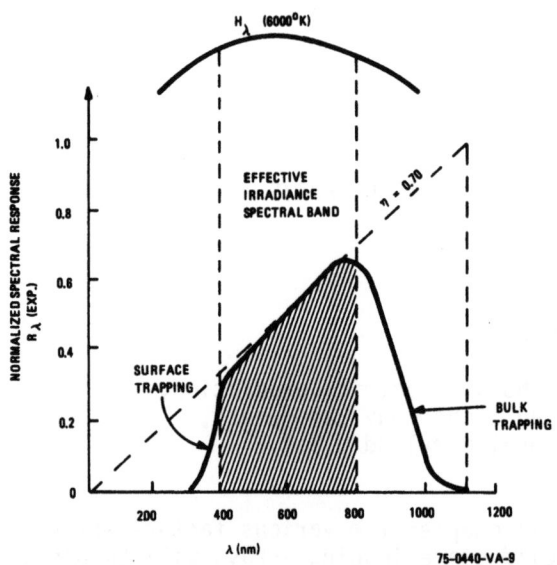

Figure 1. Spectral Response and Responsivity for a Silicon Photodiode (Dotted Curve is Ideal Spectral Response, Solid Curve is Experimental Spectral Response)

400 nm $\leq \lambda \leq$ 800 nm, the silicon photodiode approximates an ideal photon detector with a constant effective quantum efficiency determined by the silicon reflectivity,

$$\eta \simeq 1 - \left(\frac{N_{si} - N_{air}}{N_{si} + N_{air}}\right)^2 = 1 - \left(\frac{3.45-1}{3.45+1}\right)^2 = 0.70 \qquad (2)$$

where we have neglected the imaginary part of the index of refraction caused by absorption of radiation. The units of spectral response are amperes/watt or current density output per irradiance input.

$$R_\lambda = \frac{J_\lambda}{H_\lambda} \left(\frac{A/m^2}{W/m^2}\right) \qquad (3)$$

1.1 Responsivity Formulation R

The responsivity of a photosensor may be defined with respect to a spectral band (λ_2, λ_1) as follows.[1]

$$R \equiv \frac{I}{H(\text{eff.})} = \frac{\int_{\lambda_1}^{\lambda_2} R_\lambda H_\lambda d\lambda}{\int_{\lambda_1}^{\lambda_2} H_\lambda d\lambda} \quad \left(\frac{A}{W/m^2}\right) \tag{4}$$

where H(eff.) is an effective irradiance over the spectral band (λ_2, λ_1) and H_λ is an ideal blackbody irradiance given as

$$H_\lambda = \frac{2\pi c^2 h}{\lambda^5 \left(e^{hc/\lambda kT_s} - 1\right)} \quad \left(\frac{W}{m^2 \cdot \mu m}\right) \tag{5}$$

If we assume the effective quantum efficiency η is constant or slowly varying over the spectral band of interest, then we can write equation (4) as

$$R = \frac{e\eta}{kT_s} \frac{\int_{X_1}^{X_2} X^2 e^{-X} dx}{\int_{X_1}^{X_2} X^3 e^{-X} dx} \tag{6}$$

where $X = hc/\lambda kT_s \gg 1$ and T_s the specified source temperature. For a blackbody source temperature of $T_s = 6000°K$ (i.e. the "sun") and $\lambda_1 = 400$ nm, $\lambda_2 = 800$ nm, we can write for the silicon photodiode,

$$R \begin{pmatrix} \lambda_1 = 400 \text{ nm}, \lambda_2 = 800 \text{ nm} \\ \eta = 0.70 \quad T_s = 6000°K \end{pmatrix} = 0.33 \frac{A}{W} \tag{7}$$

where the watts are W - 6000°K effective within the 400 nm to 800 nm spectral band. For example, consider a silicon photodiode with area A = 15 μm x 20 μm. The detector responsivity may be written as,

$$R_D = RA = \frac{0.0984 \text{pA}}{\text{mW/m}^2} \left(\frac{616 e^-}{\mu J/m^2}\right) \tag{8}$$

with an alternate method of e- (electrons) per effective exposure density E (eff.) in $\mu J/m^2$.

1.2 Measurement of Effective Irradiance H (eff.)

The responsivity, given by equation (4), is defined in terms of a specified blackbody source (e.g. $T_S = 6000°K$) which we shall designate the "spec" source. In practice the actual "test" source may be quite different (e.g. $T_T = 2856°K$). The irradiance and spectral response of the "test" source are measured with a standard diode and narrow band filters (or a monochromator) and the current output of the detector is equated to the current output of a hypothetical "spec" source irradiated detector. Thus, we may write

$$I_{TEST} = R_\lambda(\max) \, H_\lambda(\max)_{TEST} \int_{200}^{1200} \bar{R}_\lambda \bar{H}_\lambda (TEST) \, d\lambda$$

$$I_{SPEC.} = R_\lambda(\max) \, H_\lambda(\max)_{SPEC.} \int_{400}^{800} \bar{R}_\lambda \bar{H}_\lambda (SPEC.) \, d\lambda \tag{9}$$

where $\bar{R}_\lambda$ and $\bar{H}_\lambda$ are normalized quantities. If we equate $I_{SPEC.} = I_{TEST}$ and use the $T_S = 6000°K$ radiation integrals, then we have

$$H(\text{eff.}) = \int_{400}^{800} H_\lambda(SPEC.) d\lambda = 343 \, H_\lambda(\max)_{TEST} \frac{\int_{200}^{1200} \bar{R}_\lambda \bar{H}_\lambda(TEST) d\lambda}{\int_{400}^{800} \bar{R}_\lambda \bar{H}_\lambda(SPEC.) \, d\lambda} \left(\frac{mW}{m^2}\right) \tag{10}$$

We should notice the integration limits on the "test" source (i.e. $\lambda_1 = 200$ nm, $\lambda_2 = 1200$ nm) are determined by the cut-off imposed by the sensor spectral response R_λ. For example, the photodiode and CCD detectors shown in figure 2 had effective irradiance levels of

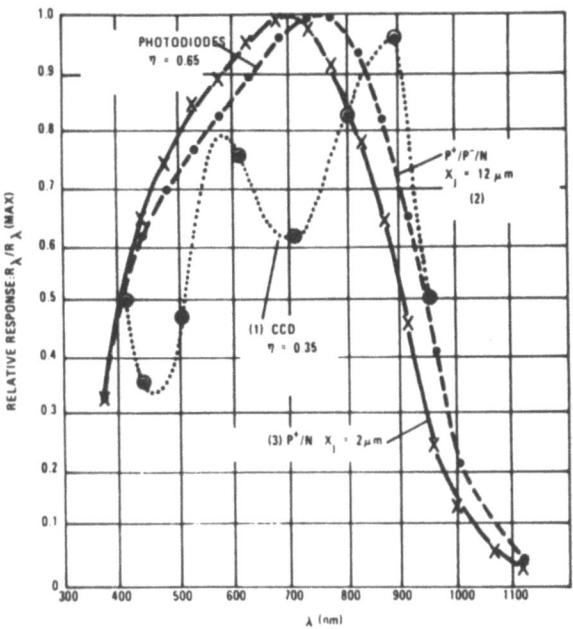

Figure 2. Phototransistor and CCD Relative Spectral Response Curves

$$H(\text{eff.})_{\text{Photodiode}} = 0.639 \text{ mW/m}^2$$

$$H(\text{eff.})_{\text{CCD}} = 0.762 \text{ mW/m}^2$$

for a particular setting of the "test" source lamp current. The above example illustrates the influence of the spectral response of the detector on the determination of effective irradiance levels. In practice, the signal-to-noise and dynamic range of the detectors are measured by attenuation of the effective irradiance level with neutral density filters.

1.3 Radiometric and Photometric Characteristics

In the preceding sections the sensor was described in terms of radiometric units (e.g. watts) and the watt was defined as the effective power from a 6000°K blackbody over the 400-nm to 800-nm spectral band. Thus, it is important to define the temperature of the source and the spectral band of interest in order to describe the type of "watt". The standard radiant flux for the human eye is the foot-candle (lumens/ft^2) or meter-candle (lumens/m^2) and it is a measure of stimulation of the human perceptual system. When the human eye is used as a standard the relative spectral response (i.e. standard "observer") curve of figure 3 is the spectral weighting function $\bar{R}_\lambda$. To describe an

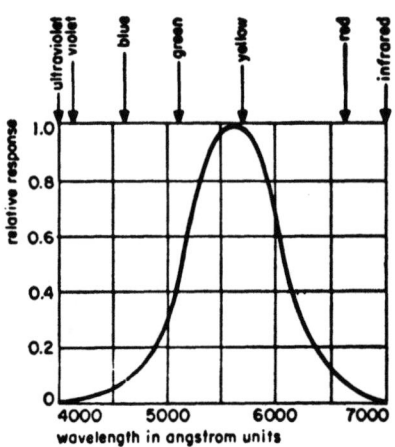

Figure 3. Spectral Response (Relative) of the Human "Eye"

electro-optical device such as silicon in terms of lm/m² is somewhat misleading since the sensor has a spectral response outside the visible range. Perhaps the best procedure is to describe the sensor response in radiometric units (A/W) and then divide this response by the luminosity of the test source (lm/W). The relative spectral response of the human "eye" is shown in figure 3 and with the standard definition of 60π lm/cm² as the total luminance from a 2042°K platinum source falling within the passband of the eye we can write

$$R_\lambda(max)_{EYE} = \frac{60\pi \times 10^4}{\int_0^\infty \bar{R}_\lambda(EYE) \, H_\lambda(2042°K) d\lambda} \quad (11)$$

The spectral luminance from a test source T_T may be written as

$$F = 680 \int_0^\infty \bar{R}_\lambda(EYE) H_\lambda(T_T) d\lambda \quad (12)$$

and the luminosity becomes

$$K = \frac{F}{\int_0^\infty f(\lambda) H_\lambda(T_T) d\lambda} \quad (13)$$

where $f(\lambda)$ is the filter characteristic of the source. K, the luminosity coefficient, is the ability of the test source to convert watts (radiant power) to lumens (luminescent power). In general, $f(\lambda)$ cuts off beyond 2.7 µm for most glasses and the upper limit of the integral is determined by the glass transmission. Thus, the number of effective blackbody watts is reduced by the glass envelope surrounding the test source. If we consider the so-called tungsten "lumen" from a T = 2856°K source, then equation (13) becomes

$$K \begin{pmatrix} \lambda_c = 2.7 \, \mu m \\ T_T = 2856°K \end{pmatrix} = 20 \, \frac{lm}{W} \tag{14}$$

The sensitivity of the sensor is calculated by dividing equations (6) and (13). In the case of the photodiode we have,

$$S = \frac{R}{K} = 0.33/20 = 16.5 \, \frac{mA}{lm} \tag{15}$$

The ideal luminosity K = 16.5 lm/W is obtained for radiation over all wavelengths. Figure 4 illustrates the luminous efficiency and luminosity of a blackbody as a function of absolute temperature.

The key point in the understanding of photometric units is that a lumen represents the amount of radiant power incident on the sensor when the luminous flux at the sensor is one lumen. The use of lumens does not provide the user any information with respect to radiant sensitivity and the specification of performance must consider the spectral distribution of the incident power. Figures 5 and 6 illustrate the natural illuminance levels on the surface of the earth and the range of these levels. The conversion of these photometric units into radiometric units requires a knowledge of the spectral distribution of the radiation. For example, if we consider the sun (T_s = 6000°K) and define the integration limits from λ_1 = 400 nm to λ_2 = 800 nm, then the luminosity becomes

$$K \begin{pmatrix} \lambda_1 = 400 \, nm, \, \lambda_2 = 800 \, nm \\ T_s = 6000°K \end{pmatrix} \approx \frac{200 \, lm}{W}$$

whereas, integration over all wavelengths yields 90 lm/W as seen from figure 4.

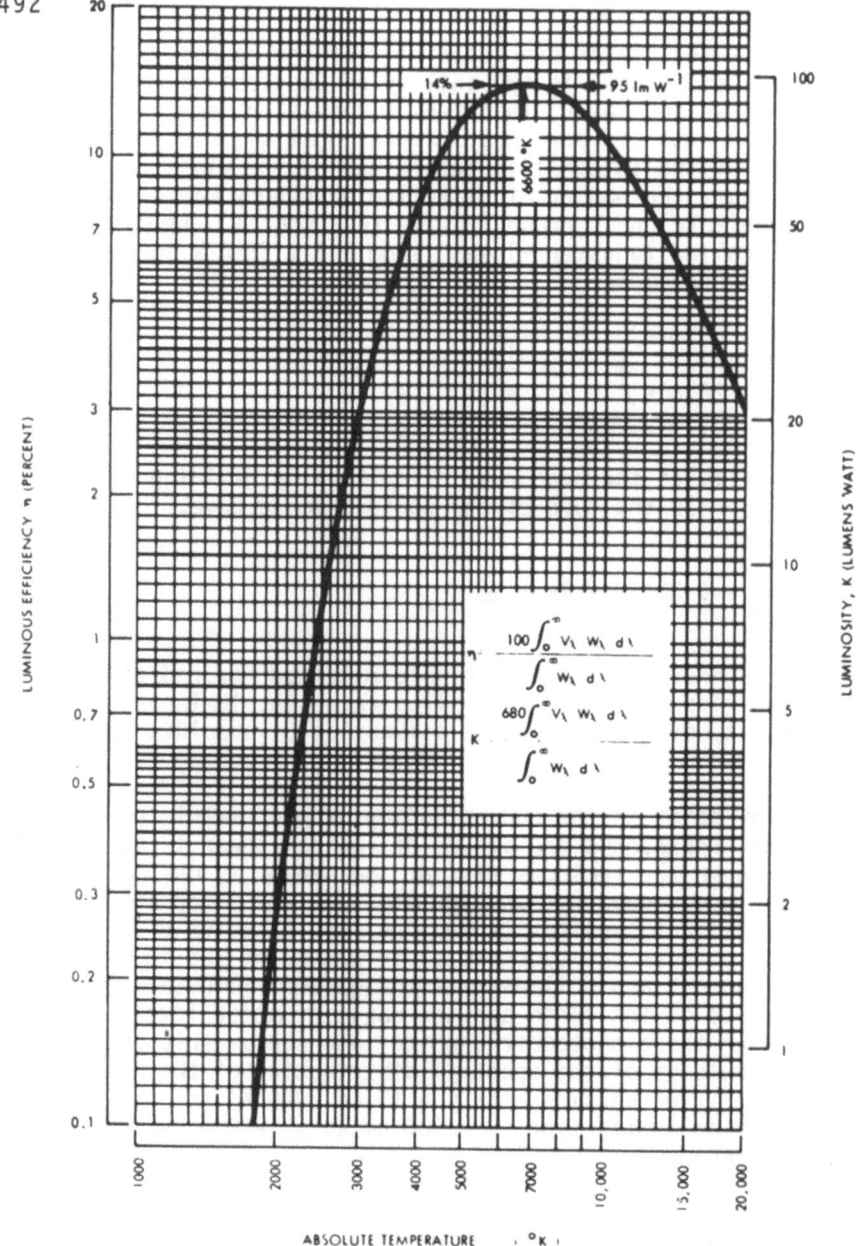

Figure 4. Luminous Efficiency η and Luminosity K of Blackbody as a Function of its Absolute Temperature where W_λ is the Spectral Radiant Emittance of the Blackbody and V_λ is the Relative Spectral Response of the Average Human Eye.[3]

493

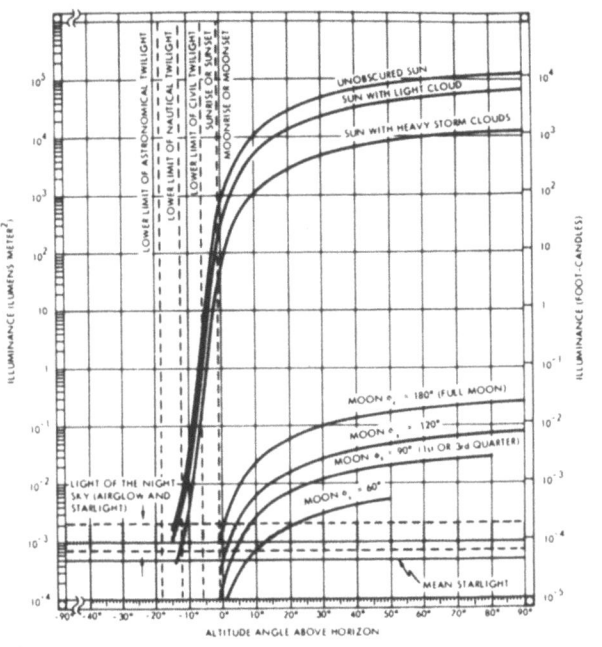

Figure 5. Illuminance Levels on the Surface of the Earth Due to the Sun, the Moon, and Sky[4]

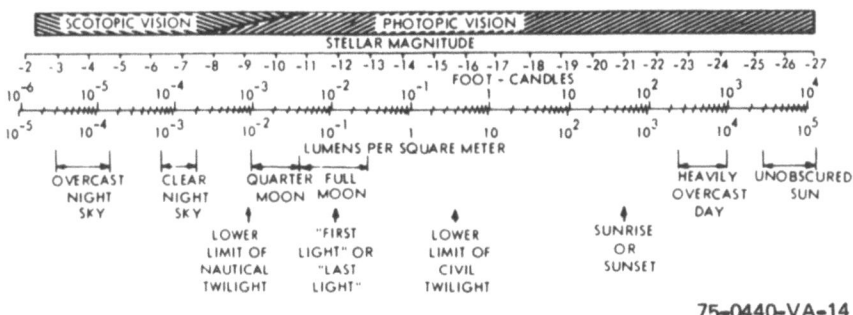

Figure 6. Range of Natural Illuminance Levels[4]

2. MODULATION TRANSFER FUNCTION (M.T.F.)

The concept of modulation transfer function (M.T.F.) has revolutionized optical system design. The M.T.F. is the system response to a sinusoidal spatial frequency input, normalized to zero spatial frequency. If we restrict the discussion to the photosensor or detector array, then the input signal is band-limited by the spacing or "pitch" of the detector elements. This follows from the Nyquist sampling theorem which states the highest spatial frequency recoverable or reconstructable in the X-direction (see figure 7) of the image has a wavelength equal to twice the pitch (i.e. $\lambda_s = 2P$). The spatial frequency, which corresponds to this limit, will be denoted as

$$\boxed{f_s(\max) = \frac{1}{2P}} \qquad (16)$$

and this is the high frequency limit. In general, the output S/N of the detector array is lowest at $f_s(\max)$ and the information content in the image is often highest. We will now proceed to derive the M.T.F. for the detector array of figure 7.

If it were possible to take discrete samples of the irradiance $H(x,y)$ with perfect resolution (i.e. $\Delta X = 0$), then the Nyquist theorem states $H(x,y)$ must be sampled at least twice during each cycle of its highest spatial frequency component $f_s(\max)$. Continuing this reasoning, if the detector element amplitude resolution (i.e. detector responsivity) were infinite, then the scene would be reproduced with no loss of information up to the Nyquist spatial frequency limit given by equation (16). Since ΔX cannot be zero in practice [i.e. a finite number of photons arrive in a finite time interval and the fluctuations in their arrival sets the ultimate signal-to-noise of the detector] we will have an "error" signal generated because of spatial integration. The integrated sample at X_0 over the detector dimension ΔX gives the mean value of $H(X_0)$ with an error signal,

$$\Delta H(X_0) = H(X_0) - \frac{1}{\Delta X} \int_{X_0 - \Delta X/2}^{X_0 + \Delta X/2} H(X) dx \qquad (17)$$

which is a measure of the uncertainty introduced for $\Delta X > 0$. Let us denote

$$H(X) = H_0 (1 + A \cos 2\pi f_s X) \qquad (18)$$

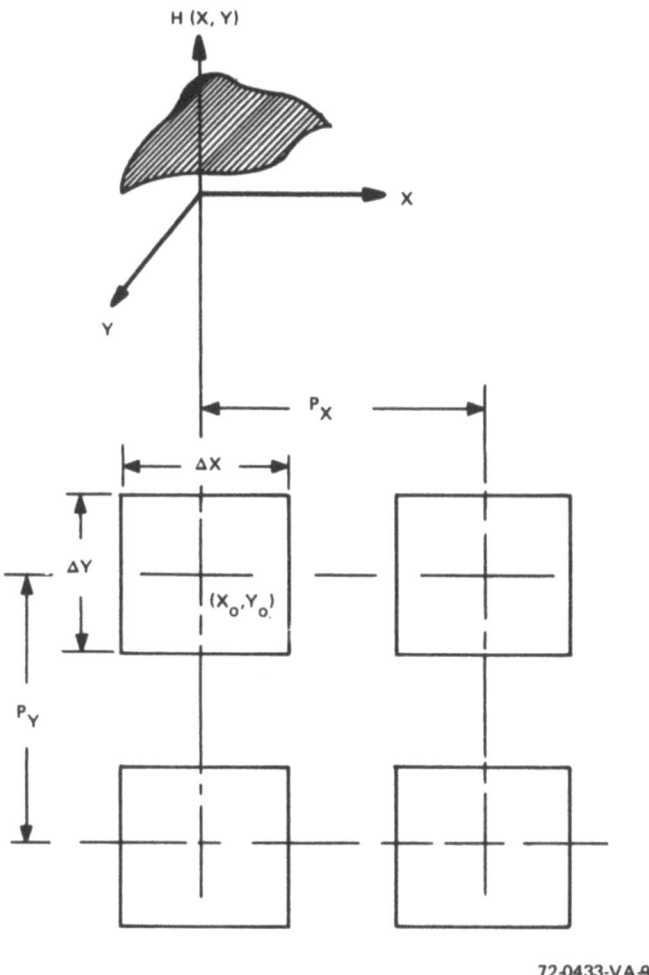

Figure 7. Irradiance on Detector Element Array

and substitution of equation (18) into (17) yields

$$\Delta H(X_0) = H_0 A \left(1 - \frac{\sin \pi f_s \Delta X}{\pi f_s \Delta X}\right) \cos 2\pi f_s X_0 \qquad (19)$$

$$= H(X_0) - H_m(X_0)$$

where the measured signal is

$$H_m(X_0) = H_0 (1 + M \cos 2\pi f_s X_0) \qquad (20)$$

with a modulation

$$M = A \frac{\sin \pi f_s \Delta X}{\pi f_s \Delta X} \qquad (21)$$

The modulation transfer function is defined as,

$$\boxed{M.T.F. \equiv \frac{M(f_s)}{M(o)} = \frac{\sin \pi f_s \Delta X}{\pi f_s \Delta X} = \frac{\sin \frac{\pi \Delta X}{2P} \frac{f_s}{f_s(max)}}{\frac{\pi \Delta X}{2P} \frac{f_s}{f_s(max)}}} \qquad (22)$$

which is shown in figure 8 for $\Delta X = 2P$ and $\Delta X = P$.

An alternate method to derive the M.T.F. of a detector is to Fourier transform the spread function of the detector $S(x)$.

$$S(f_s) = \int_{-\infty}^{\infty} S(x) e^{j2\pi f_s x} dx \qquad (23)$$

and for a spread function defined by

$$S(x) = S_0 \quad \text{a constant} \quad X_0 - \frac{\Delta X}{2} \leq X \leq X_0 + \frac{\Delta X}{2}$$

$$= 0 \quad \text{all other } X$$

we have

$$M.T.F. \equiv \frac{|S(f_s)|}{|S(o)|} = \frac{\sin \pi f_s \Delta X}{\pi f_s \Delta X}$$

as given in equation (22).

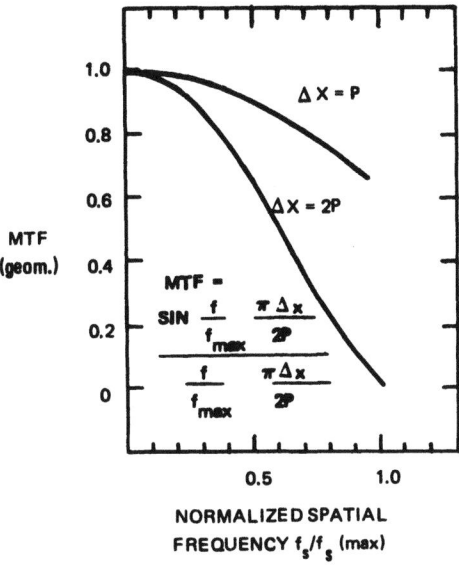

Figure 8. Modulation Transfer Function (M.T.F.) vs Normalized Spatial Frequency for $\Delta X = 2P$ and $\Delta X = P$ ($f_s(\max) = 1/2P$)

2.1 Experimental Determination of M.T.F.

The experimental determination of M.T.F. is to slide the detector element slowly and quite accurately by means of a stepping motor past a narrow slit, which is typically demagnified to less than 5 μm including effects of lens diffraction. The detector output is amplified and drives the Y-axis of a x-y recorder while the X-axis is driven from the stepping motor. The composite profile obtained contains the slit profile as well as the sensor profile. Figure 9 illustrates the basic M.T.F. determination. To remove the slit profile from the composite profile the slit profile spread function is determined by differentiation of the edge trace produced by a knife edge scan. The Fourier transform of the slit profile spread function yields the M.T.F. of the slit image. The knife edge scan is preferred to a slit measurement since the latter suffers from low signal-to-noise ratios. The slit spread function for a 0.5 mil slit demagnified by a 1:10 lens is shown in figure 10 as a function of wavelength.

The slit image spread function $S_{Slit}(X)$ and detector spread function $S_{DET}(X)$ are slid past one another and the overlapping area produces the composite output profile given as

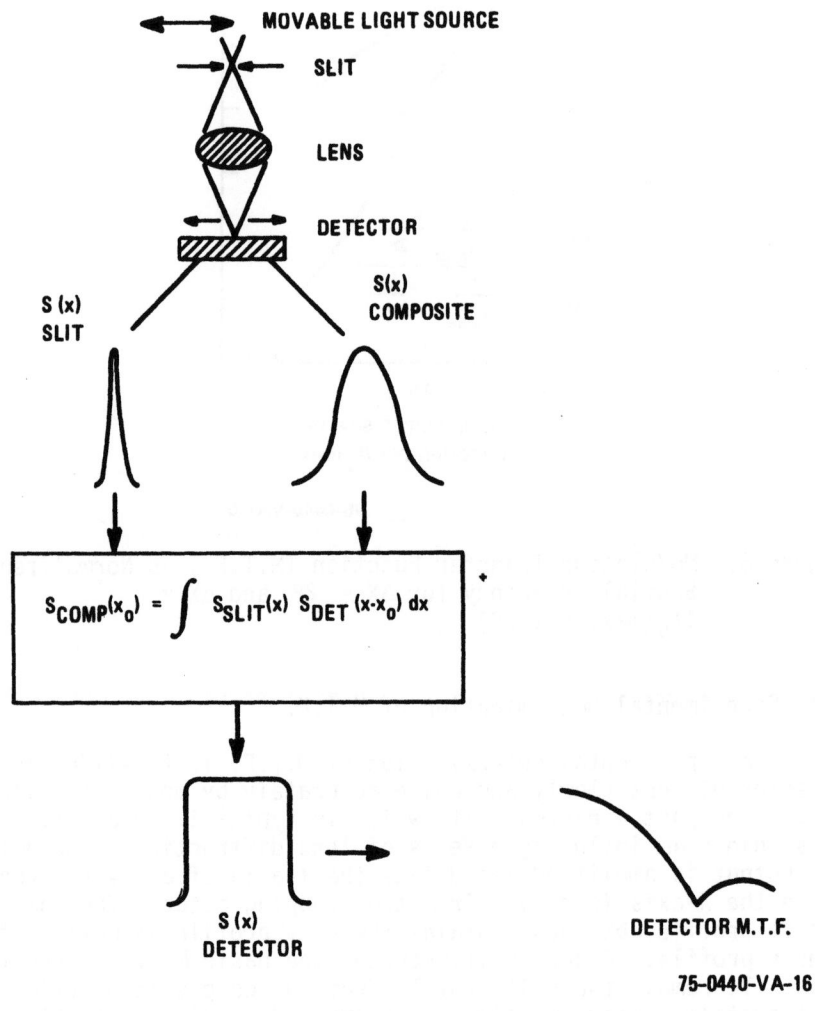

Figure 9. Modulation Transfer Function (M.T.F.) Determination

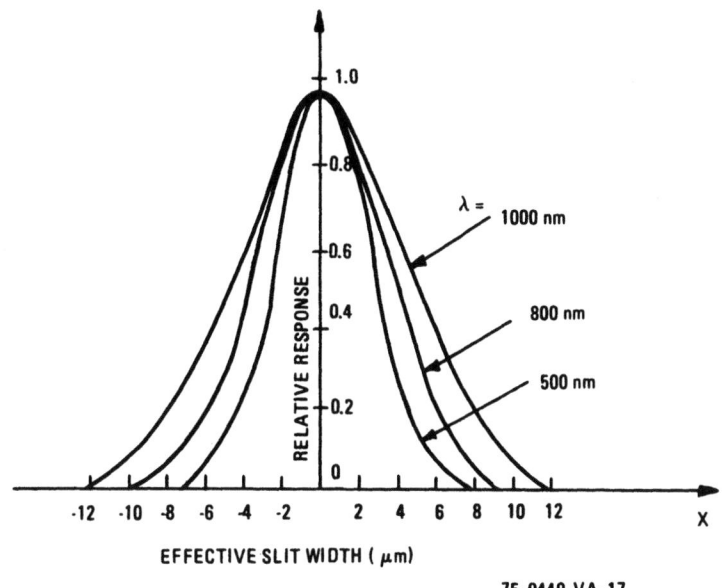

Figure 10. Slit Spread Function for a 0.5 mil Slit Demagnified by a 10X Lens Including Effects of Lens Diffraction

$$S_{COMP.}(X_o) = \int S_{Slit}(X) \, S_{DET.}(X-X_o) dx \tag{24}$$

where X_o is the coordinate of the detector center. Figure 11 illustrates the detector spread function with the effects of the slit removed through a process similar to deconvolution although equation (24) is not a convolution integral. The ideal spread function for the detector is approached at short wavelengths since the radiation does not penetrate as deep into the silicon. The effect of long wavelengths is to cause a smearing of the detector spread function and a corresponding loss in M.T.F. as illustrated in figure 12. The theoretical M.T.F. of a detector with $\Delta X = P$ at the Nyquist limit, $f_s = f_s(max)$, is given by equation (22),

$$\text{M.T.F.}\bigg|_{\substack{f_s = f_s(max) \\ \Delta X = P}} = \frac{\sin \pi/2}{\pi/2} = \frac{2}{\pi} = 0.637 \tag{25}$$

and this limit is approached as shown in figure 12.

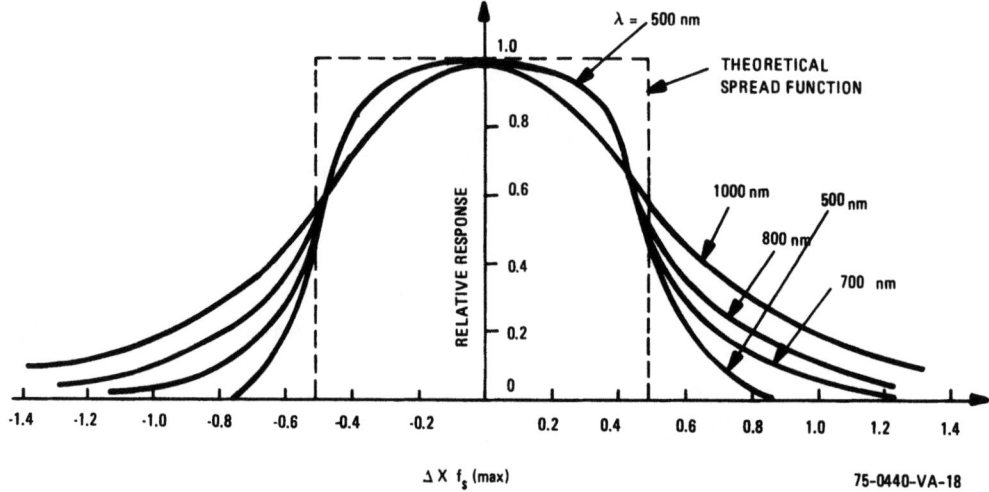

Figure 11. Detector Spread Function S_{DET} as a Function of $\Delta X f_s(max)$ and Incident Wavelength λ

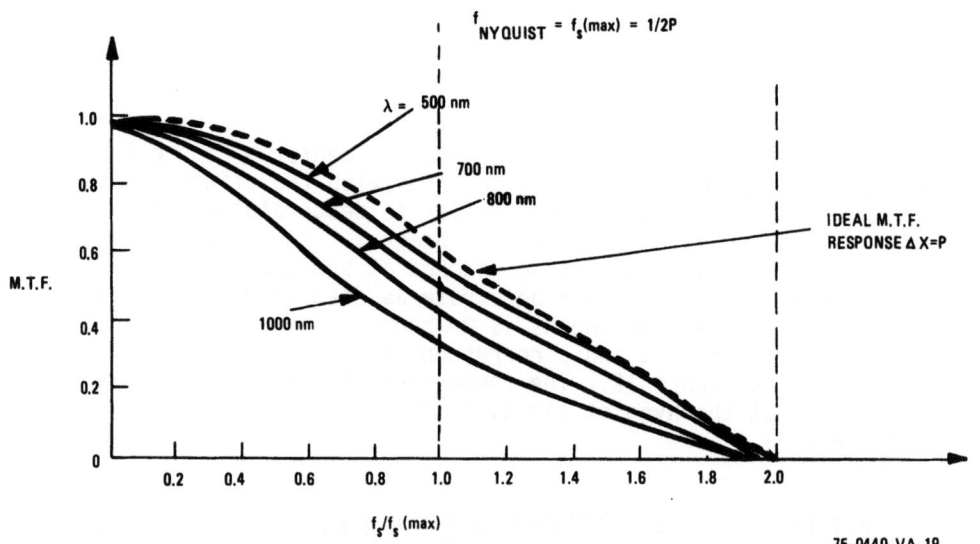

Figure 12. M.T.F. of the Detector as a Function of $f_s/f_s(max)$ and Incident Wavelength λ

3. NOISE CONSIDERATIONS

The low-light-level performance of a solid-state imaging system is determined by the noise equivalent signal (NES) which is the input exposure density ($\mu J/m^2$) that will provide a S/N = 1 at the system output. In order to realize the ultimate performance of the basic sensor elements (pixels), the noise introduced by the system through signal processing and conditioning must be minimized. Thus, present-day methods of signal processing use high-speed operational amplifiers followed by fast sample and hold circuits to provide signal reconstruction and automatic dark-level subtraction.[1] Figure 13 illustrates a test apparatus for the measurement of responsivity and noise. The analog signal processor consists of the input preamplifier followed by a sample and hold circuit to provide the video reconstruction shown in the output waveform. The sample and hold circuit combined with the roll-off of the amplifier filters the noise power spectrum. The measurement procedure involves a 10-bit A/D converter to sample the signal output from each pixel in the array 1,024 times at each irradiance level and record the mean and variance in A/D bits. The mean corresponds to the signal while the variance represents the noise. Linearity and streaking can be determined from the data and with narrow band optical filters the spectral response and streaking can be measured. The test apparatus of figure 13 is shown for the CCD sensor output; however, other sensor outputs interface in the same manner.

The noise is converted from A/D bits to an equivalent input exposure density ($\mu J/m^2$), called the NES, by multiplying the rms A/D bits by the reciprocal slope of the transfer curve at the particular exposure density. The reciprocal slope of the transfer curve is called the quantizing interval and is given as

$$Q_I = \frac{\Delta E}{\Delta(A/D \text{ bits})} \quad \frac{\mu J}{m^2 \cdot bit} \tag{26}$$

The NES becomes

$$NES = Q_I B_{rms} \quad (\mu J/m^2) \tag{27}$$

where B_{rms} the bits variance at the specified exposure density. The NES, which is measured by this procedure, consists of four principal terms:

  a. System noise from analog signal processor, power supplies, pulse jitter, mechanical vibrations, etc.

Responsivity and Noise Measurement

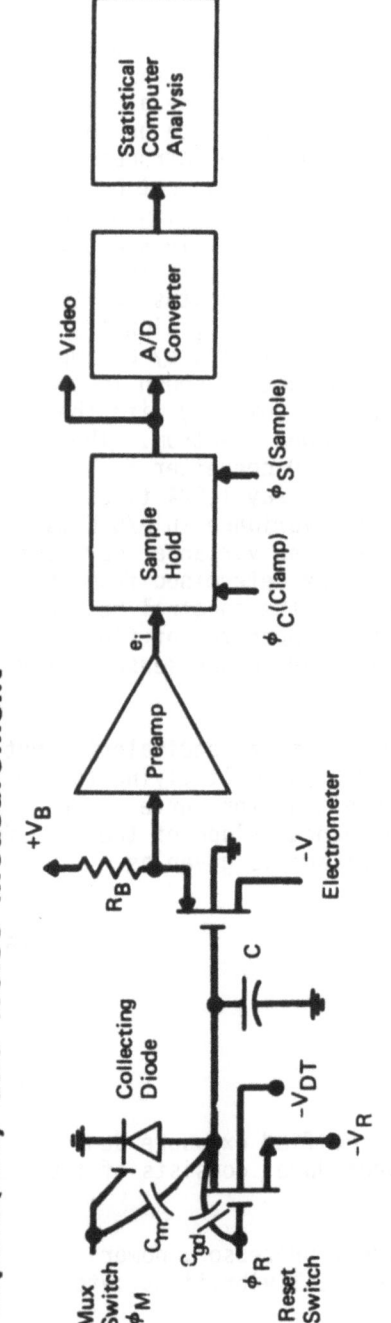
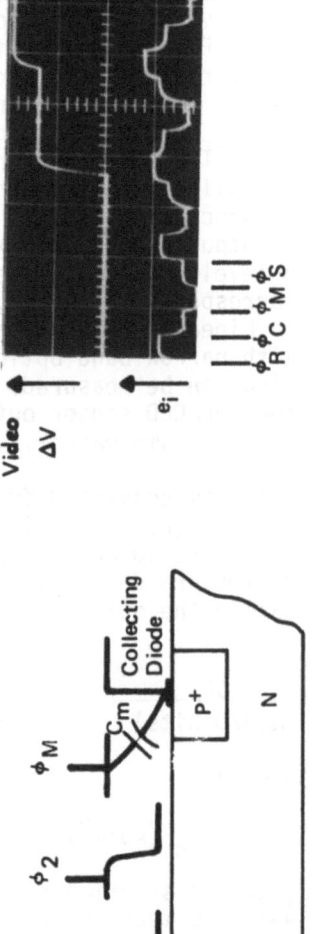
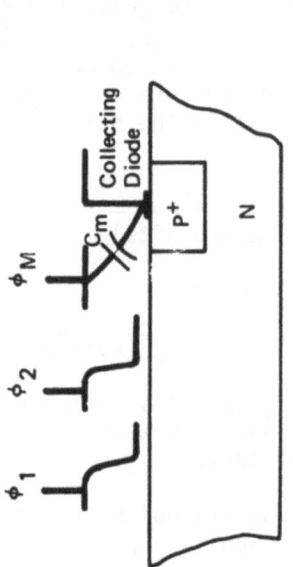

Figure 13. Measurement of Responsivity and Noise[1]

b. Chip noise from the sensor array and which is determined by the geometrical design and fabrication processes.

c. Radiation shot noise from the fluctuation in arriving signal photons.

d. Quantization noise from the uncertainty associated with the finite size of the quantizing interval Q_I.

The measured or total NES is given as

$$NES^2(total) = NES^2_{syst.} + NES^2_{chip} + \frac{eE}{R_D} + \frac{Q_I^2}{12} \qquad (28)$$

where the radiation shot noise involves the exposure density E and the quantization noise assumes an error which varies linearly with time. For a noise current i_n at the input to the preamplifier in figure 13, the equivalent noise charge at the electrometer (on-chip) input is $C\, i_n/g_m$ and the noise equivalent signal associated with the system may be written as

$$NES_{syst.} = \frac{C\, i_n}{eg_m R_D} \qquad (29)$$

where g_m is the transconductance of the electrometer. Thus, we note an important consideration in the reduction of system noise contributions is to increase the $g_m R_D/C$ ratio.

The NES_{chip} may be formulated analytically from the following three sources of noise on the chip:

a. Nyquist noise of reset and address switches with noise charge $Q_n^2 = kTC$, where C is a node capacitance or "well" capacitance for input CCD noise.

b. Shot noise associated with sensor, switches, or shift register (for CCD's) thermal leakage currents with noise charge $Q_n^2 = eI_L \tau$, where τ is the exposure time and I_L the leakage current.

c. Surface trap and bulk trap noise associated with the sensor, output electrometer, or shift register (for CCD's) and which may be written as $Q_n^2 = kTC_t$ where C_t is an effective trap capacitance.

With these terms the chip NES becomes

$$\boxed{NES_{chip} = \frac{\left[kT(C + C_t) + eI_L\tau\right]^{1/2}}{R_D}} \quad (30)$$

The above expression is modified to include an $R(1-\epsilon)^N$ for CCD's due to the finite transfer inefficiency ϵ and the number of transfers N. A principle limitation to the NES_{chip} is the Nyquist noise of the output reset switch, which gives an rms pixel noise charge of 200 e$^-$ for a 0.25 pF output capacitance. This limitation has been removed in CCD imagers with a method of signal processing called correlated double sampling[1] in which two samples are taken within a pixel time window. This method can be used to approach the limiting case of shot-noise for buried-channel CCD imagers; however, the surface channel CCD imagers are limited by the need to inject an electrical fat zero which gives a kTC contribution at the input injection circuit.

3.1 Correlated Double Sampling[1]

A method of signal processing called correlated double sampling has been developed which removes the switching transients at the output collection diode of a CCD, eliminates the Nyquist noise of the reset switch-output capacitance combination, provides dc restoration and increases dynamic range, and suppresses surface state and 1/f noise contributions.

Let us examine the four distinct timing intervals employed in the readout circuit of figure 14.

1. Reset

The n-channel MOSFET reset switch is turned on the voltage V_G across the capacitor C is reset to the reference voltage V_R with a noise uncertainty V_n. This noise voltage may be introduced through inadequate filtering of the reference supply voltage and the Nyquist noise contribution of the reset switch, [9], [10] where the latter is given by $V_n = (kT/C)^{1/2}$ or in terms of noise charge $Q_n = (kTC)^{1/2}$. The full Nyquist voltage appears across C when the electrical time constant formed by the series resistance of the reset switch and integration capacitance C is much less than the time the reset switch is on. The p-channel electrometer is connected to an operational amplifier which is the preamplifier in the CCD signal processor circuit shown in figure 15. The timing diagram for the signal processing is illustrated in figure 16. At the start of the reset interval the pixel charge is in transit to the last well (i.e., see phase ϕ_2 and figure 14) of the electrical bit adjacent to the mux gate ϕ_M and the collection diode.

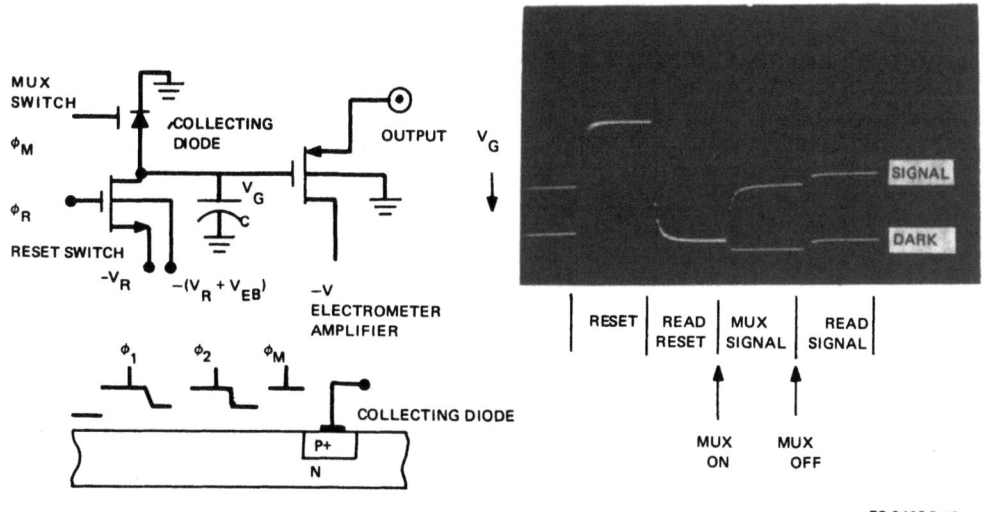

Figure 14. CMOS Correlated Double Sampling Readout Circuit[1]

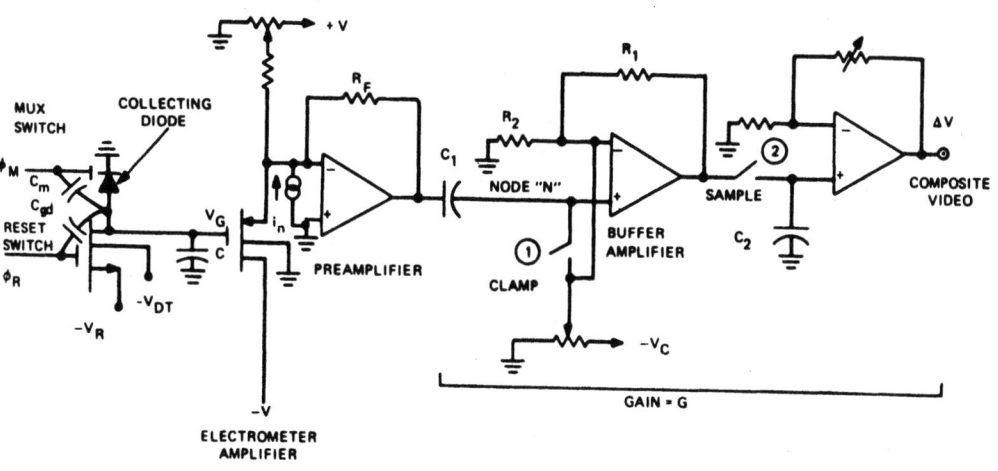

Figure 15. Schematic Diagram of a CDS Processor with Critical Capacitances, Noise Sources, and Signal Nodes[1]

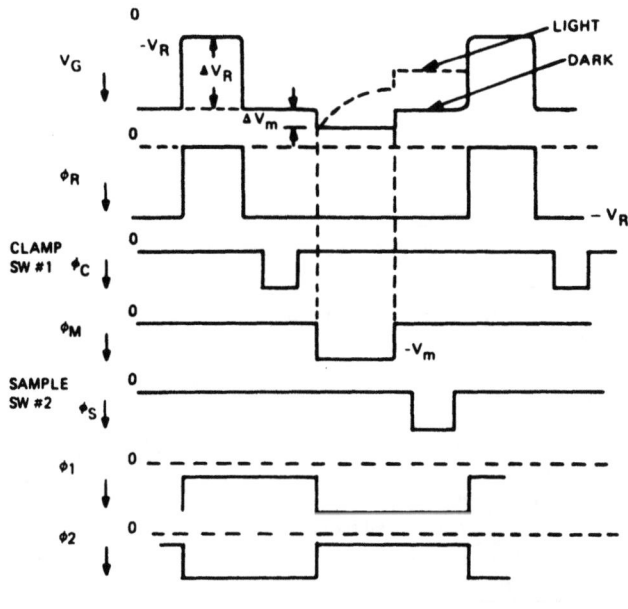

Figure 16. Timing Diagram for CCD Line Array with CDS Analog Signal Processor. The Four Steps are Separated in Sequence to Correspond to Explanation in the Text[1]

2. <u>Read Reset</u>

After the n-channel reset switch is turned off, the voltage present on the gate of the electrometer consists of a feedthrough pedestal ΔV_R and a noise voltage V_n. With the reset switch off, the gate voltage is holding on a high impedance point with a time constant of seconds. In the read reset interval, the clamp switch 1 is turned on and C_1 is charged to a voltage indicative of the voltage on the electrometer. Switch 1 is turned off and one side of the capacitor, node N, is clamped or dc restored to a reference voltage V_c, while the other side of the capacitor represents the instantaneous sample of the gate voltage. The instantaneous voltage across the clamp switch from this moment on is the differential or incremental charge caused by a change in the on-chip gate or collection diode voltage. With the clamp switch turned off, the measured reset level is holding on the high impedance node N formed by the clamp capacitor C_1, and the noninverting input of the buffer amplifier.

3. <u>Mux Signal</u>

At the start of the mux signal interval the pixel charge is raised in the storage well (ϕ_2 goes high) and ϕ_M goes low to transfer the pixel charge to the collection diode. The collection of pixel charge (minority carriers) discharges the voltage V_G as shown in the signal waveform of figure 14. If we assume, for the moment, there is no pixel charge, then the only charge transferred to the gate electrode is the feedthrough pedestal $\Delta V_m = V_m C_m/C$, where V_m is the mux voltage swing and C_m the feedthrough capacicance from the mux gate to the collection diode. The charge is removed, however, when the mux gate is turned off as shown in figure 14. A Nyquist noise of $Q_n^2 = 2kTC_m$ is introduced, which may be minimized for $C_m < 0.01$ pF for the case where the mux gate does not overlap the collection diode. Alternatively, an overlapping mux gate ϕ_m may be held at a fixed dc potential with the clock ϕ_2 transferring charge to the collection diode (see figure 14). In the absence of any optical pixel charge we would collect the leakage current from the sensor and the shift register wells.

4. Read Signal

After the mux signal is turned off, the running output voltage on node N is the time difference between the previously clamped reset level and the same reset level plus signal increment introduced by the closure of the mux switch (i.e., there is negligible leakage of the reset level between read reset (clamp) and read signal (sample) intervals). Thus, the reset noise, which includes Nyquist noise and V_R power supply noise, is correlated within a pixel time window. The signal increment, which consists of sensor and shift register leakage current added to photocharge, is amplified and passed to the output of the signal processor by the closure of the sample switch 2. The output video stream is a sequence of pixel element responses free from reset noise and proportional to the minority carrier signal increment introduced by closure of the mux switch.

The correlated double sampling (CDS) method removes switching transients similar to an earlier technique [5] which used a gated charge integrator in lieu of storing (clamping) the actual diode reset level for subsequent subtraction from the reset level plus signal increment to give the signal increment without reset noise. The Nyquist noise of the reset switch has been removed since it is correlated within a pixel time, and this means a removal of a noise charge

$$Q_n = \frac{(kTC)^{1/2}}{e} = 200 e^-$$

for a 0.25 pF capacitor. The "1/f" surface-state noise is also suppressed by the filter characteristic of the analog signal processor which is shown as follows.

The transfer function, which acts on any time-varying components of the signal between clamp and sample intervals, may be written as

$$T(s) = T_0 \frac{(1 - e^{-s\tau})}{1 + s/\omega_0} \qquad (31)$$

where T_0 is the signal gain and τ is the delay time between the end of the clamp pulse and the end of the sample pulse, and ω_0 is the bandwidth of the front-end preamplifier. Figure 17 illustrates a plot of the filter characteristic for a value of $\tau = T/2$, where T is the clock period. The important features of this filter are the "double zeros" of $|T(\omega)^2|$ at $\omega = 2N\pi/T$ ($N = 0,1,2,...$). The "double zero" at the origin ($\omega = 0$) serves to suppress 1/f and and low frequency noise arising from power supplies, pulse jitter, etc, and the double zero at even harmonics of the fundamental clock frequency also suppresses surface-state noise generation in the p-channel MOS electrometer amplifier. Thus, the signal processing does not degrade but enhances the qualities of the sensor element by removing the Nyquist noise and filtering the 1/f noise while amplifying the signal. In addition to these features, we have automatic dark level subtraction to increase the dynamic range since the video signal is clamped by the reference voltage V_c. There is no need to filter out the clock fundamental and higher order harmonics; and the video output is already in a format for image display or further data processing.

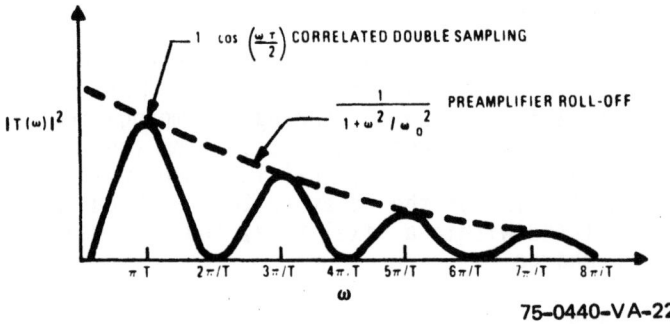

Figure 17. Filter Characteristics of CDS Analog Processor with $\tau = T/2$

4. FIGURE OF MERIT F.M. = S/N x M.T.F. (eff.)

In the design of low light level image arrays we must concentrate on two important quantities: (1) sensitivity and (2) resolution. The sensitivity can be described by the noise equivalent signal (NES) discussed in section 3 and represented by the signal-to-noise ratio given by equation (28). The resolution is described in terms of the observable spatial resolution (line pairs/mm or lines/mm) at a certain contrast level or system gain. This design parameter is called the <u>modulation transfer function</u> or M.T.F. and the use of it has revolutionalized optical system design. The M.T.F. is the modulus of the optical transfer function which is the system response to a sinusoidal spatial frequency input f_s band-limited by the sensor element spacing or pitch P. Let us examine the sensor line array shown in figure 7 with ΔY in the along-track direction and ΔX in the across-track direction. The image or scene moves in the along-track direction while the line array is scanned electronically in the across-track direction. The modulation transfer function in the across-track direction is from equation (22)

$$\text{M.T.F.}_x = \frac{\sin(\pi f_s \Delta X)}{\pi f_s \Delta X} \quad (32)$$

and in the along-track direction

$$\text{M.T.F.}_y = \frac{\sin(\pi f_s \nu \tau)}{\pi f_s \nu \tau} \cdot \frac{\sin(\pi f_s \Delta Y)}{\pi f_s \Delta Y} \quad (33)$$

where ν is the velocity of the scene across the image plane and τ the exposure (line) time (i.e., $d = \nu \tau$ is the distance the image moves during the exposure time). The additional term in equation (33) is due to the relative motion between the image and sensor during the time the output signal is generated.

If the noise is determined by the first signal preamplifier, then to a first-order approximation the noise is independent of detector area and the S/N ratio increases proportional to the detector area,

$$S/N \sim \Delta X \Delta Y \quad (34)$$

If we make the assumption the effective resolution (M.T.F.) is the geometric mean of its components, then a figure-of-merit may be defined as[6]

$$\boxed{\text{figure-of-merit (F.M.)} \equiv \Delta X \Delta Y \left[(M.T.F._x)(M.T.F._y)\right]^{1/2} \sim S/N \; M.T.F. \; (\text{eff.})} \quad (35)$$

The F.M. may be maximized at the Nyquist limit $f_s(\text{max}) = 1/2P$ and letting $\nu\tau = P$ to conserve bandwidth and provide two samples in the along-track direction. With these conditions the F.M. becomes,

$$F.M.\bigg|_{f_s=f_s(\text{max})} = \Delta X \Delta Y \left[\frac{\sin\left(\frac{\pi\Delta X}{2P}\right)}{\left(\frac{\pi\Delta X}{2P}\right)}\right]^{\frac{1}{2}} \left[\frac{2}{\pi} \frac{\sin\left(\frac{\pi\Delta Y}{2P}\right)}{\left(\frac{\pi\Delta Y}{2P}\right)}\right]^{\frac{1}{2}} \quad (36)$$

Differentiation of equation (36) with respect to ΔY yields

$$\tan\frac{\pi\Delta Y}{2P} = -\frac{\pi\Delta Y}{2P} \quad (37)$$

with a solution

$$\Delta Y = 1.292P \quad (38)$$

as the optimum value for the along-track dimension. Equation (36) is plotted in figure 18 with the value $\Delta X/P$ varied and $\Delta Y = 1.292P$ held constant at its optimum value. To the left of the peak the detector has a high M.T.F. but a low S/N with the converse true to the right of the peak. For $\Delta X/P > 1$ a bilinear array is required, rather than a true linear array, with an offset in the along-track direction as shown in the 2-P offset of figure 19. The offset must be an integer multiple of P in order to reconstruct the image on a uniform Cartesian grid. In situations where aliasing is expected to present a problem the bilinear array offers optimum performance since at $\Delta X = 2P$ there is no response and aliasing effects are minimized. Another consideration in the selection of a bilinear, maximum-performance array is the need to avoid excessive image misregistration in either the x or y components due to angular deviations from perpendicularity to the array of the image velocity vector. This would necessitate small spacings between the two rows of detectors.

4.1 Equiresolution Bilinear Array

In applications, such as cartography, equal resolution is required in both the across-track and along-track directions. Thus, we can set the M.T.F.'s equal at the limiting spatial frequency $f_s(\text{max})$ and let $\nu\tau = P$ to conserve bandwidth to find,

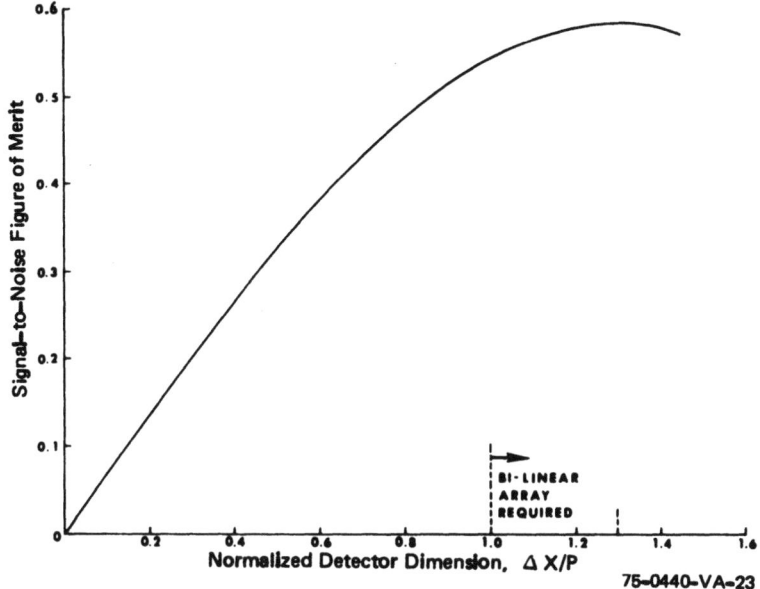

Figure 18. Figure-of-Merit[6] as a Function of $\Delta X/P$

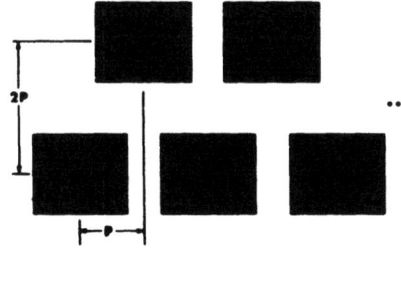

Figure 19. Bilinear Detector Element Array for Maximum Performance

$$\frac{\sin\left(\frac{\pi \Delta X}{2P}\right)}{\left(\frac{\pi \Delta X}{2P}\right)} = \frac{2}{\pi} \frac{\sin\left(\frac{\pi \Delta Y}{2P}\right)}{\left(\frac{\pi \Delta Y}{2P}\right)} \tag{39}$$

Equation (39) provides the relationship between $\Delta X/P$ and $\Delta Y/P$ to obtain equal resolution in both x and y directions. Thus, the F.M. becomes

$$\left. F.M. \right|_{f_s(max)} = \frac{4 \Delta X P}{\pi^2} \sin\left(\frac{\pi \Delta Y}{2P}\right) \tag{40}$$

subject to the constraint of equation (39). The equiresolution bilinear array has a maximum F.M. with the values

$$\Delta X = 1.45P \quad \text{(preamplifier noise-limited)} \tag{41}$$
$$\Delta Y = 1.17P$$

as shown in figure 20. The M.T.F. of $f_s(max)$ is 0.33 for the conditions above and if we desire $P = 15\ \mu m$, then $\Delta X = 22\ \mu m$ and $\Delta Y = 18\ \mu m$ are logical selections to achieve a maximum performance

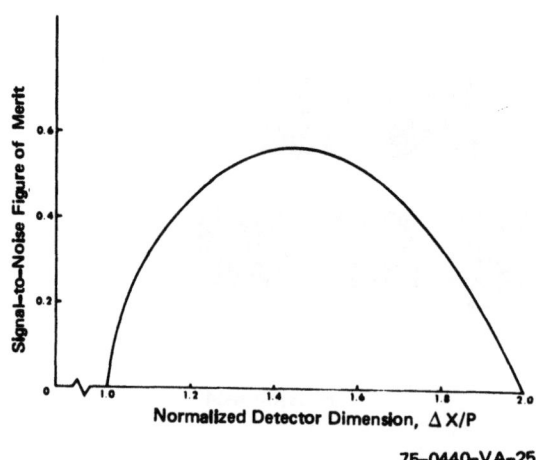

75-0440-VA-25

Figure 20. Figure-of-Merit of Equal Resolution Bilinear Array[6]

linear detector array. The M.T.F. deviation between along and cross track is minimal as shown in figure 21. Figure 22 illustrates a 2P bilinear CCD array (P = 15 μm) with the design indicated by equation (41) and defined by an aluminum light shield.

If the detector array has very low noise and the detector area limits the noise through such factors as thermal "shot" noise $(eI_L \tau)^{1/2}$, Nyquist noise $(kTC_D)^{1/2}$ or radiation "shot" noise $\left(\frac{eE}{R_D}\right)^{1/2}$, then the signal-to-noise ratio becomes

$$S/N \sim (\Delta X \Delta Y)^{1/2} \qquad (42)$$

and the optimum set of values are

$$\Delta X = 1.26P \quad \text{(detector area noise-limited)} \qquad (43)$$
$$\Delta Y = 0.85P$$

which yields a M.T.F. = 0.46 at f_s(max). For P = 15 μm the dimensions of the low-noise bilinear array are ΔX = 19 μm and ΔY = 13 μm. Such an array could be designed with a 1-P offset in the along-track direction to minimize misregistration of the image as discussed previously. In this design the aliasing will be decreased since the spatial sampling frequency is higher in the along-track direction.

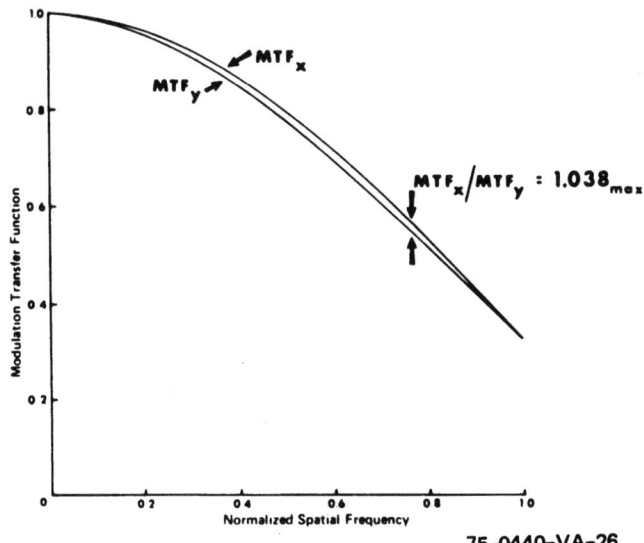

Figure 21. Illustration of Similarity of x and y Modulation Transfer Functions for Bilinear Equiresolution Array[6]

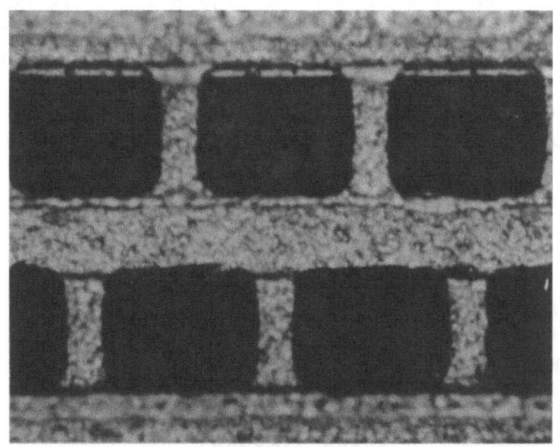

Figure 22. 2P - Bilinear Offset CCD Array P = 15 μm; ΔX = 22 μm; ΔY = 18 μm

4.2 Maximum Performance Linear Array

We have discussed the bilinear detector array with 1-P and 2-P offsets in the along-track direction for bandwidth conserving $v\tau = P$. In this section we will examine the zero offset case (i.e., 0-P) which corresponds to a linear array of elements, Since $\Delta X = P$ the M.T.F.$_x$ is constrained and we need an adjustable parameter to obtain equiresolution at the Nyquist limit $f_s(max)$. An examination of equations (32) and (33) indicates for $v\tau = P$ we must have $\Delta Y \rightarrow 0$ for equiresolution. Alternately, if we select $\Delta Y = P$, then $v\tau \rightarrow 0$ is required for equiresolution. In the first case there will be appreciable signal reduction and in the second case the bandwidth requirements are considerable. If we simply constrain $v\tau = P$ and formulate the figure-of-merit F.M., then from equations (32), (33) and (35) we have,

$$\left. F.M. \right|_{f_s(max)} = P^2 \left(\frac{2}{\pi}\right)^{3/2} \left[\left(\frac{\Delta Y}{P}\right) \sin\left(\frac{\pi \Delta Y}{2P}\right)\right]^{\frac{1}{2}} \tag{44}$$

for the preamplifier noise-limited case. The maximum F.M. occurs with the selection

$$\Delta X = P \quad \text{(preamplifier noise-limited)}$$
$$\Delta Y = 1.292P \quad (45)$$

with a M.T.F.$_x$ = $2/\pi$ (0.637) and M.T.F.$_y$ = 0.28. For the case of detector noise-limited performance the $_y$F.M. becomes,

$$F.M.\Big|_{f_s(max)} = P\left(\frac{2}{\pi}\right)^{3/2} \sin^{\frac{1}{2}}\left(\frac{\pi \Delta Y}{2P}\right) \quad (46)$$

with maximum value at

$$\Delta X = \Delta Y = P \quad \text{(detector noise-limited)} \quad (47)$$

and a M.T.F.$_y$ = 0.41. Thus, for $\Delta X = P = 15 \mu m$ the value of ΔY may range from 15 μm to 20 μm, the exact value determined by the partitioning of noise between the detector and the preamplifier. Figure 23 illustrates a 0-P photodiode array with $\Delta X = P = 15 \mu m$ and $\Delta Y = 20 \mu m$.

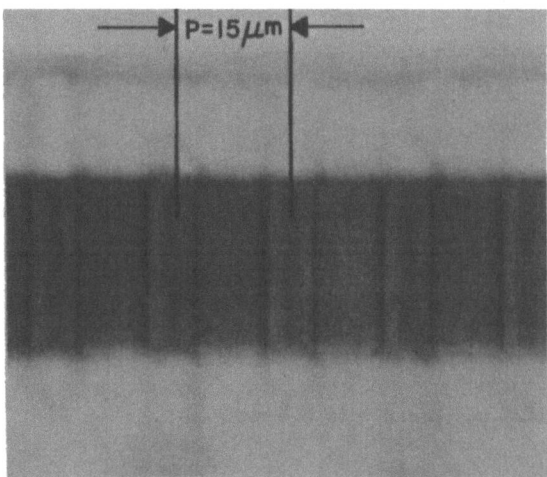

75-0440-PA-28

Figure 23. Linear Photodiode Array $\Delta X = P = 15 \mu m$; $\Delta Y = 20 \mu m$;

4.3 Other M.T.F. Considerations

In general, there are three M.T.F. factors in sensor design:

- geometrical M.T.F.
- diffusion M.T.F.
- transfer inefficiency M.T.F.

We have discussed the geometrical M.T.F. associated with the aperture definition of the sensor. The diffusion M.T.F. is a result of the penetration of the incident radiation into the silicon with subsequent diffusion of the charge carriers from the point of generation to the depletion region of the reverse-biased diode.

Figure 24 illustrates a cross section of a photodiode sensor array with pitch P. The long wavelength radiation penetration into the silicon is described by the incident flux density at a given wavelength λ as

$$N_\lambda = \frac{N_0(\lambda)}{2} \left(1 + \cos 2\pi f_s x\right) \left(1 - r(\lambda)\right) e^{-\alpha(\lambda)y} \tag{48}$$

where N_0 is the peak photo flux, $r(\lambda)$ the reflection coefficient of the silicon and $\alpha(\lambda)$ the absorption coefficient as shown in figure 25. The spatial variation with x as indicated in equation (48) is for the calculation of the M.T.F. The diffusion equation, which describes the movement of photogenerated carriers may be written as,

$$-D\nabla^2 p + p/\tau = G(x,y) = \frac{\partial N_\lambda}{\partial y} \tag{49}$$

where $p(x,y)$ is the excess minority carrier density, D the diffusion coefficient, and τ the bulk recombination lifetime. The current density for the photodiode may be written in the form[7]

$$J_\lambda = \frac{eN_0(\lambda)}{2} \left[n(o) + n(f_s) \cos 2\pi f_s x\right] \tag{50}$$

where $n(o) = n(f_s = o)$ and the M.T.F. is,

$$\text{M.T.F. (diffusion)} = \frac{n(f_s)}{n(o)} \tag{51}$$

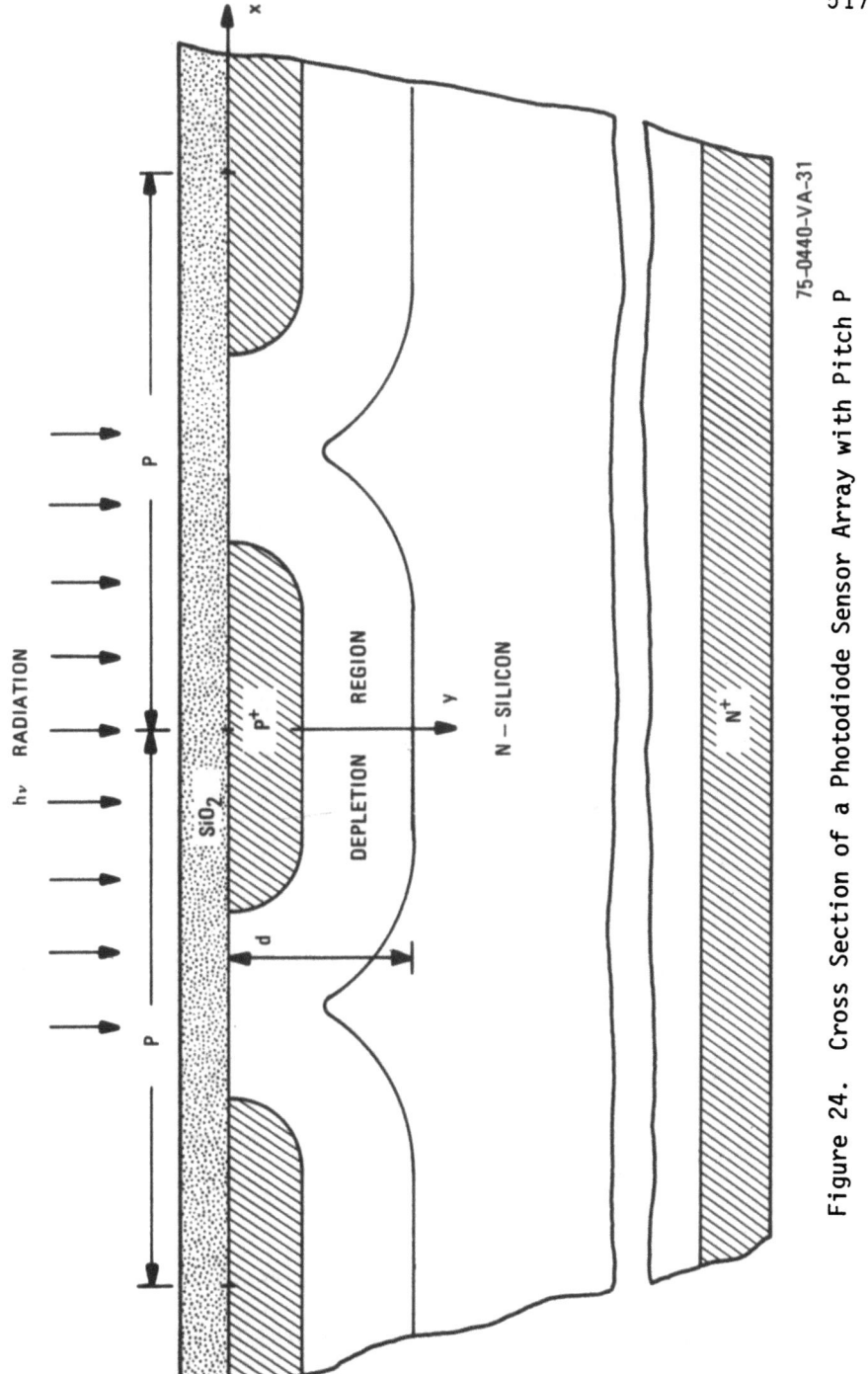

Figure 24. Cross Section of a Photodiode Sensor Array with Pitch P

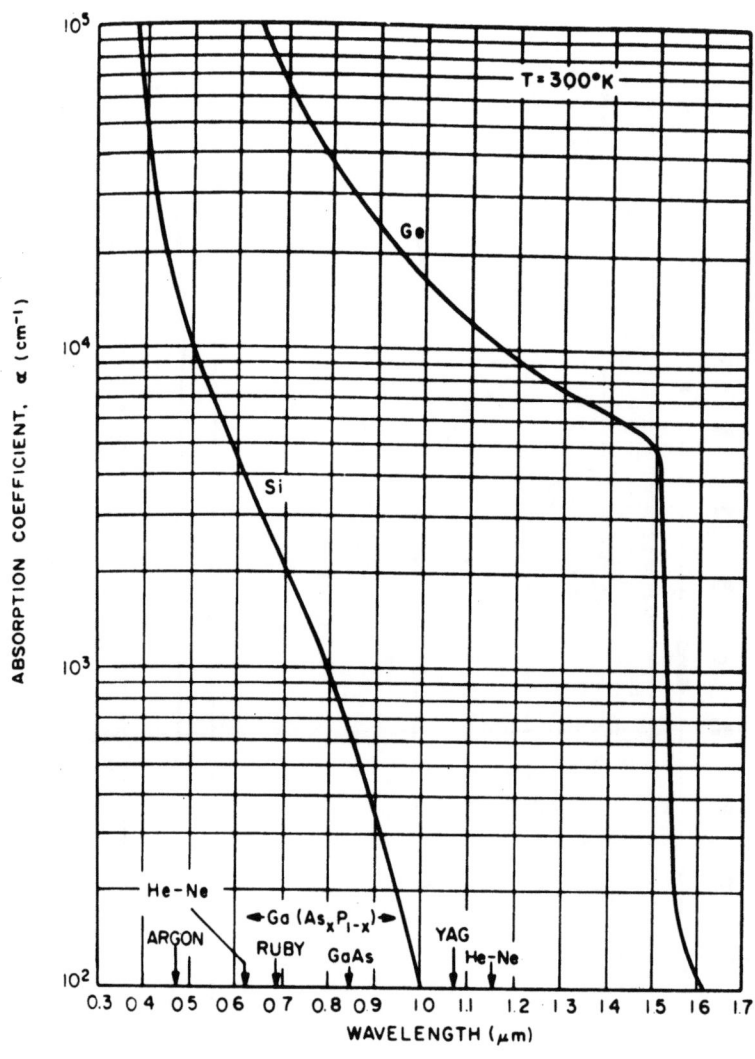

Figure 25. Absorption Coefficient vs Wavelength for Ge and Si at 300°K [After W.C. Dash and R. Newman, "Intrinsic Optical Absorption in Single Crystal Germanium and Silicon at 77°K and 300°K," Phys. Rev., 99, 1151, (1955)]

If we assume the junction depth d is thin, such that negligible trapping of photogenerated charge carriers occurs in the P^+ region, then we may solve the boundary value problem of a depletion region with

$$P = 0 \quad \text{at } y = d \text{ and } y = \infty \tag{52}$$

The hole flux entering the P^+ region $J_\lambda(x)$ may be calculated by evaluating the hole diffusion current density that enters the depletion region and adding to this quantity the photogenerated holes absorbed in the depletion region width d. Essentially we assume all the photogenerated carriers within the depletion region are collected efficiently and no excess carrier buildup is possible since the high electric field sweeps the carriers out of this region. Thus, we need only concern ourselves with the photogenerated carriers outside the depletion region and their wavelength and spatial frequency dependence. The value of M.T.F. becomes,

$$\boxed{\text{M.T.F (diffusion)} = \frac{1 - \dfrac{e^{-\alpha d}}{1+\alpha L}}{1 - \dfrac{e^{-\alpha d}}{1+\alpha L_0}}} \tag{53}$$

where $\dfrac{1}{L^2} = \dfrac{1}{L_0^2} + (2\pi f_s)^2$ and $L_0^2 = D\tau$. Typical values for a photodiode sensor array are $d = 5\ \mu m$, $L_0 = 50\ \mu m$ and $P = 15\ \mu m$, where $f_s(\max) = 1/2P$. Figure 12 illustrates the influence of $\alpha(\lambda)$ on the M.T.F.

The CCD sensors have a M.T.F. degradation due to the finite transfer inefficiency ϵ associated with each transfer. The transferred charge is added to trailing samples which causes a dispersive effect on the output signal. We can derive the M.T.F. due to tranfer inefficiency through the discrete recursive relation[8] for the signal charge in a delay line

$$q_s(X,t) = \epsilon q_s(X,t-1) + (1-\epsilon) q_s(X-1,t-1) \tag{54}$$

where X and t are normalized to the CCD cell pitch P and clock time T_c, respectively. Since equation (54) is a discrete set of signal values in the time domain, we can transform the signal to the Z-domain.

$$Q_s(X,Z) = Z^{-1}\left[\epsilon Q_s(X,Z) + (1-\epsilon) Q_s(X-1,Z)\right] \tag{55}$$

where $Z = e^{ST_c}$ and S is the complex frequency. We can calculate the transfer function after N transfers as,

$$\left[\frac{Q_s(X,Z)}{Q_s(X-1,Z)}\right]^N = \left[\frac{(1-\epsilon)Z^{-1}}{1-\epsilon Z^{-1}}\right]^N \simeq e^{-N\epsilon(1-Z^{-1})} \qquad (56)$$

and the M.T.F. is determined with the substitution $Z = e^{j2\pi f_s/f_c}$ where $f_c = 2f_s(max)$. The M.T.F. becomes

$$\boxed{\text{M.T.F. (transfer inefficiency)} \doteq e^{-N\epsilon\left[1-\cos \pi f_s/f_s(max)\right]}} \qquad (57)$$

which is illustrated in figure 26.

The actual M.T.F. used in the design approach must be the composite of all three M.T.F.'s discussed in this section. Thus, for a CCD imager we must write the M.T.F. as,

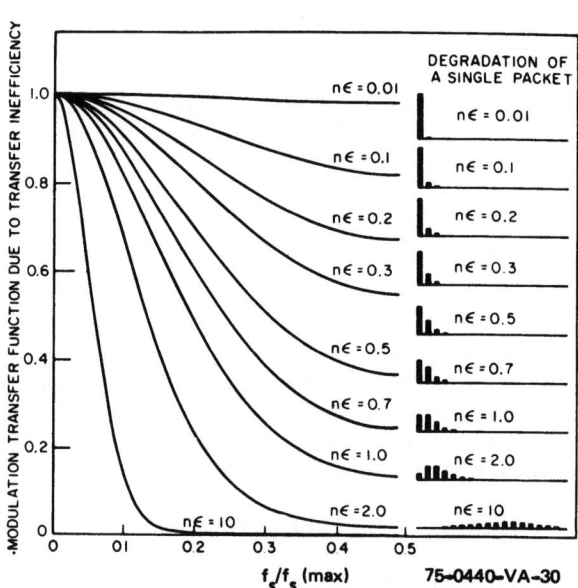

Figure 26. Degradation in M.T.F. Due to Various Values of Nϵ as a Function of Normalized Frequency[10]

$$\text{M.T.F.}_x = e^{-N\epsilon[1-\cos \pi f_s/f_s(\max)]} \cdot \frac{1 - \frac{e^{-\alpha d}}{1+\alpha L}}{1 - \frac{e^{-\alpha d}}{1+\alpha L_0}} \cdot \frac{\sin \frac{\pi f_s}{2Pf_s(\max)}}{\frac{\pi f_s}{2Pf_s(\max)}}$$

(58)

instead of the single geometrical M.T.F.

4.4 Area Array Design

The design of an area array proceeds along the same direction as discussed previously for the line array. The F.M. is formed for the sensor,

$$\text{F.M.} = \Delta X \Delta Y \, (\text{M.T.F.}_x \, \text{M.T.F.}_y)^{1/2}$$

where the S/N $\sim \Delta X \Delta Y$ and preamplifier noise-limited performance is assumed. The area will not have the relative image velocity correction term since the array is electronically scanned in both x and y directions. In TV applications the format of the array is limited to a 4:3 aspect ratio to reconstruct the image for the observer. Furthermore, the video bandwidth restrictions combined with the need to provide a "flicker-free" picture for human viewing (i.e., 30 frames/sec interlaced 2:1), determine the number of picture elements (pixels) in the array. Further restrictions are placed by horizontal and vertical "retrace" intervals which limit the number of usable TV lines/raster height and horizontal resolution.

5. STREAKING CONSIDERATIONS

Streaking in the image may be caused by the following:

- temperature - nonuniform increase in leakage current with temperature fluctuations
- responsivity - nonuniformity in gain from element to element (effective quantum efficiency and amplifier gain)
- spectral   - nonuniformity in spectral response

The effect of streaking is the same as fixed pattern noise and it is highly objectionable in the image. If the sensor output is amplified and passed directly to a display, then solid-state imagers must possess highly uniform characteristics across the array. In some applications, compensation techniques may be used to remove streaking of the image.

REFERENCES

1. M.H. White, D.R. Lampe, F.C. Blaha, and I.A. Mack, "Characterization of Surface Channel CCD Imaging Arrays at Low Light Levels," IEEE Trans. J. Solid-State Circuits, SC-9, 1, 1974.

2. M.H. White and D.R. Lampe, "Noise Considerations in Solid-State Imagers," IEEE Intercon 74, New York City, N.Y., March 1974.

3. Optical Engineering Handbook, ed. J.A. Mouro, General Electric Co., Syracuse, N.Y., 1966.

4. Electro-Optics Handbook, RCA Defense Electronic Products, Burlington, Mass., 1968.

5. J.D. Plummer and J.D. Meindl, "MOS Electronics for a Portable Reading Aid for the Blind," IEEE J. Solid-State Circuits, SC-7, 111, 1972.

6. F.C. Eliot, "Geometric Design of Linear Array Detectors," IEEE Trans. Electron Devices, ED-21, 613, 1974.

7. M.H. Crowell and E.F. Labuda, "The Silicon Diode Array Camera Tube," B.S.T.J. 48, 1481, 1969.

8. W.B. Joyce and W.J. Bertram, "Linearized Dispersion Relation and Green's Function for Discrete Charge Transfer Devices with Incomplete Transfer," B.S.T.J. 50, 1741, 1971.

9. D.F. Barbe, "Imaging Devices Using the Charge-Coupled Concept," Proc. IEEE, 63, 38, 1975.

10. M.F. Tompsett, J. Vac. Sci. Technology, 9, 1166, 1972.

INTERLACING IN SOLID-STATE IMAGE SENSORS

C. H. Séquin

Bell Laboratories
Murray Hill, New Jersey 07974, U. S. A.

ABSTRACT. Solid-state image sensors to be used for commercial broadcast television or for most closed-circuit TV systems have to supply the video information in a 2:1 interlaced timing format. The implementation of such a readout scheme requires special considerations for the case of frame transfer devices. Practical solutions will be reviewed, and the consequences for the modulation transfer function will be discussed.

1. INTERLACED TELEVISION FORMATS

Commercial broadcast television and most closed-circuit TV systems for direct viewing reduce the apparent flicker to the observer by using a 2:1 interlaced format. Two fields are displayed in each complete frame period which typically lasts 1/30 or 1/25 s. In the first field all odd lines are displayed and during the second field all even lines are written in between. Thus, if the eye of the observer is focused on a certain area covering several scan lines on the display tube, it will receive a flash of light every 1/60 or 1/50 s when the electron beam sweeps through this area, or twice as often as in the case of a normal consecutive readout of all the lines in the complete frame. The frequencies of 50 or 60 Hz are above the flicker sensitivity of the human eye, and the display will therefore appear to be of steady brightness.

It is natural to ask what would happen if an even higher degree of interlacing were employed. It turns out that most other schemes are objectionable to the observer. A 3:1 interlacing scheme, for example, tends to capture the focus point of the observer and move it up or down the screen in a 3-phase action, depending on the sequence in which the three fields are presented. This effect has been called "line crawling". Because of such physiological effects, TV systems for direct viewing do not employ more complicated interlacing schemes.

2. INTERLACED READOUT OF AREA IMAGE SENSORS

To be compatible with existing systems, solid-state TV cameras have to supply their video output in this interlaced format. Unless some kind of information reorganization is employed, such as interpolation schemes which reduce the bandwidth required for transmission, the solid-state sensor in the TV camera has to be read out in a manner that produces the video signal directly in the proper format, i.e., all the lines of one field in consecutive order.

2.1 X-Y-Addressed Sensors

The implementation of such a readout sequence normally presents no difficulties in x-y-addressed sensor arrays, which permit, at least in principle, a random readout sequence of individual sensor elements. In practice, however, such devices normally comprise two shift registers, which select subsequent lines or columns in a straightforward sequence. It is then practical to use the vertical scan register to address the lines in pairs and to use an additional multiplex gate to determine whether line A or B in each pair actually gets addressed (Fig. 1). During one field the multiplex switch remains in the same state, while the vertical address register is operated in a normal manner to select every other line. This approach has the advantage that it needs only one scan register stage for every two lines in the device and thus leaves more space for the design of each stage.

2.2 Line-Addressed Charge Transfer Devices

An image sensor may consist of an array of linear charge transfer registers running in the horizontal direction [1]. Each line is addressed and read out individually into a common output diode or into a serial output register by applying proper driving clock pulses to that particular line. This is done under the control of a vertical address register. The same considerations thus apply as in the case of the x-y-addressed sensor. Each stage of the vertical shift register serves a pair of lines, among which the proper one is selected by a suitable field multiplexor.

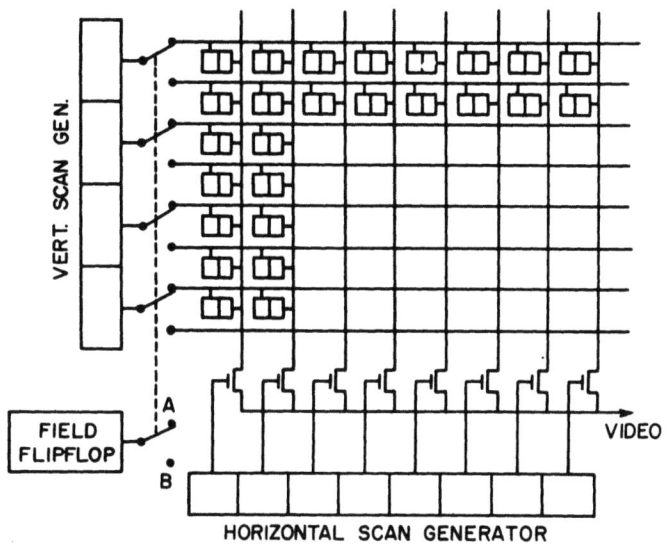

Fig. 1   The use of multiplex switches to achieve an interlaced readout in x-y-addressed structures.

2.3 Interline Transfer Devices

Charge transfer devices with vertical transfer registers represent a separate case since they move the whole pattern of charge in unison and thus have no addressing circuitry to select individual lines. If one field has to be read out while the other one is kept in place, the two sets of charge patterns corresponding to the two fields have to be maneuvered around each other. This is not possible in a normal frame transfer device, and this case will be treated separately in the following section.

In interline transfer devices [2] this maneuver presents no big problem. The set of charge packets which corresponds to the field to be read out is laterally transferred into the vertical storage registers, while the other set remains in the sensor sites and continues integration.

In principle two transfer gates are required so that one field can be selectively transferred to the storage registers while the other one is retained. To avoid the topological problems with the design of such a device, it is more practical to use the transfer phases of the vertical registers themselves to provide this gating function [2]. This can readily be achieved by using a 2-phase approach with ion-implanted barriers to produce the directionality of the transfer electrodes. The same kind

of barrier is also used in each electrode to give lateral
isolation toward the integration sites. In order to transfer the
integrated charge packet from this site into the storage registers, a combined action of the photogate that covers all the
integration sites and of the receiving electrode in the storage
register is required. A charge packet can be transferred over
the implanted barrier only if the photogate is turned low at the
same time that the receiving electrode is high. Thus, in each
field one phase of the transfer register is kept low and the
other one high, thereby permitting only one set of charge packets
to be transferred laterally.

2.4 Frame Transfer Devices

Frame transfer structures [3] do not lead naturally to a
properly interlaced readout of the two charge populations since
the whole charge pattern has to move in unison. Alternate lines
could still be read out in the proper sequence, if an additional
store having half the size of the normal storage area is added at
the bottom of the device (Fig. 2a). This store collects every
second line that emerges from the storage area during the vertical line steps and stores them for an additional field time, while
the first set of lines is read out directly. The stored lines are
then read out via a second serial register in the subsequent
field. The active area of such a device would be more than 5/2
times the size of the effective imaging area, and the region of

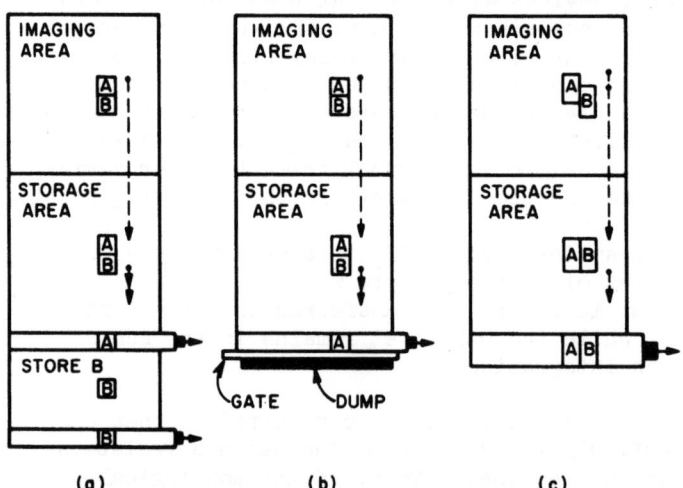

Fig. 2 Three ways of achieving a 2:1 interlaced readout from a
frame transfer imaging structure: (a) use of an extra
store, (b) discarding every other line, and (c) adding
adjacent lines. (Ref. 4)

the first horizontal readout register would require a highly complex electrode and interconnection arrangement.

In a simpler approach, every second line could be discarded by dumping it into a drain diode running along the serial register (Fig. 2b). In most applications, however, the waste of half the signal and the corresponding loss in signal-to-noise ratio and sensitivity are not admissible.

Rather than discarding every second line, it is preferable to combine it electrically with one of its neighbors. Interlacing is then achieved by using different combinations of line pairs for the two fields (Fig. 2c). Such a mixing scheme results in a somewhat reduced vertical response for the highest spatial frequencies owing to the overlap of the combined "lines" of subsequent fields. From the device point of view such a mixing scheme is preferable since it requires a less complicated structure and does not result in a loss in sensitivity. Furthermore, the mixing of neighboring lines can be performed in the imaging area during integration itself, and thus the number of lines in the storage area can be reduced to the number of lines in a single field. This results in a considerable reduction in the number of necessary features in the device.

3. ELECTRONIC ALTERATION OF THE INTEGRATING WELL

3.1 Frame Transfer Devices

In frame transfer devices the position of the integrating potential wells depends on which electrodes are held at a high potential during the integration period (Fig. 3). The location of the effective sensor elements can thus be altered from field to field by using a different set of electrode potentials for both fields, and it can then be matched to the interlaced location of the scan lines in the display [4].

In 2-phase structures this is achieved quite naturally by forming the integrating potential wells alternately under the Phase 1 or Phase 2 electrodes. In order to obtain the proper offset in 3-phase structures, one field is integrated under a single set of electrodes (Fig. 3a) and the other field jointly under the other two electrode sets (Fig. 3b). In this manner the center of the charge collecting well is shifted by exactly half a cell dimension as is required for best results. Normally, there is no difference in the sensitivity of the two fields, because the same number of minority carriers are collected in both cases.

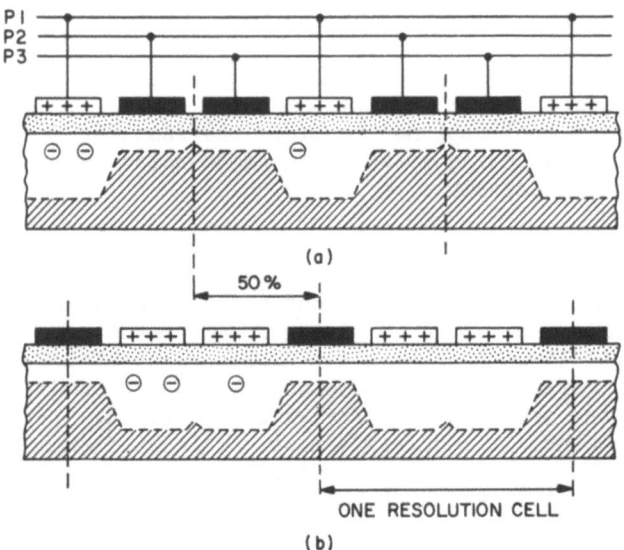

Fig. 3 Vertical cross sections through imaging area of a 3-phase frame transfer device and potential distribution during the integration of two 2:1 interlaced fields: (a) field A integrated under a single set of electrodes and (b) field B integrated jointly under two electrode sets. (Ref. 7)

3.2 Experimental Results

Different readout and interlacing modes have been compared on a 3-phase frame transfer device, fabricated with a single-level metal. Close-up photographs of the resolution wedges reproduced on the display are shown in Fig. 4. The expression A/B under each display denotes symbolically under which electrodes the two fields have been integrated.

1/N: Sixty-four lines are integrated underneath the Phase 1 electrodes and are displayed by 64 scan lines in a noninterlaced fashion. The picture quality is clearly impaired by the low number of scan lines.

1/1: The lines are always integrated underneath the Phase 1 electrodes but are displayed as two interlaced fields containing the same information. All the features in the display show up in pairs, and since they are displayed alternately in the two

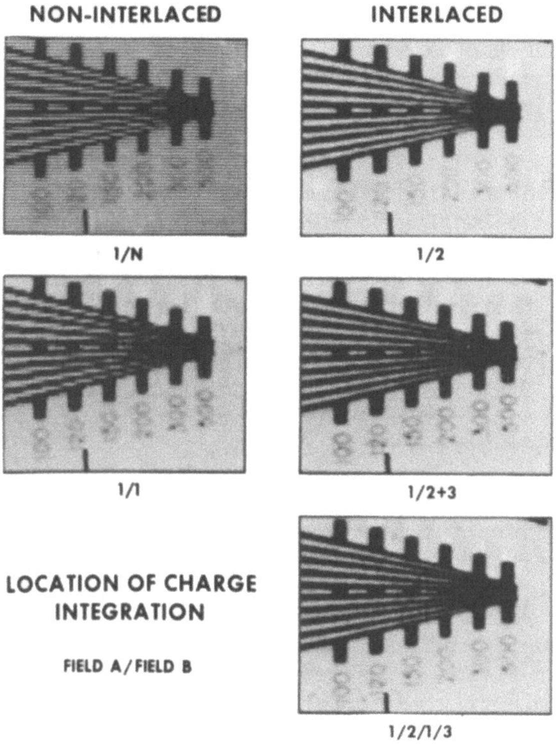

Fig. 4   Resolution wedges as reproduced by various integration schemes collecting the charge for subsequent fields in succession underneath different electrodes A/B/... as indicated below each picture. (Ref. 4)

different fields, the whole picture seems to jump up and down at a rate of 30 Hz.

1/2: The integration sites are interlaced and are located alternately underneath the Phase 1 and Phase 2 electrodes for fields A and B, respectively. The vertical resolution of the system is improved over the earlier schemes. The fact that the integration sites are slightly paired results in an unevenness of the edges of the slanted lines. It also produces some jitter in the display since the wedge pattern seems to jump up and down by about one-third of the scan line separation.

1/2+3: The integration sites are truly 2:1 interlaced. The charge is accumulated alternately by the electrodes of Phase 1 and of Phases 2+3 jointly for fields A and B, respectively. The vertical resolution is further enhanced over the 1/2 scheme. The unevenness of the edges of the bars and the jitter of the display have disappeared. No degrading effects from the different shape of the depletion region during integration of the two fields could be observed.

1/2/1/3: A double interlacing scheme is employed. In succession the electrodes belonging to Phases 1, 2, 1, 3 serve as integration sites. Thus, as a time average, the center of charge collection for field B falls between the electrodes of Phases 2 and 3. Indeed, the resulting photograph looks the same as for the previously treated case (1/2+3). However, if observed in real time, the display shows an objectionable flicker. This stems from the fact that fields B' and B", integrated by the electrodes of Phases 2 and 3, respectively, contain different information, which alternates at a rate of 15 Hz.

Out of the five modes discussed, the scheme 1/2+3 is the most preferable. It provides the most accurate reproduction without any objectionable flicker of the display.

The displayed white point defects, which are attributable to localized structural nonuniformities in the silicon crystal, will also vary between the two fields. The mechanism that generates the excessive amount of minority carriers is activated only if the defect lies in the depleted region underneath an electrode that is biased to a certain minimum potential. Thus, for fields integrated underneath different electrodes, a different pattern of white defects is observed. In a truly interlaced mode two sets of white defects alternate at a rate of 30 Hz. Generally, the mode 1/2+3 will show more white defects than the mode 1/2 since during a full frame time the defects underneath all three sets of electrodes will become activated. This effect can be compensated to a certain extent by integrating field B with a lower potential on the electrodes of Phases 2 and 3.

3.3 Effect on the Modulation Transfer Function

The above approach to interlacing, in which one vertical transfer cell is used to generate the video signals for two lines in the display, is one of the major advantages of the frame transfer devices over the line-addressed devices which need a separate channel for each display line. The interline transfer structure can also make use of vertical transfer cells being shared between two fields, but there are significant differences. Firstly, the actual sensor areas of the two fields do not overlap and thus the modulation transfer function in the vertical direction has a slower roll-off. Secondly, the total integration time for each individual sensor is a full frame time. This results in a somewhat slower response of the sensor to fast moving objects and may result in a break-up of the image when the camera is panning too swiftly. For applications in which this could be objectionable, both sets of integration sites could be read out in each field by using two lateral transfers during two subsequent half-cycles of the clocks driving the vertical registers. By suitably pairing the elements in both fields, an interlaced readout with the same vertical MTF as that of the frame transfer device can be obtained.

In an image sensor with two distinct sets of integration sites corresponding to the two interlaced fields, the overall modulation transfer function (MTF) of the sensor is not affected by the sequence in which the sensor information is read out. However, in a frame transfer device the display looks quite different, as illustrated by the experimental results discussed in the previous section, when the integration sites are either kept stationary in the same place or when they are electronically interlaced. But the differences in the resulting vertical modulation transfer functions are entirely due to the different numbers and positions of the samples taken in the two modes. Although the MTF produced by one field of integration sites is not altered by the interlaced readout format, proper interlacing which effectively doubles the number of samples reduces drastically the aliasing problems caused by the discrete nature of the sensor sites and the scan lines in the display. The Nyquist theorem, applicable to all sampled data systems, requires that the information be properly band limited to half the sampling rate before the sampling process is executed in order not to cause aliasing. An ideal sharp cutoff indicated by the dashed line in Fig. 5 can normally not be achieved in an image sensor. The roll-off of the MTF, which is caused by the optical system or by the processes in the image sensor which occur before the generated minority carriers reach the integration sites, does not normally meet this requirement. The total MTF that acts upon the incoming optical information (curve AB) is the product of the MTF of the lens (curve A) and the device itself (curve B). This curve will then be reflected around the harmonics of the sampling frequency $f_s = 1/L$, where L is the spatial period of the sensor cells in the vertical direction. The dominant reflection around the fundamental of the sampling frequency is shown as curve $(AB)^*$ in Fig. 5. In the 2:1 interlaced format this reflection occurs at a frequency which is twice as high, and the corresponding curve is labeled $(AB)^{**}$. Thus the aliasing components appearing in the baseband are much lower and will produce fewer Moire effects. In an arbitrary manner we can define the useful range of spatial frequencies in the vertical direction to be limited by the point where the reflected part of the spectrum equals 50% of the magnitude of the fundamental branch. For the example shown in Fig. 5,

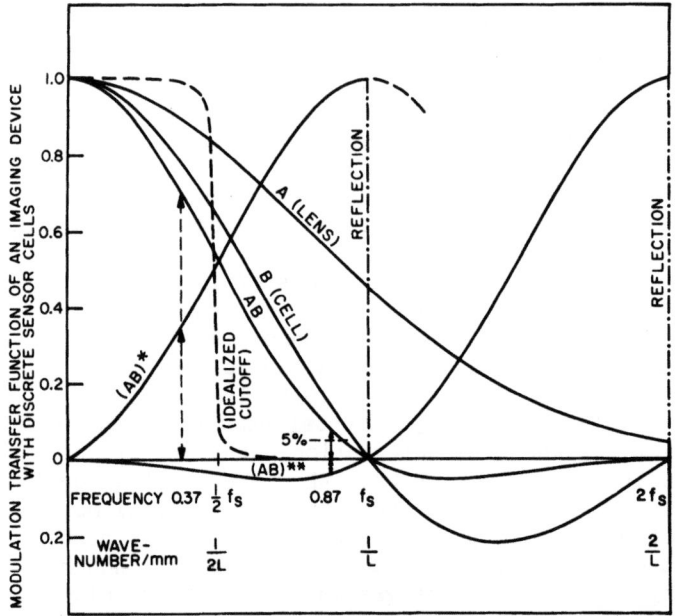

Fig. 5 Effect of the sampling process due to the discrete nature of the resolution cells on the vertical modulation transfer function. (Ref. 4)

this occurs at a frequency of $0.37f_s$ for the noninterlaced case and at $0.87f_s$ for the 2:1 interlaced case. Thus the useful frequency range has been more than doubled. For most applications there can thus be no doubt that a given frame transfer device should be used in the interlaced format.

The trade-offs are more complicated when the case of the 2:1 interlaced frame transfer device is compared to the interline transfer device with two discrete sets of sensors which are contiguous in the vertical direction. The ensuing trade-offs between resolution and aliasing will be treated in a separate lecture.

4. INTERLACING IN LINEAR ARRAYS

Originally some linear charge-coupled image sensors were built which integrated the incident light directly in the transfer channel [3,5]. For long devices with many elements this approach is impractical because the optical smearing during readout would become unacceptable. All newer devices employ therefore one or two readout registers with separate, laterally attached integration sites [3]. These sites are typically outlined by a channel stopping diffusion or by a thick field oxide. Their position can thus not be altered electronically, and the overall modulation transfer function and resolution, measured after all elements have been read and the video signal has been reassembled, will not depend on the sequence in which the individual charge packets were read out and transmitted.

A possible benefit from an interlaced readout of the sensor sites may result from the fact that the readout register needs fewer transfer elements than when all elements are read out in the same scan. As described for the case of the interline transfer device, two sets of sensors might be fitted to the two phases of a 2-phase readout register, and alternate gating during the lateral transfer would then be achieved by proper biasing of the receiving transfer electrodes. However, the same sensor density with the same number of elements in the readout registers can also be obtained if a bilinear approach is used in which interdigitated integration sites are attached to two separate readout registers. Thus, the use of an interlaced readout scheme seems only justifiable in order to achieve increased resolution density if at least four sensor elements can be fitted into the linear period of a readout cell and if the interlaced readout is combined with a bilinear organization.

Another approach, which increases the resolution of a linear sensor system but which keeps the solid-state sensor simple, uses a mechanical wobble of the sensor or of the optical system, so that the same sensor cells effectively detect different interlaced locations in subsequent lines. This scheme has been used to improve the resolution in spectrum sensors in astronomical applications [6]. Since the data of such systems are normally not intended for direct observation, interlacing formats of a high order are possible and are actually used to improve resolution and to avoid aliasing problems.

REFERENCES

[1] Weimer, P. K., Kovac, M. G., Shallcross, F. V., and Pike, W. S., "Self-Scanned Image Sensors Based on Charge Transfer by the Bucket Brigade Method," IEEE Trans. on Electron Devices ED-18, 996-1003 (1971).

[2] Amelio, G. F., "Physics and Applications of Charge Coupled Devices," IEEE INTERCON, New York, Digest, Vol. 6, paper 1/3 (1973).

[3] Tompsett, M. F., Amelio, G. F., Bertram, W. J., Buckley, R. R., McNamara, W. J., Mikkelsen, J. C., and Sealer, D. A., "Charge Coupled Imaging Devices: Experimental Results," IEEE Trans. on Electron Devices ED-18, 992-996 (1971).

[4] Séquin, C. H., "Interlacing in Charge Coupled Imaging Devices," IEEE Trans. on Electron Devices ED-20, 535-541 (1973).

[5] Séquin, C. H., "Experimental Investigation of a Linear 500-Element 3-Phase Charge-Coupled Device," Bell Syst. Tech. Jour. 53, 581-610 (1974).

[6] Beaver, E. A., Burbridge, E. M., McIlwain, C. E., Epps, H. W., and Strittmatter, P. S., "Digicon Spectrophotometry of the Quasistellar Object PHL 957," Astrophysical Jour. 178, 95-103 (1972).

[7] Séquin, C. H., and Tompsett, M. F., "Charge Transfer Devices," Suppl. #8 to "Advances in Electronics and Electron Physics," Academic Press, Inc., N. Y. (1975).

AMPLIFIER AND AMPLIFIER NOISE CONSIDERATIONS

James A. Hall
Westinghouse Electric Corporation
Advanced Technology Laboratory
Baltimore, Maryland U.S.A.

ABSTRACT. Principal noise sources in analog preamplifiers for conventional television cameras are the load resistor Johnson noise, and the first stage transistor noise. When optimized, these contributions are equal and $i_n = \sqrt{\frac{8}{3} \pi^2 C_S^2 \epsilon_n^2 \Delta f_v^3}$. Typical values are C_S = 20 pF, ϵ_n = 1.5 x 10^{-9} V/$Hz^{1/2}$, Δf = 4.5 x 10^6 Hz, and i_n = 1.47 x 10^{-9} amp or q_n = 2040 electrons per Nyquist sample. In a CCD sensor with simple sampled and reset amplifier input, C_S is two orders of magnitude smaller and Johnson noise in the reset switch dominates at 280 electrons per Nyquist sample for C_S = 0.25 pF. This noise can be removed by correlated double sampling, leaving amplifier noise

$q_n = \sqrt{\epsilon_n^2 \Delta f_n C_S^2}$ = 60 electrons per Nyquist sample for

C_S = 0.25 pF and Δf_n = 1 MHz. For lower noise, especially at TV data rates, one should consider the distributed floating gate amplifier described in the following chapter.

1. INTRODUCTION - AMPLIFIERS FOR CONVENTIONAL TELEVISION

All television camera tubes in use today have signal output electrodes which can be electrically represented as a current generator shunted by a capacitance to ground. Signal current levels for vidicons, silicon diode array target tubes, SEC camera tubes and the like range from a few nanoamperes or less to about 1000 nA in a video frequency band extending from low frequencies to several MHz. The noise produced by the camera tube load circuit and by the first preamplifier stages is of the same order as the signal at low image irradiance values, hence the amplifier noise characteristic is a principal factor limiting television camera performance at low light levels, and TV amplifier design has been extensively studied, at least for entertainment television applications.

1.1 Characteristics of The Signal - The Keyed Clamp

As shown in figure 1, the scanning beam of a camera tube is blanked during the horizontal and vertical retrace intervals so that no beam electrons can land on the target to erase the stored charge image. Thus the signal current from the camera tube is as shown in figure 2, varying during each active scan line time as a function of the exposure distribution in the previous frame time, but falling to zero during the beam blanking interval. The time scale in figure 2 is that standardized for U.S. broadcast television, and must be modified for European or other television standards. To reproduce a television picture fully, one could use a dc-coupled amplifier whose passband extended below the low frame frequency of the television system, but there are important advantages instead in using an ac-coupled amplifier and in clamping the output signal to a reference voltage before beginning each scanning line. As shown in figure 2, a clamp placed after one or several high gain low noise ac-coupled amplifier stages can be keyed on during the blanking interval to establish the voltage level corresponding to zero signal current so that synchronizing and display blanking signals can be added independently of picture content. As shown, the inserted display blanking level is usually "blacker" than the zero signal current level to ensure that retrace lines will not be visible in remote display devices. The difference between display blanking and the zero signal current level is known as the pedestal.

Use of a line frequency zero level keyed clamp has another advantage, since the essential video signal then has no frequencies below f_h, 15,750 Hz for U.S. broadcast standards. Thus excess low frequency noise from input preamplifier stages and power supply or other drift phenomena with frequencies below f_h are effectively removed.

The maximum required video frequency is selected to reproduce the finest detail required in the system in the direction parallel to the scanning lines. In the television industry, as shown in figure 3, it is standard practice to express the spatial frequency of image structure in terms of television lines or half cycles of video information per pattern height, even though the spatial frequency is measured along the scan lines, i.e., in the

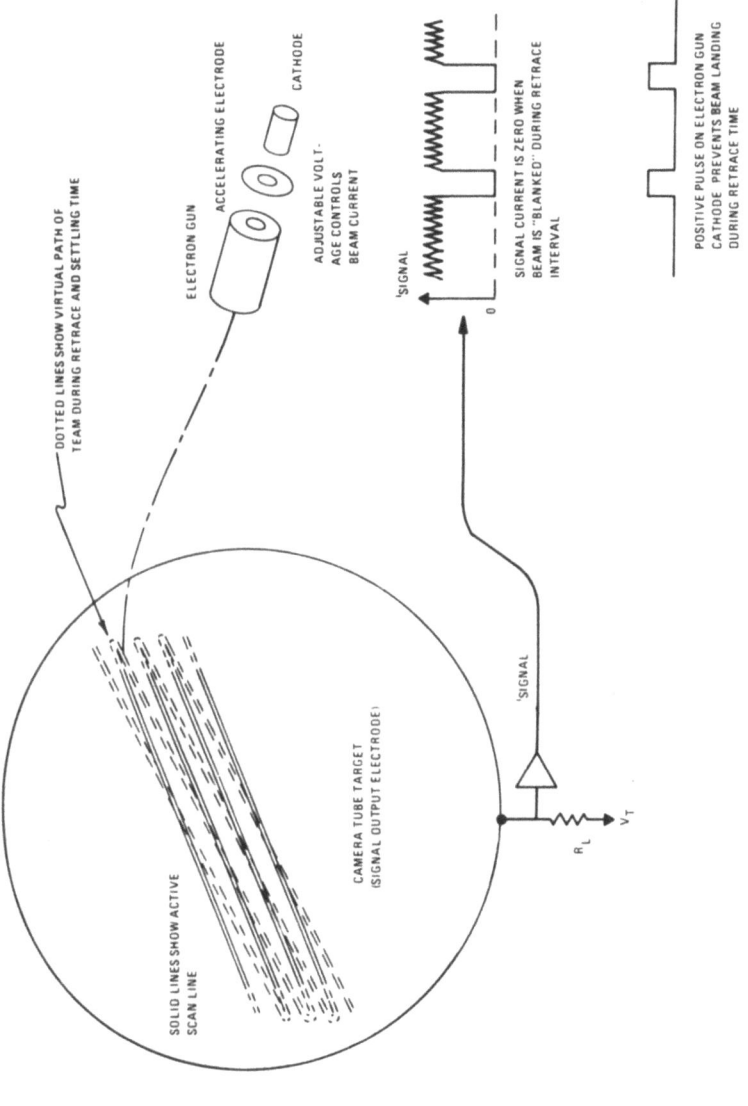

Figure 1. Electron Beam is Blanked During Retrace and Reads Stored Charge Only During Active Scan Times. Zero Signal Current During Retrace Provides Convenient Reference Level.

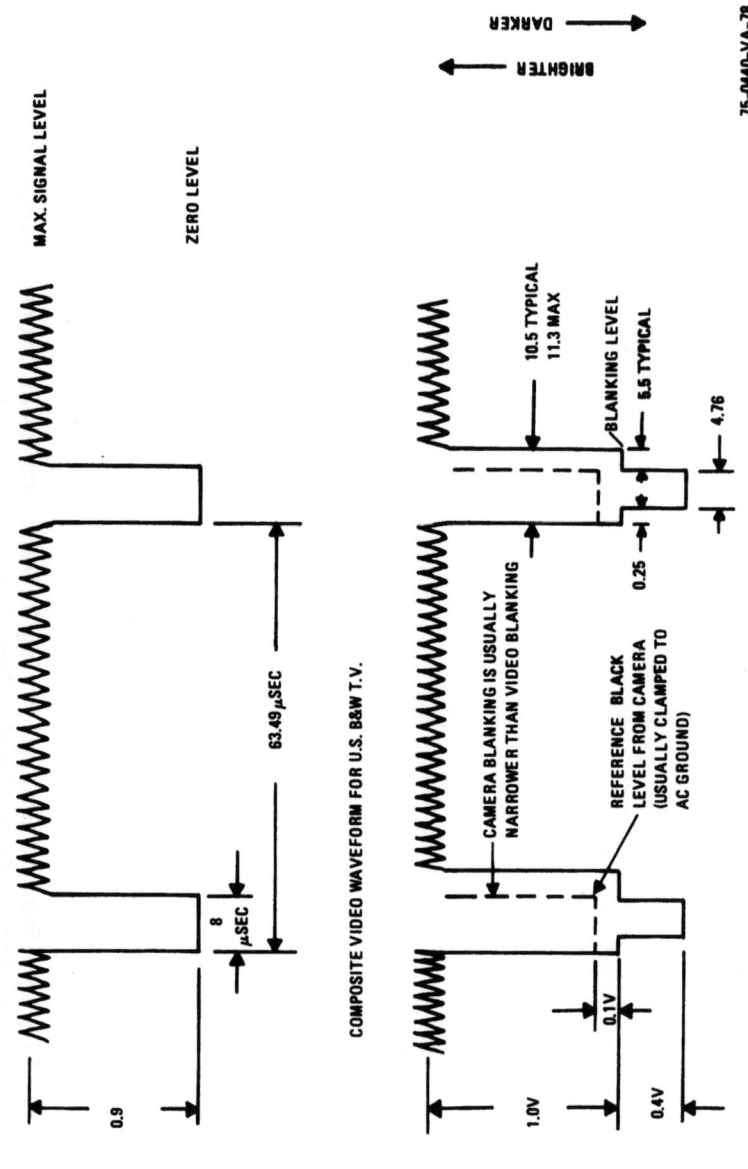

Figure 2. Television Signal Includes Added Blanking and Sync Pulses Referenced to Black Zero Current Level of the Video Signal from the Sensor.

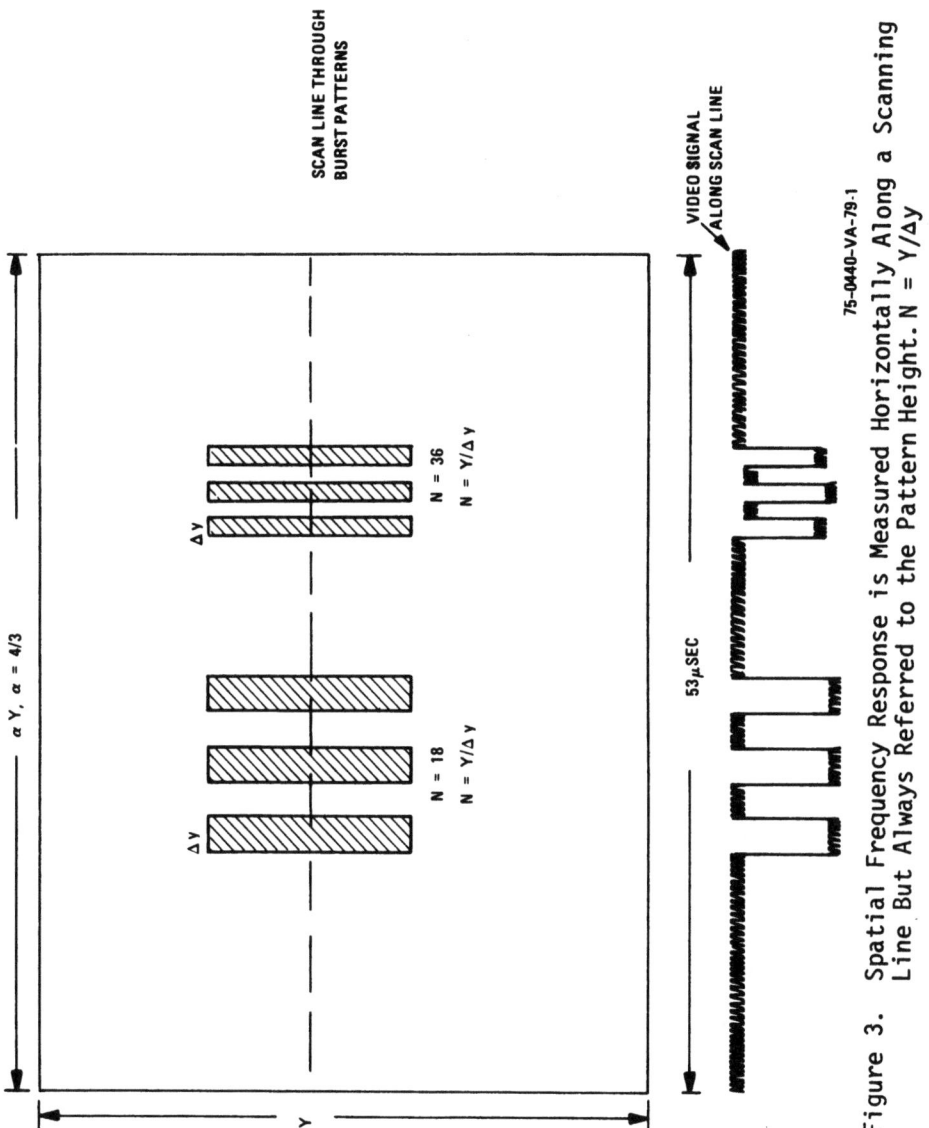

Figure 3. Spatial Frequency Response is Measured Horizontally Along a Scanning Line But Always Referred to the Pattern Height. $N = Y/\Delta y$

direction of picture width. For the usual 4x3 aspect ratio scanning pattern of commercial television, the upper required video frequency is

$$f_v = \frac{4}{3} \times \frac{N_H}{2} \times \frac{1}{t_h} \text{ Hz} = 1.26 \times 10^4 \, N_v \text{ Hz} \qquad (1)$$

t_h is the active line scan time, 53×10^{-6} sec for U.S. broadcast standards, N_H is the maximum required horizontal image spatial frequency expressed in half cycles per pattern height, and 4/3 is the aspect ratio. For a "350 TV line" system, f_v = 4.41 MHz, approximately the video signal bandwidth allocated for U.S. broadcast television.

For all camera tubes in use today target or mesh structures have finer texture than the finest image detail which the system can produce using commercial scanning standards. Thus the generation of the video signal along each scanning line is strictly an analog process. No sampling is involved so there is no first order moire problem, no preferred phase for a test pattern in the horizontal direction, and incidentally no structure defined picture element size for a television camera tube. Resolving power, or MTF, along the lines is simply a function of the aberrations in electron optical focus, the diameter of the scanning electron beam, and the details of the signal reading and erasing process at the camera tube target.

In the vertical direction a sampling structure is provided by the scanning lines of the raster, rather than by any physical structure. Even if the MTF were high at the raster line spatial frequency, the sampling would limit the highest meaningful image spatial frequency to one half cycle or "TV line" per raster line, the Nyquist frequency in the vertical direction. But response is high only when the image detail is in phase with the raster lines. Averaging over all possible phases, the maximum effective vertical spatial frequency in the image N_v is $N_A/\sqrt{2}$, where N_A is the number of active scanning lines in the raster. For U.S. broadcast standards N_A = 490 and N_v = 346 half cycles or TV lines per raster height. Thus $N_v = N_H$, the U.S. broadcast standards are intended to give balanced resolution, with the number of resolvable half cycles in the horizontal direction equal 4/3 the number in the vertical direction, corresponding to the 4:3 aspect ratio.

In this discussion it is important to differentiate between TV lines (per pattern height), a measure of spatial frequency in the image, and the scanning or raster lines one sees in the display. They are not the same. Note also that resolving power for a television system is best expressed in terms of TV lines or half cycles per scanned pattern height rather than in cycles per millimeter, since the former figure will apply equally throughout the system, from the small image at the sensor input to the various larger images reproduced on display devices, and because resolving power is often system limited rather than sensor limited.

1.2 Analog Preamplifiers

The design philosophy for television preamplifiers was set in the late 1930's on psycho-physical grounds by the same people who worked out the television transmission standards. In a reproduced picture, low frequency noise produces moving streaks along scanning lines, medium frequency noise produces "snow particles," and high frequency noise produces fine dancing scintillations. The human observer integrates information in space and time and seems in a sense to set his own bandwidth to match the scale of the detail being viewed. Thus fine textured high frequency noise is usually ignored when one views a television image of moderate quality, and medium frequency snow can be tolerated by moving away from the display and visualizing only gross image features, but noise becomes subjectively more objectionable as the frequency decreases. Thus if a choice exists, for a man-viewed system, a television preamplifier should improve the low frequency signal to noise ratio as far as possible, even at the cost of degrading the high frequency signal to noise ratio.

There are three principal camera tube load circuit and preamplifier designs in use today, shown in figures 4, 5, and 6, with a number of minor variations. All are similar in many respects, and the earliest circuit shown in figure 4 is a good starting point. The current signal from the camera tube appears as a voltage across the load impedance formed by R_L in parallel with C_S. To maximize the signal at low frequency, R_L is made large, usually from 10^4 to 10^5 ohms for broadcast scanning rates and up to 10^8 ohms in slow scan cameras. C_S is made up, as shown in figure 4, of the camera tube signal electrode capacitance to its surroundings, C_T, ranging from 10 to 30 pF depending on tube size and structural details, the capacitance of interconnections, C_W, which can be kept to 2 or 3 pF by careful design, and the input capacitance of the first amplifier stage, C_G, shown here as a field effect transistor.

Thus the signal voltage at the preamplifier input is

$$v_s = \frac{R_L}{\left(1+\omega^2 R_L^2 C_S^2\right)^{1/2}} i_s \qquad (2)$$

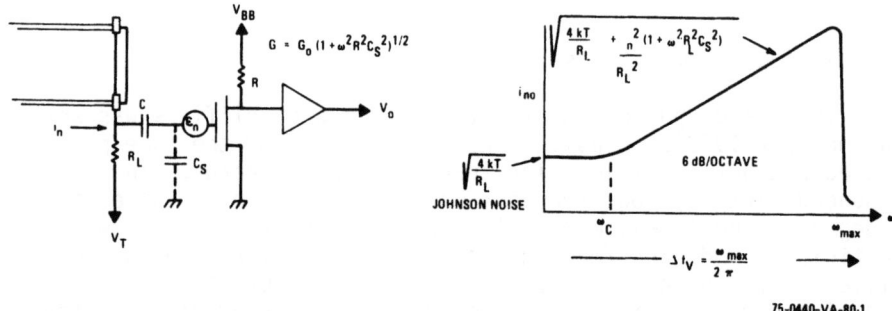

Figure 4. Classic TV Camera Analog Preamplifier Uses High Peaking in Later Stages to Provide Flat Signal Frequency Response. First Stage Transistor Noise is High Peaked But Has Low Visibility to Human Observer.

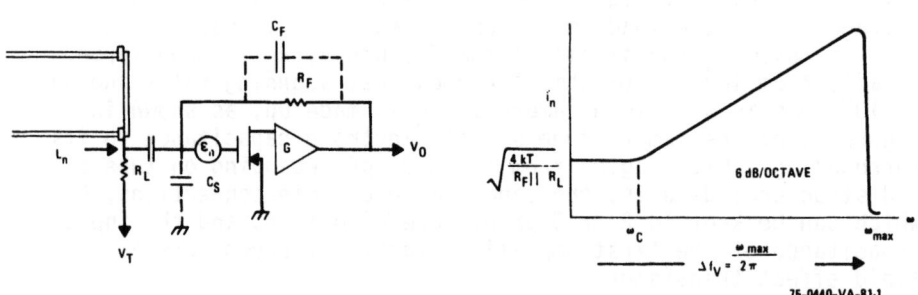

Figure 5. Current Mode Operational Amplifier Provides Flat Frequency Response for Signal by Effectively Reducing R_L to $(R_F/1+G) \parallel R_L$, But First Stage Transistor Noise is High Peaked by Falling Impedance From Signal Electrode to Ground Caused by C_S.

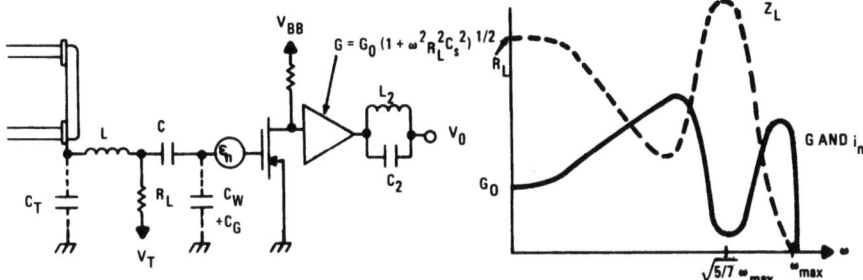

Figure 6. Series Peaking Increases Signal Power at Preamplifier Input for ω Near ω_{max}. Notch Filter Flattens Signal Frequency Response and Reduces Integrated Noise Current by Up to 6 dB.

To provide an overall video signal gain which is independent of frequency through the selected video passband, a later stage of the preamplifier is designed for a rising gain vs frequency characteristic so that amplifier gain is

$$G = G_0 \sqrt{1+\omega^2 R_L^2 C_S^2} \qquad (3)$$

The noise sources in this circuit are the Johnson Nyquist noise of R_L and the noise of the first and following stages of the preamplifier. As shown, for a field effect transistor input stage the transistor noise can be represented by a voltage generator internal to the transistor in series with the gate. Factors determining this noise voltage are discussed below.

Since the camera tube acts like a current source, it is convenient to describe the noise contributions as equivalent noise currents at the amplifier input terminal. The noise spectral density can then be written as

$$\underset{\text{Johnson Noise of } R_L}{\underbrace{\frac{d(I_n^2)}{df} = \frac{4kT}{R_L}}} + \underset{\text{Preamplifier Noise}}{\underbrace{\frac{\varepsilon_n^2}{R_L^2}\left(1+4\pi^2 f^2 R_L^2 C_S^2\right)}} \tag{4}$$

since the internal preamplifier noise is high frequency peaked by the rising gain vs frequency characteristic. The overall noise is found by integrating this expression over the video passband. The result is written as the equivalent rms input noise current

$$I_n = \left[\left(\frac{4kT}{R_L} + \frac{\varepsilon_n^2}{R_L^2}\right)\left(f_{max}-f_{min}\right) + \frac{4}{3}\pi^2 C_S^2 \varepsilon_n^2 \left(f_{max}^3 - f_{min}^3\right)\right]^{1/2} \tag{5}$$

For most cases of interest $f_{max} \gg f_{min}$ and the f_{min} terms may be neglected. For U.S. broadcast standards, for example, $f_{max} = 4.5 \times 10^6$ Hz and $f_{min} = 1.57 \times 10^4$, and neglecting f_{min} makes the noise current estimate high by less than 1/3 percent. Thus

$$I_n = \sqrt{\left(\frac{4kT}{R_L} + \frac{\varepsilon_n^2}{R_L^2}\right)\Delta f_v + \frac{4}{3}\pi^2 C_S^2 \varepsilon_n^2 \Delta f_v^3} \tag{6}$$

where Δf_v is the upper frequency limit of the video passband at the 3 dB point. This form of the noise expression shows why R_L should be made large to minimize the overall rms noise current contribution. The phycho-physical observations mentioned above are an additional and independent reason for maximizing R_L.

In attempting to minimize I_n, the designer should make C_S as small as possible by minimizing interconnect capacity, by careful choice of input transistor, and by choosing a camera tube with a small output capacitance if possible. He should keep the video frequency as low as possible, perhaps by using slower scan rates if systems requirements allow this, and should select an input transistor for low input noise, remembering however that low noise field effect transistors often have large gate capacitances which would tend to raise C_S. This question is discussed below. In addition he will make R_L as large as feasible. The limits to making R_L larger are first, that the camera tube signal electrode current of up to 1 or 2×10^{-6} amperes for broadcast scan rates must pass through R_L with at most a volt or so drop, to avoid changing the tube operating point with light level, since the signal electrode impedance is in fact finite. This limits R_L to about 5 x 10^5 Ω in broadcast type equipment. Second, the following preamplifier stages must produce a gain which rises at 6 dB per octave above the "corner frequency," $f_c = \frac{1}{2\pi R_L C_S}$. For a given C_S, a large R_L means a low corner frequency, hence a large high frequency gain is required, without objectionable phase shifts in the amplifier. It is customary to make R_L large enough so that the first Johnson noise term equals the third amplifier noise term. The second term is then usually negligibly small, and the noise equation becomes

$$I_n = \sqrt{2\left(\frac{4}{3}\pi^2 C_S^2 \epsilon_n^2 \Delta f_v^3\right)} \tag{7}$$

Once Δf_v is set by system considerations and the chosen sensor has established C_T, the designer's task is to choose an input transistor and circuit to minimize both C_G and ϵ_n. These requirements are mutually contradictory, however.

1.3 First Stage Transistor Noise

At this time discrete component junction field effect transistors selected for television preamplifier use could include the 2N4391, 2N4392, and 2N4393 n-channel JFET's intended for switching applications, as well as low noise amplifier JFET's like the 2N4139 and the microwave transistor 2N4417. A problem with most "low noise" JFET's is their high gate capacitances, which vary from 4 to 20 pF at V_{GS} = 0 for the units listed. The following table includes data measured at Westinghouse on the 2N4139, 2N4392, and 2N4417.

	ϵ_n at Midband nV/$\sqrt{Hz}$	C_G at V_{GS} = 0 pF	g_m eff
2N4392	1.2	18	10 mmho
2N4139	1.8	20	5 mmho
2N4417	1.7	4	5 mmho

This table shows the 2N4417 has exceptionally low capacitance and low noise, probably because this unit has little stray back gate to substrate capacitance.

These values may be compared with noise models for field effect transistors used by Motchenbacher and Fitchen[1], by Cobbold[2], or by Leventhal[3] which agree that the input noise should be related to the channel resistance, and hence to the transconductance as

$$\epsilon_n = \sqrt{\frac{4kT}{g_m} A} \qquad (8)$$

where A is a function of operating conditions and transistor design and processing which can in theory be as low as 2/3. This expression applies for mid band noise for both JFET's and MOSFET's, although the value of A may be slightly larger for the latter.

The transconductance for MOSFET's in the common source mode under saturation conditions is related to input capacitance as

$$g_m' = \frac{C_o^* \mu_n^*}{L^2} (V_{GS} - V_T) \text{ for MOSFET's} \qquad (9)$$

where:

C_o^* = the active gate to channel capacitance

μ_n^* = effective carrier mobility at the semiconductor surface

L = channel length between source and drain diffusions

Note that C_o^* excludes gate to source and gate to drain overlap capacitances and gate to substrate capacitance outside the channel area. These parasitic capacitances are lumped with wiring capacitance and must be minimized for good low noise performance.

The transconductance for a JFET with both gates active is more complex but similar:

$$g_m' = \frac{K \mu_n C_{A,0}}{(V_{PO}+\psi)L^2} \left(1 - \left(\frac{V_{GS}+\psi}{V_{PO}+\psi}\right)^{1/2}\right) \qquad (10)$$

where K is a constant depending on doping profiles and levels, and $C_{A,0}$ is the active capacitance from gates to channel at $V_{GS} = 0$. Any capacitance from gate to substrate, capacitances to drain or source outside the channel region, or lead capacitance are parasitics to be lumped with wiring capacitance and must be minimized.

Thus equations 8, 9, and 10 agree that

$$\varepsilon_n = \frac{\text{const}}{\sqrt{g_m}} = \frac{\text{const}}{\sqrt{C_{A,0}}} \qquad (11)$$

for a given channel length and a given voltage above threshold or above pinch off. The fact that the tabulated data does not follow this relationship means primarily that the gate capacitance values include extensive stray capacitance from the back gate to substrate on at least the 2N4139.

Now to minimize

$$i_n = \sqrt{\frac{8}{3} \pi^2 (C_T+C_W+C_{GA})^2 \frac{\text{const}}{C_{GA}} \Delta f_v^3} \qquad (12)$$

one must minimize $\frac{(C_T+C_W+C_{GA})^2}{C_{GA}}$ which requires that $C_{GA} = C_T+C_W$.

We conclude that for either MOSFET or JFET input amplifier stages, the transistor structure should be designed so that the <u>active</u> gate capacitance equals the sum of tube signal electrode and wiring capacitance, including parasitic capacitance within the transistor, after the wiring and parasitic capacitances have been minimized. Unfortunately manufacturers of purchased components do not specify the active gate capacitance. Instead, for selection of stock discrete JFET's, one must balance low noise voltage ε_n in the noise equation against C_G, the total transistor input capacitance.

$$i_n = \sqrt{\frac{8}{3} \pi^2 C_S^2 \varepsilon_n^2 \Delta f_v^3} \qquad (13)$$

where $C_S = C_T+C_W+C_G$. Typical values of C_T+C_W range from 13 to 30 pF. For a Westinghouse WX32,719 EBS tube with $C_T = 22$ pF, with $C_W = 3$ pF, the transistor choice should minimize $(25 + C_G) \varepsilon_n$. The choices from the table are the 2N4392 and the 2N4417 or two such units in parallel:

	C_G	$25+C_G$	$\times \varepsilon_n$		
2N4392	18	43	x 1.2	=	51.6
2N4417	4	29	x 1.7	=	49.3
2 x 2N4417	8	33	x 1.2	=	39.6

and the best choice would be a pair of 2N4417 transistors as an input stage.

Once the first stage transistor or transistors have been selected, the amplifier/load circuit noise current is a function only of the video frequency bandwidth. If the application permits, a narrower bandwidth will reduce the noise current significantly, but it will also reduce resolving power in the along scan direction unless the scanning frequencies are reduced proportionally. Figure 7 lists the noise current for an optimized preamplifier vs the video frequency bandwidth for the components mentioned above.

Figure 7 also shows q_n, the amplifier noise expressed as an rms charge fluctuation per Nyquist sample referred to the amplifier input. Roughly speaking the exposure during a frame time must be sufficient to produce this many carriers at the scanned surface of the target per Nyquist sample to achieve a signal equal to amplifier noise. However, the concept is useful principally for large area image details which are completely resolved. For point images the signal charge is spread over several Nyquist samples and the signal charge in each sample must equal the noise fluctuation.

The example chosen illustrates the rule of thumb that for U.S. broadcast scanning standards in a well designed system, i_n is between 2 and 3 nanoamperes, and q_n is of the order of 1000 to 3000 electrons per sample.

To use television camera tubes at lower exposure levels, one can provide prescanning gain by using a photoemissive cathode and accelerating the electrons through 10 kV or so before they strike a silicon target. A carrier pair is produced and collected for roughly every 3.5 eV of primary photoelectron energy. If the prescanning gain is G, the equivalent noise charge referred to the photocathode is q_n/G, so that at 10 kV with a gain of 2800, single photoelectron events can be distinguished from amplifier noise at slow scan rates using the lower video frequencies. The advantages and limitations of prescanning gain are discussed in another chapter comparing all solid-state and electron beam scanned imaging devices.

Variations on the basic preamplifier design are shown in figures 5 and 6. The current mode input operational amplifier shown in figure 5 reduces the effective load resistance from R_L to $R_L \parallel R_F/G+1$, and when $R_F/(G+1) \ll R_L$ provides an output voltage $i_{sig} Z_F \frac{G}{G+1}$ within its passband; where $Z_F = R_F/\sqrt{1+\omega^2 R_F^2 C_F^2}$.

$$i_n = \sqrt{\frac{8}{3} \pi^2 C_S^2 \epsilon_n^2 \Delta f_V^3}$$

Assume

$$C_S = C_T + C_W + C_G = (22 + 3 + 8)\ \text{pf} = 33\ \text{pf for parallelled 2N 4417 FET's}$$

$$\epsilon_n = 1.2 \times 10^{-9}\ \text{V/Hz}^{1/2} \quad \text{or} \quad R_{eq} = 90\ \Omega$$

Then noise current depends on Δf_V which depends on the scan rate

Δf_V	10^4	10^5	10^6	4.5×10^6	10^7	Hz
i_n	2.03×10^{-13}	6.42×10^{-12}	2.03×10^{-10}	1.94×10^{-9}	6.42×10^{-9}	Amp
q_n	63.5	201	635	1350	2010	Electrons

75-0440-VA-83-1

Figure 7. Optimum Analog Preamplifier Design Gives $q_n \propto \Delta f_v^{1/2}$ But $i_n \propto \Delta f_v^{3/2}$

For many purposes, C_F, the shunt stray capacitance of the feedback resistor, is small enough so that $\omega^2 R_F^2 C_F^2$ is negligibly small throughout the video passband and there is no need for a later high peaked amplifier stage. Surprisingly, the noise equation for this amplifier is essentially identical to that of the amplifier shown in figure 4. At low frequencies ε_n appears at the output in opposite phase but equal amplitude to the input, and

$$\frac{d(I_n^2)}{df} = \frac{4kT}{R_F} + \frac{\varepsilon_n^2}{R_F^2}.$$ At higher frequencies as $1/\omega C_S$ becomes small, $$\frac{d(i_n)^2}{df} = \frac{4kT}{R_F} + \frac{\varepsilon_n^2}{R_F^2}(1 + \omega^2 R_F^2 C_S^2)$$ as before, high peaked because of the shunt capacitance.

To eliminate need for separate high peaking, the open loop gain of this amplifier must be large enough so that $\omega R_F C_S/G+1 \ll 1$ throughout the passband. Since in any feedback amplifier phase shifts must be held below 180 degrees throughout the range where the open loop gain is greater than unity to avoid oscillation, and since this condition is easier to satisfy with amplifiers having lower mid band gain, an open loop gain of about 100 is customary. This limits R_F to about 47 KΩ for a 10 MHz video circuit with C_S = 33 pF as in the foregoing example and at this value Johnson noise in R_F is 1.8 nA by itself. Thus the operational amplifier is usually used where advantages like insensitivity to wiring and component capacitance variations outweigh a slightly larger Johnson noise contribution than in the basic circuit of figure 4. Note that the common characteristic of the analysis so far is that shunt capacitance sets noise performance and must be minimized.

Certain camera tubes, for example the CPS-Emitron used for some time in studio cameras by the BBC, have even larger output electrode capacitances, and investigators have asked whether inductive peaking of the load circuit could not, despite the large C_S, increase the available signal before the input stage noise was added. Shunt peaking was used in some early work at RCA, and a successful series peaking embodiment, shown in figure 6, was disclosed by Percival[4] and described by James.[5] The Percival/James pi-network in a sense splits the shunt capacitance into its components and places a "Percival coil" between C_T and C_W+C_G to resonate with the latter at about 85 percent of the upper video frequency limit.

In James' analysis for the circuit of figure 6,

$$C_S = C_T + (C_W + C_G)$$

$$a = \frac{C_W + C_G}{C_S} \tag{14}$$

$b = \frac{f_0}{f_r}$ where f_0 = video bandwidth

f_r = resonant frequency for L and $C_W + C_G$

$$f_r = \frac{1}{2\pi\sqrt{L(C_W + C_G)}}$$

and optimum results are claimed when R_L is large and $b = \sqrt{7/5}$. The transfer impedance of the lead circuit varies approximately as shown in figure 6, and to produce a flat overall response one uses an LC series notch filter, also shown in figure 6, in a later amplifier stage as well as normal RC high peaking. With such a notch filter the noise equation becomes

$$I_n = \sqrt{\left(\frac{4kT}{R_L} + \frac{\varepsilon_n^2}{R_L^2}\right)\Delta f_v \left(1 - \frac{2b^2}{3a} + \frac{b^4}{5a^2}\right) + 4\pi^2 C_S^2 \varepsilon_n^2 \Delta f_v^3 \left(\frac{1}{3} - \frac{2b^2}{5} + \frac{b^4}{7}\right)} \tag{15}$$

This circuit analysis was verified by measuring the noise output of a JFET input preamplifier with $\varepsilon_n = 1.2 \times 10^{-9}$ when

C_T = 27 pF a = 1/3 (approx)

$C_W + C_G$ = 18 pF $b = \sqrt{7/5}$

$\Delta f_v = 7 \times 10^6$ Hz $\alpha = 1 - \frac{2b^2}{3a} + \frac{b^4}{5a^2} = 1.728$ (16)

$R_L = 10^5 \, \Omega$ $\beta = \frac{1}{3} - \frac{2b^2}{5} + \frac{b^4}{7} = 0.0533$

and $I_n = 2 \times 10^{-9}$ amp, compared to 3.7×10^{-9} amp without a pi-network. Inspection suggests the pi-network actually makes the Johnson noise term slightly larger but acts to greatly reduce the amplifier noise term by reducing the effective capacitance.

Other examples show an improvement factor of $1/\sqrt{2}$ to 1/2, a worthwhile improvement if C_S must be high due to the choice of camera tubes. For camera tubes with more modest input capacitances peaking is little used, however, because of the need to individually adjust both the peaking and notch filter networks in any case of component replacement and because the noise reduction is in the higher video frequencies where the human observer is remarkably noise tolerant. A more fruitful application of this technique will be for imaging systems with automatic output where, for example, a thresholding circuit passes signal locations to a computer memory and the human observer is no longer a link in an analog system. There the old design criteria no longer apply, and use of series or shunt peaking should give a significant performance improvement.

2. NOISE PERFORMANCE WITH SAMPLED AND RESET INPUT

Camera tubes with magnetic deflection of the electron scanning beam are most easily operated with continuous beam scanning motion and therefore with an analog output circuit using a load resistor to develop an output signal voltage depending on the rate of reading stored charge. Array sensors are operated by addressing each element in turn using switches or charge transfer electrodes, the signal charge from an element is presented to the amplifier as a packet, and there is every reason not to use a load resistor. A typical output circuit for either a charge coupled device or a diode array sensor is shown in figure 8. The gate of the input amplifier is reset by closing, then opening the reset switch after the signal has been read from the previous element. The signal charge from the next element is then transferred to the gate node by actuating the transfer gate, or in a diode array by closing an address switch, and the signal is then sampled by operating a series switch at the preamplifier output and held for read-out on a shunt capacitor.

Consider first the case of the CCD sensor. The signal voltage at the amplifier input is q_s/C_S. To maximize response, the first amplifier transistor is usually built on chip to minimize C_S, and is always a field effect transistor for high input impedance, and will be called an electrometer amplifier. Since the amplifier measures signal charge, it is simplest to express noise in terms of the equivalent rms charge fluctuation at the gate of the electrometer amplifier. In this circuit, excluding the noise sources within the CCD,

$$q_n = \sqrt{AkTC_S + \varepsilon_{n1}^2 \Delta f_{n1} C_S^2 + \frac{i_{n2}^2 \Delta f_{n2} C_S^2}{g_m^2}} \qquad (17)$$

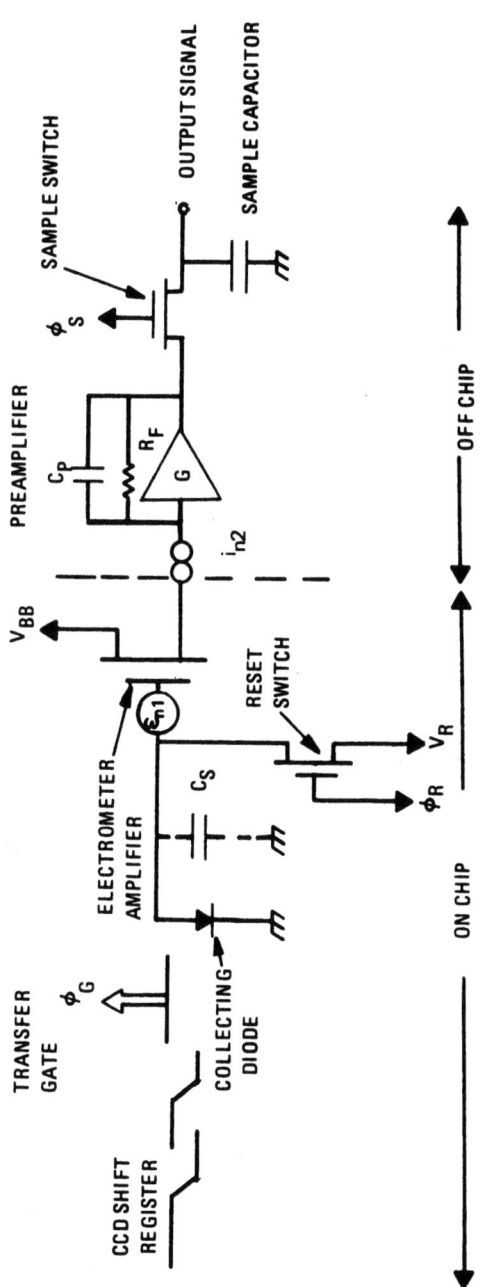

Figure 8. For a Simple Sampled and Reset Amplifier Input, the Principal Noise Source is Johnson Noise in the Reset Switch.

The three noise terms are:

a. The Charge Uncertainty Left After Resetting the Gate Node Capacitance C_S

   While the reset transistor switch is closed it forms an RC circuit in which the channel resistance parallels C_S. The channel resistance Johnson noise spectrum is band limited by the parallel combination. The instantaneous value of the noise charge at the moment the switch opens is left on the capacitor to be added to the next signal charge transferred to the electrometer gate node. If the reset switch has no excess noise, $q_n = \sqrt{kTC_S}$. Measurements indicate typical MOS transistor switches may have up to $\sqrt{2}$ times this value. This term corresponds directly to the load resistor Johnson noise in the camera tube preamplifier.

b. The First Stage Amplifier Noise

   The first electrometer amplifier stage is invariably a field effect transistor, whose noise can be well represented as shown by a noise voltage generator, ε_{n1} internal to the transistor in series with the gate. This noise voltage appears amplified at the off chip preamplifier output, band limited at Δf_{n1} by the preamplifier, and its instantaneous value is added to the signal and sampled and held when the sample switch is opened. The preamplifier gain is assumed to be large enough so that noise in the sample switch is insignificant. The mean square of the equivalent input noise charge is $\varepsilon_{n1}^2 \Delta f_{n1} C_S^2$ and if there is no significant excess 1/f noise in the amplifier passband, ε_n is simply the mid band noise voltage. If the passband includes a significant 1/f contribution, one can evaluate the

   integral $\dfrac{1}{2\pi \varepsilon_{n1}^2} \displaystyle\int_{2\pi f_1}^{2\pi f_2} \varepsilon_n^2(\omega)\, d\omega = \Delta f_n$ to determine an

   effective noise bandwidth.

   This term corresponds exactly to the first stage transistor noise term in the camera tube amplifier analysis except that high peaking is not needed here since the signal depends on the amount of charge, not the rate at which the charge is read. Again, rms electrometer amplifier noise is directly proportional to C_S, which must be minimized, usually by keeping the first stage on chip close to the CCD output.

c. The Noise Contribution of the Following Off Chip Stages

This term was simply lumped with the first stage noise for the camera tube case, but here is expressed separately because the amplifier is usually off chip and can be measured separately, and because the on chip electrometer amplifier stage will often have a low enough transconductance so that off chip noise contributions are significant. This term is meaningful for a specific amplifier configuration, and here we will assume a current mode input operational amplifier with an effective input noise of 2.6×10^{-12} amp. Again C_S must be minimized to minimize this noise component.

To consider a specific example, consider a CCD chip with a P-channel MOSFET on chip input amplifier. Let:

$$C_S = 0.25 \text{ pF}$$

$$\epsilon_{n1} = 20 \times 10^{-9} \text{ V/Hz}^{1/2}$$

$$\Delta f_{n1} = \Delta f_{n2} = 1 \times 10^6 \text{ Hz}$$

$$A = 2$$

$$i_{n2} = 2.6 \times 10^{-12} \text{ amp/Hz}^{1/2}$$

$$g_m = 150 \text{ }\mu\text{mho} \tag{18}$$

Then $q_n = \sqrt{2 \times 10^{-33} + 2.5 \times 10^{-35} + 1.85 \times 10^{-35}}$ coul $\doteq 283$ electrons

Thus most of the equivalent input noise is reset switch Johnson noise, at least at this comparatively low data rate.

To remove most of the reset noise contribution, Westinghouse engineers have developed correlated double sampling.[6] Unlike most noise contributions, reset noise is sampled and held on C_S at the moment the reset switch is opened. If leakage to the electrometer gate node is small, it is possible to read this reset noise before the next signal charge packet is transferred and to separate signal from noise. A mechanization is shown in figure 9. After the reset switch has opened, a clamp switch closes at the output of the preamplifier to store the negative of the amplified reset noise charge on the coupling capacitor. The clamp switch then opens, sampling the electrometer and preamplifier transistor noise and holding this sample also on the

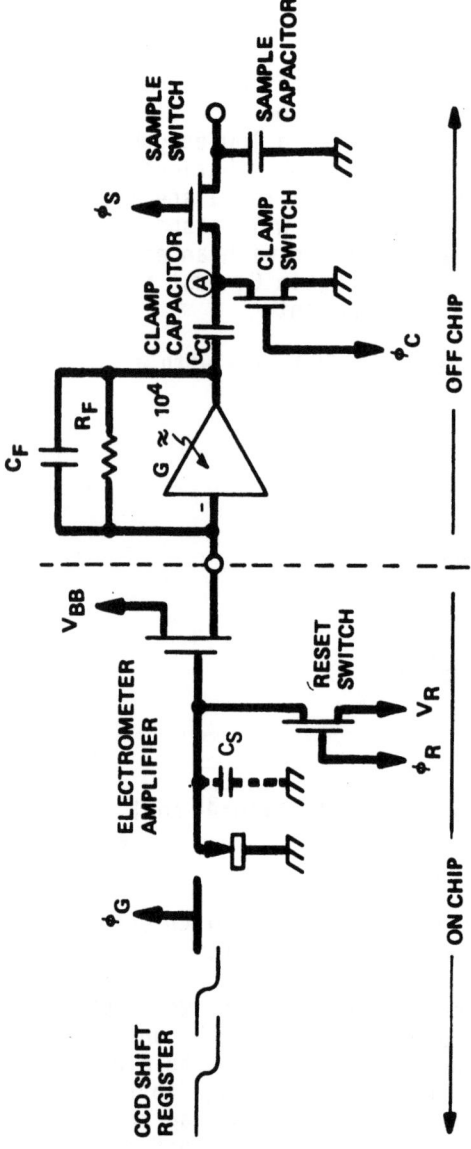

Figure 9. Correlated Double Sampling Removes Reset Noise at CCD Output.

clamp capacitor. The next signal charge packet is then transferred to the gate node of the on chip electrometer amplifier. The voltage at the preamplifier output is now the amplified signal plus the amplified reset noise plus the sampled transistor and amplifier noise plus the on going time varying transistor and amplifier noise. The voltage at the output of the clamp capacitor is the sum of amplified signal plus sampled electrometer transistor plus amplifier noise plus the on going transistor and amplifier noise. The reset noise has been removed. The transistor sample switch is then closed and opened, holding at the output the sum of the amplified signal plus two independent samples of electrometer transistor and preamplifier noise. The second amplifier noise sample was taken at the instant the sample switch was opened. As shown in the example, the reset noise was an order of magnitude larger than the amplifier noise and its removal is important despite the double sampling of amplifier noise.

Correlated double sampling has other advantages, however, since the clamp and sample operations are performed within a single element time, usually only one or a few microseconds apart. The system output voltage is the difference in instantaneous preamplifier output voltage between these two samples, which effectively provides ac coupling to neglect any frame rate or quasi-dc level shifts at the chip output. The effective frequency response of this system for a 2 microsecond delay between the opening of the clamp and sample switches is shown in figure 10. The response at zero frequency is zero, and is very low at low frequencies to greatly reduce the effect of excess 1/f noise in the electrometer amplifier, the off chip preamplifier, or the power supplies.

With correlated double sampling the amplifier noise equation becomes:

$$q_n = \sqrt{\varepsilon_{n1}^2 \, \Delta f_{n1} C_s^2 + \frac{i_{n2}^2 \, \Delta f_{n2} C_s^2}{g_m^2}} \qquad (19)$$

where Δf_{n1} is now roughly twice the analog bandwidth of the operational preamplifier to represent the two noise samples. For the same on chip components used above

$\varepsilon_{n1} = 20 \times 10^{-9}$ V/Hz$^{1/2}$ $i_{n2} = 2.6 \times 10^{-12}$ amp/Hz$^{1/2}$

$\Delta f_{n1} = 2 \times 10^6$ Hz $\Delta f_{n2} = 2 \times 10^{-6}$ Hz

$C_s = 0.25 \times 10^{-12}$ fd $g_m = 150$ μmho

$$q_n = \sqrt{5 \times 10^{-35} + 3.76 \times 10^{-35}} = 58 \text{ electrons.} \qquad (20)$$

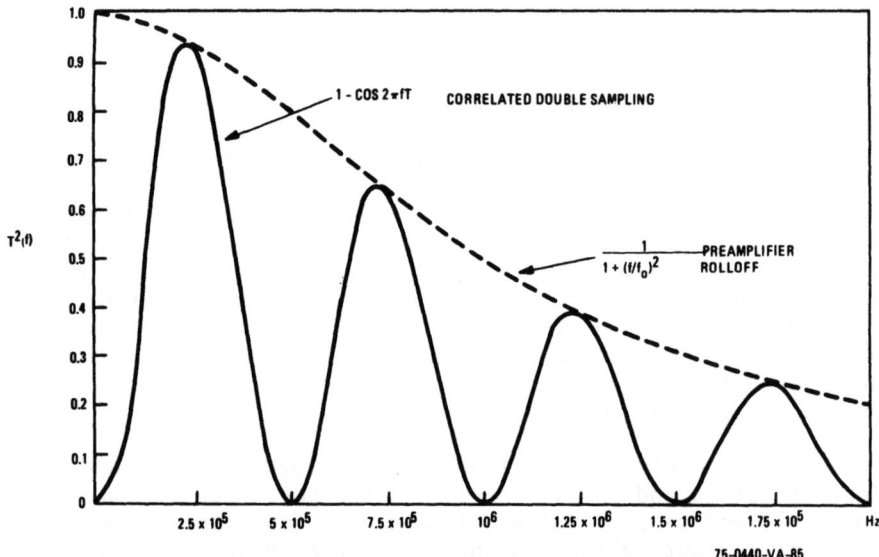

Figure 10. Effective Frequency Response of a CDS System with 2 µsec Between Opening of Clamp and Sample Switches.

Thus with correlated double sampling a realistic CCD output circuit and amplifier embodiment should give noise of less than 60 electrons with an amplifier noise bandwidth of 1 MHz, raised to an effective bandwidth of 2 MHz by the double sampling of amplifier noise. Since $q_n = \text{const} \times \Delta f^{1/2}$, correlated double sampling can give even lower noise at lower data rates. On the other hand, at data rates of 5 to 10 MHz required for full resolution television imaging with brute force interlacing, $g_n \approx 131$ to 185 electrons, large enough to rule out desirable low light level performance.

What can be done to reduce this figure? As shown earlier in this chapter, $\varepsilon_n \alpha g_m^{-1/2}$ and $g_m \alpha C_{AG}$, the active gate capacitance. As before, the active gate capacitance should be matched to the capacitance of the CCD output circuit, after that has been minimized as far as possible. But the collection diode and reset switch diffusion and the interconnecting and overlap capacitances total about 0.1 pF even with careful design, and at 0.25 pF C_S seems within a factor of two of the bare minimum. ε_n cannot be made much smaller for an electrometer transistor with such a small active gate capacitance. The approach used in camera tube type equipment was to parallel input transistors to increase g_m and lower ε_n. If this could be done without increasing C_S

proportionally, one could reduce first stage noise significantly, and also provide enough first stage signal gain so that the effect of off chip preamplifier noise was significantly reduced. The distributed floating gate amplifier described in the next chapter in a sense accomplishes this goal.

REFERENCES

1. Motchenbacher, C.D. and Fitchen, F.C., <u>Low Noise Electronic Design</u>, Wiley Interscience, New York, 1973.

2. Cobbold, R.S.C., <u>Theory and Application of Field Effect Transistors</u>, Wiley Interscience, New York, 1970.

3. Leventhal, E.A., "Derivation of 1/f Noise in Silicon Inversion Layers from Carrier Motion in a Surface Band," Solid-State Electronics, $\underline{11}$, pp 621-677, 1968.

4. Percival, W.S., British Patent 528,179 (1939).

5. James, I.J.P., "Fluctuation Noise in Television Camera Head Amplifiers," Journal (Brit.) IEEE, No. 20, Part IIIA, pp 796-803, 1952.

6. White, M.H., et al, U.S. Patent 3,781,574.

PRINCIPLES OF LOW-NOISE SIGNAL EXTRACTION FROM PHOTO-
DIODE ARRAYS

Robert R. Buss, Satoru C. Tanaka, and Gene P. Weckler

Reticon Corporation, 910 Benicia Avenue, Sunnyvale, CA.

ABSTRACT. Recent developments in both array design and in signal extraction techniques, as they pertain to linear photodiode arrays, are considered. Signal extraction techniques as well as device design can differ depending on the ultimate use of the information. Approaches to signal extraction include a simple current amplifier and various forms of charge amplifiers. Of particular importance for low-noise performance is a sequence of a discrete-time differentiator followed by a discrete-time integrator. Measured performance is illustrated and the circuit techniques discussed for an 1872-element linear photodiode array. Also included are tradeoffs on device design and their effects on performance.

INTRODUCTION

Self-scanned photodiode arrays are adaptable to performance at very low light levels, but to accomplish such performance requires careful attention to the various sources of contaminating noise, most of which derive from the external circuitry. Low-noise components in noise-suppressing configurations are required.

Reason for the need of a low-noise amplifier

The video-line output of a photodiode array is effectively a capacitance, large compared to the capacitance of the individual reverse-biased photodiodes which are sequentially connected to the video line. There is thus a substantial capacitive attenuation of the signal voltage before its appearance at the external terminal of the array. The signal, as it appears at the terminal, must compete with the noise of the first amplifier stage, as a minimum, and must also compete with any thermal noise generators connected to the first stage. This factor is even more evident as the length of the array increases since the output capacitance of the video line for a typical general-purpose linear photodiode array is approximately 0.2N pf, where N is the number of diodes. A factor-of-two reduction is possible for very-low-noise applications at the expense of a somewhat less general-purpose device. Even if this reduction is possible, the trend of the array technology has been to increase the length of the arrays. This progress in the development of self-scanning photodiode arrays can best be seen by the plot of Figure 1, which plots the number of elements in linear arrays versus the approximate date that they became commercially available. Although this curve shows

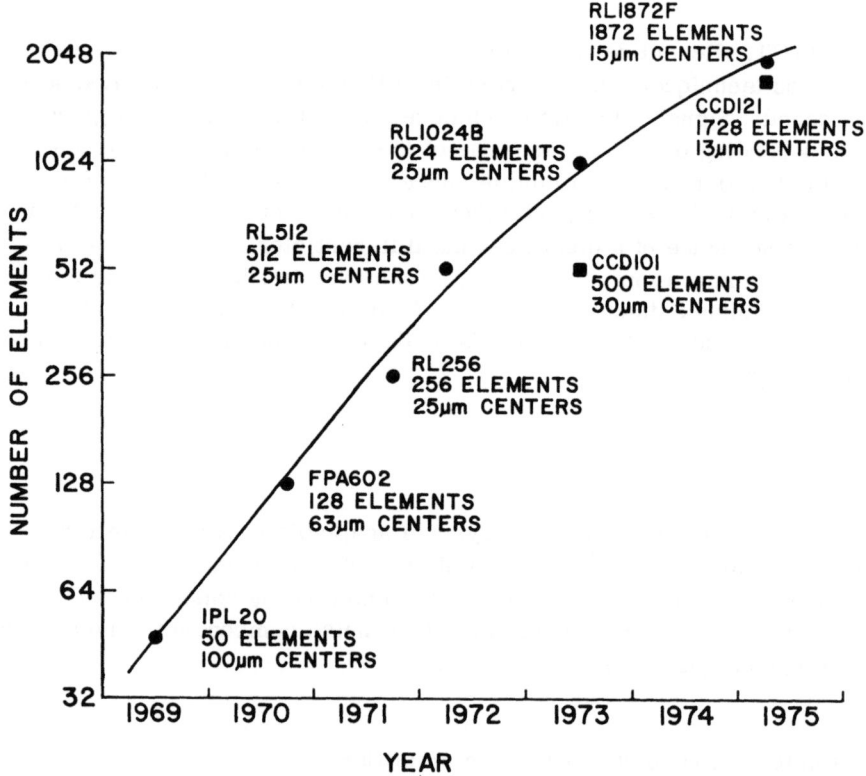

Figure 1. Progress in development of self-scanning photodiode arrays - number of elements vs. year of introduction.

steady progress toward longer arrays, it is evident that this curve is just about to saturate. It may well be further extended if and when high-purity four-inch or larger-diameter wafers become a reality, at which time low-noise signal extraction problems become even more acute.

GENERAL APPROACHES TO SIGNAL EXTRACTION

General approaches to signal extraction will be discussed in this section. The details will be discussed in a later section.

The extraction of signals from photodiode arrays has been developed through an evolutionary process to the point where RMS noise levels of a few hundred electrons are seen. This combined with the high quantum efficiency of the photodiode results in a noise equivalent irradiance in the neighborhood of $1\mu joule/M^2$ (2870° K tungsten source).

The various techniques of signal extraction were slow in evolving. This was in part due to the lack of availability of low-noise FETs and in part due to lack of necessity. It may be surprising to some that very few actual applications really need ultra-low-noise capability. Of course, low noise is always nice to have; however, the vast majority of the applications for solid-state imaging devices find it is far easier just to turn up the light level than to fight the problems of low-noise circuitry.

Also, many potential users of photodiode arrays have been discouraged by theoretical dissertations on how the large output capacitance sets a theoretical limit on their performance. These articles failed to point out that there existed above this limit four decades of useful dynamic range. But because of its importance it is video-line capacitance and its effects that form the basis of the low-noise amplifier discussion that is to follow.

At this point, before proceeding, let us establish a common definition for dynamic range and for noise levels.

Dynamic range is a chameleon*

Everyone uses the term dynamic range with complete confidence and understanding; however, no two individuals should ever enter into a discussion of dynamic range until after they have established a common

---

* A small lizard whose color changes to match the environment.

definition for it. Let us take this advice and with the aid of Figure 2 establish an understanding of the various dynamic ranges of concern.

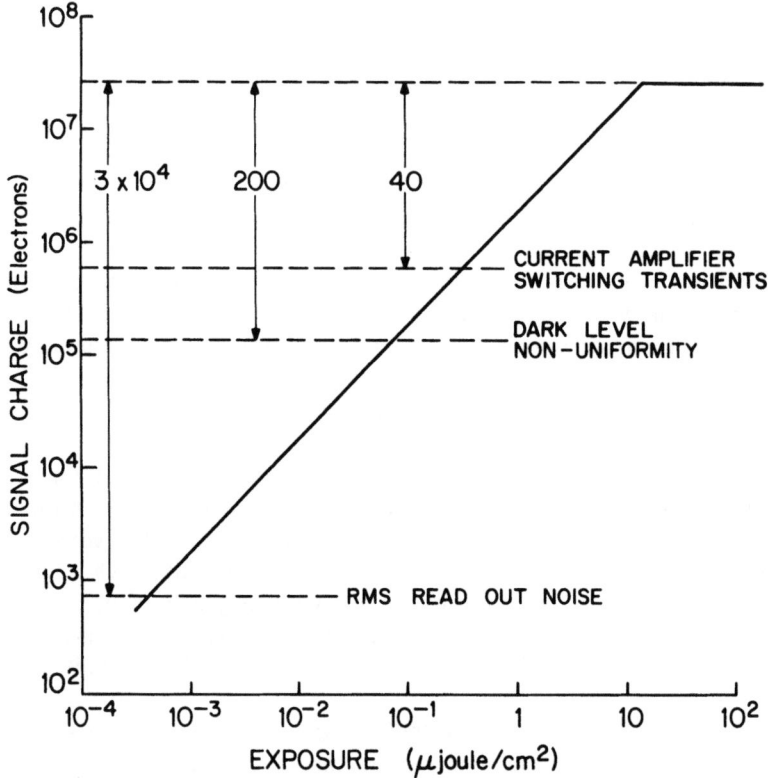

1872 ELEMENT PHOTODIODE ARRAY – 15μm x 16μm pixel size
SCAN RATE : 400 KHz (100 KHz/VIDEO LINE)
INTEGRATION TIME : 33 msec
TEMPERATURE : 23°C
ILLUMINATION : 2870°K TUNGSTEN LAMP

Figure 2. Factors affecting the definition of dynamic range.

There are basically two definitions that should be clarified. The first is usually applied to a sampled-and-held output. In this definition both switching transients and fixed-pattern noise are disregarded. The output from each pixel is individually considered and its dynamic range is the output at saturation divided by the RMS noise on that pixel. Using this definition, a dynamic range of greater than 3×10^4 was measured on an 1872-element linear array operating at a 100 KHz sampling rate. The measured RMS noise level corresponds to 770 electrons. Design philosophies relating to this type of performance will be presented in a later section. The above definition applies to applications requiring high-quality imaging such as aerial reconnaissance, where the ability of a computer exists to massage out the effects of both nonuniform sensitivity and fixed-pattern noise.

For most applications the dynamic range is limited either by the fixed pattern or by the nonuniformity of the fixed pattern, and for some applications the range is limited by switching transients. This definition is sometimes referred to as a machine-readable dynamic range, since this is the dynamic range one could obtain with a simple comparator. For the simple case of the current amplifier, the minimum level for which a threshold can be set is determined by the peak value of the switching transient and, as shown in Figure 3, this gives a machine-readable dynamic range of typically 40:1.

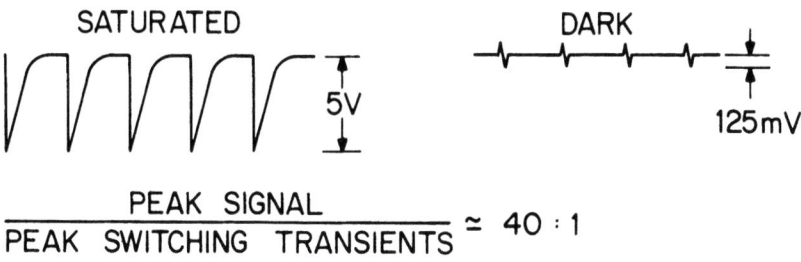

Figure 3. Waveforms for a simple current-amplifier circuit.

The switching transient is largely bipolar and so can be reduced by integration to the extent that the fixed-pattern noise becomes the limit to minimum threshold. As shown in Figure 4, the peak-to-peak amplitude of the fixed-pattern noise in the dark is typically 25 mv, thus making possible a machine-readable dynamic range of 200:1. This is a typical figure for currently available linear photodiode arrays. Techniques such as differential cancellation have been successfully incorporated to substantially reduce the fixed-pattern noise. Experimental devices using a new shift-register design have shown promise of reducing the effects of the fixed-pattern by a factor of about five.

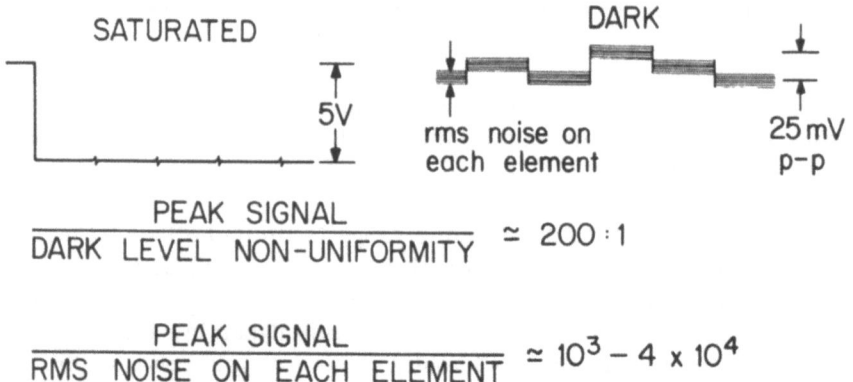

Figure 4. Waveforms for an integrate, sample-and-hold circuit.

However, it is not necessarily the fixed-pattern dark noise that determines the ability of an array to distinguish one level from another. Suppose we wish to divide the dynamic range into N bands, N-1 thresholds. What determines the maximum value of N? For most applications it is not the fixed-pattern noise in the dark, but rather the fixed pattern that appears due to nonuniform photosensitivity. These variations can be as high as ±10% with broadband illumination. This variation is typically a slow variation along an array and its effect can be minimized by using a floating threshold. This effect can also be greatly reduced by limiting the illumination spectrum, i.e., filtering out long wavelengths near the band edge of the silicon.

The type of noise and the corresponding nature of the application has much to do with the method of signal extraction to be used. For instance, if a quantitative measurement of irradiance is to be made, it is best to measure the charge emanating from each pixel since the p-n junction operating in the charge-storage mode is basically a photon counter. For this application, thermally generated noise, the well-known

KTB limit, becomes a decisive factor. Surprising as it may seem, however, most applications require only detection of the existence of contrast change; therefore, it is usually simpler and more economical merely to monitor the recharge current pulse to each pixel. A simple block diagram is shown in Figure 5. The approach, although simple, is the most susceptible to switching noise since the largest bandwidths are necessary for this technique.

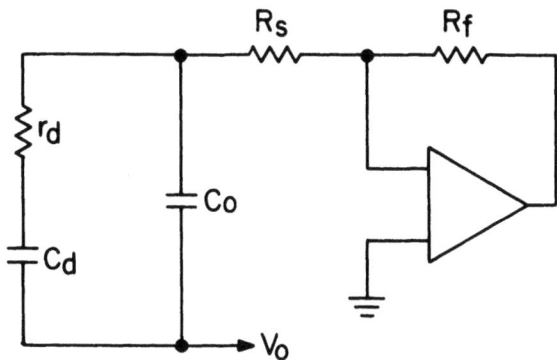

Figure 5. Block diagram for a simple current amplifier.

In other applications switching noise may be of prime consideration such as in a signal processor where a level comparator is used. The major source of switching noise is the capacitance between the access switches and the output video line. As the nth access switch is turning off, the $(n+1)^{st}$ access switch is turning on; thus as one drives a charge into the video line, the other is removing a charge. If both the coupling capacitances between the switches and the video line and the voltage excursions of the switches are the same, then the net charge remaining on the video line is zero; however, if one is looking at current, then any lack of absolute coincidence of the two switch edges in time or shape will result in a switching transient in the output video.

With this background on the origin of fixed-pattern and switching noise on the video line, let us proceed to the methods available for signal extraction.

Signal extraction using a charge amplifier or current integrator

The advantage offered by the detection of charge is immediately obvious; however, there are several ways to implement the function of integration which is necessary to measure the accumulated charge. The function of integration in itself is quite straight forward; however, it becomes complicated by practical considerations. Since integration restricts bandwidth, it seems desirable to perform the functions as soon in the processing scheme as possible, thus restricting the entry of noise. But the necessary resetting process may itself introduce unacceptable noise.

It is possible to make the first amplifier stage a resettable integrator; but the practical problems which were previously alluded to begin to appear. If a voltage gain of unity is to be maintained, i.e., Δe on the photodiode produces Δe out of the amplifier, it requires the integration capacitor be equal to the photodiode capacitance which usually is slightly less than a picofarad. The use of such a small capacitance cannot be ruled out for laboratory environments; however, such a small value often is not practical for general use. Further, it is necessary periodically to reset or discharge this capacitor. The capacitance of most switches suitable for this purpose is not negligible, also the switch capacitance generally is non-linear as well; thus switch-introduced noise may become dominant.

It is often better to use an initial current amplifier to increase the signal level and decrease the effective source impedance, all with an amplifier which discriminates against low-frequency noise, and then to integrate the resulting current pulse to obtain a charge-related signal.

There are two approaches to resetting the integrator. The simplest approach is to reset after each pixel. This works well at the higher signal levels, improving the machine-readable dynamic range typically to several hundred to one; however, as the signal level is reduced, the energy injected by the switching process becomes dominant and its peak value can, if one is not careful, overdrive the amplifier. The second approach resets the integrator once per line during the flyback or holding time between scans. The output of the integrator is a stair-step, the difference between the nth and the $(n+1)^{st}$ step corresponding to the signal from the $(n+1)^{st}$ pixel. This signal is first amplified then ac coupled and dc restored which turns it into a pulse train just as if the capacitor were reset after each pixel. But now the integrating amplifier must handle the sum of all charges of the linear array, so the problems of overload become extreme.

To examine the relevant factors, let us first consider just the monitoring of the recharge current required by each pixel. This is the simplest of all signal-extraction techniques and is shown schematically in Figure 5. The accompanying Figure 3 shows the typical output waveshapes for both saturation and dark conditions and, as one might expect, there exist switching transients. Since the net charge resulting from these switching transients is ideally zero, it is reasonable that to eliminate these transient spikes one should integrate through both signal and transient to obtain only the signal charge.

There are two techniques employed in integrating the signal current pulse to obtain charge. One method is that of the simple charge amplifier and the other method is that of forming a pseudo sample-and-hold on the array video line, i.e., let the charge switch out onto the video line and integrate the signal from each pixel on the line with reset just prior to the output from the following pixel. Both types of circuits will be discussed in the following sections concluding by briefly pointing out the salient properties to compare the two circuits.

Signal extraction using a current amplifier

A simple equivalent circuit of the video output line is shown in Figure 6.

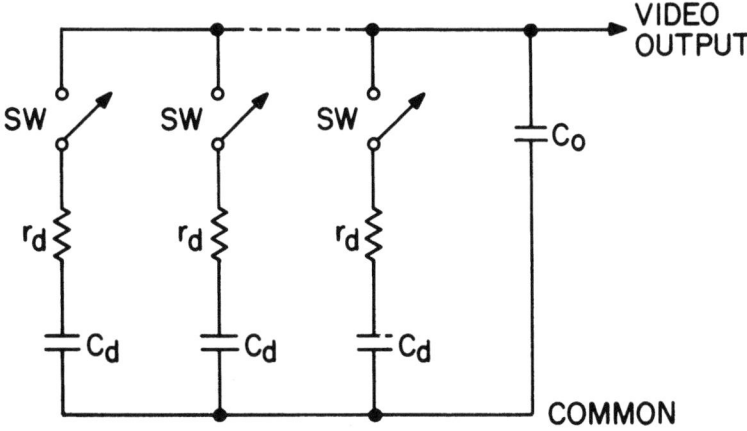

Figure 6. Equivalent circuit of array.

The resistor r_d is the resistance of the MOS FET switch and C_d is the photodiode's junction capacitance. For a single diode under access, the output circuit is further simplified as shown in Figure 7.

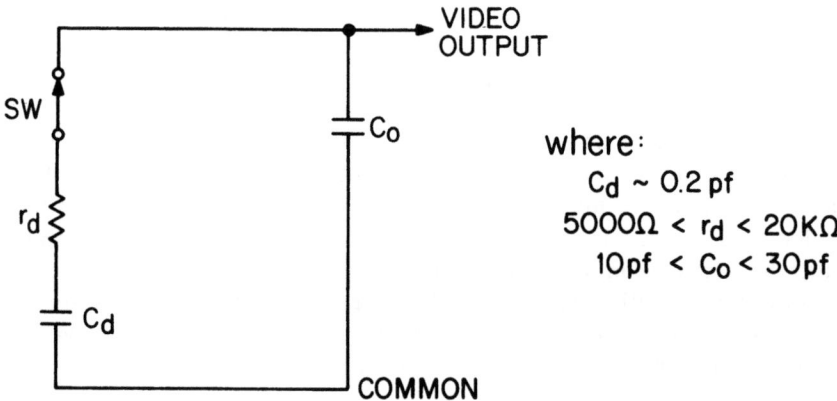

Figure 7. Equivalent circuit for the single diode being accessed.

The array common is normally biased to +5 volts, thus providing not only compatibility to TTL peripheral circuitry, but also automatically referencing the output video to ground with suitable bias on the photodiode.

When SW is closed the capacitor, C_d, will charge almost instantly (from the video output line) with a time constant of $r_d C_d$. These conditions make a current amplifier a suitable preamplifier.

If the current amplifier is ideal with $A \to \infty$, then the output voltage, e_o, has the same shape as the charging current under short-circuit conditions and would be a true reproduction of the charging current. However, there are three major factors which contribute to substantial deviation from an ideal current amplifier. These factors are:

1. amplifier noise
2. fixed residual pattern
3. amplifier stability.

Amplifier noise must be considered because the maximum signal is derived from the light-generated charge with a saturation charge level of approximately 2 picocoulombs. To obtain the minimum amplifier noise requires a series resistor, R_s, between the array output and the amplifier input, as shown in Figure 8 at (a) or a shunting feedback capacitor, C_f, as shown at (b) to restrict the noise bandwidth of the amplifier. The second factor, the fixed residual pattern, is a consequence of the segmented sampling process. This source of contaminating signal is also reduced by insertion of the resistor, R_s, or capacitor, C_f, because it slows down the network response and integrates the sum of the switch pulses derived from the leading edge of the pulse accessing a diode switch and the lagging edge of the previous pulse disabling its switch.

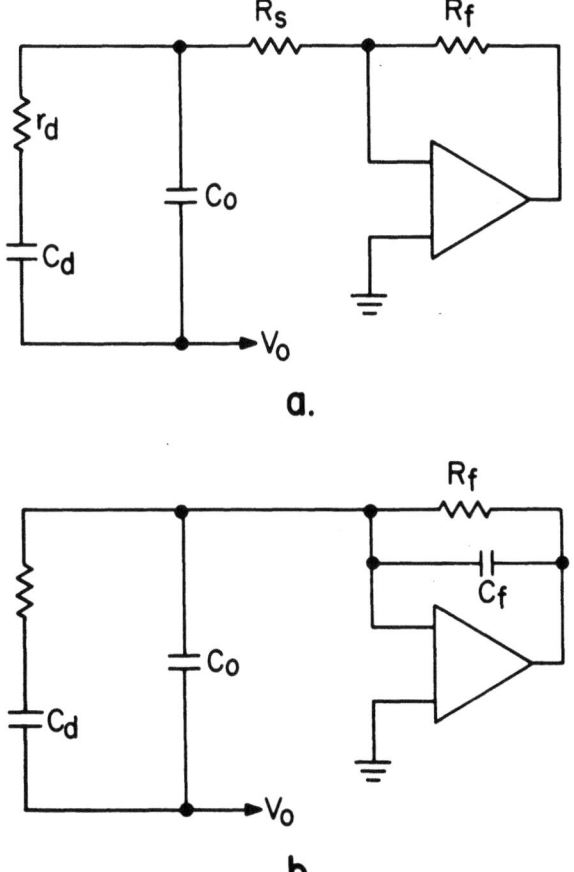

Figure 8. Equivalent circuit for two forms of current amplifier; a) With resistive stabilization, b) With capacitive stabilization.

The integration process can be seen by referring to Figure 9.

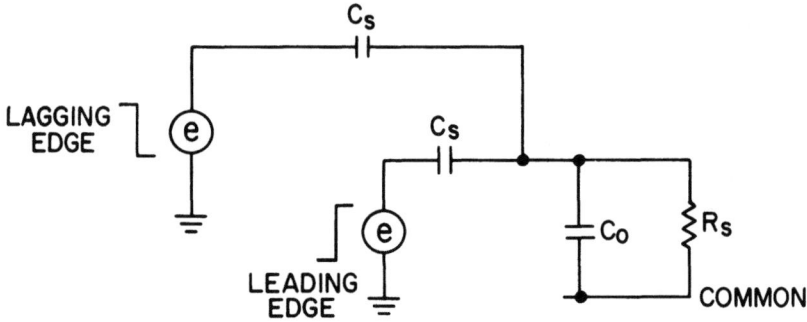

Figure 9. Equivalent circuits showing the feed-through coupling paths for switch energy.

The switch pulse is essentially a differentiated current pulse through C_s, where $C_s \ll C_o$. If R_s is very large, very little switch current flows through the resistor and most of the current charges C_o and is held while the switch energy from the trailing edge charges C_o in the opposite polarity, and aids in cancelling the switch energy. Any residue discharges through R_s with a time constant determined by $C_o \cdot R_s$. However, if $R_s \rightarrow 0$, the switch energy begins to flow through R_s and, hence, is pulsed out with the video. If the leading and lagging edges of the switch pulses are not perfectly coincident, then switching spikes appear at the amplifier output with amplitude a function of the value of R_s. So the addition of R_s reduces the switch spikes by integration on C_o.

The third factor can be attributed to the wideband fast-response characteristic needed for the current amplifier. The bandwidth must be greater than 10 MHz with slew rates greater than 100 v/μsec. Amplifiers with these characteristics exhibit instability when their inputs are terminated with capacitance. Since the array output is essentially a capacitor, a series resistor, R_s, or a shunt feedback capacitor, C_f, as shown in Figure 8 is required to stabilize the amplifier.

Inspection of the Bode plot of a typical amplifier, shown in Figure 10, shows why a series resistor or shunt feedback capacitor may be necessary to stabilize the amplifier configuration of Figure 8. Curve A is the open-loop plot and Curve B is the closed-loop plot with $R_s=0$ or $C_f=0$ and with breakpoint ⓐ approximately determined by $R_f \cdot C_o$. This is an unstable configuration because the "closure rate" between curve A and B is > 12 db/octave. To insure a stable configuration it should be less than 12 db. By adding R_s or C_f a new corner is added with the modified closed loop response of curve C. This reduces the closure rate to $\leq$12 db/octave, and, hence, insures stability.

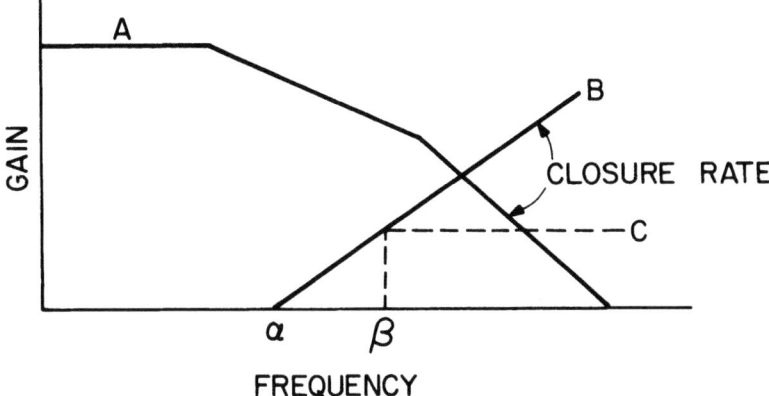

Figure 10. Open-loop gain plot for current amplifier.

Introducing the series resistor or shunt feedback capacitor reduces the detrimental effects of the aforementioned factors, but at the expense of the sampling rate.

An equivalent circuit of a diode under access is shown in Figure 11. The values of the components are typical of a small array and are given to illustrate the formation of the video pulse and exemplify the effects of R_s.

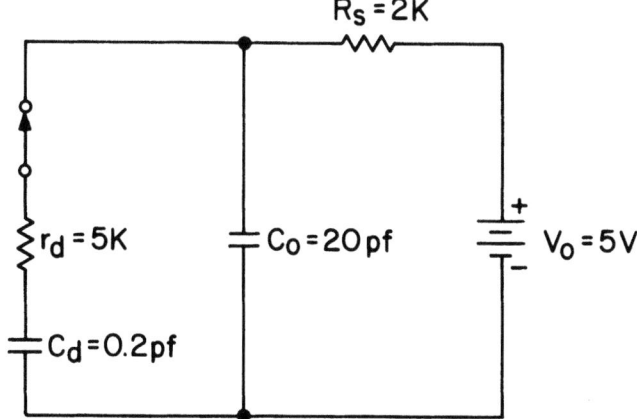

Figure 11. Equivalent circuit of diode under access, with circuit stabilization.

Assume that C_d is completely discharged prior to switch closing and with an ideal clock pulse. When the switch closes C_d charges almost instantly in a time approximately determined by $r_d \cdot C_d$=1 nanosec. To replenish the charge taken from C_d, C_o charges through R_s at a much slower rate which is approximately determined by $C_o \cdot R_s$=40 nanosecs. Figure 12 depicts the voltage and currents of C_o and R_s respectively.

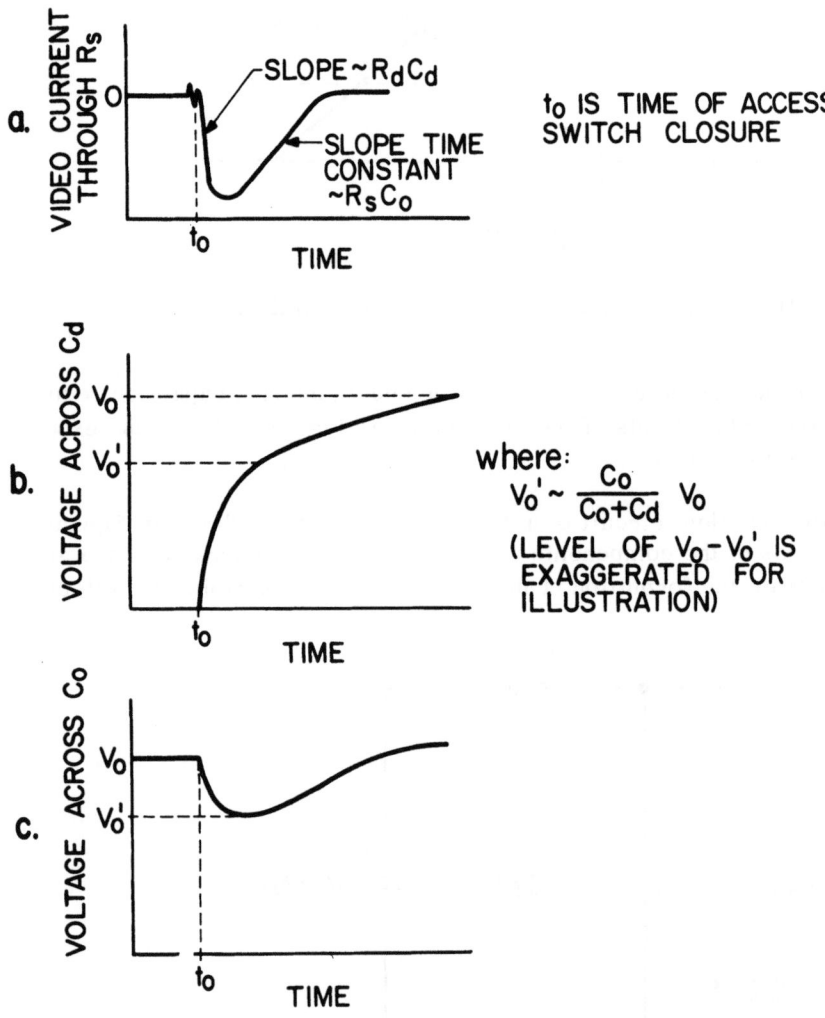

Figure 12. Current and voltage waveforms for a current amplifier.

Since the trailing slope is the dominant factor in determining the pulse width and its time constant is mainly determined by $R_s \cdot C_o$, it is readily seen that the speed of signal extraction and consequently the sampling rate is reduced by using R_s. An exactly similar effect occurs with the alternate use of C_f. The configuration (Figure 8b) gives results equal to that of the configuration of Figure 8a when $R_f C_f = R_s C_o$, but the R_s of 8a introduces thermal noise whereas C_f of 8b does not; therefore 8b is to be preferred if possible. It is interesting to point out that if $5R_s C_o = 1/f_s = 5R_f C_f$ then $f_{3db} = 1/2 \pi RC = 5f_s/2\pi = .8f_s$ which is between the sampling frequency and the nyquist frequency.

Signal extraction using a low-noise voltage translator (charge amplifier)

Another method for signal extraction is simply to buffer the output video line with a high-impedance low-noise FET and a shunt switch to provide a reset path for the video line. This forms a pseudo sample-and-hold for each pixel of information sampled on the video line.

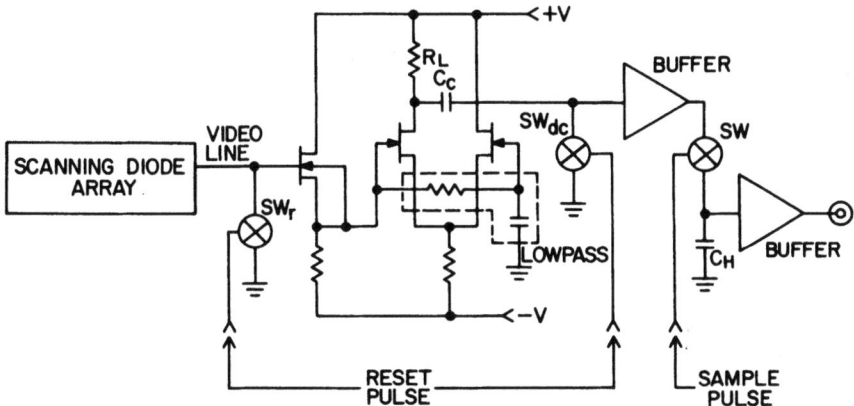

Figure 13. Partial circuit of a low-noise voltage translator.

Figure 13 depicts a partial schematic of this voltage interface circuit. Briefly, the circuit operates as in the following description. As each single storage cell is interrogated by the multiplexing switch the charge flows onto the video line and is temporarily stored on the video line and its associated capacitance. While held in temporary storage it is amplified and sampled by the sampling switch onto a hold capacitor, C_H. The coupling capacitor C_c, and switch, SW_{dc}, form a dc restoration circuit which translates the low-frequency noise out of the information band. The SW_{dc} is synchronously clocked with SW_r.

The differential amplifier following the buffer stage aids to stabilize any dc off-sets which may occur with the array, reset switch, or the buffer amplifier. This amplifier provides a differential current mode at dc and very low frequencies through a low-pass filter connecting the buffer output to the opposite gate of the differential FET. Since the bias current is essentially a constant current at the sources of the differential pair for a small dc voltage variation, the voltage seen across R_L remains con-relatively constant. However, the signal pulse voltage is amplified by the first FET of the differential pair but not the second and is seen across R_L. A point to note is that if the current source for the differential pair were a transistor or other constant-current device, rather than a resistor, the bias changes across R_L would have even less effect as dc voltage changes take place at the output node of the buffer amplifier. A further discrimination against dc and very low-frequency signals is obtained by use of SW_{dc} and the coupling capacitor C_c, as discussed further in a following section.

For small signal levels on the video line, less than 50 mv, this voltage-mode interface circuit has two distinct advantages over the current amplifier: (1) it can be made simpler; (2) it has superior signal-to-noise performance; however it is currently limited in speed to less than 5 MHz because of sampler and clamp limitations. Since it forms a pseudo sample-and-hold circuit for clocking frequencies greater than 2 KHz sampling rate, the circuit inherently is a zero-order-hold, or "box car" signal processor, whereas the current-amplifier interface requires an additional sample and hold circuit; therefore the circuitry is markedly simpler and also takes maximum advantage of the pixel energy.

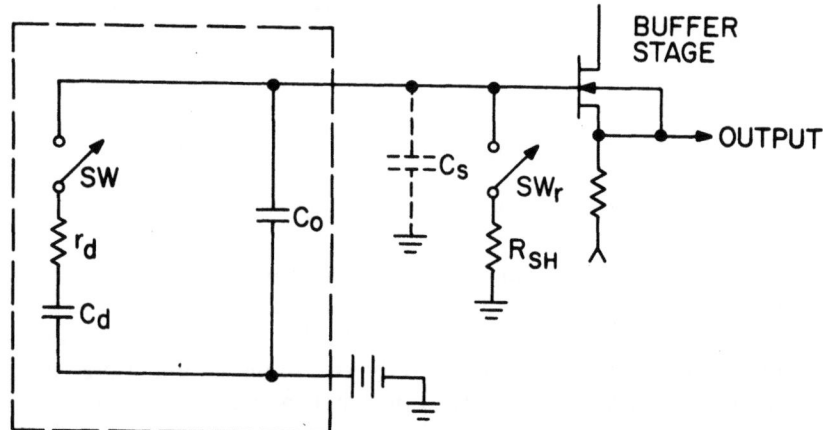

Figure 14. Equivalent circuit showing the voltage translator and a single diode cell.

Another advantage is the superior signal-to-noise performance over the current amplifier. Since the limiting noise is in the interface circuits and not in the array, obtaining low-noise devices for the interface circuits is of the utmost importance (noise in this discussion is the uncorrelated thermal and shot noise). The voltage-hold interface circuit offers this advantage. A low-noise, high speed FET buffer is readily available, whereas obtaining a low-noise, high-speed, high-gain operational amplifier at low cost is not as readily achieved. The most significant factors which make the voltage-mode video interface circuit an inherently lower-noise system are the sampling process and the high-impedance interface.

Let us proceed in a manner similar to that in the analysis for the current-mode extraction. Figure 14 depicts a single cell under interrogation where the closing of SW_r is taken as occurring at t = 0. As previously discussed, r_d is the switch resistance and C_d is the storage capacitance of a single one of the array diodes. SW_r is the shunt switch and R_{SH} is its associated series resistance. The diode's voltage discharge time constant, $r_d C_d$, as explained for the current-mode form, is less than 50 nanoseconds. This video-line voltage is amplified and sampled onto a hold capacitor after the voltage has reached a pseudo equilibrium state. The waveforms and relative sample-pulse timing are shown in Figure 15. The pseudo equilibrium condition which exists following charge transfer can be held for any desired length of time with limiting conditions imposed by the FET's bias and offset current. The deleterious effects are negligible for a hold duration of 2 μsec or less when a relatively good FET with pico-amp bias and offset currents is used.

To refresh the video-line capacitor and to recharge the photodiode capacitor to full charge, a reset switch, SW_r is provided. At the instant this switch is activated, the memory capacitor, C_d, continues to charge up to the full battery voltage, V_0, through a time constant $\tau = r_d C_d$, because the shunt switch resistance $R_{sh} \ll r_d$ (see Figure 15e); however, the video-line capacitor charges with a time constant of $R_{sh} C_0$, which is, again, extremely fast because the typical values for a small array are $C_0 \cong 15$ pf and $R_{sh} \cong 200$.

With a good shunt switch the time to reset the charges is typically in the order of 200 nsec. Again referring to Figures 15f and 15g, the currents during the refresh period for the photodiode and the video-line capacitor are shown just after the reset pulse activates the shunt switch. This reset pulse relative time and pulse duration is shown in Figure 15b.

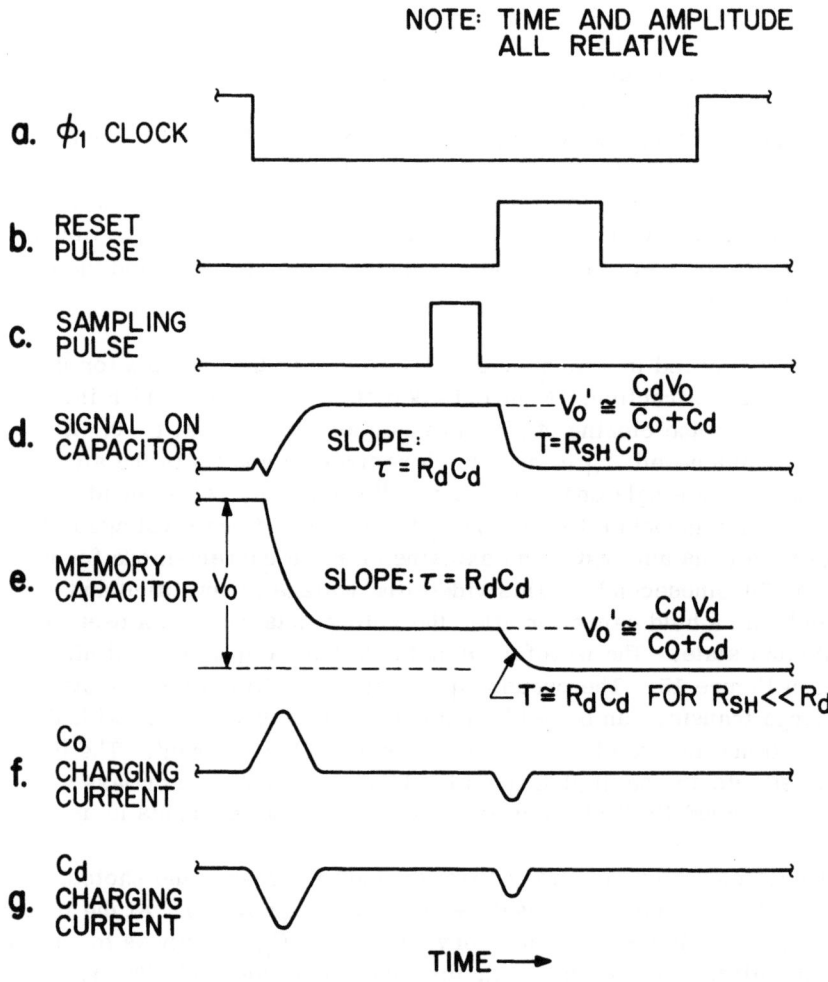

Figure 15. Relative timing diagram for the voltage translator.

System bandwidth and noise consideration

Because this is a sampled data system it possesses an inherent ability to discriminate against low-frequency noise. Therefore, the bandwidth must be carefully considered for information and noise spectral density.

It is well known that in order to reproduce a pulse with a fixed pulse width, the bandwidth must be such as to recover the frequency spectrum which is a $\sin(\omega/2)/(\omega/2)$ function, with ω the frequency in radians. If, however, a 500 nanosec pulse is postulated, a passband which accepts the spectrum to the sixth zero crossing is 30 MHz, a demand which can exorbitantly increase the cost of the system. Fortunately, most of the energy is in the central section and a reasonable reproduction of the pulse can be achieved by making a system with a bandwidth only as wide as the first zero crossing. This approximation yields a reasonable bandwidth of 2 MHz for a 500 nanosec pulse. This design also allows enough bandwidth to operate a dc restoration circuit to return the baseline to its proper level at a 1 MHz sampling rate.

Although the amplifier bandwidth is from 0 to 2 MHz, the noise bandwidth is not the same; the noise bandwidth is reduced by the sampling process, as shown below.

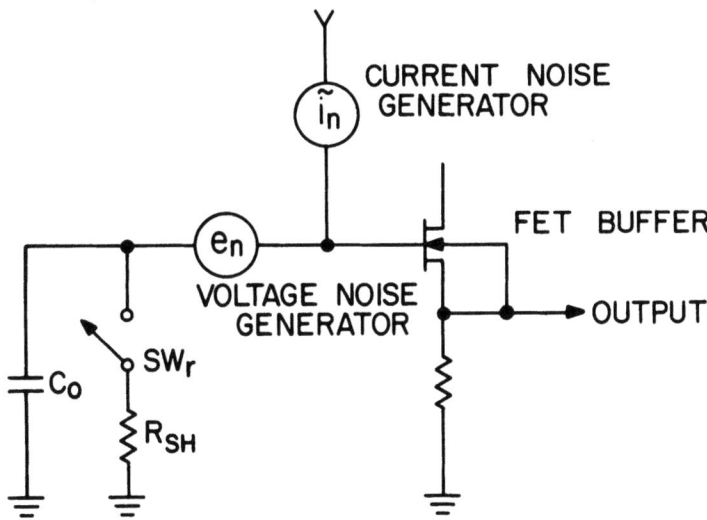

Figure 16. Noise equivalent circuit for the voltage translator.

Figure 16 depicts the buffer stage with the noise generators, the sampling switch, and C_o, the video line capacitance. By selecting a low-noise FET the noise current and voltage generators can be as small as 0.005 pa/$\sqrt{Hz}$ and 5 nanovolts/$\sqrt{Hz}$ respectively. The current-generator contribution is negligible because of the shunting effect of C_o, and because of the action of the sampling switch, as shown below.

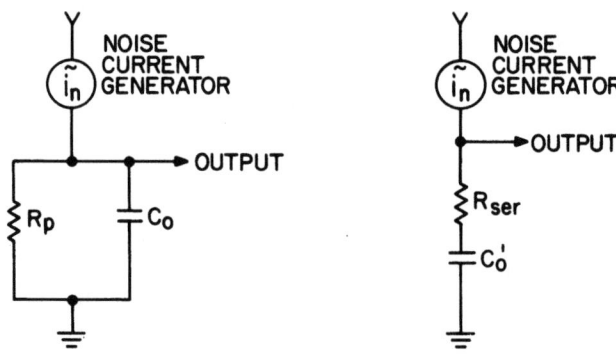

a. PARALLEL EQUIVALENT LOAD FOR THE NOISE CURRENT GENERATOR

b. SERIAL EQUIVALENT LOAD FOR THE NOISE CURRENT GENERATOR

Figure 17. Equivalent circuits to illustrate the impedances important to the noise current component.

Figure 17 shows the equivalent circuit in both its parallel and its equivalent series resistance form, but without the shunt switch. The series form may make the behavior more intuitively obvious. Without the shunt switch, the noise voltage seen across this network can be very large at low frequencies, especially if $R_p > 10^{10} \Omega$, which is readily realized with a good FET, because the voltage spectral noise density is

$$e_n(\omega) = i_n(\omega) R_p / \left| 1 + j\omega R_p C_o \right|$$

where i_n is the noise generator spectral density and the bar lines in the denominator indicate magnitude only.

It is readily seen that as R_p becomes very large the noise voltage at low frequencies grows without bound. Thus if the network did not have the shunt switch the noise could be extremely large near dc even with 0.005 pa/√Hz as a current noise generator. However, this current generator is shunted every clock cycle; therefore the circuit effectively samples the current generator. This makes the low-frequency noise current component near dc translate up to near the sampling frequency and to form a noise power spectrum around the modulation products of $(\omega_s \pm \omega_n)$, where ω_s is the sampling frequency and ω_n is the noise spectral frequency. It can readily be seen that with the noise current generator's low-frequency bandwidth translated around the sampling frequency, $\omega_s \pm \omega_n$, (see Figure 18) the noise current essentially

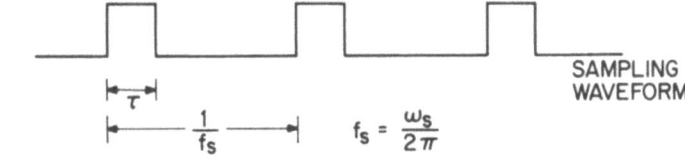

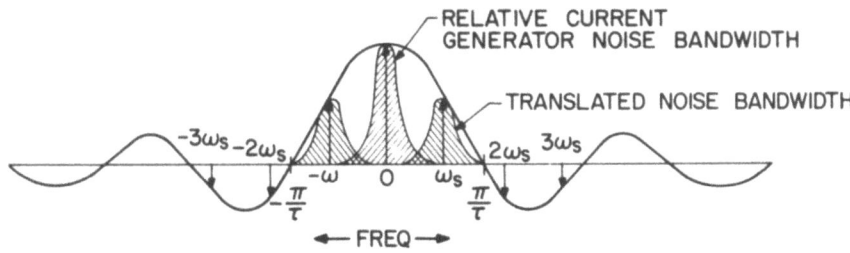

Figure 18. Spectrum for a sampled noise current generator.

contributes no noise component because C_o, the video line capacitance, shunts the generator's signal to ground, i.e., if

$C_o = 50$ pf, $f_s = 1$ meg, $\overline{i_n} = 0.005$ pa$/\sqrt{Hz}$

then $\overline{e_n}$ = (voltage across the network) = 16 pico volts$/\sqrt{Hz}$.

With this low impedance presented by C_o, even if the noise source were integrated over a narrow bandwidth around the sampling frequency the quantity is negligible compared to the noise voltage generator. It can be seen that this system also effectively reduces the i/f component of any current source. The thermal noise of the resistance is a current generator of comparable magnitude $\overline{i_n} = (4kT/R_p)^{1/2} \sim 10^{-15}a/\sqrt{Hz}$ and thus likewise negligible.

The dominant noise source is the voltage generator source. Here again, as in the current source analysis, the data sampling reduces the low-frequency noise from this generator.

Figure 19 shows the equivalent circuit with the noise generator in series. As the switch samples the generator the noise spectrum is translated up just as with the current generator to $\omega_s \pm \omega_n$ except this time this spectrum is amplified and processed, and the noise is not reduced at this point. But its effective noise bandwidth (in part) is reduced by the sampling switch at the hold capacitor. At this point the sampling is synchronous with the shunt switch and operates with a window only during the inactive time of the shunt switch. Between samples the value is held constant. This technique, known as zero-order-hold, reduces the high-frequency contribution. The low-frequency component of the 1/f type is reduced by maintaining a small sampling window and then processing through a sampling differentiating network which is also used for dc restoration. The analysis of this network is discussed in the following sections under low-frequency noise suppression with a discrete-time differentiator.

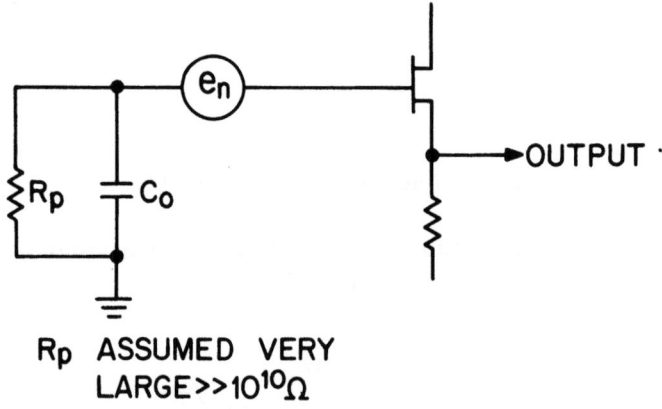

Figure 19. Equivalent circuit for the noise voltage generator.

A RECOMMENDED CIRCUIT FOR SIGNAL PROCESSING

If the fixed-pattern noise and the switching noise can be ignored, as in some applications, then the next limiting noise source of the system is the uncorrelated noise. In this case it is very important to minimize the noise in the signal extraction circuits; however, there is a limit imposed by the devices used in the circuit. Even with the best device available, either discrete or integrated amplifier, there is noise associated with it. See Figure 20 for a typical-noise-versus frequency plot for a typical semiconductor amplifier found on the market today. The general characteristic of the noise response is the increasing noise on each end of the spectrum. The low-frequency noise increases with decreasing frequency, and at the high end of the spectrum the noise increases with increasing frequency. There is a central zone of the

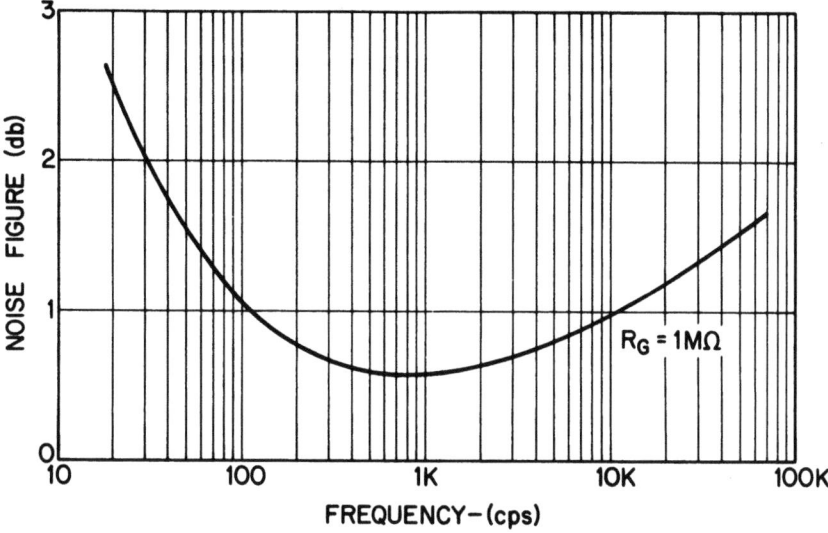

Figure 20. Typical noise figure plot for a solid-state amplifier.

frequency spectrum which is relatively flat and lower in noise amplitude. In general this is true for all devices; different devices differ in the width, slope and magnitude of the noise spectrum, but the overall response shape is similar.

This noise characteristic is the basis for recommending the following signal-processing circuit and forms the basis for its noise analysis. See Figure 21.

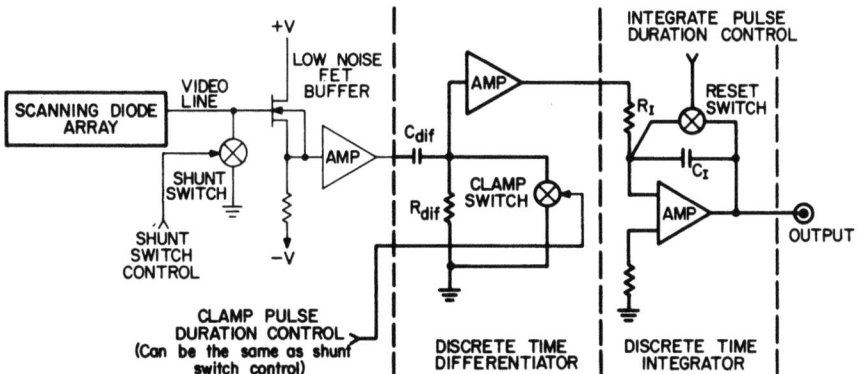

Figure 21. Low-noise signal processor for a self-scanned photodiode array.

This circuit under discussion is broken into two main sections which involve processing the signal and assumes that the signal has been removed via a buffer stage with a synchronous reset switch, as shown in Figure 14, or by a current amplifier which converts the charge to a voltage on the output as discussed previously. From here, the signal is processed through a low-frequency discrete-time noise-differentiating network (which is incidentally used to restore dc) and the discrete-time integrating circuit. An equivalent circuit is discussed in the following sections.

Low frequency noise suppression with a discrete-time differentiator

Essentially the low-frequency noise component in the apparent signal bandwidth is suppressed by the high pass or discrete-time differentiation network. This is possible because all of the video information coming out of the array is translated up in frequency and is centered about the sampling rate, ω_s

A high pass filter is formed with a sampled RC differentiating network. See Figure 22. By operating the switch SW_1 with a sampling waveform as shown in Figure 23, a discrete-time difference network is realized.

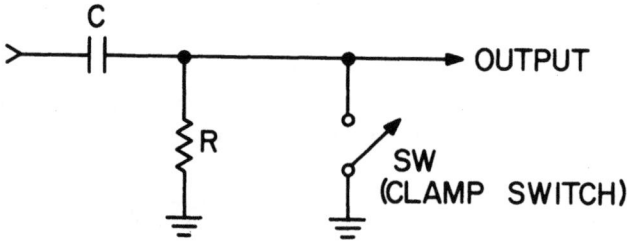

Figure 22. Circuit for a discrete-time differentiator.

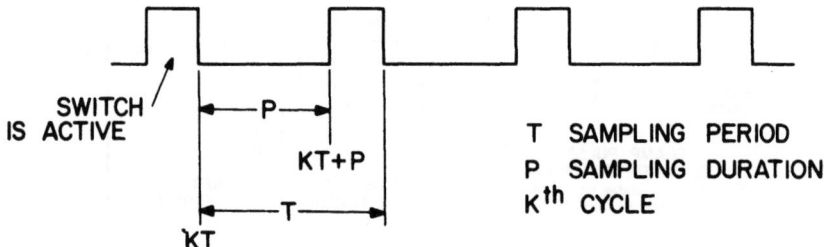

Figure 23. Sampling Switch waveform.

An example of this network operating in a discrete-time mode is shown in Figure 24. The clock is operated synchronously with the discrete step of the discrete-time triangular wave such as that which may be seen from a full-scan integrated video output from an array. The capacitor, C, charges to the value of the previous sample, KT, and subtracts that value from the next step sample $KT + T$, where T is sample duration and K is the Kth incremental time, and thus yields only a differentiated or the difference output of the two adjacent steps.

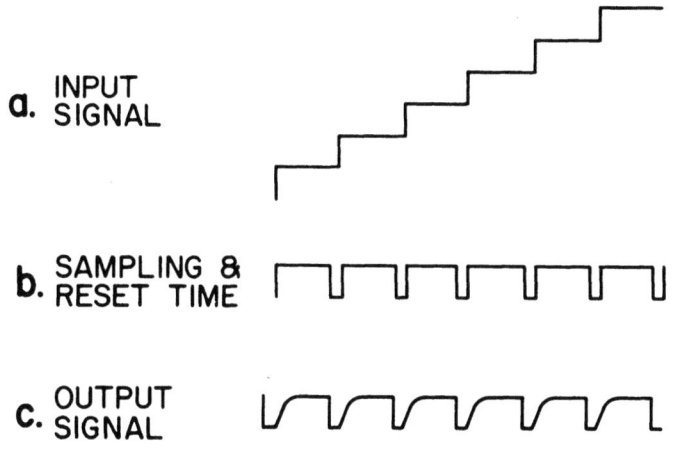

Figure 24. Example with a stepped triangular charge input wave.

By using this sampling technique the RC differentiation network has its passband shifted in frequency with the zero at the origin translated to ω_s, the sampling frequency; as in the classical sense of a sampled data system, this function repeats itself periodically with ω_s, the sampling frequency. See Figure 25.

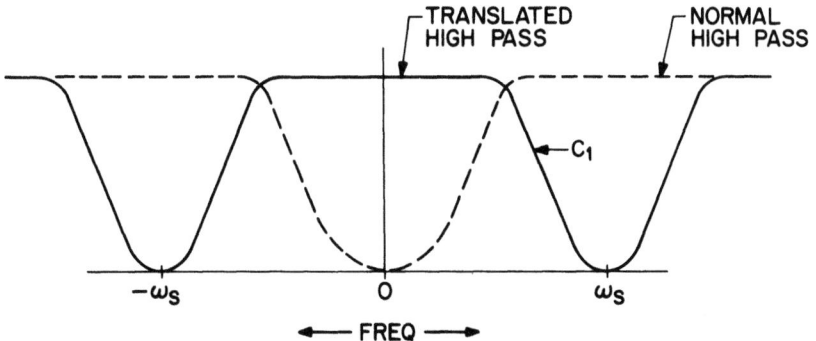

Figure 25. Translation of a high-pass filter's passband through sampling.

The frequency translation occurs because the sampling wave form time-multiplies with the input signal which shifts the input spectrum up around each harmonic of the sampling-frequency spectrum.

The multiplying waveform is redrawn in Figure 26(b) along with the input signal waveform 26(a), and the associated output waveform, 26(c) of the differentiating network.

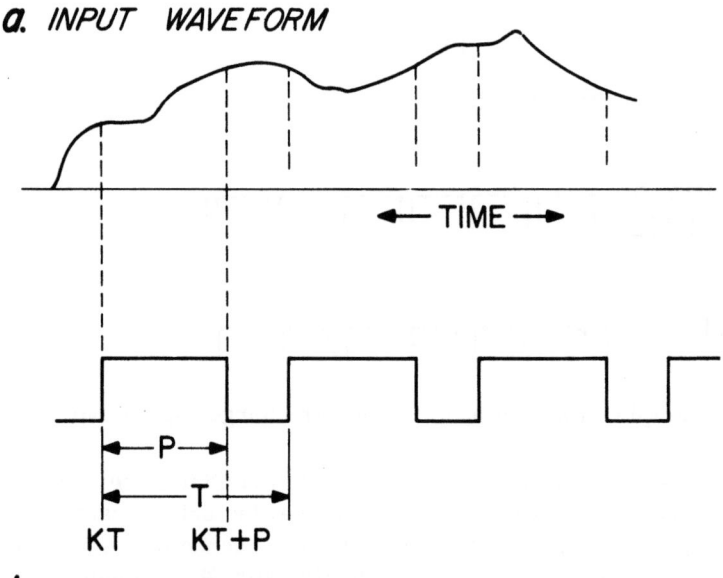

a. INPUT WAVEFORM

b. INPUT TIME WINDOW

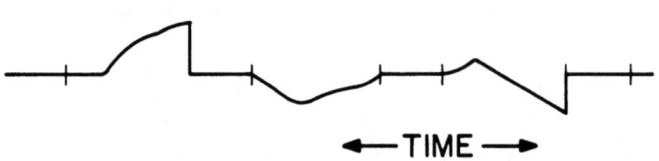

c. DIFFERENTIATED OUTPUT

Figure 26. Signal multiplied by a time window.

Since the initial charge for each pixel held by the capacitor is charged at the time, kT, the beginning of the sample duration, its charged voltage subtracts from the input voltage until the time, t = KT + P, and repeats after the switch opens with a new charge on the capacitor. This switching is performed on the incoming network, hence the input signal is time multiplied and applied to the differentiating network.

To simplify the switching of the incoming signal without the need to consider initial stored charge each transition of the switch, the circuit of Figure 22 is redrawn into its equivalent current-source circuit. See Figure 27.

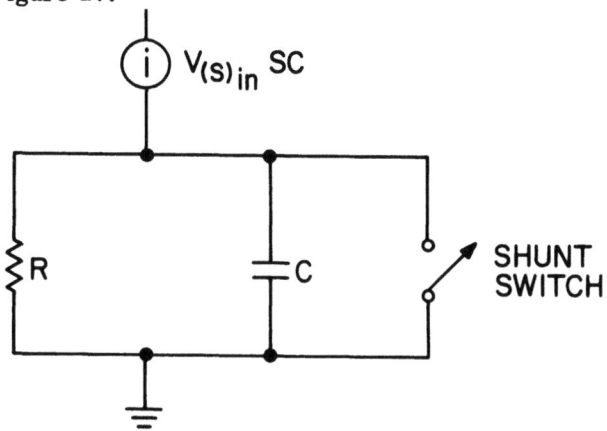

Figure 27. Equivalent current-source circuit.

This then yields a transfer function

$$H_{(s)} = R/(SCR+1)$$

with a current generator $i = V_{(s)in} SC$

where $V_{(s)in}$ = Laplace-transformed input signal.

The sampling waveform is applied to the shunt switch. This multiplies the input waveform. To multiply the input function, the Fourier series of the waveform is obtained, then time multiplied with the input time function. Then the product's Laplace transform is obtained and multiplied with the transfer function.

The sampling waveform has the Fourier series

$$P_{(t)} = \sum_{n=-\infty}^{n=\infty} C_n e^{jn\omega_s t}$$

where
$$C_n = \frac{P}{T} \frac{\sin(n\omega_s P/2)}{n\omega_s P/2} e^{-jn\omega_s P/2}$$

T = sampling period
P = sampling duration
$\omega_s = 2\pi/T$

Taking this function and multiplying with the incoming signal $f_{(t)}$ yields the modified sampled signal,
$$f^*_{(t)} = P_{(t)} \times f_{(t)} = \sum_{n=-\infty}^{n=\infty} C_n e^{jn\omega_s t} f_{(t)}$$

or by the shift theorem, the Laplace transform is
$$\mathcal{L}\left[e^{jn\omega_s t} f_{(t)}\right] = F(s + jn\omega_s t)$$
where $\mathcal{L}\left[f_{(t)}\right] = F_{(s)}$

then
$$f^*_{(s)} = \sum_{n=-\infty}^{n=\infty} C_n F(s + jn\omega_s)$$

so that the sampled signal is a sum of the products of the original spectrum shifted by the sampling frequency and its integral multiples.

The incoming spectrum of the current generator is by substitution
$$F^*_{(s)} = \sum_{n=-\infty}^{n=\infty} C_n V_{(s + jn\omega_s)} (s + jn\omega_s)C$$

Multiplying this spectrum with the transfer function given earlier yields the output response; that is
$$R_{(s)} = F^*_{(s)} \times H_{(s)}$$

$$R_{(s)} = \sum_{n=-\infty}^{n=\infty} C_n V_{(s+jn\omega_s)} \frac{(s + jn\omega_s) CR}{(s + jn\omega_s) CR + 1}$$

The output response for the first coefficient, C_1, and $S = \pm j\infty$ has the response as shown in Figure 28 as a solid line. This response corresponds to the original high pass shown in the dotted line before sampling. The region of interest now is that of the two minima about $\pm \infty_s$.

a. SAMPLING FREQUENCY SPECTRUM C_N

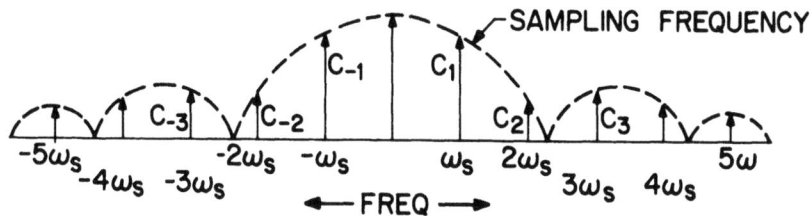

b. ORIGINAL IMPULSE RESPONSE $H_{(s)}$

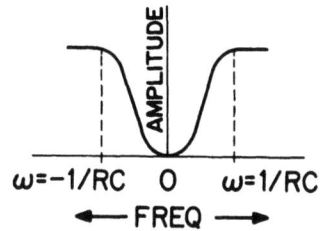

c. SAMPLED RESPONSE $H_{(s)}$

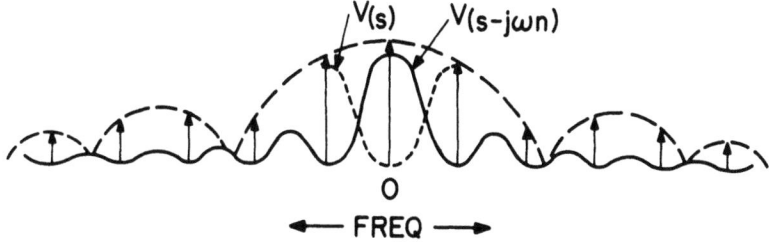

Figure 28. Frequency translation for a discrete-time differentiator.

The point to note is the translation of the low-frequency response to the vicinity of the sampling frequency; this translation greatly reduces the noise contribution from a low-frequency noise generator.

The translated network response has the same response as before only shifted. Therefore, in design, just consider the RC high pass connected without the switch and then translate the frequency by ω_s, the sampling frequency. See Figure 28 for a pictorial representation of the frequency translation.

Another point in the design consideration is to maintain the percentage sampling time as large as possible. This minimizes the width for the $(\sin x)/x$ amplitude envelope of the sampling frequency, ω_s, and its harmonics (Figure 28(a).) This reduction tends to reduce the effects of aliasing of the spectral energy from about the adjacent $n\omega_s$.

The discrete-time integrator

In actual use an operational amplifier used as a feedback or Miller integrator is a more accepted method, but for this analysis a simple RC integrator will suffice in its place just as well.

The integrator aids in reducing the noise from frequency regions both above and below the sampling frequency. This band narrowing also is to be attributed to the sampling system. Figure 29 shows a simple low-pass filter and an associated switch to reset the capacitor every sampling period.

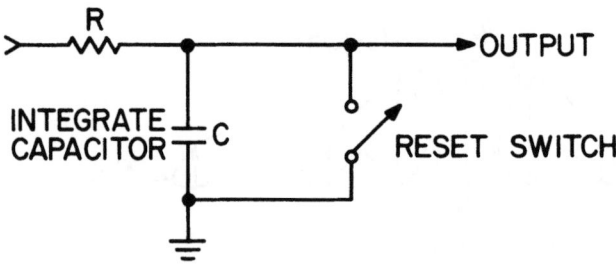

Figure 29. Discrete-time integrator.

Figure 30 shows the relative timing diagram; 30(a) shows the multiplying switch function; 30(b) shows the modified input signal voltage as seen by the resistor and capacitor in series when the switch is open; 30(c) shows the integrated output.

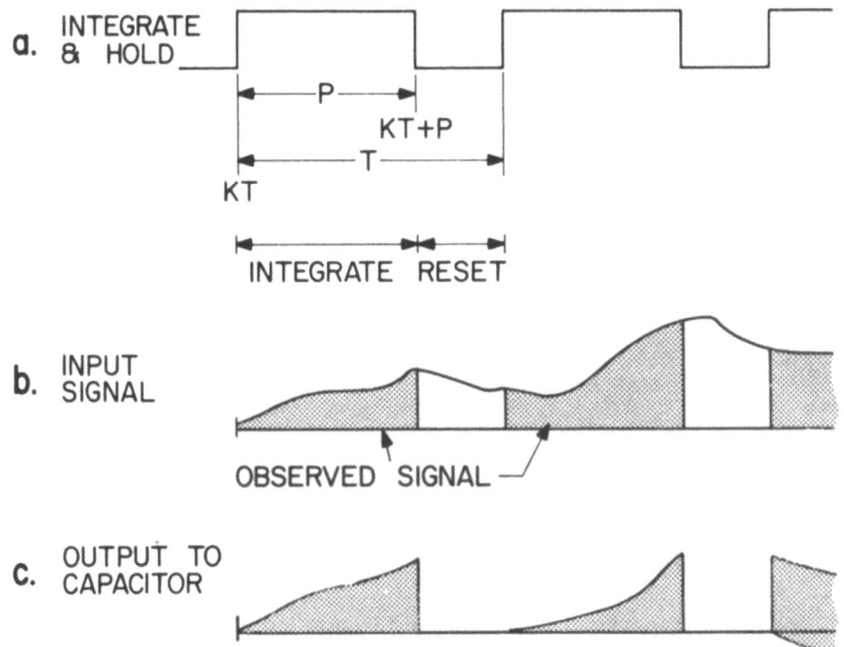

Figure 30. Relative timing diagram for the discrete-time integrator.

Again, to determine the frequency response of this system, referring to Figure 31, although the filter is low pass, the sampling effectively band limits the noise frequency spectrum around sampling frequency.

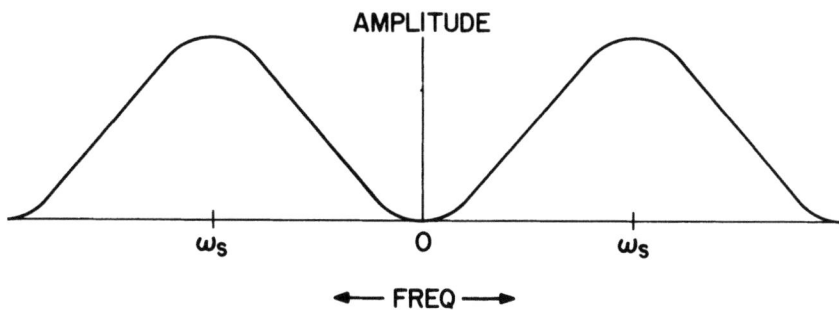

Figure 31. Frequency response of the integrator.

The analysis as before is based on the time-multiplied input signal and the switching function. This result is readily seen because for each pixel undergoing integration during $KT \leq t \leq KT+P$ (refer to Figure 30(a)) the input signal voltage is applied to the low pass network. As before the sampling time function

$$P_{(t)} = \sum_{n=-\infty}^{n=\infty} C_n e^{jn\omega_s t}$$

is multiplied by the input time function $f_{(t)}$ to obtain the sampled time function

$$F^*_{(t)} = \sum_{n=-\infty}^{n=\infty} C_n e^{jn\omega_s t} f_{(t)}$$

Again as before, the function is transformed to

$$F^*_{(s)} = \sum_{n=-\infty}^{n=\infty} C_n F_{(s+jn\omega_s)}$$

Then the output response $R_{(s)}$ is obtained by multiplying $F^*_{(s)}$ with $H_{(s)}$, the low-pass transfer function.

$$R_{(s)} = \sum_{n=-\infty}^{n=\infty} C_n F(s+jn\omega_s) \cdot \frac{1/RC}{(s+jn\omega_s)+1/RC}$$

or in terms of $s = j\omega$

$$R_{(j\omega)} = \sum_{n=-\infty}^{n=\infty} C_n F(j\omega+jn\omega_s) \cdot \frac{1/RC}{(j\omega+jn\omega_s)+1/RC}$$

The bandpass is centered around each $n\omega_s$ frequency with the three-db points about $(\omega - n\omega_s) = 1/RC$. However, again, the sampling frequency spectrum reduces the magnitude by the $(\sin x)/x$ function.

593

Combined effect of both filter responses upon noise reduction

Figure 32 shows the responses of the two network transfer functions as discussed earlier and the combined output from the array.

a. TRANSLATED HIGH PASS

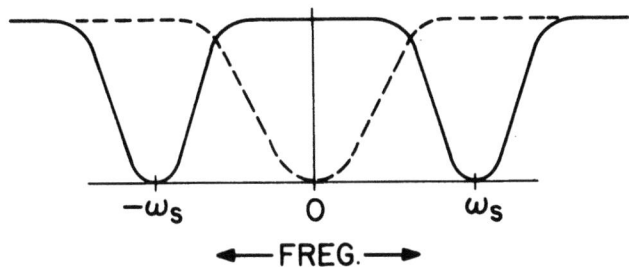

b. TRANSLATED LOW PASS

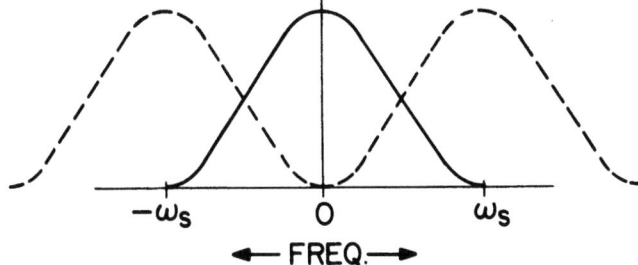

c. OUTPUT RESPONSE

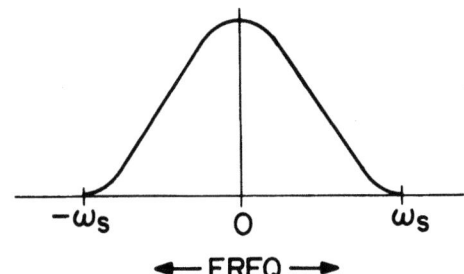

Figure 32. Representation of the response from the combined filters.

For the noise components, the combined transfer response is the product of the solid curve of Figure 32(a) with the dotted curve of 32(b), or a curve which has a zero about $n\omega_s$ and also a zero at the origin, so both high-frequency and low-frequency noise components are attenuated. Maximum response to noise is to components in the frequency region of $\omega_s/2$.

But the desired signal is itself derived from a sampled system, so there is a new translation which gives an output signal response as in Figure 32(c), where in this case the frequency represents an amplitude-modulation frequency for any individual pixel. A constant level at a picture element gives maximum attainable output for that level, but a level changing at the sampling frequency gives equal positive and negative input to the integrator with zero resultant output.

Thus the mechanization for noise reduction can be seen as an effective bandwidth reduction. The salient property which makes this noise reduction possible is that the video information is amplitude modulated on the sampling frequency and is shifted up in frequency in a process analogous to a suppressed carrier system, where the signal products formed are $\omega_s \pm \omega_v$, where ω_s is sampling frequency and ω_v is the video information.

This factor basically makes this signal processing technique an ideal system for high signal-to-noise ratio for the following reason: the system bandwidth may be reduced by band limiting around the sampling frequency in such a way as to preserve the dc component of the video information. This technique is similar to the video transmission and reception of tv signals.

Because the array's video information is translated up around the sampling frequency, the low-frequency response of the amplifier may be reduced, thus rejecting the noise otherwise introduced by the devices used in the signal extraction process such as the FET used for a buffer amplifier stage. The sampled differentiating and reset integrating networks fulfill this requirement, referring again to Figure 32. Note the dotted line in (a) which is the original spectrum of the high pass or differentiating network without sampling; however, this effective response is translated around the sampling frequency, ω_s, which effectively leaves a slot in the frequency spectrum around the sampling frequency where the incoming information signal is carried. This is shown as a solid line in Figure 32(a).

595

The integrate and hold circuit in effect translates the signal down from the sampling frequency, ω_s, to 0. This circuit reduces the high frequency end of the spectrum, because the filter's response multiplies the sampling spectrum. See Figure 32(b). A low pass filter is in effect sampled with its characteristic translated up to form a relatively narrow band pass around the sampling frequency. This reduces the noise spectrum on the upper end of the noise spectrum. The reset also reduces low-frequency noise.

Although this circuit forms a band pass against the uncorrelated noise spectrum, it convolves with the input video signal spectrum and translates the video information's center frequency down to dc and the low pass spectrum as in Figure 32(c). The frequency convolution takes place because the sampling switch of the low pass synchronously time multiplies the incoming samples of the video. The frequency domain was introduced for signal processing because it lends itself to a simpler noise analysis using band-limiting techniques.

EXPERIMENTAL MEASUREMENT

This low-noise circuit configuration was developed specifically to measure the detectable level of irradiance input to a Reticon self-scanned array. Its noise performance characteristic was measured with a thermally cooled RL-1872 array. By using a calibrated light source the equivalent source noise was measured. A comparison was made with the noise generated by a similar amplifier connected to a dummy array. Noise was considered to be effective over an equivalent amplifier bandwidth of 250 KHz.

General experimental configuration

The configuration of Figure 33 was used to obtain the experimental data. The sections are depicted in more detail in Figures 34 and 35. The intermediate section is shown as part of the integrator in Figure 35.

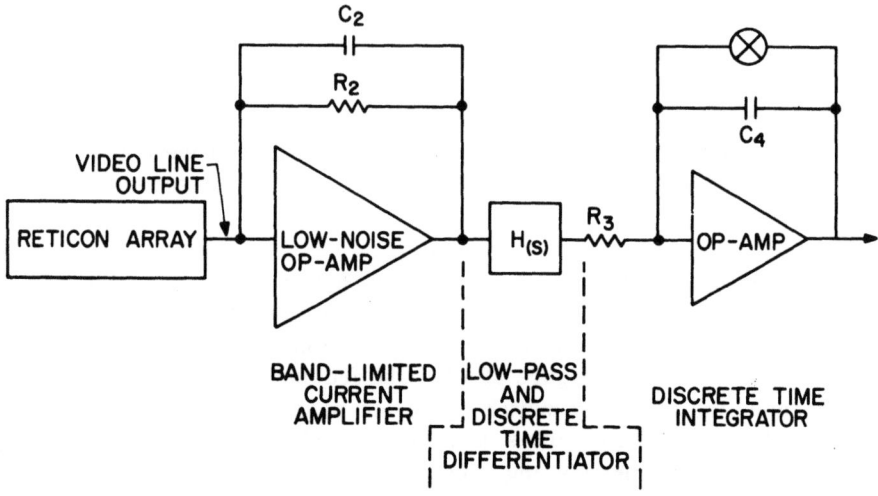

Figure 33. Block diagram of a low-noise amplifier.

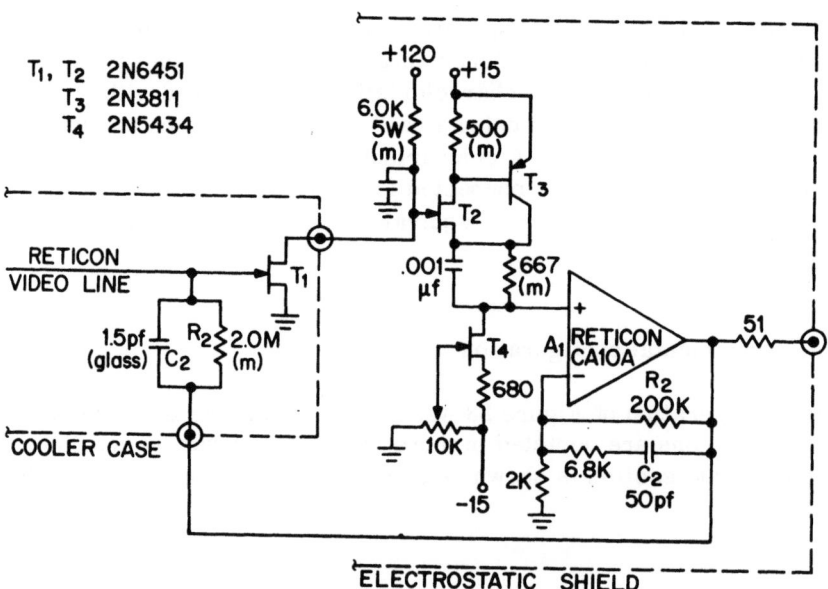

Figure 34. Schematic circuit of the current amplifier.

Figure 34 shows the schematic circuit for the transimpedance current amplifier as evolved for the tests. The source-follower following the input stage serves now only as a mechanism for control of the dc offset voltage introduced by the FET input stage. The source follower was adapted to a voltage-translation buffer by supplying an adjustable constant current by means of T_4. T_2 and T_3 are both selected for their low-noise characteristics.

The time constant $R_2 C_2$ determines the rate of decay of the output voltage as each cell contributes its charge to the video line. The time constant thus is a compromise between providing the maximum energy (area) in each output pulse, yet not carry crosstalk to the next bit period. R_2 contributes a substantial amount of thermal noise, diminished by increasing R_2. C_2 is already at a practical lower limit. $R_2 C_2$ can reasonably be as large as 1/3 the clock period (3.3 µsec) since for such a ratio only 5% of the initial charge transferred to the video line at each clocking still remains on the video line at the end of the clock period. Therefore, R_2 as shown is at its practical limit. Note that a proper noise budget must be considered, since so many separate sources contribute to the overall noise.

At low light levels, additional flat gain may be desirable between the input transimpedance amplifier (Figure 34) and the discrete time integrator (Figure 35). For some of the experimental work a commercial Tektronix low-noise differential amplifier was used (single ended) for this function when needed. It has a minimum calibrated gain of 100 and bandwidth dc to 1 MHz.

Discrete time integrator

The area of the output pulse as derived from the input amplifier is proportional to charge delivered by each photodiode, so long as each pulse transient is essentially complete by the end of the clock period. A discrete-time integrator thus may be used to give a proportional output that is generally flat-topped toward the end of each bit period, where the current amplifier's transient output amplitude has nearly vanished, so that a sampling pick-off becomes noncritical, and discrimination against high-frequency noise becomes optimized.

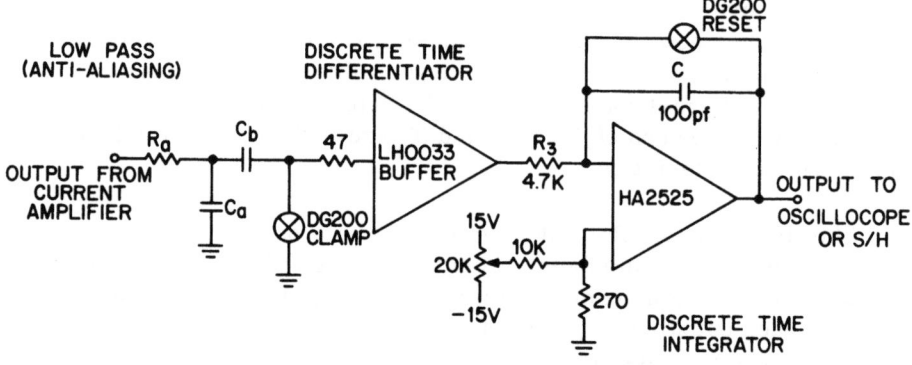

Figure 35. Block diagram for the discrete-time differentiator and integrator.

This circuit, also shown in Figure 35, along with the dc restoration network, performs the integration function. This circuit is the actual Miller integrator referred to in the analysis of the discrete time integrator and essentially resets every clock cycle, after integrating the pulse at the output of the buffer stage shown as LH0033 **buffer**.

Qualitatively, the actual input to the integrator is an analog input (continuous noise) plus the discrete-time signal input, from which is subtracted the sampled version of the continuous analog input. The discrete-time component is virtually unaffected. The result is to discriminate against low-frequency noise.

Discrete-time differentiating network

This network is also shown in Figure 35 at the input to the LH0033 buffer. This network not only performs as the differentiator as described in the analysis under the section, "Low Frequency Noise Suppression With a Discrete-Time Differentiator", it is a dc restoring circuit as well.

In addition to the differentiating network, there is a pre-filter which performs an anti-aliasing function by band limiting the noise. As in so many situations the design of the input filter and differentiation network is a compromise. The clamp is held for a fraction of the bit period, so that time constants while clamped should be short compared to the clamped period. The DG 200 clamp has an "on" resistance less than 100 ohms, so the overall time constant, 100 ohms times 0.5nf, is adequately short, while the unclamped filter time constant, 560 ohms times 1 nf, gives a high-frequency cutoff of 280 KHz without severe loading on the operational amplifier. But complete readjustment of the interior capacitor charge, while clamped, through 560 ohms cannot occur, so the result is a compromise between a short time constant for discharge while clamped and a longer one to limit the low frequency.

Array configuration

The array, mounted on the thermoelectric cooler, was located in a defined position approximately one meter directly below a ribbon-filament incandescent light source (100 W), operating at a color temperature of 2870°K. The enclosing box restricted the angular output by means of a collimating adjustable iris approximately 2 cm maximum diameter located 10 cm below the filament.

Radiant flux measurements were made by locating a calibrated detector at the array position. Attenuation of the radiant flux was accomplished by placing neutral-density filter(s) over the exit aperture so that light was uniformly diminished without any change in distribution. A United Detector Technology PIN-10 CAL/DF detector was used to take prime data.

Electrical readings were taken on a calibrated oscilloscope.

Data

Table I shows a set of data taken with the circuit configuration as described. This data was compiled to obtain noise-equivalent charge, i.e., how many electrons per pulse constitutes the noise charge equal to the signal charge. This was calculated from the data of Table I and is 770 electrons.

The minimum detectable signal charge (where the signal charge equals noise charge) is 770 electrons, which is the same as 1.23×10^{-16} coul./pulse. Then for any known clocking or scan time the noise equivalent current can be calculated.

Taking the scan time equal to 20.48 milliseconds as in line two, yields a noise equivalent current of 0.006 pico amps. This is approximately a factor of 20 times lower than the minimum detectable signal level with the current amplifier with a bandwidth of 250 KHz without the signal processing circuit.

TABLE I

Measurement of Charge-Deflection Ratio
(Room Temperature)

Scan Period msec.	(20.5 K) Video mv Illumined	No Start	Current NA	UDT Light	Dark	$\mu J/m^2$	Defl. mv.
33.33 (90% sat)	1.578	0.185	63.4	17.62	0.141	48900	10000
20.48	1.616	0.185	69.8			30100	6200
10.34	1.631	0.185	70.6			15050	3100
5.12	1.630	0.185	70.5			753	1600

Peak-to peak noise deflection   1.5 mv

Deflection ratio = $(6200 \text{ mv})/(30100 \mu J/m^2) = 0.21 \text{ mv}/\mu J/m^2$

Video line current = (video mv operation - video mv, no start)/20.5K

Average current, well below saturation   70 nA

(Charge/pixel) x 468 = Q/scan = I x scan period.

For 20.48 msec. $q = (I \times 20.48 \times 10^{-3})/468 = (70 \text{ nA} \times 20.48 \times 10^{-3} \text{sec})/468 = 3.06 \times 10^{-12}$ coulombs = 1.91×10^7 e for 6200 mv = 3.08×10^6 e/volt.

For 11V, $q = (11/6.2) \times 1.91 \times 10^7 = 3.39 \times 10^7$ electrons, or 5.42×10^{-12} coulombs.

Common line at 5.94 volts. For a normal 5.0 volts on the common line, $q_{sat} = (5.00/5.94) \times 3.39 \times 10^7 = 2.85 \times 10^7$ electrons.

Noise equivalent charge = $(1.5 \text{ mv}/(6 \times 11000 \text{ mv})) \times 3.39 \times 10^7$
= 770 electrons.

CONCLUSION

From the array user's standpoint the signal processing circuit can be very useful to optimize the signal-to-noise ratio with respect to cost and implementation by keeping these key factors in mind:

1. The information signal from the array is essentially an amplitude-modulated suppressed-carrier system, i.e., it is translated up around the sampling rate.

2. The use of a current amplifier alone introduces a broadband noise which can be significantly reduced by processing with the discussed system.

3. A voltage-translating scheme produces an inherent sampling system which provides for an easy implementation of the discrete-time differentiation and integration.

With this circuit implementation an excess of 80 db of dynamic range is possible when the saturated signal output is referred to thermal equivalent noise of the amplifier.

BIBLIOGRAPHY

<u>Analysis and Synthesis of Sampled-Data Control Systems</u>, Benjamin C. Kuo, 1963, Prentice-Hall, p. 8.

<u>Signal Systems and Communication</u>, B. P. Lathi, John Wiley & Sons, Inc.

ELECTRICAL MANUFACTURING, "Sampled Data Systems", Carl O. Carlson, November 1959.

<u>Information Transmission Modulation & Noise</u>, Mischa Schwartz, McGraw-Hill Books Inc., 1959.

DISTRIBUTED FLOATING GATE AMPLIFIER

G. F. Amelio, R. H. Dyck
Research and Development Laboratory
Fairchild Camera and Instrument Corporation
4001 Miranda Avenue, Palo Alto, Ca. 94304

ABSTRACT. In order to extract the full performance capability from a buried channel CCD image sensor, it is essential that a low noise on-chip amplifier be used. The distributed floating gate amplifier fills this need. By repetitively sampling the video signal and coherently summing the result, the signal-to-noise ratio is improved by the square root of the number of stages. In this paper, the theory behind the distributed floating gate amplifier is presented and results from laboratory devices are presented.

1. INTRODUCTION

In a charge-coupled imaging device which is designed for low light level imaging applications, it is essential to include an on-chip amplifier which is capable of providing a low noise-equivalent-signal (NES) level. Since such a charge-coupled device can efficiently transport charge packets of the order of ten electrons or less in size, it is desirable to achieve an NES level which is also of the order of ten electrons or less. A distributed floating gate amplifier (DFGA) has been developed which exhibits an NES of less than 20 electrons at a noise bandwidth of 3 MHz.

2. Non-Destructive Signal Detection with a Floating Gate Amplifier (FGA)

A floating gate will respond in potential to

the amount of signal charge in the CCD channel and
this change in potential can be used to modulate the
current in an FET. The signal charge is not mixed
with any other charge in this process and can therefor
be recovered perfectly intact. The input circuit of
the FGA can have the desirable features of having a
very low input capacitance and freedom from reset
noise, i.e., from the fundamental capacitance noise
of a reset switch.

The design and operation of the FGA have been
described in detail by Wen (1974). A less detailed
treatment of the FGA has been incorporated with a
description of the DFGA as reported by Amelio (1974)
and these are reproduced below.

3.   THE DFGA

A DFGA was developed as part of a 190 x 244-
element area imaging device. The following is a description of the concept, design, and performance of
the DFGA in this context.

In addition to the more conventional gated
charge detector preamplifier, the 190 x 244 array incorporates a type of preamplifier capable of amplifying charge, a new concept in the area of semiconductor
devices. The DFGA concept is based on the property of
charge-coupled devices that signal charge can be
passed under a sensing electrode and then further
transported to other sensing electrodes with no signal
degradation. By sensing the signal charge repeatedly,
it is possible to improve the signal-to-noise ratio
in power relative to a single stage amplifier by the
number of times the signal is sensed. By summing
the amplified signals in a second CCD register, proper
reconstruction of the signal in the time domain is
automatically obtained.

A schematic diagram of the DFGA incorporated
in the 190 x 244 array is shown in Figure 1. It consists of four functional parts: an input register, a
bank of charge amplifiers with floating gate inputs,
an output register, and an output amplifier. This
particular DFGA has 12 stages. It uses inverting amplifiers between the two registers and has a floating
gate output amplifier. A 4-phase register clocking
scheme is used to obtain maximum clocking flexibility;
the two registers are driven by the same set of clocks.

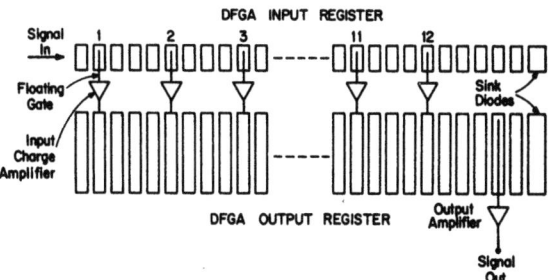

Figure 1 Schematic of the Distributed Floating Gate Amplifier (DFGA)

The DFGA, shown in the Figure 2 photograph, occupies an area of approximately 250 x 400 μm. A single DFGA stage consists of a source, a floating gate, a bias electrode, a control gate, and a gate that serves to minimize clock coupling from the control gate to the floating N+ diffusion, as shown in Figure 3. It can be seen from the potential well profiles in Figure 3 how charge control is achieved. During the period that a signal charge packet in the input register is under the floating gate, the control gate is pulsed on for a precise time interval. During this time interval, a small fixed charge and a signal-dependent current flows; the larger the intial charge packet, the less current flows. The effect of signal charge level on the voltage of a floating gate is determined analytically by a capacitance network. A cross-sectional view of a floating gate in Figure 4 identifies the several capacitances involved. The responsivity is given by

$$\frac{dV_{FG}}{dQ_s} = \frac{1}{C_2 + C_4 + C_{in} + \frac{C_3}{C_1}(C_1 + C_2 + C_4 + C_{in})} \quad (1)$$

This responsivity by design can be on the order of 5 μV/electron.

The charge gain for small signal is given by

$$A_Q = \frac{dV_{FG}}{dQ_s} g_m t \quad (2)$$

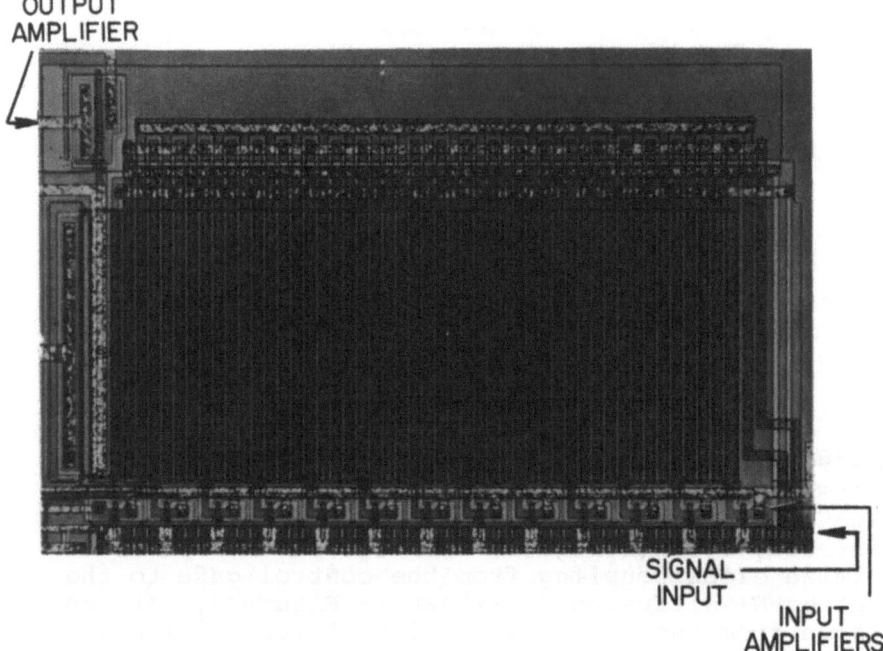

Figure 2 Photograph of the Distributed Floating Gate Amplifier (DFGA)

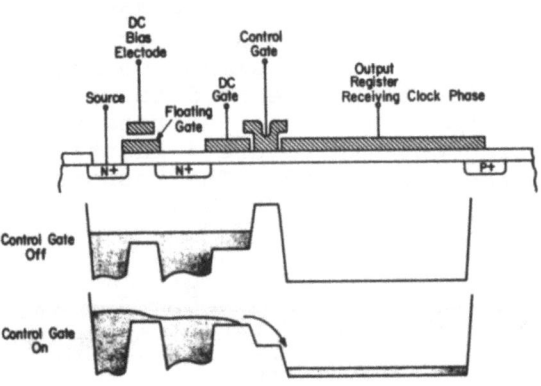

Figure 3 Cross-Section of One Input Charge Amplifier in the DFGA

where g_m is the MOS transistor transconductance and t is the time that the control gate is on. Since g_m is in general a function of the current level of the MOS transistor and since the noise level is also a function of the current level, it is desired to establish the current level which gives the optimum signal-to-noise ratio at the output for small signal level. It can be shown that shot noise in the MOS transistor is the dominant noise source of the DFGA. Therefore, the output of a single stage in electrons is

$$n_S = \frac{dV_{FG}}{dQ_S} \frac{g_m t Q_S}{q} \qquad (3)$$

The noise in rms electrons is

$$n_N = \sqrt{\frac{I_D t}{q}} \qquad (4)$$

where I_D is the drain current. This gives a signal-to-noise ratio

$$\frac{n_S}{n_N} = \frac{dV_{FG}}{dQ_S} g_m \sqrt{\frac{qT}{I_D}} Q_S \qquad (5)$$

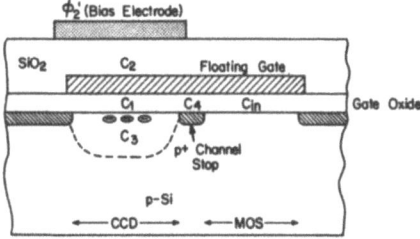

Fig. 4 Cross-Sectional View of the Floating Gate Input Structure showing the several capacitances used in the analysis

By assuming g_m is proportional to I_D and by considering that I_D and C_{in} (from Equation 1) are both proportional to the channel width of the MOS transistor, it is possible to optimize the MOS transistor geometry. Then, from I_D vs V_{FG} and the source bias for the particular transistor design, it is possible to optimize I_D. Figure 5 shows how I_D and the signal-to-noise ratio (SNR) change with gate voltage. The curve of I_D vs V_{FG} was measured and the SNR curve was calculated using this data. It may be noted that the optimum occurs close to the end of the exponential range of the FET.

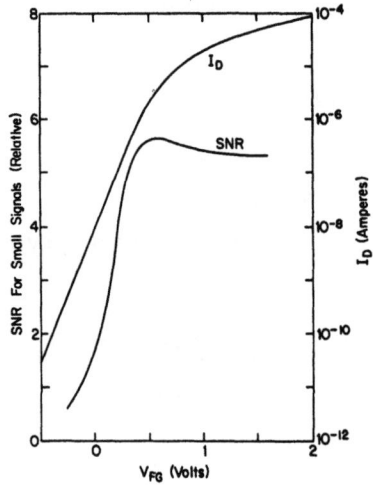

Figure 5 The Dependence of Drain Current and Signal-to-Noise Ratio on the Voltage of the Floating Gate in the DFGA

The size of a CCD output-register stage is determined by the saturation input signal level of the DFGA and by the number of stages in the DFGA. In this device, over half of the DFGA area is used by the CCD output register.

There is no clearly defined optimum number of stages. Since the fundamental principle behind the DFGA concept shows that SNR in voltage improves as the square root of the number of stages and since the size of each output-register stage must be increased linearly as the number of stages increases, SNR increases as the one-fourth power of the output register area.

For the 12-stage DFGA on the 190 x 244 area image sensor, the following performance has been observed. The dynamic range referred to the input is determined by a saturation level of approximately 5×10^4 electrons per picture element and by an rms noise level in the dark of approximately 10 electrons. The resulting dynamic range, defined as the ratio of saturation signal to minimum usable signal, may be estimated as follows:

$$\text{D.R.(input)} = \frac{n_{sat}}{n_{min}} = \frac{n_{sat}}{5n_{rms}} = \frac{5 \times 10^4 \text{ electrons}}{50 \text{ electrons}} = 10^3 \quad (6)$$

This performance was observed at a 1 MHz element sampling rate at 25°C ambient. However, the measurement was made in such a way that the element sampling rate could have been increased to approximately 7 MHz without altering either the output signal or output noise level.

The relationship between input charge and output voltage is shown in Figure 6. The NES for the unit tested was 10-20 electrons.

An image obtained with the 190 x 244 array and this DFGA is shown in Figure 7 together with a corresponding image obtained using the FGA output from the array. This image demonstrates the quality of signal which has been obtained at higher light levels using the DFGA output.

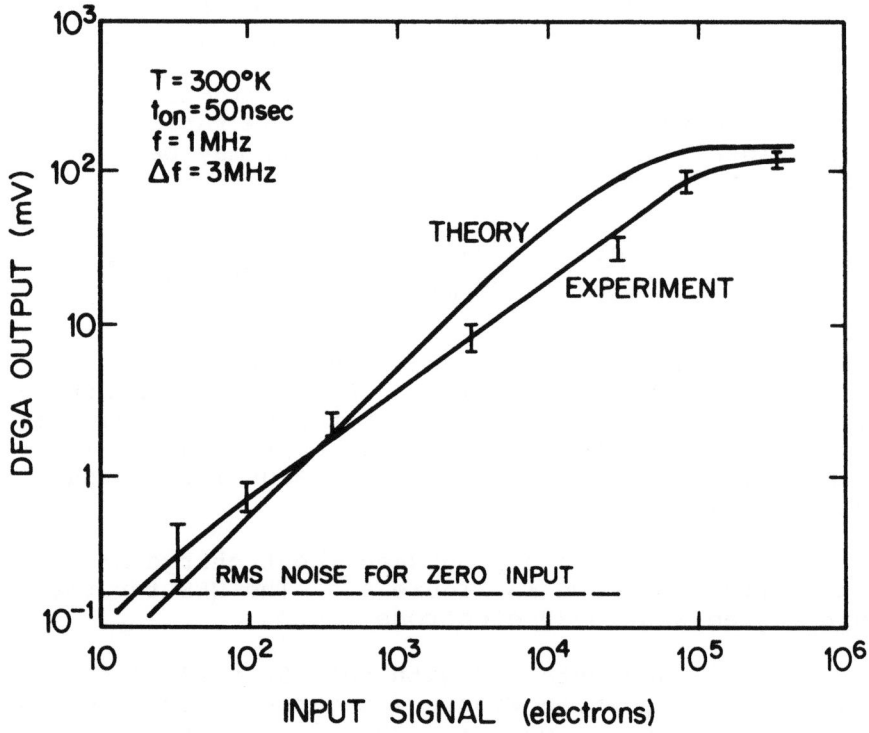

Figure 6 The DFGA Transfer Characteristics predicted by theory and measured for one device

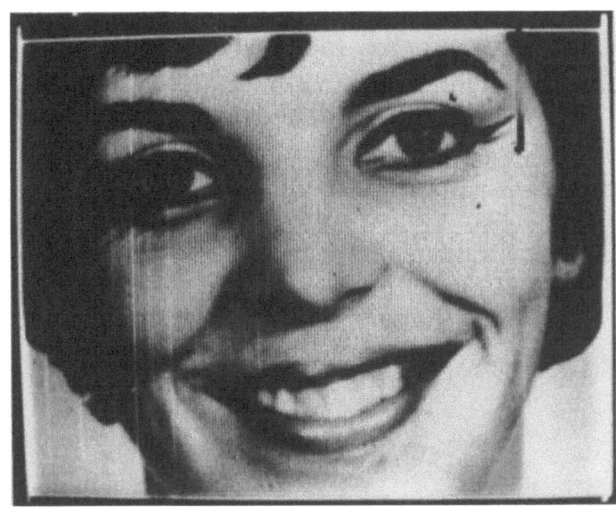

(a)   FGA Output

(b)   DFGA Output

Fig. 7   Imagery taken with the 190 x 244 array

REFERENCES

1. D. D. Wen, "Design and Operation of a Floating Gate Amplifier", IEEE J. of Solid-State Circuits, SC-9, 410-14 (Dec. '74)

2. G. F. Amelio, "The Impact of Large CCD Image Sensing Area Arrays," Proc. of the International Conf. on the Tech. and Appl. of Charge-Coupled Devices, Edinburgh, Sept. 25-27, (1974)

FIXED PATTERN NOISE AND COOLED PHOTOSENSOR ARRAYS

G. F. Amelio, R. H. Dyck
Research and Development Laboratory
Fairchild Camera and Instrument Corporation
4001 Miranda Avenue, Palo Alto, Ca. 94304

ABSTRACT. Imposed on the video signal from a photosensor array are in general, undesirable signals resulting from a number of parasitic effects. These effects include clock interference, leakage currents, random noise, and fixed pattern noise. Cooling reduces some of these effects and improves the video signal-to-noise ratio. This paper will discuss the effects of random and fixed pattern noise on photosensor array performance and the benefits of cooling. Data obtained by cooling actual devices will be presented.

1. INTRODUCTION

Most imaging devices suffer in performance from an assortment of effects that appear within the video bandwidth of the output signal. The subject of this paper is this class of effects as they are found in charge coupled imaging devices; charge transfer inefficiency effects are excluded.

2. TYPES OF FIXED PATTERN NOISE (FPN)

2.1 Dark current non-uniformities

Dark Current Spikes. A single element with a high dark current is said to produce a dark current spike. The spike amplitude depends heavily on processing. In general, it only depends slightly on channel potential. In most cases, the temperature dependence is that of the intrinsic carrier concentration of silicon. Fig. 1 shows an image from a 100 x 100 element array, the CCD201, which contains dark current spikes (from Dyck and Jack, 1974).

<u>Gaussian-shaped bands in linear arrays.</u>
This type of FPN has been seen in some reject CCD101 units (unpublished Fairchild data). It is of interest because it may be a dark current phenomenon which exists in other devices as well. In general, the band is 400-600 μm in width. Model: minority carriers are injected into the substrate from a point outside the active area by some type of charge pumping mechanism. These carriers diffuse away from the point of generation to a characteristic radius determined by $\sqrt{D\tau}$ and the substrate thickness. Some of these minority carriers are collected in the sensor array, producing the gaussian band in the video output signal. The origin of the minority carriers can be controlled by design.

<u>Excess edge dark signal in area image sensors.</u>
Effects related to the gaussian bands described above can occur near edges. A defective channel stop on the perimeter of the array can leak charge into the channel. This effect is dependent on the perimeter structure used in the design.

2.2   Clock Feed-thru into the video output signal

In the simplest CCD scanner, i.e., in a linear imaging device with a single CCD output register, the only clock feedthrough within the video bandwidth is the random noise on the clock waveforms or parasitics from count-down logic circuits which might be present within the system. Linear arrays with two CCD output registers can have an odd-even effect. Fig. 2 shows this effect in the output of such an array, the Fairchild CCD110. In this device, it has been shown that the effect is primarily due to capacitive coupling in the device package rather than in the monolithic circuit itself.

2.3   Fat Zero Fixed Pattern Noise

In an SCCD imager where a fat zero is generally required, this charge may be injected electrically or optically. Since it is in general desirable to maximize the dynamic range of an imager, it is

Figure 1   Imaging with the CCD201 at 27°C and at 10% of saturation illumination level

desirable to obtain the lowest possible fixed pattern noise in the fat zero.

With electrical injection from a single injection element, very stable bias supplies are required to provide a constant fat zero charge-pulse train; the level of stability required is approximately the same as is required for some of the preamplifier bias supplies. However, certain types of area image sensors are not well suited to the use of a single electrical injection element.

In an area imager of the frame transfer type, it is advantageous to have one injection element for each vertical CCD register. In this case, a large fixed-pattern-noise can exist due to variations in control-gate threshold voltage from element to element.

Optical injection can be achieved simply by the use of a small amount of uniform illumination. However, in order to maximize dynamic range, the degree of uniformity must be set at a very high level.

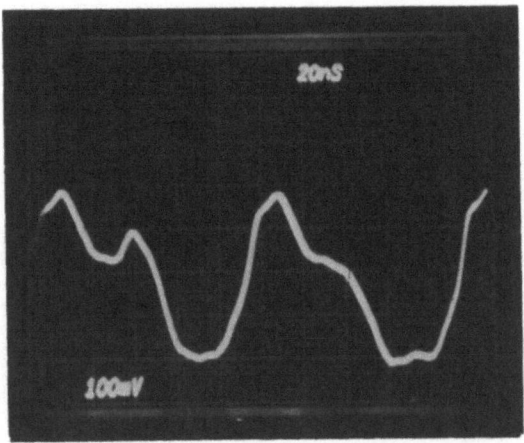

Fig. 2 The video output waveform for two elements of a CCD110 at a 10 MHz element output rate and under uniform illumination. The reset baseline is at the top and the video signal pulses are downward-going. An odd-even effect distorts the waveform but has virtually no effect on the peak amplitudes of the signal pulses.

For example, if a dynamic range of 200-to-one is desired and a 20 percent fat zero is required to obtain satisfactory transfer efficiency, then the non-uniformity must be 2.5 percent or less.

3. Temperature Dependence of Fixed Pattern Noise

From simple theories it is not possible to predict precise temperature dependences. This is because the dark current parameters, τ_0 and S_0, in the equation

$$j_D = \frac{qn_i}{2} \left(\frac{W_D}{\tau_0} + S_0 \right)$$

contain defect cross-section factors and these can be temperature dependent. Empirically, it has been found that τ_0 and S_0 are effectively constants. This is the case for a variety of silicon devices from a little above room temperature down to the order of $-40^\circ C$ and possibly lower.

Some CCD imaging devices follow n_i vs. T down to $-40°C$ within experimental error. Some have a smaller temperature dependence at the lower temperatures. Fig. 3 shows the dependence for such a device. Whereas some units decrease in dark current by 1000 times on cooling to $-40°C$, this unit only decreased by 250 times. Fig. 6 of the chapter entitled "Buried Channel CCD's" shows images at $-40°C$ for the same device used to produce the image in Fig. 1. This illustrates the reduction of dark current spike amplitudes by greater than 100 times on cooling from $27°C$.

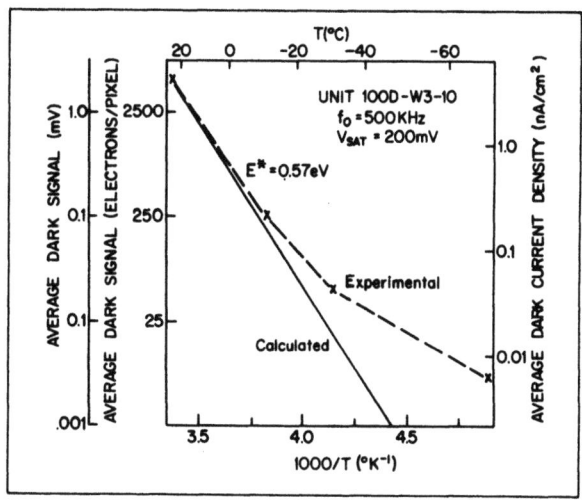

Fig. 3 Experimental and Calculated Curves of Dark Current vs temperature. The experimental data was taken with a unit exhibiting an average dark current of 4 nA/cm^2 at room temperature. The two ordinates on the left give the dark output and the corresponding dark charge per pixel. The ordinate on the right gives the calculated dark current density

REFERENCES

1. R. H. Dyck and M. D. Jack, "Low Light Level Performance of a Charge-Coupled Area Imaging Device," *Proceedings of International Conf. on the Tech. and Appl. of Charge-Coupled Devices*, Edinburgh, Sept. 25-27, 1974

Section VII

SYSTEMS

ALIASING AND MTF EFFECTS IN PHOTOSENSOR ARRAYS

D. F. Barbe

Naval Research Laboratory
Washington, D. C. 20375

S. B. Campana

Naval Air Development Center
Warminster, Pennsylvania 18974

ABSTRACT. Analyses of the sampling process in the space domain and in the spatial frequency domain are presented. The effects of aliasing which accompany response beyond the Nyquist limit and the effects of prefiltering to reduce the response beyond the Nyquist limit are discussed. Empirical results, obtained with CCD imagers, which illustrate the tradeoffs between aliasing and MTF are discussed.

1. THE SAMPLING PROCESS IN THE IMAGE PLANE

The process by which a charge configuration, $Q(X,Y)$, is formed in an imaging array in response to an irradiance distribution, $H(X,Y)$, is illustrated in Fig. 1. For simplicity, only on dimension (X) is discussed. An array can be described mathematically by the repetition of a basic cell with repeat distance d. Let $R(X)$ denote the response as a function of X across the basic cell. For simplicity this is shown in Fig. 1 as a rectangular response function of width ΔX. The absolute value of the Fourier Transform of the basic response function is called the modulation transfer function (MTF) of the cell. Let $H(X)$ denote the irradiance as a function of X on the image plane. The Fourier Transform of $H(X)$ describes the irradiance in terms of its spatial frequency content. Before the sampling process in the array, $H(f)$ is filtered (multiplied) by $R(f)$. This corresponds to the convolution of $R(X)$ and $H(X)$ in the X-domain. The array then samples the result. In the X-domain, this means $H(X) * R(X)$ is multiplied by a periodic train of delta functions with periodicity d. The result is $Q(X)$ shown in Fig. 1D. In the spatial frequency domain the sampling process consists of convolving H(f) R(f) with S(f) which is the Fourier Transform of S(X). The result is the repetition of H(f) R(f) at intervals $f_s = 1/d$. Mathematically, the complete process is given by Eqs. (1) and (2) in the X-domain and the spatial frequency domain respectively:

$$Q(X) = \left[H(X) * R(X)\right] S(X), \qquad (1)$$

where $S(X) = \sum_{i=-\infty}^{\infty} \delta(X-id)$

and

$$Q(f) = \left[H(f) R(f)\right] * S(f), \qquad (2)$$

where

$$S(f) = f_s \sum_{i=-\infty}^{\infty} \delta(f - if_s),$$

$f_s = \frac{1}{d}$, and Q(X) and Q(f) are a Fourier pair.

2. ALIASING

If the sampling frequency, $f_s = 1/d$, is not at least twice the highest frequency component present in H(f) R(f), then when H(f) R(f) is convolved with S(f), the repeated spectra H(f) R(f) will overlap each other as shown in Fig. 1D. Since natural scenes are not band limited, the spectra will always overlap unless H(f) is prefiltered. Overlapping of the spectra causes frequencies higher than the Nyquist limit ($f_n = \frac{1}{2} f_s$) to appear in the passband ($-f_n \leq f \leq f_n$) as lower frequency components - thus the term "aliasing". Fig. 1D is redrawn and expanded in Fig. 2. The solid curve represents the true frequency response H(f) R(f), and the broken curves represent aliased spectra. Branch (1) is the true response to frequencies from zero to f_n; branch (2) is the response in the passband to frequencies from f_n to 2 f_n; branch (3) is the response in the passband to frequencies from 2 f_n to 3 f_n; and branch (4) is the response in the passband to frequencies from 3 f_n to 4 f_n. For example, Fig. 2 shows that the frequency 1.5 f_n would give a response at 0.5 f_n.

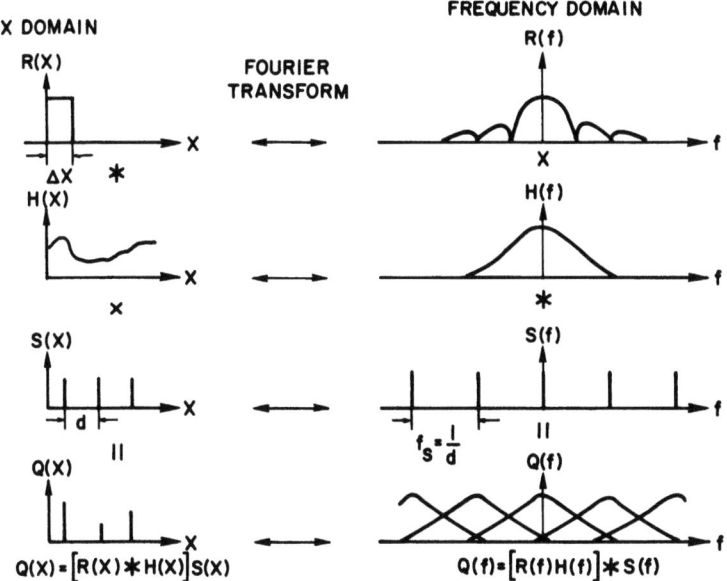

Fig. 1 Sampling process in the X-domain and in the spatial frequency domain.

This can also be shown in the X-domain via Fig. 3. Fig. 3A shows that the sampling gives a true reproduction of the frequency component f_n. Fig. 3B shows that the frequency component 1.5 f_n appears as (or is aliased to) 0.5 f_n. Fig. 3C shows that 2 f_n is aliased to zero.

3. PREFILTERING

From the preceeding discussion it would appear that aliasing is an extremely serious problem. Now let us investigate how aliasing can be reduced. Fig. 4A shows a response function which has high values beyond the Nyquist frequency. It is of course desirable to have high MTF up to f_n for the array to respond to high spatial frequencies. At the same time, it is desirable to sharply cut off the response at f_n with a prefilter function as shown in 4B to eliminate aliasing. Unfortunately such an optical filter function is not physically realizable. A more realistic optical prefilter function is shown in 4C. Fig. 4D shows that the result of prefiltering (to minimize aliasing) is to reduce the response to high spatial frequencies in the passband. Therefore, there is a tradeoff between high MTF near the Nyquist frequency and aliasing.

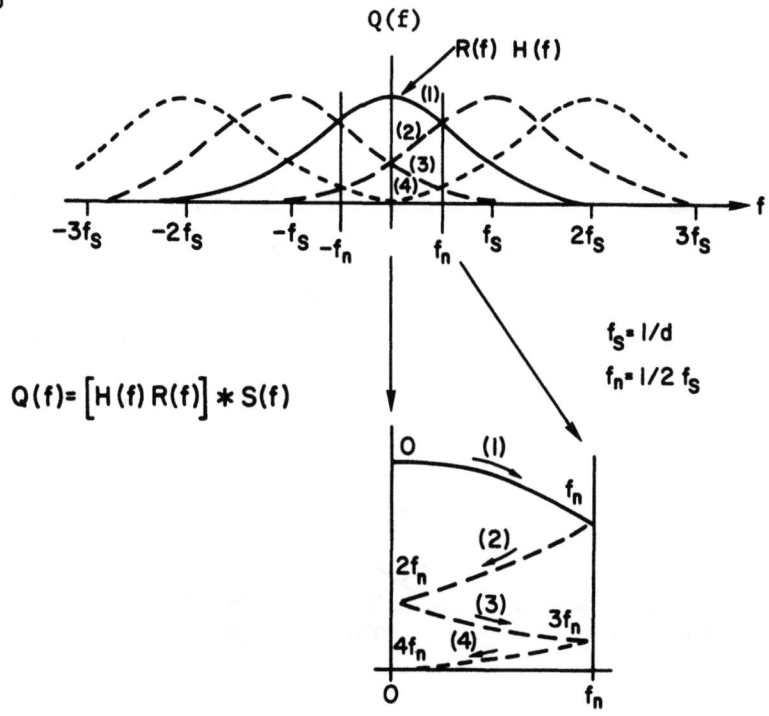

Fig. 2 Frequency domain analysis of sampling process. Branch (1) is the desired response and branches (2), (3), and (4) represent response to $f > f_n$ aliased into the passband.

4. ALIASING VERSUS MTF IN CCD's

CCD (Charge Coupled Device) technology offers great flexibility in the design of two-dimensional imagers. Barbe and White [1] have examined the major differences between the two basic CCD imager designs (interline transfer and frame transfer) emphasizing responsivity and MTF (Modulation Transfer Function) within the spatial frequency passband defined by the pixel (picture element) spacing. It has been suggested, however, that since CCD's sample the image plane and are, therefore, subject to aliasing, that the spatial frequency response beyond the sampling limit is a major design consideration. It has been argued that since response beyond that limit serves only to introduce spurious signals into the allowable passband, this response should be eliminated by filtering the scene before sampling. In practice, however, this cannot be done perfectly. Some response to frequencies within the passband must be sacrificed if spurious response

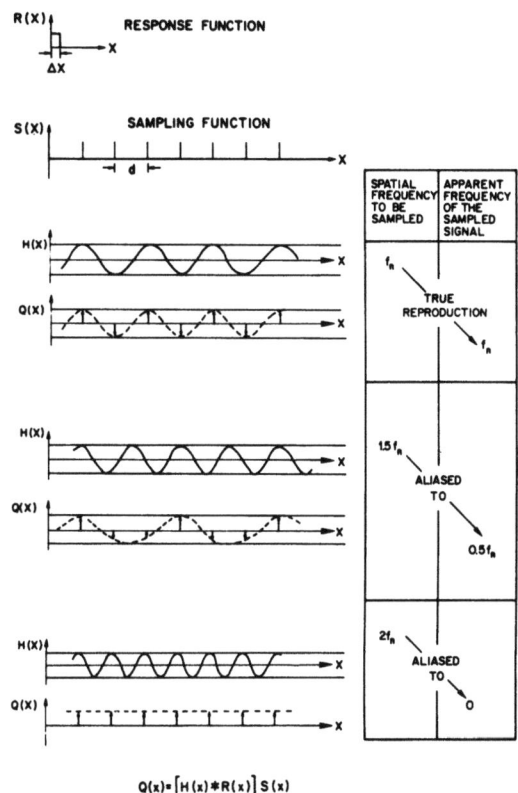

Fig. 3 X-domain analysis of sampling illustrating the true response to the spatial frequency f_n and the aliased response to spatial frequencies above f_n.

is to be avoided. Thus, in the design and use of sampling imagers there is a tradeoff to be made between the adversity of aliasing and the inevitable loss in image sharpness incurred in reducing aliasing. Although several authors [2,3,4] have offered arbitrary guidelines for making this tradeoff, there is no firm theoretical basis for trading off desirable response to reduce spurious response. It is, therefore, necessary to resort to an empirical approach, utilizing representative scenes reproduced with an imaging system for which true and spurious response is known. To this end Campana and Barbe [5] utilized a 100 x 100 element CCD imager as a vehicle to produce a wide range of imagery under controlled conditions of spatial frequency response.

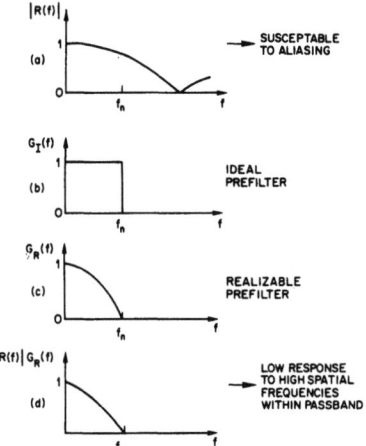

Fig. 4 (A) Response of sensor having high MTF,
(B) ideal prefilter characteristic,
(C) realizable prefilter characteristic, and
(D) prefiltered response

4.1 The experiment

A Fairchild CCD-201 area imager was used to make the pictures reproduced in the remainder of this chapter. This device has 100 columns, spaced at 1.6 mils, and 100 rows at 1.2 mils. The array utilizes the interline transfer structure, with more than 50% of the image area obscured by the aluminum on the vertical scanning registers, as shown in Fig. 5. In the vertical direction, sensing elements are contiguous rather than overlapped as in the interlaced vertical frame transfer structures. The non-contiguity in the horizontal direction makes this structure particularly vulnerable to aliasing, as is evidenced by the imagery shown in Fig. 6. Each of the nine views shown are of a different set of vertical bars extending over the entire sensing area. The frequency of the bars is expressed in multiples of the sampling frequency, f_n. Since the horizontal cell spacing is 1.6 mils, f_n is 12.3 cycles per millimeter. Thus, only the top row of Fig. 6 represents true imagery. The remaining six views are Moire' patterns produced by the interaction of the CCD structure and the bars of the test chart. Note that the aliased frequencies correspond closely to those predicted by the frequency domain analysis of Fig. 2.

In order to reduce the strength of spurious signals, it is necessary to filter the input scene before sampling. In a CCD

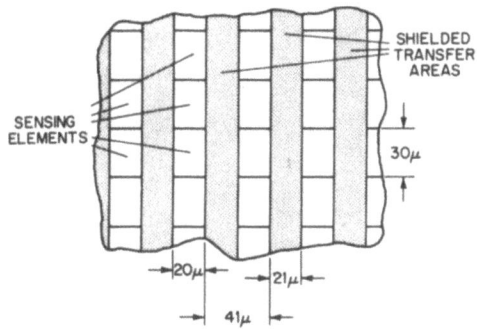

Fig. 5   Structure of the 100 x 100 element CCD imaging chip used to perform the aliasing experiments.

this can be accomplished at the image plane by varying the size of the sensing element relative to the element spacing. For example, Fig. 7 shows the MTF's of arrays having (A) non-contiguous apertures, (B) contiguous apertures, and (C) overlapping apertures. Note that (A) would have good response to high spatial frequencies in the passband but would be susceptible to aliasing. The overlapping apertures of (C) would provide the prefiltering to reduce aliasing but would result in poor response to high spatial frequencies in the passband.

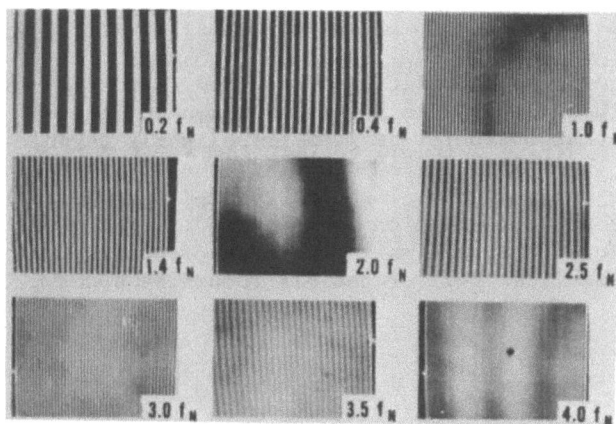

Fig. 6   Bar pattern imagery of a 100 x 100 element CCD. Fundamental frequency is given relative to the Nyquist frequency f_n.

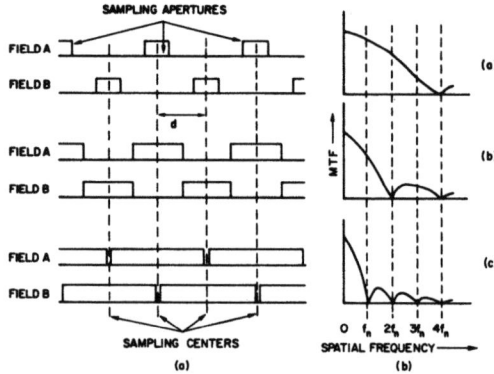

Fig. 7 Responses of sensors having
(A) non-contiguous pixels,
(B) contiguous pixels, and
(C) overlapping pixels

Further prefiltering can be accomplished using an optical low-pass filter at the input to the lens, [6] or by simply defocussing the lens. Since in this experiment the object distance was fixed, it was convenient to utilize the latter effect. MTF was measured at several focus positions of the lens. The measurements showed that at each focus position the additional MTF degradation produced by defocussing the lens had the form $[\sin(\pi f/f_c)/(\pi f/f_c)]$, where f is the spatial frequency, and f_c is a parameter dependent on focus. Thus, it was possible to closely simulate various conditions of cell size/cell spacing by merely rotating the lens focus ring to preselected positions.

Three different focus conditions were used. Condition "A" is that of best focus, where the MTF is as shown in Fig. 8A. Note that the null of the horizontal response is slightly beyond 4 f_n. In condition "B" the MTF is degraded so that the first null of the horizontal response occurs slightly beyond 2 f_n (Fig. 8B). This condition is an approximation to the case of contiguous elements. Condition "C" approximates the case of overlapped elements, where the first null of the MTF occurs at about 1 f_n (Fig. 8C).

4.2 Test imagery

Figs. 9, 10, 11, and 12 show imagery produced by the 100 x 100 CCD imaging chip for each of the three prefiltering conditions. Condition "A", of course, represents the worst case for aliasing, while condition "C" produces imagery with the least aliasing. The aperiodic scenes in Fig. 9, 11, and 12 show no deliterious effects due to aliasing for any of the conditions, while the effect of prefiltering seriously reduces the high spatial frequency content and reduces the sharpness of the imagery. Figs. 10 A and B illustrate the noticeable effect of aliasing on the vapor trails of three evenly spaced (periodic) jet engines. Fig 10C shows the Moire' is effectively removed by prefiltering. In Fig. 11 the CCD is used to image an aerial photograph of an offshore platform and Fig. 12 is a lunar landscape. In both scenes, prefiltering has an obvious deliterious effect. As in Fig. 11, small details such as the legs of the platform, and small craters are lost in Condition "C".

Other scenes, too numerous to reproduce in this chapter, were imaged as described above. In all cases the result was the same. Condition "A" produced sharp imagery with no obvious aliasing problem other than the occasional dropout of small details due to the shielding between photosites. Condition "B" produced slightly less sharp imagery with no dropout of small details. Condition "C" produced images that were generally inferior to those of "A" and "B".

To answer the question "When does it pay to prefilter" a very unusual scene was set up. This scene consisted of a set of vertical bars at 1.6 f_n as a background, with various aperiodic forms superimposed. Fig. 13 contains the CCD images of this scene. The scene contains a pair of vertical bars, one white and one black, which are positioned so that they are aligned with the light and dark bars of the Moire' produced by the interaction of the imager and the periodic background. In view C these bars stand out quite clearly, since prefiltering has greatly reduced the spurious response. However, in view A the bars are quite difficult to see since the spurious signal has a high contrast. Note, however, that if the bars are not aligned with the Moire' pattern, as with the large black bar at the left, they stand out quite well as an interruption of the Moire' pattern. In fact, several large gray objects in both scenes are visible in views A and B but not in view C where they blend in with the uniform gray produced by prefiltering. Thus, in this very artificial situation, aliasing in the form of Moire' can be detrimental or beneficial, depending on the exact conditions of size, orientation, and intensity.

5. DISCUSSION OF RESULTS

It seems evident from the imagery presented that the "aliasing problem" is not as severe as has been supposed. Indeed the prefiltering required to significantly reduce the response beyond the Nyquist limit appears to do more harm than good in most instances. This result may seem surprising considering the large degree of spurious response in the sensor as indicated in Figs. 8 A and B. Intuitively it would seem that all of the scene energy

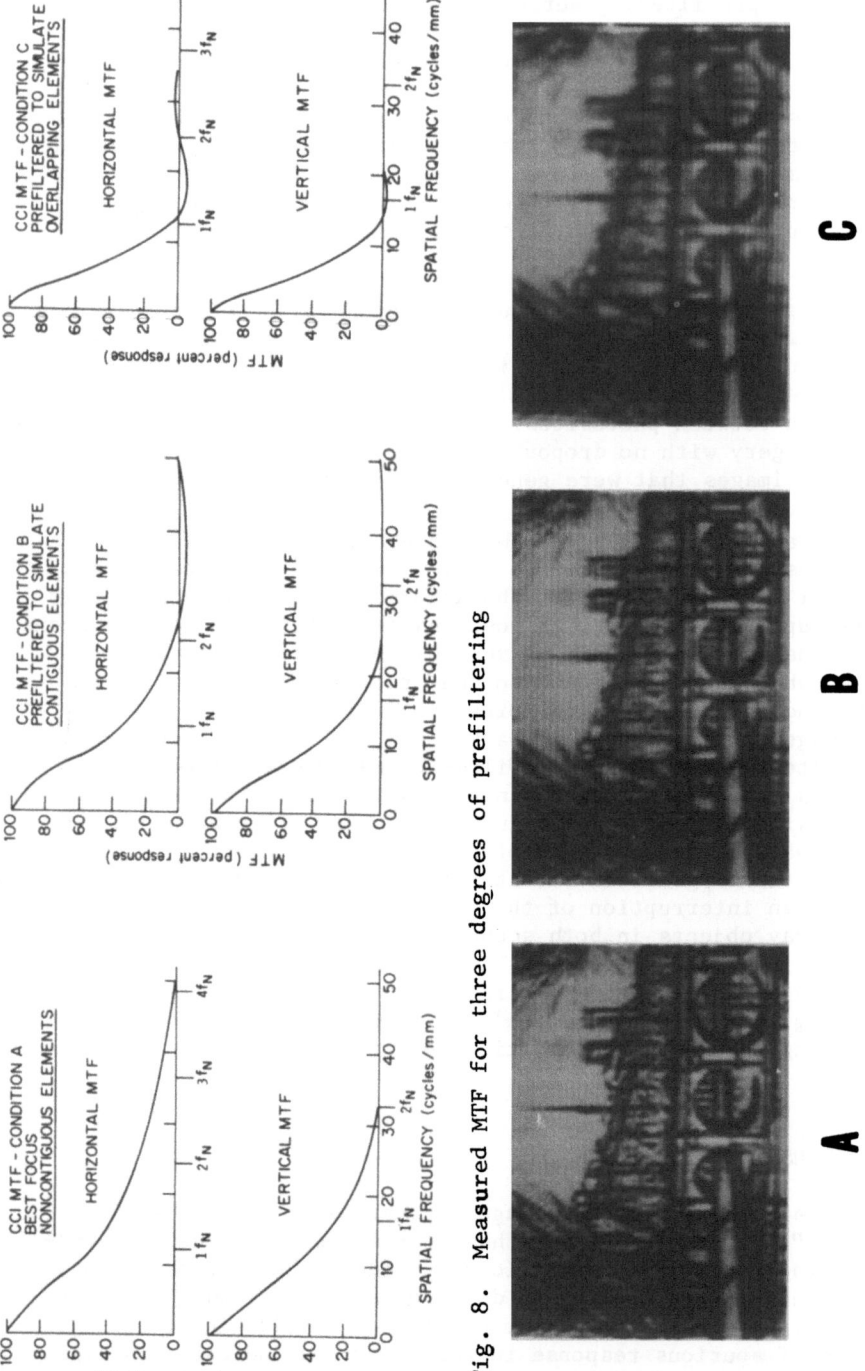

Fig. 8. Measured MTF for three degrees of prefiltering

Fig. 9. CCD imagery using an aperiodic scene

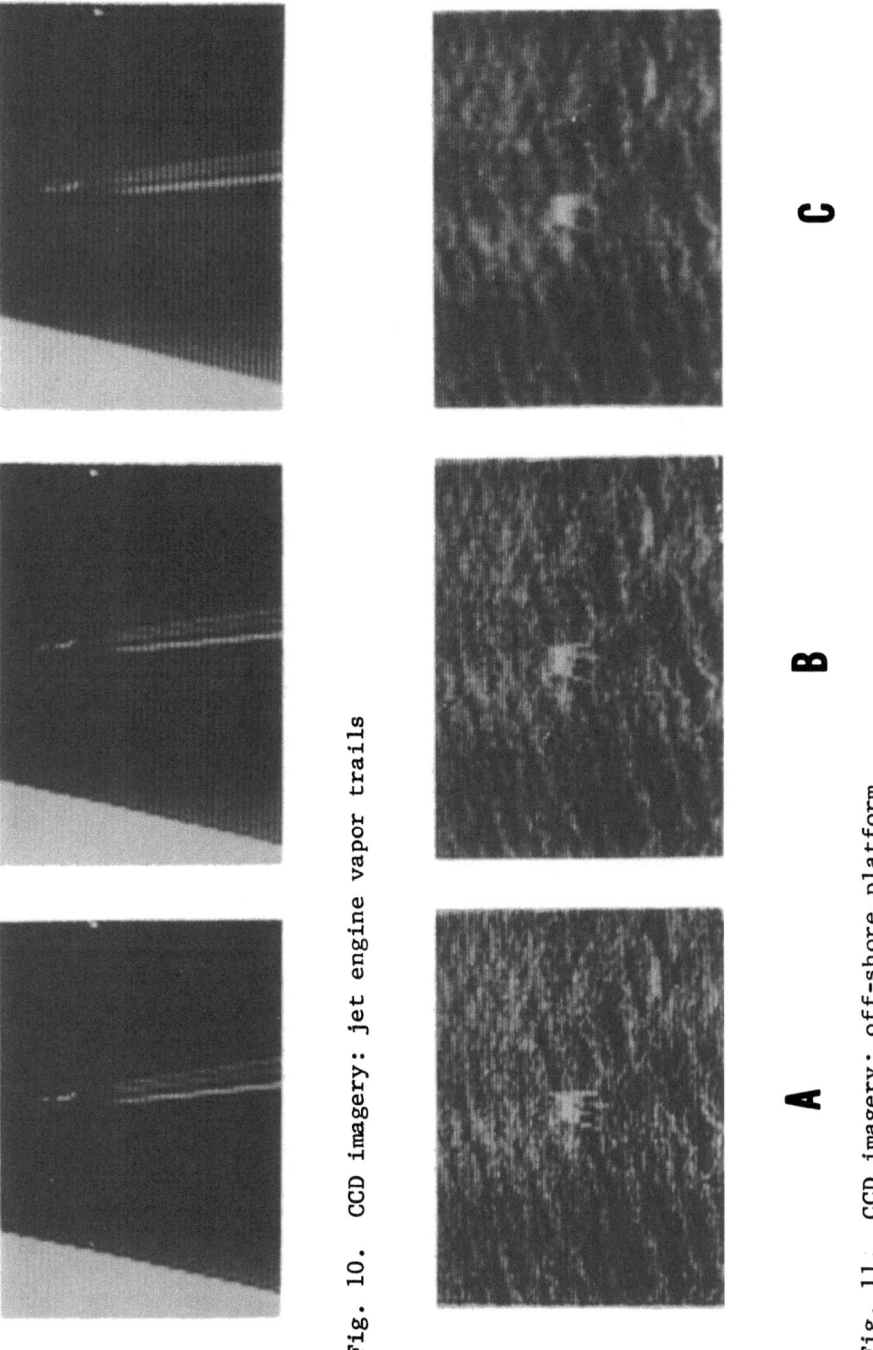

Fig. 10. CCD imagery: jet engine vapor trails

Fig. 11. CCD imagery: off-shore platform

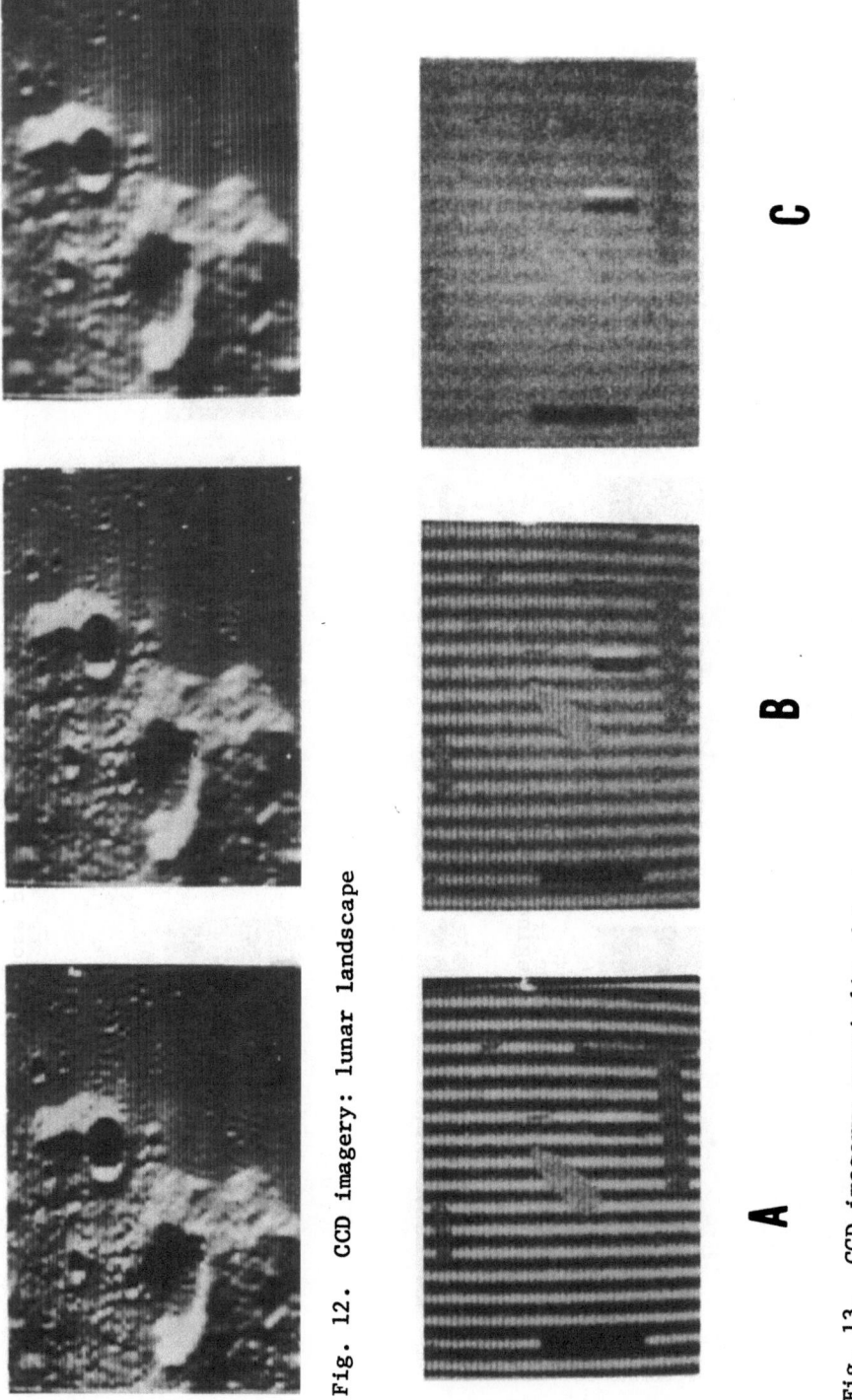

Fig. 12. CCD imagery: lunar landscape

Fig. 13. CCD imagery: aperiodic objects on periodic background

contained in the two-dimensional frequency domain beyond the sampling limit would be reproduced as a spurious signal, or noise occupying the same frequency space as the signal, thereby tending to obscure it. The problem with this analysis is that the viewer is influenced primarily by the spatial content of the reproduction rather than the frequency content. Although a Moire' pattern might occupy the same frequency domain as the true signal, it does not occupy the same space in the image plane as is the case with additive noise. This is illustrated graphically in Fig. 13, where the Moire' pattern extends up to, but not through, the objects of interest; i.e., <u>Moire' in the display is limited to the extent of the scene causing the Moire'</u>.

A similar argument can be used in the case of aperiodic objects. Suppose the scene contains many sharp edges and lines. The frequency spectrum of these scene elements may extend far beyond the sampling limit, and will, therefore, be aliased into the useful passband. In the image plane these elements will be reproduced as sharp but broken edges and lines. However, this distortion does not extend beyond those elements subtending the original edges and lines, and therefore, does not affect the appearance of scene elements elsewhere in the scene. While prefiltering may render the edges and lines as continuous, it also spreads part of the energy of these scene elements to adjacent areas, thereby blurring the image and reducing the edge contrast. Thus, it appears that frequency domain analysis is not adequate for judging the effect of aliasing at the image plane.

If the sensor is to be used to detect subresolution sized objects, such as point images of stellar objects, it is a definite disadvantage to have unresponsive intensities at the image plane. In these instances it would be preferable if the sensors were contiguous, or if the scene were prefiltered to that degree. The former solution would be preferable for those applications where sensitivity is a prime consideration. Although further prefiltering will eliminate the abrupt jumps of the point image as it moves from element to element, it is questionable whether defocussing to Condition "C" is merited. If in a given application it becomes clear that such effects are a real problem additional prefiltering external to the sensor is easily implemented [6].

In summary, the effects of aliasing are not as severe as might be predicted from frequency domain analysis. An excellent compromise between aliasing and image sharpness is obtained when sensing elements are contiguous. Further prefiltering should be considered only if it has been experimentally demonstrated that aliasing will be a real problem.

6. REFERENCES

1. D. F. Barbe and M. H. White, "A Tradeoff Analysis for CCD Area Imagers," in Proc. 1973 CCD Applications Conf., pp. 13-20.
2. O. H. Schade, "Image Reproduction by a Line Raster Process," <u>Perception of Displayed Information</u>, L. M. Biberman, editor, Plenum 1973.
3. R. Legault, "The Aliasing Problems in Two Dimensional Sampled Imagery,", Ibid.
4. C. H. Sequin, "Interlacing in Charge-Coupled Imaging Devices," IEEE Trans. Electron Devices, <u>ED-20</u>, pp. 535-541, 1973.
5. S. B. Campana and D. F. Barbe, "Tradeoffs Between Aliasing and MTF," in Proc. 1974 CCD Applications Conf., pp. 168-176.
6. M. Mino, and Y. Okano, "Optical Low-Pass Filter for Single-Vidicon Color TV Camera," Journal SMPTE, <u>81</u>, pp. 282-285, 1972.

SIGNAL AND NOISE IN THE DISPLAY OF IMAGES

James A. Hall

Westinghouse Electric Corporation,
Advanced Technology Laboratory,
Baltimore, Maryland U.S.A.

ABSTRACT. Schade, Coltman and Anderson, Rosell, and others have shown that the ability of an observer to perform tasks of detection, classification, or identification using electronic imaging devices or systems depends on SNR_D, the Display Signal to Noise Ratio. Since the eye averages noise over its own effective integration time of 0.1 to 0.2 seconds, and since the observer uses the entire area of the image of each significant detail in performing detection or identification, not just the area of a sensor or display element, the human observer can obtain useful information in many cases where the electrical video signal to noise ratio or element signal to noise ratio is far less than unity. SNR_D is defined after Rosell and equations given so that the designer or user can predict observer performance for electronic viewing systems in realistic situations at the time systems are being designed. Results of recent human interface studies are included.

1. INTRODUCTION

Students of the visual process beginning with Barnes and Czerny[1] in 1932 have postulated that fluctuations or noise in the visual signal from the eye set the threshold to seeing with lower illumination levels or lower contrasts. When electronic aids to vision were first considered for critical applications, investigators like Coltman[2] in 1954 suspected that the signal to noise ratio in the electronically displayed image would determine the visibility of the image to the observer, and devised experiments to measure how detectable contrast and spatial frequency were related to signal to noise ratio for simple bar patterns.

The question of observer performance is appropriate for this book since the majority of solid-state imaging systems being considered today will present an image to a human observer who must perform a task, make a decision, or simply enjoy a television program based on the information transferred to his eye and brain from the display. Very truly, the observer's eye and brain are the final components in any man-viewed system, and the action he takes as a result of what he sees is the true system output. Fortunately, the subject of information transfer from an electronic display to an observer has been well and recently treated in "Perception of Displayed Information," edited by L.M. Biberman, Plenum Press, New York and London, 1973. This chapter will introduce the concept of display signal to noise ratio, show how it may be calculated, and give a few examples of the display signal to noise ratio required to perform some simple tasks. Anyone who becomes involved in the design or analysis of man-viewed systems should refer to the Biberman text. This chapter will follow some of the notations of Rosell and Willson in Chapter 5 of "Perception of Displayed Information" to aid in use of that book.

2. SNR_D FOR AN APERIODIC TEST OBJECT

Assume an optical system forms an image of a simple scene on the photosensitive surface of an electronic imaging system. Assume for the present a system like an intensified CCD where the elements are much smaller than the image detail, with sufficient gain so that individual photoelectrons cause visible scintillations or flashes or light on the display, whose element structure is also much smaller than the detail in the displayed image. If the displayed image is sufficiently large and sufficiently bright, it should be the signal to noise ratio in the display rather than fluctuations in the observer's eye which determines his ability to see, and the signal to noise ratio in the display, not yet defined, should be related to the signal to noise ratio in the video signal in a manner which depends on the properties of the observer's eye. Rosell and Willson[3] consider the case shown in figure 1 of a simple test pattern containing as a test object a small disc or rectangle whose area "a" in the image at the sensor is slightly brighter or darker than the uniform surrounding background. At least for images which are large enough to be clearly resolved by the imaging system, they postulate, following Coltman, that the input "signal" is the difference between the product of the area and sensor exposure in area "a", E_0, and the product in an equal area of background E_b, or better is the difference between the average number of photoelectrons $\bar{n}_0$ generated from area "a" and the average number, $\bar{n}_b$, from an equal area of uniform background during some exposure time t. But this difference fluctuates from sample to sample because the

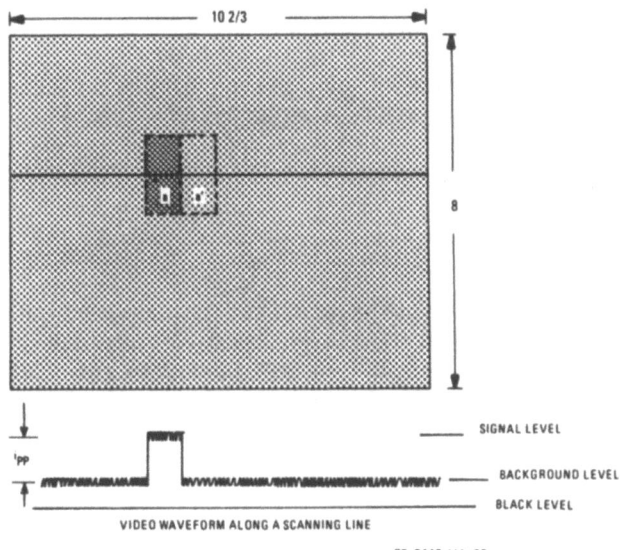

Figure 1. The Eye Compares the Brightness Averaged Over the Image of Test Object "a" with the Same Area of Background to Constitute a Signal. The Video Signal Current for a Flat Field Equals the Current from the Entire Scanned Area, hence $q_r = i_{pp} \frac{a}{A} t_e$.

emission of photoelectrons is a random process to which Poisson statistics apply, so that $\tilde{n} = \sqrt{\bar{n}}$, the rms deviation in n equals the square root of the average value. Hence, the signal to noise ratio of the difference signal from the first photosurface in the sensor is

$$\text{SNR} = \frac{\overline{n_o} - \overline{n_b}}{\sqrt{\overline{n_o} + \overline{n_b}}} = \frac{2M \, \overline{\dot{n}}_{av} \, at}{\sqrt{(\overline{\dot{n}_o} + \overline{\dot{n}_b})at}} \qquad (1)$$

where:

$\overline{n_o}$ = average number of electrons or charge carriers generated from object area "a" in time t

$\overline{n}_b$ = average number of electrons generated from background area "a" in time t

$\dot{\overline{n}}_o$ = average rate of generation of photoelectrons from the object area, electrons cm-2 sec-1

$\dot{\overline{n}}_b$ = average rate of generation of photoelectrons from background area, electrons cm-2 sec-1

M = modulation contrast = $\dfrac{\dot{\overline{n}}_o - \dot{\overline{n}}_b}{\dot{\overline{n}}_o + \dot{\overline{n}}_b} = \dfrac{\dot{\overline{n}}_o - \dot{\overline{n}}_b}{2\,\dot{\overline{n}}_{av}}$

Rosell and Willson then postulate that this same signal to noise ratio will appear in the displayed image of an ideal imaging system, although area "a" and time t have not yet been defined. To relate them to the characteristics of the observer, we note that the eye acts to integrate information in both space and time. Area "a", therefore, is the area of the simple test object, large enough so that it is fully resolved but small enough so the observer at comfortable viewing distance from the array treats area "a" as a unit and does not break it up into sub-areas. Based on flicker experiments, the eye at normal television display brightness has been shown to have an effective integration time t_e of 0.1 to 0.2 seconds, longer at lower display brightnesses. If the imaging system is being scanned rapidly enough to produce a flicker free display, the eye should integrate brightness fluctuations or noise over several frame times, up to its own integration time, and over area "a" to establish the signal to noise ratio which affect the observer. Thus "a" is the area of the pertinent image detail, referred to the sensor input, if it is neither too small nor too large, and $t = t_e$, the integration time of the eye, which is actually a function of display brightness and ambient illumination at the viewer's position but will be treated as a constant in this analysis.

Finally, Rosell and Willson convert the signal and noise from electron emission rates to currents, and refer these currents to those which would be generated from the entire equivalent scanned area of the photosensitive surface, it it were irradiated at the level of the detail or background area in question, then add a term for system noise, to derive their first key equation:

$$SNR_D = \left[2\, t_e\, \Delta f_v\, \dfrac{a}{A}\right]^{1/2} \cdot SNR_V \qquad (2)$$

where:

SNR_D = display signal to noise ratio

t_e = integration time of the eye, 0.1 to 0.2 sec

Δf_v = video frequency bandwidth

a = area of (small) detail being observed

A = area of the entire scanned area

SNR_v = video signal to noise ratio, defined in accordance with television engineering usage as peak-to-peak signal divided by rms noise

Note that a and A were originally referred to the dimensions of the image at the sensor input, but the ratio a/A is unchanged if referenced to the display, or to the corresponding areas in the scene being viewed.

This definition of display signal to noise ratio in terms of video signal to noise ratio was first suggested by Coltman and Anderson[4] in 1960. It applies when the detail area "a" is large enough so that the signal amplitude is a function of sensor irradiance only, not of the object size; and SNR_v is derived from the signal above background from the test area "a". Thus SNR_v is a function of contrast and of the average sensor irradiance.

Note that the coefficient of SNR_v is just what one might have expected. $2t_e \Delta f_v = t_e/(2\Delta f_v)^{-1}$ is the ratio of the eye's sampling time to the Nyquist sampling time of the system. The video signal is referred to unit area by dividing by A, to the area in question by multiplying by "a", and the signal to noise ratio, originally expressed in terms of signal shot noise, varies as the square root of the exposure.

3. SNR_D FOR PERIODIC TEST OBJECTS

System performance is normally measured with periodic, usually sinusoidal signals, although bar charts are more customary as signal sources for imaging systems. The same form of equation for display signal to noise ratio should apply to a completely resolved bar chart image if for area "a" one substitutes the area of one black or white bar, although the value of SNR_D to recognize a bar pattern might be lower than for a simple isolated test object of the same area. In television practice, the spatial frequency of a bar pattern is usually expressed in terms of N TV lines or half cycles per pattern height Y, as shown in figure 2. If a bar of width Δy has a length to width ratio ϵ, its area is $\epsilon \Delta y^2$. For an image with aspect ratio α, $A = \alpha Y^2$. Hence $a/A = \epsilon \Delta y^2 / \alpha Y^2$ and

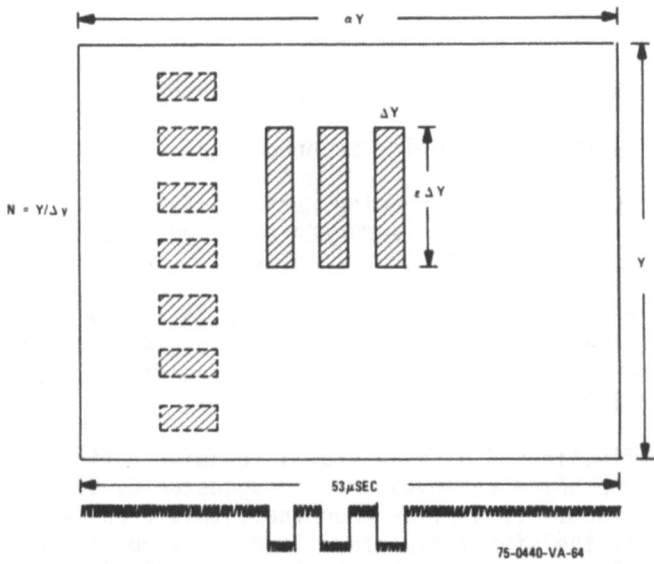

Figure 2. Spatial Frequency for a Bar Chart Test Object is N Half Cycles per Pattern Height. For Example Shown, $N = 15$. $a/A = \epsilon \Delta y^2 / \alpha Y^2 = \frac{\epsilon}{\alpha} \frac{1}{N^2}$.

$$SNR_D = \left(2t_e \Delta f_v \frac{a}{A}\right)^{1/2} \cdot SNR_V = \left(2t_e \Delta f_v \frac{\epsilon \Delta y^2}{\alpha Y^2}\right)^{1/2} \cdot SNR_V$$

and since $N = Y/\Delta y$ TV lines per pattern height,

$$SNR_D = \left(2t_e \Delta f_v \frac{\epsilon}{\alpha}\right)^{1/2} \cdot \frac{1}{N} \cdot SNR_V, \qquad (3)$$

where SNR_V is the video signal to noise ratio for the signal from the completely resolved test pattern, and the equation again applies only for test patterns coarse enough so that they are completely resolved by both the optical and electronic imaging system, i.e., when the MTF is still 100 percent at least for the fundamental of the bar chart spatial frequency. It was exactly this case which was investigated by Coltman in his original experiments.

4. EFFECTS OF REAL MTF'S - APERIODIC OBJECTS

In any real imaging system the MTF falls as spatial frequency increases from zero because of optical aberrations, electron optical aberrations, carrier diffusion in image sensing or storage members, or if all else were ideal, simply because of diffraction effects from finite optical apertures. For simplicity, Rosell and Willson refer to each image forming element as a finite aperture, producing a blurred image of each point in object space because of diffraction. Actually any aberration effects are included by inference.

Consider the profile of the image of a bar object through a real lens as a function of object width, as shown in figure 3. The image of the wide bar has the same amplitude as the object and the same full width at half amplitude, but the corners of the image are rounded as shown. This is an example of the completely resolved case mentioned above. Although the image is in a sense broader than the object, a theory ignoring image broadening is at least approximately correct. Note that the imaging process assumed is nondissipative, loss free. In the optical case this means we have ignored reflection, absorption, and scattering losses. Note also that the object waveform is referred to image space, shown as it would be if imaged by a perfect lens of very large diameter.

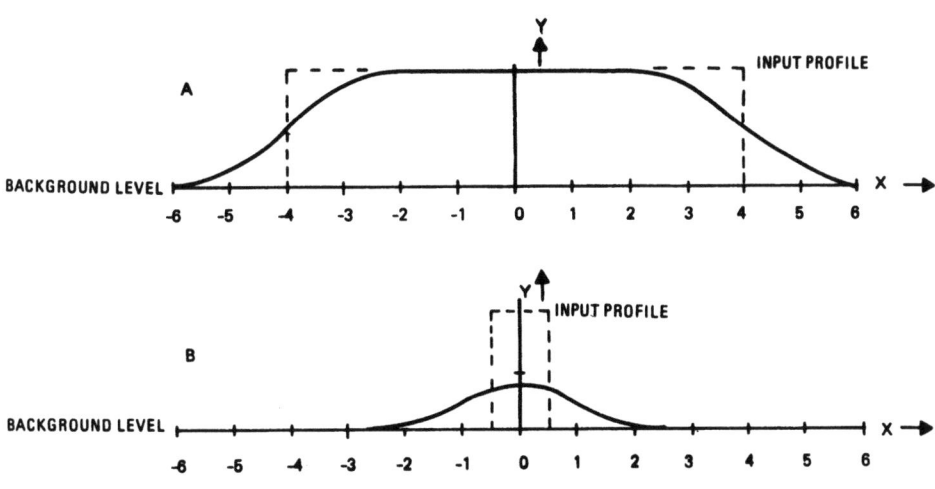

Figure 3. Image Profile of an Isolated Bar Object After Filtering Through the MTF of a Real Imaging Device. Wide Bar is Nearly Completely Resolved. Visual Signal is Area Under Curve Above Background and is Unchanged by Imaging Process.

Figure 3 also shows the image of a narrow bar which is not completely resolved. The amplitude in the image above background is substantially decreased and the effective image width, either full width across the skirts or full width at half amplitude, is greatly increased. However, the radiant power in the image is unchanged, that is the areas above background under the object and image curves are equal. Now experiment shows that the eye simply expands its integration area to include all of the aberrated image. Hence, as defined above, the signal, the difference in the sensor between the photoelectron flux or the total charge from the bar image and the charge from the same area of background, remains unchanged by dissipationless imaging, even with aberrations. The noise, however, is greater, since the larger area includes more background flux, hence more shot noise. In fact, since any noise source in the imaging system can be referred to the input photosensitive surface as an equivalent shot noise, even for the zero background exposure case, the broadening of the test object's image causes the eye to include a larger area of noise, hence reduces SNR_D. The effect is well described in figure 4. The effect for aperiodic images is quantitatively discussed by Rosell and Willson, but is omitted here to concentrate on the effect with periodic test objects.

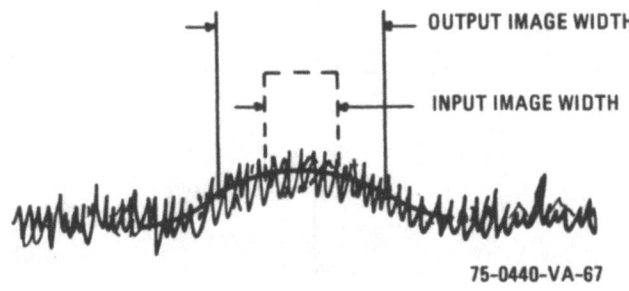

Figure 4. Broadening of Image Area Causes Eye to Include a Larger Sample of Noise with the Image.

5. EFFECTS OF REAL MTF'S - PERIODIC OBJECTS

For periodic objects falling MTF does not act to increase the sampling area "a" whose width is fixed at one half the spatial period by the nature of the bar chart. Instead, aberrations and diffraction effects from each focusing structure act to reduce signal amplitude by redistributing some of the white bar energy into the space of the black bars. The variation of signal amplitude as a funcion of spatial frequency for a bar chart input is described by $R_{SQ}(N)$, the square wave response function, shown in figure 5. The video waveform is a square wave of 100 percent amplitude at low spatial frequencies, degrades through a rounded top waveform like that in figure 6 at intermediate spatial frequencies, and is virtually a pure sine wave of reduced amplitude at high spatial frequencies. The falling signal amplitude proportionally reduces SNR_V which is defined as the ratio of peak-to-peak signal to rms noise by standard TV usage. A basic assumption for the SNR_D model, however, was that the signal sensed by the eye was the total flux from each sampled area, the product of average current density times area. To apply here, equation (3)

$$SNR_D = \left(2t_e \Delta f_v \frac{\epsilon}{\alpha}\right)^{1/2} \cdot \frac{1}{N} \cdot SNR_V \qquad (4)$$

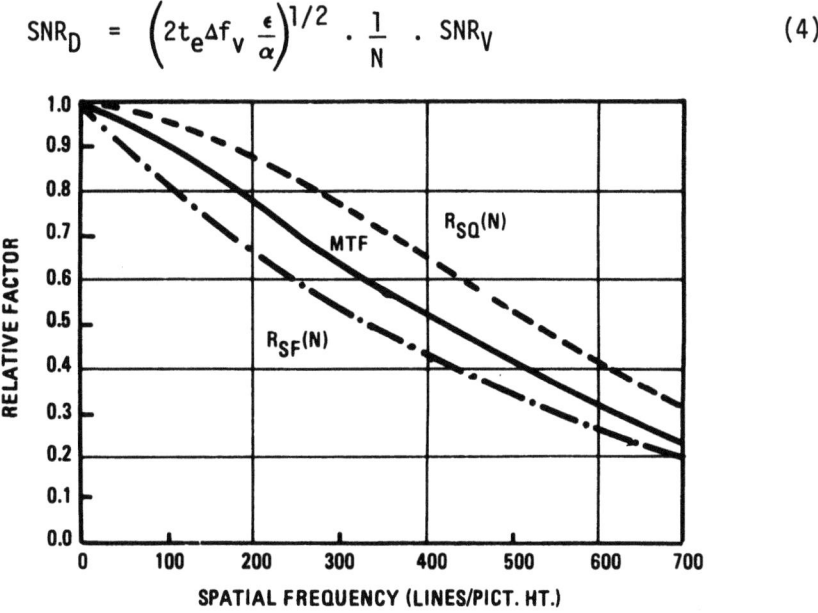

Figure 5. System Resolving Power is Best Described as MTF (N), P.P. Response to a Sine Wave Input. P.P. Response to a Square Wave Bar Chart is $R_{SQ}(N)$. Average Signal Response to a Square Wave Bar Chart is $R_{SF}(N)$.

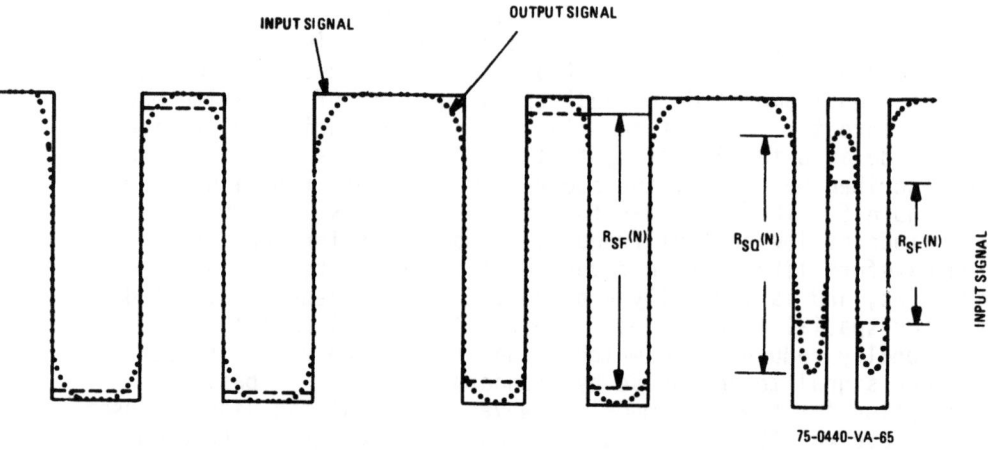

Figure 6. Peak-to-Peak Amplitude of Actual Video Waveform with Bar Chart (Square Wave) Input is Used for Computing $R_{SQ}(N)$. Detection Model Depends on Average Signal for Each Sampling Area (Half Cycle) $R_{SF}(N)$, the Square Wave Flux Factor.

must be modified to base SNR_V on the average signal amplitude rather than the peak signal amplitude as shown in figure 6. In Rosell and Willson's investigation, SNR_V was measured in the usual way for each spatial frequency, N, as $\Delta i_{pp}/I_n$, where I_n is primarily an added noise. SNR_V was then multiplied by $R_{SF}(N)/R_{SQ}(N)$ to calculate the equivalent square wave video signal amplitude having the same area under each half cycle as the actual video signal. The waveforms and the definitions of $R_{SF}(N)$ and $R_{SQ}(N)$ are shown in figure 6. Thus in terms of the measured pp signal:

$$SNR_D = \left(2t_e \Delta f_V \frac{\epsilon}{\alpha}\right)^{1/2} \cdot \frac{1}{N} \cdot \frac{R_{SF}(N)}{R_{SQ}(N)} \cdot \frac{\Delta i_{pp}}{I_n} \qquad (5)$$

where:

t_e = integration time of the eye, 0.1 sec

Δf_V = video frequency bandwidth

ϵ = length to width ratio of bars in test pattern, usually different for each spatial frequency N

α = aspect ratio of scan pattern = 4/3

N = spatial frequency in half cycles (TV lines) per pattern height

$R_{SF}(N)$ = square wave flux factor

$R_{SQ}(N)$ = square wave response

Δi_{pp} = signal current referred to sensor output for test chart group at frequency N

I_n = rms noise current referred to sensor output

For the camera tube, a 1-1/2-inch high resolution vidicon, used in Rosell and Willsons' experiments $R_{SF}(N)/R_{SQ}(N)$ varied from unity for gross images to 2/3 at higher spatial frequencies as shown in figure 5. An approximate value for SNR_D may be calculated by setting $R_{SF}/R_{SQ} = 1$, that is simply by using Δi_{pp} as measured to calculate SNR_V for use in equation (4) or equation (2). The calculation of optical signal as peak electrical signal x area obviously gives too large a result except at very low frequencies.

One qualification is in order. Equation (5) as developed here applies when the principal noise originates in the amplifier after the scanning process as it does in a vidicon camera or in some solid-state array imagers. For EBS camera tubes and for intensified charge coupled devices, aberrations in focusing can in effect occur in the gain process by which each photoelectron typically gives rise to 1000 to 4000 charge carriers in the array. Since these charge carriers may diffuse to any of several sensor sites, each photoelectron results in a spread function distribution of charge on the array. In this case the aperture, the imaging process, acts as a low pass filter band limiting the noise spectrum. This important case is neglected here as beyond the scope of this chapter, but it is treated by Rosell and Willson and others.

6. OBSERVER PERFORMANCE - THE REQUIRED SNR_D

To establish what display signal to noise ratio a man needs to recognize detail, Rosell and Willson measured observer performance first in detecting the images of small rectangular test objects in a uniform but noisy field. The rectangular image was generated electronically, hence there was no MTF degradation, and the equation used to define SNR_D was

$$SNR_D = \left[2\Delta f_v \, t_e \, \frac{a}{A}\right]^{1/2} \cdot SNR_V \qquad (6)$$

The rectangle was generated in one of four quadrants of the display, the signal strength and added gaussian noise were varied in many combinations so that the image was just detectable or just not detectable, and the observer asked to choose the quadrant in which the rectangle appeared. The video bandwidth was also changed in this experiment.

More important, the dimensions of the rectangle were changed from 4x4, 4x64, 4x128, and 4x180 scan lines, based on a U.S. 525 line system, to check whether the eye could still use the entire area of the rectangle in detecting the image, as had been assumed in the model. These corresponded to 0.13° x 0.13° to 0.13° x 6.02° at the observer's eye. The results shown in figure 7 are that threshold $SNR_D = 3$ as defined by equation (6) if the eye integration time is assumed to be 0.1 sec, for all rectangle shapes. Throughout their work, Rosell and Willson measured video signal to noise ratio and calculated display signal to noise

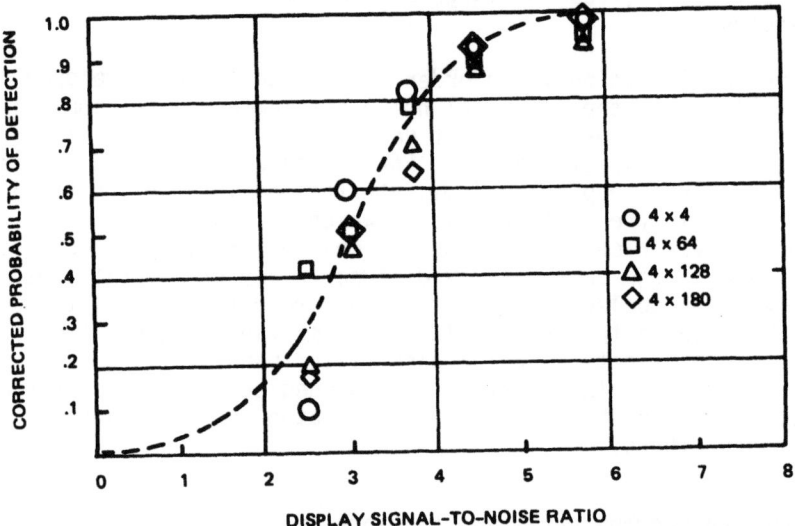

Figure 7. Measured Probability of Detection vs SNR_D for an Isolated Rectangular Image on an 8 x 10-2/3 inch Display with 490 Active Scanning Lines. Image Size is Given in Scan Lines.

ratio using equation (6). Next, the rectangle length was held at 96 scan lines and its width varied from 4 to 32 scan lines. Here SNR_D required for 50 percent probability varied from about 3 for the narrow two rectangles to about 5 for the widest rectangle, as shown in figure 8. Apparently the eye fuses image areas into single samples up to an angular subtense of about 0.5°. Long thin samples are mostly edges and the eye is known to be an edge sensor. Finally a series of squares from 8x8 to 64x64 scan lines was added, 0.06 to 2.14° at the eye, with results shown in figure 9. Again the eye appears to fuse the full area into a single signal only up to about 0.5° x 0.5° angular subtense. This effect was further checked by doubling the distance from the observer to the display.

Periodic bar patterns were then used as test objects, using a high resolution vidicon camera as the signal source but adding gaussian noise as before. The observer stated whether he could resolve the pattern in the image as SNR_D was randomly varied. Since the MTF of this real imaging system modified the results, display signal to noise ratio was defined as:

$$SNR_D = \left(2t_e \Delta f_v \frac{a}{A}\right)^{1/2}. \quad SNR_V = \left(\frac{2t_e \Delta f_v \epsilon}{\alpha}\right)^{1/2} \frac{1}{N} \frac{R_{SF}(N)}{R_{SQ}(N)} \frac{i_{pp}(N)}{I_n} \quad (7)$$

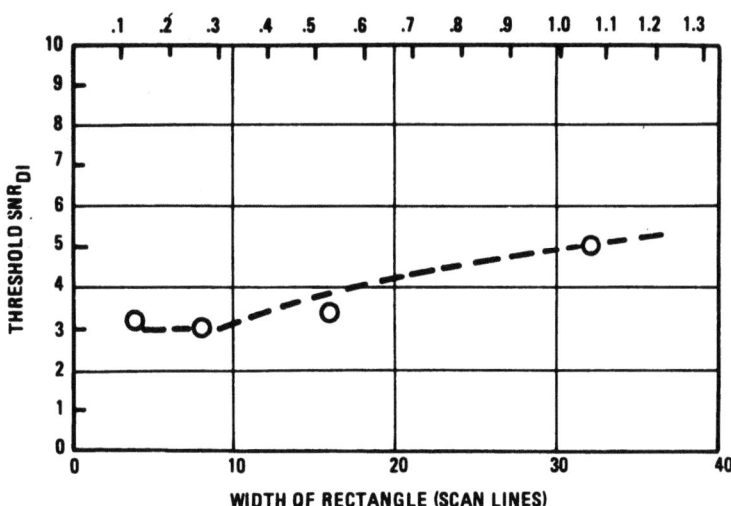

Figure 8. Threshold SNR_D is Approximately Constant at 3.2 to 3.5 When Rectangle Width Subtends Less than 0.5° at the Eye. Rectangle Length = 96 Scan Lines for All Tests in this Series.

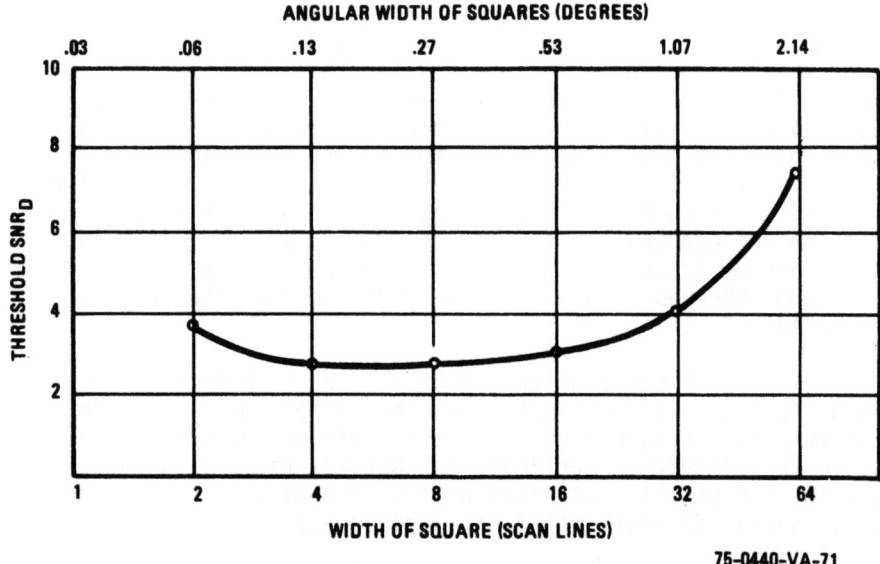

Figure 9. Threshold SNR_D for Square Images Rises Above 0.5° Angular Width Because Eye Fails to Treat Larger Areas as a Unit. Rise at Lower Widths Probably Caused by Limited Acuity or by MTF of Display.

based on the area of one black or one white bar in the image as sample area "a". The quantity $\frac{\epsilon}{\alpha N^2} = \frac{a}{A}$. This expression was derived above in Section 5.

Human observer measurements on bar chart images included spatial frequencies from 104 to 635 half cycles per picture height and length to width ratios of 5:1 to 20:1. As shown in figure 10, threshold signal to noise ratio varied from about 4 to 2 as N varied from 100 to 500 half cycles per picture height, while ϵ, the length to width ratio, was kept constant. Some of this variation may be due to the fixed 28-inch viewing distance from the 8 x 10-2/3 inch display. If the observer was allowed to move back from his 28-inch viewing distance in front of the 8 x 10-2/3 inch display, SNR_D dropped at the lowest frequencies for ϵ = 5:1 bars so that SNR_D varied from 2.7 at N = 100 to 2.0 at N = 500.

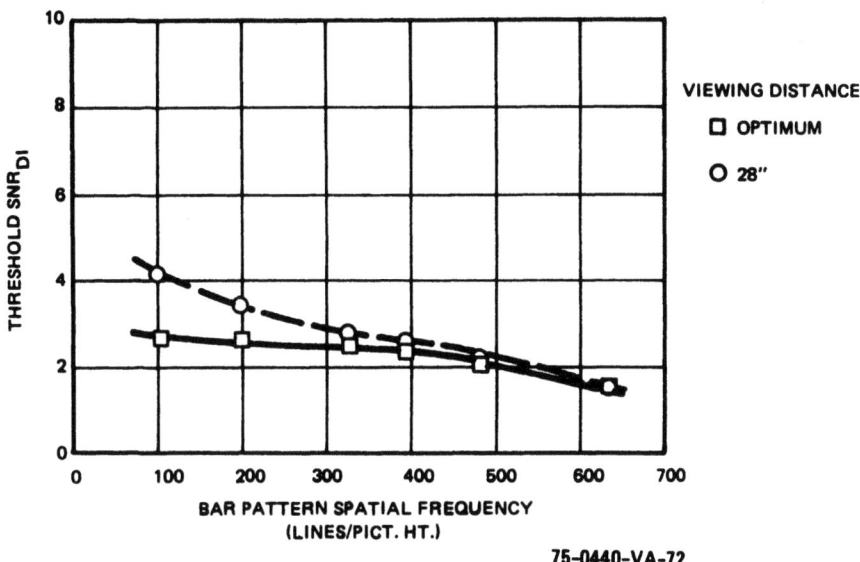

Figure 10. Threshold SNR_D vs Bar Pattern Spatial Frequency Varies Much Less if Observer Can Adopt an Optimum Viewing Distance at Lowest Spatial Frequencies.

7. THE JOHN JOHNSON CRITERIA FOR REAL VISUAL TASKS

Many man-viewed systems, especially for the military, are used to extend the observer's ability to detect, recognize, or identify objects at a distance, rather like night binoculars. Imaging systems are often evaluated in terms of reproduction of bar chart test objects. John Johnson[5] of the U.S. Army's Night Vision Laboratories in 1958 related observed performance against bar charts to performance on real visual tasks. The real objects used in his tests were military vehicles against background, and in viewing a scene containing a vehicle or other object of interest, Johnson established four levels of observer performance:

Performance Levels	Meaning
Detection	An object is present.
Orientation	The object is approximately symmetric, or it is asymmetric and its orientation may be discerned.
Recognition	The object may be classified as a house, truck, ship, man, etc.
Identification	The object can be described to the limit of the observer's knowledge as a pickup truck, F-4 aircraft, M-48 tank, etc.

Johnson tested imaging systems against test charts and against military objects so that conditions were the same, i.e., the irradiance and contrast produced at the sensor by the bar chart and that produced by the military object were the same, varied the spatial frequency of the bar charts, and compared the observer's ability to resolve the bar chart with his ability to detect, recognize, or identify the military object. He described his comparison in terms of the number of bar chart half cycles which must be resolved in a distance equal to the narrowest object dimension in order for the observer to perform his task as suggested in figure 11.

Performance Level	Perceived Bar Chart Resolution per Minimum Object Dimension, Half Cycles	
Detection	2	+1 / −0.5
Orientation	2.8	+0.8 / −0.4
Recognition	8.0	+1.6 / −0.4
Identification	12.8	+3.2 / −2.8

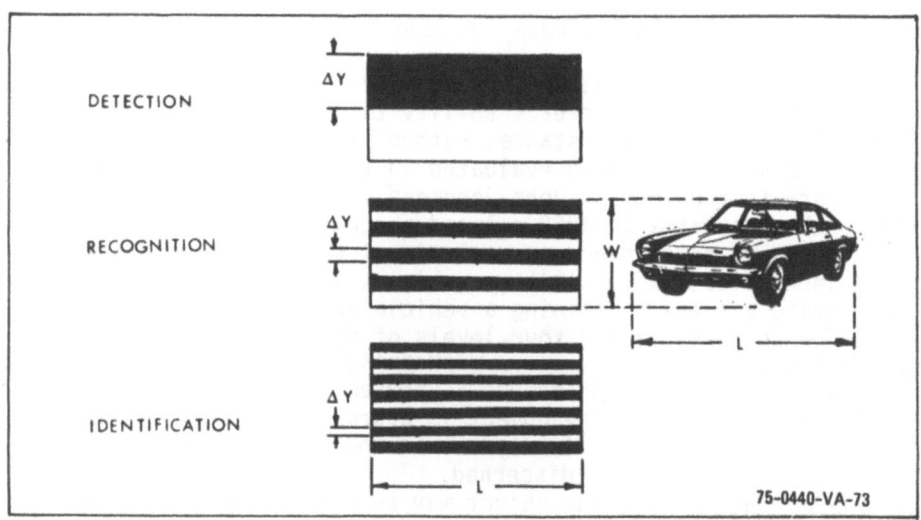

Figure 11. Johnson's Criteria Specify Eight Resolvable Half Cycles on an Equivalent Bar Chart Test Object to Recognize a Car as a Car.

Thus to identify a partially submerged submarine by the shape of its sail will require about 13 resolvable elements across the height of the sail, under the same viewing conditions. This implies that the system has enough resolution capability to provide this many elements. For example, a "350 line" broadcast TV system with 525 scanning lines would require the narrow dimension of the object to be identified occupy about 4 percent of the linear field of view. The entire field would be only 25 sail heights high.

Rosell and Willson checked Johnson's recognition criteria by viewing transparency images of military vehicles with a high performance 875 line vidicon camera, adding noise, and measuring the video signal to noise ratio at which recognition probability was 50 percent. The vehicles, shown in figure 12, were a tank, a van truck, a half tracked vehicle with a top mounted radar antenna, and a bulldozer with a derrick. The vehicle images occupied approximately 17x34 scan line widths on the 875 line raster (825 active lines), and the pictures were taken at 45° depression angle so that the sides and tops of the vehicles were imaged. Display signal to noise ratio at the recognition level was calculated from the measured video signal to noise ratio for the vehicle image against the background, by associating SNR_D with the area of a bar whose length matched the length of the vehicle and whose

75-0440-VA-75

Figure 12. These Photographs Were Used for Recognition Experiments: Upper Left, Tank; Upper Right, Van Truck; Lower Left, Half-Truck with Antenna; and Lower Right, Bulldozer with Derrick.

width was 1/8 of the vehicle height. That is, Johnson's criterion was assumed to apply, and the display signal to noise ratios were calculated using it, then compared with previously measured results on rectangles and bar charts. Results shown in figure 13 gave SNR_{DT} = 3.3. Bar patterns of 7, 9, and 11 bars with each group matching the vehicle size required SNR_{DT} = 2.9. The results are certainly in fair agreement, since the task of recognizing a vehicle by its outline is not exactly the same as that of detecting the brightness modulation which defines a bar pattern. The feature on each vehicle which was used for recognition, like the derrick on the bulldozer, seemed to have a width about 1/8 the vehicle height, and it certainly was not periodic, hence SNR_{DT} results closer to those for isolated rectangles do not seem surprising. The experiment was then repeated by superposing the vehicle transparencies on a transparency of a real background, with increases to 3.8, 4.1, and 5.0 depending on background complexity. Since the required SNR_{DT} for all these cases varied by less than 2:1, Johnson's recognition criterion was shown useful for designing man-viewed systems for a given observer performance, although Rosell and Willson suggest some refinements.

The identification experiments used pictures of M47, M48, Stalin, Panther, and Centurion tanks, also taken from an angle 45° above the horizontal as shown in figure 14. Two images sizes

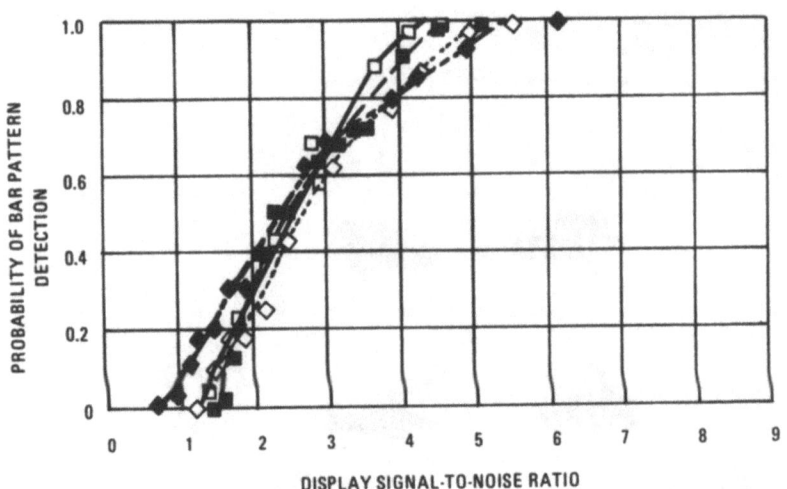

Figure 13. Probability of Recognition vs SNR_D for a (o) Tank, (◇) Radar Half-Track, (□) Van Truck, and (●) Derrick Bulldozer. Background was Uniform.

Figure 14. Photographs of Tank Models Used in Identification Experiments: Upper Left, M47; Upper Right, M48; Center Left, Stalin; Center Right, Panther; Lower Center, Centurion.

were used, 69x138 scanning line widths and 108x216 scanning line widths on the 825 active scanning line raster, much larger than for recognition since identification depends on seeing smaller details. Video signal to noise ratio was varied by adding gaussian noise, and SNR_{DT} at the 50 percent probability of identification point was converted to display signal to noise ratio by referring the measured SNR_{VT} to the area of a bar as long as the tank image and 1/13 as wide as the tank was high to agree with Johnson's identification criterion. Against a uniform light background SNR_{DT} for the smaller images averaged 5.2, for the larger images, 6.8.

While these results are not in as good agreement with measured performance on bar patterns as the recognition values were, they would still give an estimate of the exposure level required to perform a given identification task which is within a factor of two, based on the bar chart values, and this is certainly helpful to the system designer. Put another way, for critical tasks like identification of vehicles the imaging system must provide 13 to 26 resolvable elements across the smallest dimension of the vehicle as measured with a bar chart under sensor exposure and contrast conditions like those produced by the vehicle, including the possible contrast degrading effects of the optical path through the atmosphere and the optical system.

7.1 Effect of Sensor Element Size

Real solid-state imaging systems use discrete sensing elements which may not be negligibly small compared to image detail. For square elements the geometric MTF of such sensors in one dimension is

$$MTF = \frac{\sin\left(\frac{\pi \Delta x}{2P} \frac{f_s}{f_{s\ max}}\right)}{\frac{\pi \Delta x}{2P} \frac{f_s}{f_{s\ max}}} \tag{8}$$

where:

f_s = spatial frequency of sinusoidal test pattern at the sensor

$f_{s\ max} = \frac{1}{2P}$ = the Nyquist sampling frequency

P = the array pitch, the center to center distance between sensing elements

Δx = width of the active sensing element

At the Nyquist frequency $f_s = f_{s\ max}$, and for a 100 percent area effective array $\Delta x = p$. Then

$$MTF_{f_{s\ max}} = \frac{\sin \pi/2}{\pi/2} = 0.637 \tag{9}$$

Actual measured MTF data on CCD imagers ranges from 0.31 down to about 0.15, mostly because of lateral carrier diffusion in the silicon, which is not of course included in the geometric model. These low values will essentially eliminate aliasing problems in the array, although the data is still sampled, and produce an image which is below the Nyquist band limited frequency. The observer in turn should place himself far enough from the display so that the element structure, if there is one, is not resolved by his eye. Then the foregoing criteria should apply. The aliasing problem is addressed in Chapters 6 and 7 of the book "Perception of Displayed Information."

8. THE EFFECT OF ELEMENT TO ELEMENT NONUNIFORMITIES

One last point is that solid-state array imaging systems are rarely completely uniform. If one views a uniformly bright scene, the resulting display may have a slight or more serious pepper and salt appearance. This variation in element to element brightness in the display may be more prominent at low sensor irradiance levels, due to dark current variations or other variations in element dark offset, or may be prominent at high levels due to variations in element response or gain. The resulting pattern is not noise, since it is stationary, but the observer's eye is similarly affected. In fact early work by Rose[6] related to the display signal to noise ratio concept was done with photographs of noisy electronic displays in which the time varying character of the noise had been removed. Thus the same SNR_D values as those listed above also apply to element to element nonuniformities, and the rms deviation in element to element brightness in the display may be treated as a (stationary) noise at the Nyquist frequency, and designated $\tilde{B}$. In general, if an image detail in the display has area "a", and the elements in area "a" have average brightness B_0, the signal presented to the eye will be $(B_0-B_b)a/p^2$, where B_b is the average brightness of elements in the adjacent background, p is the element pitch, and a/p^2 is the number of elements in sample area "a". The rms noise will be $\tilde{B}(a/p^2)^{1/2} = \tilde{B}(n_a)^{1/2}$ and

$$SNR_D = \frac{(B_0-B_b)a/p^2}{\tilde{B}(a/p^2)^{1/2}} = \frac{B_0-B_b}{\tilde{B}}(n_a)^{1/2} \qquad (10)$$

and this ratio must be at least 2 to 3 for the detail to be seen through the sensor and display nonuniformities. Further, the frozen noise can, at least in theory, be combined with the time varying noise by adding noise powers on an element by element basis, to obtain an inclusive display signal to noise ratio for mosaic type arrays.

REFERENCES

1. Barnes, R. and Czerny, M., (1932), Zeits, f. Phys. <u>1932</u>: 79.

2. Coltman, J.W., (1954), "Scintillation Limitations to Resolving Power in Imaging Devices," J.O.S.A. <u>44</u> (3), 234-237.

3. Rosell, F.A. and Willson, R.H., "Recent Psychophysical Experiments and The Display Signal to Noise Ratio Concept," Chapter 5 in "Perception of Displayed Information," Ed. L.M. Biberman, Plenum, New York.

4. Coltman, J.W. and Anderson, A.E., (1960), "Noise Limitations to Resolving Power in Electronic Imaging," Proc. I.R.E. <u>48</u> (5): 858-865.

5. Johnson, John (1958), Image Intensifier Symposium, Fort Belvoir, Va., Oct. 6-7, 1958, AD 220160.

6. Rose, Albert, "The Sensitivity Performance of the Human Eye on an Absolute Scale," J.O.S.A., Vol. 38, No. 2, 1958, 196-208.

TIME DELAY AND INTEGRATION IMAGE SENSORS

D. F. Barbe

Naval Research Laboratory
Washington, D. C. 20375

ABSTRACT. The tradeoff between signal and resolution in the "pushbroom" imaging mode is described. The extension of the pushbroom concept using CCD's in the time delay and integration (TDI) mode to provide increased exposure time, and therefore increased signal, without sacrificing resolution is discussed. The organization of CCD TDI imaging arrays and noise limited resolution are discussed. Finally, MTF effects in TDI arrays are described in detail.

1.0 INTRODUCTION

In certain applications, it is important to view the object plane from a platform which has linear motion relative to the object plane. In such circumstances, the object/sensor motion may be used to generate one dimension of the image field. Typical examples of this type of image generation include strip and panoramic modes of aerial reconnaissance. This mode of image generation is commonly called the "pushbroom" mode. Figure 1 shows a schematic diagram of an electro-optical system using a linear array of sensor elements in the "pushbroom" mode. As the sensor array moves across the scene due to the relative motion, the scene is synthesized into a series of strips. If a linear array is used, there is a tradeoff between signal-to-noise ratio and resolution in the along-track direction. The use of CCD's in the TDI mode has allowed the "pushbroom" concept to be extended to provide additional sensitivity (signal) without sacrificing resolution. The purpose of this chapter is to discuss the use of CCD's in the TDI mode.

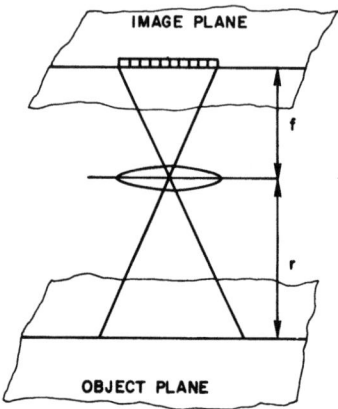

Fig. 1 Schematic diagram of an imaging system.

2.0 THE PUSHBROOM IMAGING MODE

When the sensor and the scene to be viewed have a constant relative velocity, a linear array of sensor elements can be used for imaging as shown in Fig. 2. The minimum geometrically resolvable dimension in the along-track direction in the object plane, d_{geom}, is determined by the speed of the sensor projected onto the object plane via the optical system, V_o, and the integration time of the sensor, T; i.e.,

$$d_{geom} = V_o T \qquad (1)$$

where d_{geom} is the along-track distance covered by a sensor element projected onto the object plane in an integration time. Using the magnification relation, the speed of a point in the object plane projected onto the image plane, V_i, is

$$V_i = (f/r) V_o \qquad (2)$$

where f = focal length of the optical system.
r = object plane to image plane distance.
V_o = speed of a point in the image plane projected onto the object plane.

The output signal is proportional to the input irradiance, H, and the integration time; i.e.,

$$S \propto HT. \qquad (3)$$

Therefore, the product of the geometrical resolution and the signal out of the array is independent of T. This represents a basic tradeoff for the pushbroom mode; i.e., T can be reduced to decrease d_{geom}, but S also decreases.

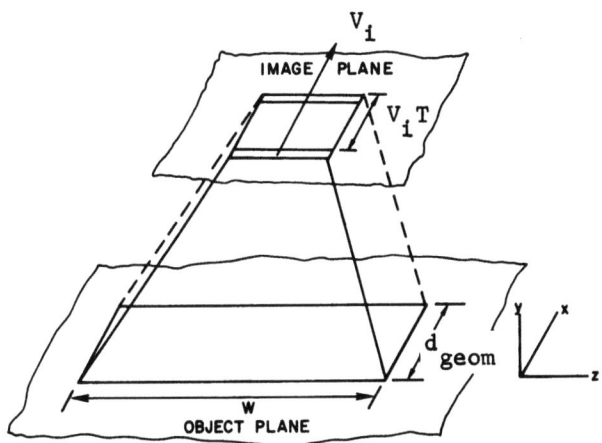

Fig. 2 Pushbroom imaging concept (optics not shown).

3.0 THE TIME DELAY AND INTEGRATION (TDI) MODE

If the image plane were composed of a number (M) of elements contiguous in the x-direction to form an array of columns, if the columns were electrically delayed in the x-direction at the same speed, V_i, as the scene is scanned across the image plane, and if the outputs from all elements in a given column were added, then the output signal would be M times larger than that from a single line array of equal elemental dimensions. This situation is shown in Fig. 3 and is called the time delay and integration mode. Other names are delay-and-add and image-motion compensation modes. The basic result is that the signal in Eq. (3) is increased by the factor M while geometrical resolution in Eq. (1) is unchanged. Thus, the new equations analogous to Eqs. (1) and (3) are

$$d_{geom} = V_o T, \qquad (4)$$

and

$$S \propto HMT .\qquad(5)$$

A major advantage of the TDI mode is that the exposure time is increased by the factor M without affecting the geometrical resolution. This improves the low-light-level capability without affecting the resolution or data rate.

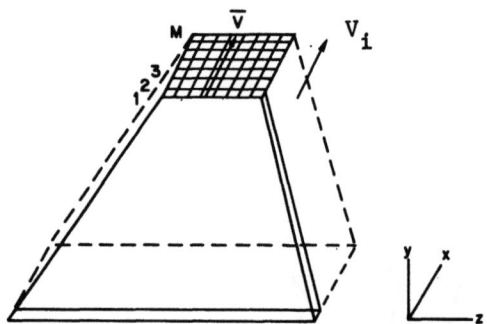

Fig. 3 Time delay and integration (TDI) concept (optics not shown).

4.0 TDI ARRAY ORGANIZATION

The natural TDI CCD chip organization, shown in Fig. 4, is basically a parallel-serial design. The parallel imaging columns are composed of M delay-and-add stages (CCD stages). These N columns are multiplexed into an N-stage CCD serial shift register for readout. There is no need for separate frame or line storage with this mode of operation; i.e., all of the columns are optically active. Only the horizontal output register is shielded from light.

5.0 NOISE LIMITED RESOLUTION

For low-light levels, the TDI imager would operate in the noise limited regions; i.e., the minimum resolvable dimension in the along-track direction is given by

$$d_{min} = V_o T \left[\frac{CSAMHT/e}{(SAMHT/e + MN_1^2 + N_2^2)^{\frac{1}{2}}} \right]^{-1} ,\qquad(6)$$

where

C	=	scene contrast,
S	=	responsivity in mA/watt,
A	=	optically active area of sensor row,
M	=	number of TDI stages,
T	=	integration time,
H	=	irradiance on the image plane,
e	=	electronic charge,
N_1	=	noise (in number of electrons) introduced at each TDI stage, and
N_2	=	noise (in number of electrons) introduced by the output amplifier [1].

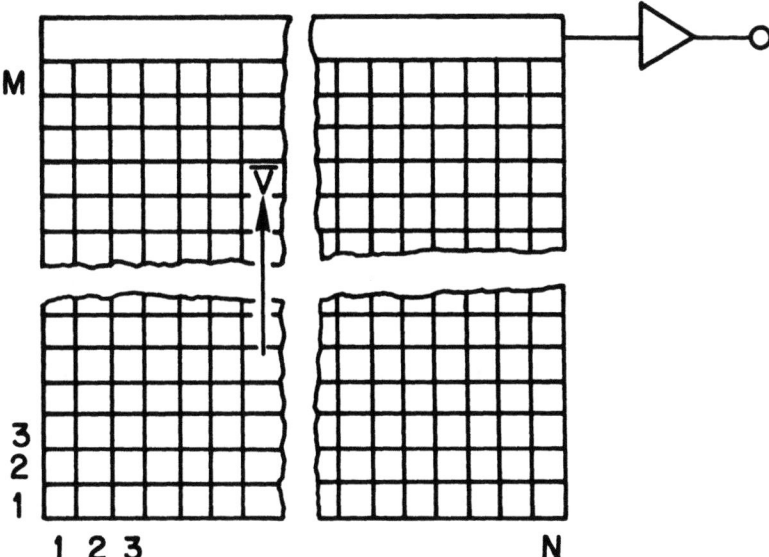

Fig. 4   Time delay and integration array organization.

There are three cases to be examined with regard to the functional dependence of d_{min} on M: (1) if photon noise dominates, then d_{min} is proportional to $M^{-\frac{1}{2}}$; (2) if N_1 dominates, the d_{min} is proportional to $M^{-\frac{1}{2}}$; and (3) if N_2 dominates, then d_{min} is proportional to M^{-1}.

6.0 SPATIAL FREQUENCY RESPONSE

In any approach to low-light-level imaging it is necessary to be able to predict, by the use of appropriate theory, the ability of the system to resolve detail. In low-light-level imaging the maximum quantity of information transmitted is frequently not of great interest, and it is often necessary to give up information of a certain type in order to increase the quantity of information of other types. However, in order to make comparisons it is necessary to know the relationships of the image transfer properties of individual system components to the equivalent property of the complete system. In addition, it is necessary to know how well each component transfers image details and contrasts.

Use of the modulation transfer function (MTF) makes it possible to compute the effects on image rendition of the different elements through which an image is transmitted between the scene and the observer. Actually the information continues on to the observer's brain, and some consideration as to what is required to recognize an object is necessary. Any mechanism that operates on or collects information modifies it in some way, so that information delivered in hard-copy form is different from the original scene information. These differences can be predicted to a great extent by the use of MTF's. Manipulation of the MTF's permits the description of the difference between the input information and the output information without an exact knowledge of the nature of the information itself. Because the effects of each element that transmits the information can be accounted for separately and the total effect determined by the multiplication of each of these functions, their use is extremely advantageous. The close relationship between spatial frequency and fineness of detail, which preserves intuitive clarity, makes the MTF a powerful tool in the analysis of imaging systems.

The effects which cause the imaging array response (signal) to decrease at high spatial frequencies are: (1) the geometry of the integration aperture, (2) the degree to which the average speed of the charge packets is matched to the speed of the scene moving across the array, (3) the discrete nature of the charge motion, and (4) the charge transfer inefficiency. These effects are treated separately in the remainder of this chapter. Since any scene can be analyzed into Fourier components in object space (x,y), a general treatment of the response of a sensor array reduces to an analysis of the response to a sinusoid of arbitrary spatial frequency.

6.1 Integration MTF

If the distance between samples on the image plane in the x-direction is p, then the Nyquist frequency is

$$f_n = 1/2p \ . \tag{7}$$

The Fourier component of the irradiance on the array at spatial frequency f is

$$H_1 = H_0 \left[1 + m \cos (2 \pi f x) \right]. \qquad (8)$$

This can also be written

$$H_1 = H_0 \left[1 + m \cos (\pi \frac{f}{f_n} \frac{x}{p}) \right]. \qquad (9)$$

If the integration aperture is Δx in width, then the output charge pattern is that due to a different intensity pattern H_2

$$H_2 = \frac{1}{p} \int_{x_i - \frac{1}{2}\Delta x}^{x_i + \frac{1}{2}\Delta x} H_0 \left[1 + m \cos (\pi \frac{f}{f_n} \frac{x}{p}) \right] dx. \qquad (10)$$

Integrating Eq. (1) and simplifying gives

$$H_2 = \frac{H_0}{p} \left[1 + m \; \frac{\sin \left(\frac{\pi}{2} \frac{f}{f_n} \frac{\Delta x}{p} \right)}{\frac{\pi}{2} \frac{f}{f_n} \frac{\Delta x}{p}} \; \cos (\pi \frac{f}{f_n} \frac{x}{p}) \right]. \qquad (11)$$

Therefore, the MTF of the integration process is

$$MTF_{integ} = \frac{\sin \left(\frac{\pi}{2} \frac{f}{f_n} \frac{\Delta x}{p} \right)}{\frac{\pi}{2} \frac{f}{f_n} \frac{\Delta x}{p}} \qquad (12)$$

In the TDI CCD array organization, the integration aperture Δx is equal to the distance between samples, p, (this is also true in the y-direction). Therefore, the picture elements are contiguous in both directions. This is shown in Fig. 5 for a 3-phase system. Then Eq. (12) reduces to

$$MTF_{integ} = \frac{\sin \left(\frac{\pi}{2} \frac{f}{f_n} \right)}{\frac{\pi}{2} \frac{f}{f_n}}. \qquad (13)$$

The integration MTF in the z-direction will be given by the same expression with the appropriate modification if p_z is different from p_x.

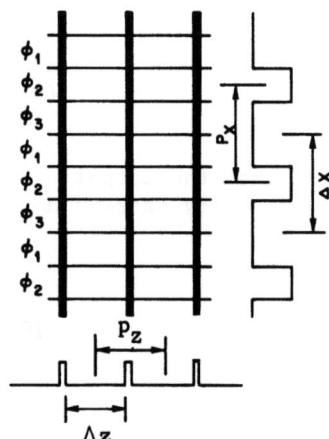

Fig. 5 Schematic diagram showing integration aperture dimensions.

6.2 Synchronism MTF

If the average speed of the charge packets is not exactly equal to the speed of the motion of the scene across the image plane, then the response will be degraded [2]. If the difference between the average speed of the charge packets ($\bar{V}$) and the speed of the scene on the image plane (V_i) is ΔV, then after M TDI stages the charge packets will be displaced from where they should be if the synchronism were perfect, by the distance $MP(\Delta V/V)$. In effect the array sees a traveling wave instead of a fixed Fourier component representing each spatial frequency. The MTF degradation due to this effect can be determined by noting the equivalence between (1) an infinitesimal aperture sampling a traveling wave with relative speed ΔV for a time MP/V and (2) an aperture of width $MP(\Delta V/V)$ sampling a stationary wave as shown in Fig. 6. Therefore the form of this MTF is Eq. (13) with Δx replaced by $MP(\Delta V/V)$ i.e.,

$$(MTF)_{\Delta V} = \frac{\sin\left(\frac{\pi}{2} \frac{f}{f_n} M \frac{\Delta V}{V}\right)}{\frac{\pi}{2} \frac{f}{f_n} M \frac{\Delta V}{V}} \tag{14}$$

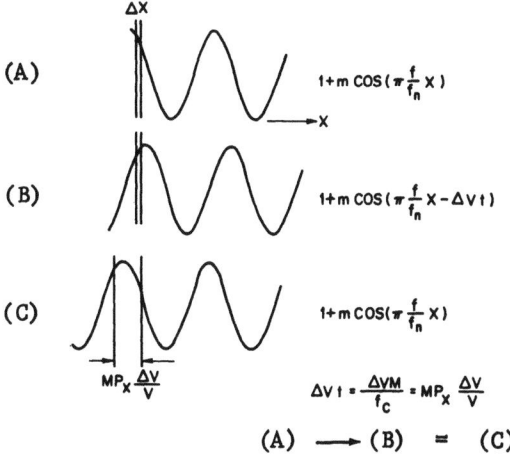

Fig. 6 Diagrams indicating the effect of non-synchronism on aperture size.

Figure 7 shows curves of $(MTF)_{\Delta V}$ vs. f/f_n with $M(\Delta V/V)$ as the parameter. For $M(\Delta V/V) = 2$, the MTF is 0.64 at $f/f_n = \frac{1}{2}$ and zero at $f/f_n = 1$. Further MTF degradation due to higher $M(\Delta V/V)$ would be intolerable. This is an important design criterion for TDI CCD arrays. If it is assumed that a practical value of $\Delta V/V$ is 1%, then using the criterion that $M(\Delta V/V) \leq 2$, gives $M \leq 200$. Therefore for $\Delta V/V \simeq 1\%$, more than 200 delay-and-add stages is impractical. <u>The inequality $M(\Delta V/V) \leq 2$ is the primary criterion used to choose the number of TDI stages</u>.

6.3 Discrete charge motion MTF

Even if the <u>average</u> speed of the charge packets is equal to the speed at which the image moves across the sensors, there will still be a loss of response due to the discrete nature of the charge transfer compared with the uniform motion of the scene across the sensor [3]. Between two successive transfers of charge, the image will move a distance $d = p/n_\phi$, where p is the center-to-center spacing between CCD stages (not electrodes) and n_ϕ is the number of transfers per stage (number of phases). This motion will change the sensitivity function for an element in the direction of motion from $S_1(x)$ to $S_2(x)$ where

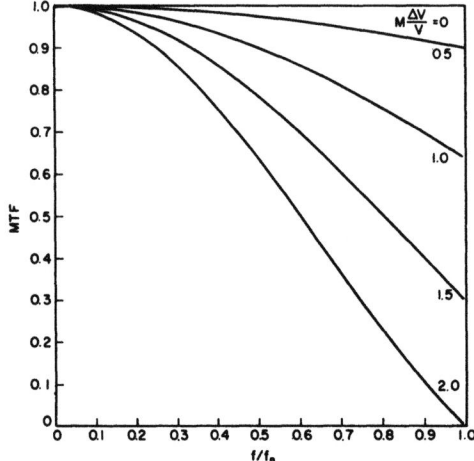

Fig. 7 MTF due to non-synchronism versus spatial frequency.

$$S_2(x) = \frac{1}{d} \int_{d/2}^{d/2} S_1(x-x')\, dx' \quad . \tag{15}$$

The response will thus be changed from

$$R_1(f) = \int_{-\infty}^{\infty} S_1(x) e^{i2\pi fx} dx \tag{16}$$

to

$$R_2(f) = \int_{-\infty}^{\infty} S_2(x) e^{i2\pi fx} dx \quad . \tag{17}$$

Substituting Eq. (15) into Eq. (17) gives

$$R_2(f) = \frac{1}{d} \int_{-\infty}^{\infty} \left(\int_{-d/2}^{d/2} S_1(x-x')\, dx' \right) e^{i2\pi fx} dx \tag{18}$$

Making the substitution $w = x-x'$ gives

$$R_2(f) = \frac{1}{d} \int_{-d/2}^{d/2} \left(\int_{-\infty}^{\infty} S_1(w) e^{i2\pi fw} dw \right) e^{i2\pi fx'} dx'$$

$$= \frac{1}{d} \int_{-d/2}^{d/2} R_1(f) e^{i2\pi fx'} dx'$$

$$= R_1(f) \frac{\sin(\pi fd)}{\pi fd} \tag{19}$$

Using $f_n = 1/2p$ and $d = p/n_\phi$ in Eq. (19), and recognizing that the $\sin(x)/x$ factor as the MTF due to the discrete nature of the charge transfer gives

$$MTF_{disc.} = \frac{\sin\left(\frac{\pi}{2} \frac{f}{f_n} \frac{1}{n_\phi}\right)}{\frac{\pi}{2} \frac{f}{f_n} \frac{1}{n_\phi}} \tag{20}$$

At the Nyquist frequency the MTF is

0.900 for $n_\phi = 2$,
0.955 for $n_\phi = 3$,
and 0.974 for $n_\phi = 4$.

Therefore a four-phase device gives the highest MTF. This is because the charge motion is more nearly continuous.

6.4 Transfer inefficiency MTF

The gain and phase shift due to transfer inefficiency are [3]

$$G_n = \exp\left[-n\epsilon(1-\cos 2\pi fp)\right], \tag{21}$$

and

$$\Delta\phi_n = n\epsilon \sin(2\pi fp), \tag{22}$$

where n is the number of transfers and ϵ is the charge transfer inefficiency. Eq. (22) is the transfer MTF in arrays having separate integration and transfer; however, it does not apply to the x-direction of TDI imagers. This is because charge packets added at different points along the columns see different numbers of transfers and different phase shifts. For the x-direction of a TDI imager, the transfer MTF can be determined by computing the output signal resulting from a sinusoidal irradiance function of spatial frequency f; i.e.,

$$H_1(x) = H_0 (1 + m \sin 2\pi f x) . \qquad (23)$$

After N transfers during which the image is moved down the array, the output charge pattern will be that due to a different intensity pattern $H_2(x)$:

$$H_2(x) = \frac{1}{N} \sum_{n=1}^{N} H_0 \left[1 + n G_n \sin (2\pi f x + \Delta\phi_n) \right] , \qquad (24)$$

where G_n and $\Delta\phi_n$ are given by Eqs. (21) and (22). The cumulative transfer MTF is given by the ratio of the output modulation to input modulation. Expanding Eq. (24), using the approximation

$$\sum_{n=1}^{N} g(n) \simeq \int_{0}^{N} g(n) \, dn , \qquad (25)$$

and manipulating gives:

$$(MTF)_{\epsilon,x} = \frac{1}{N(a^2 = b^2)} \left(\left\{ a - \exp(-Na) \left[a \cos (Nb) - b \sin (Nb) \right] \right\}^2 + \left\{ b - \exp(-Na) \left[a \sin (Nb) + b \cos (Nb) \right] \right\}^2 \right)^{\frac{1}{2}}, \qquad (26)$$

where

$$a = \epsilon \left[1 - \cos (\pi f/f_n) \right] \quad \text{and}$$
$$b = \epsilon \sin (\pi f/f_n) .$$

Consider, for example, a 100 stage TDI imager using four-phase clocking and having n = 400 and $\epsilon = 10^{-4}$. Then at the Nyquist limit, $(MTF)_{\epsilon,x} = 0.961$.

Transfer in the horizontal direction (2) is not accompanied by TDI in that direction; therefore, the $(MTF)_{\epsilon,z}$ is given by Eq. (21).

In summary, the MTF in the direction of TDI is the product:

$$(MTF)_x = (MTF)_{integ,x} \times (MTF)_{\Delta V}$$
$$\times (MTF)_{disc.} \times (MTF)_{\epsilon,x} \quad . \qquad (27)$$

The MTF in the horizontal direction is the product:

$$(MTF)_z = (MTF)_{integ,z} \times G_n \quad . \qquad (28)$$

7.0 REFERENCES

1. S. B. Campana, "Charge-Coupled Devices for Low Light Level Imaging," in Proc. 1973 CCD Applications Conf., pp. 235-245.
2. H. V. Soule, <u>Electro-Optical Photography at Low Illumination Levels</u>, John Wiley and Sons, New York, 1968, pp. 332-333.
3. "Moving Target Sensors," Texas Instruments, Navy Contract N00039-73-C-0070, Final Report, 1973.

SOLID STATE INFRARED IMAGING

D. F. Barbe

Naval Research Laboratory
Washington, D. C. 20375

ABSTRACT. The uses of charge coupled devices (CCD's) and charge injection devices (CID's) in infrared focal plane arrays are discussed. The characteristics of infrared radiation and the differences between visible imaging and infrared imaging are discussed. The InSb CID, accumulation mode extrinsic silicon CCD, and extrinsic silicon photoconductor CCD sensors are described. Staring mode, parallel scanned, serial scanned, and serial-parallel scanned focal plane arrays are discussed. The use of CCD's and CID's in a focal plane array is discussed in detail, and the effect on range is illustrated.

1. INTRODUCTION

It is common practice to describe electromagnetic radiation by its wavelength or frequency as shown in Fig. 1. Other chapters of this volume have delt with imaging in the spectral region in which silicon is useful, primarily the visible (0.4 to 0.75 microns) and into the near infrared ($\leq 1.1\mu$). Strictly speaking, the infrared region includes the range 0.75 to 1000μ; however, this chapter will deal with imaging and detection in the middle and far infrared regions (3 to 15μ). Since this chapter deals with a different part of the electromagnetic spectrum, which has quite different imaging parameters from the visible region, a brief review is in order.

2. LAWS OF BLACKBODY RADIATION

The basic law which describes the spectral radiation from a blackbody is Planck's law:

$$W(\lambda) = \frac{2\pi c^2 h}{\lambda^5} \frac{1}{\exp(ch/\lambda kT)-1} \,, \tag{1}$$

where $W(\lambda)$ = spectral radiant emittance in W cm$^{-2}\mu^{-1}$

λ = wavelength in μ (μ is used to denote microns),
h = Planck's constant = 6.6256×10^{-34} W sec^2,
T = absolute temperature in $^{\circ}$K,
c = speed of light = 2.9979×10^{10} cm sec^{-1},
k = Boltzmann's constant = 1.38×10^{23} W sec $^{\circ}$K^{-1}.

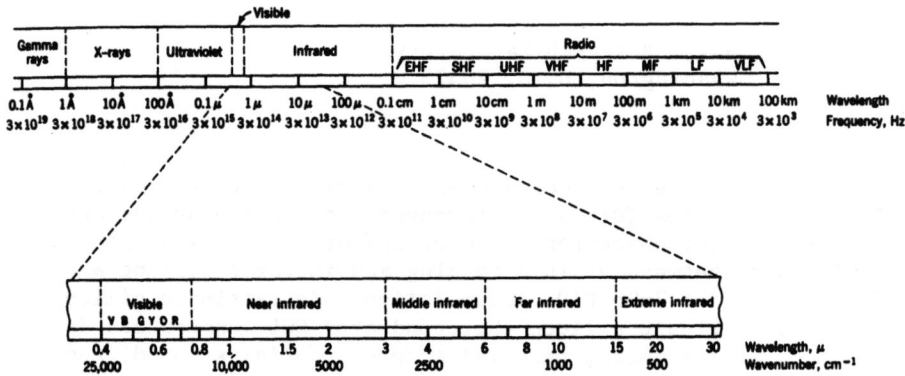

Fig. 1 The electromagnetic spectrum (after Hudson [1]).

Dividing by the photon energy (hc/λ) gives the spectral photon flux in cm$^{-2}\mu^{-1}$sec^{-1}

$$Q(\lambda) = \frac{2\pi c}{\lambda^4} \frac{1}{\exp(ch/\lambda kT)-1} \,. \tag{2}$$

Differentiating Planck's law and solving for the wavelength of maximum spectral radiant emittance gives Wein's law:

$$\lambda_{max} T = a, \tag{3}$$

where

λ_{max} = wavelength of maximum spectral radiant emittance and
a = 2897.8μ $^{\circ}$K.

Fig. 2 illustrates Equations (1) and (3).

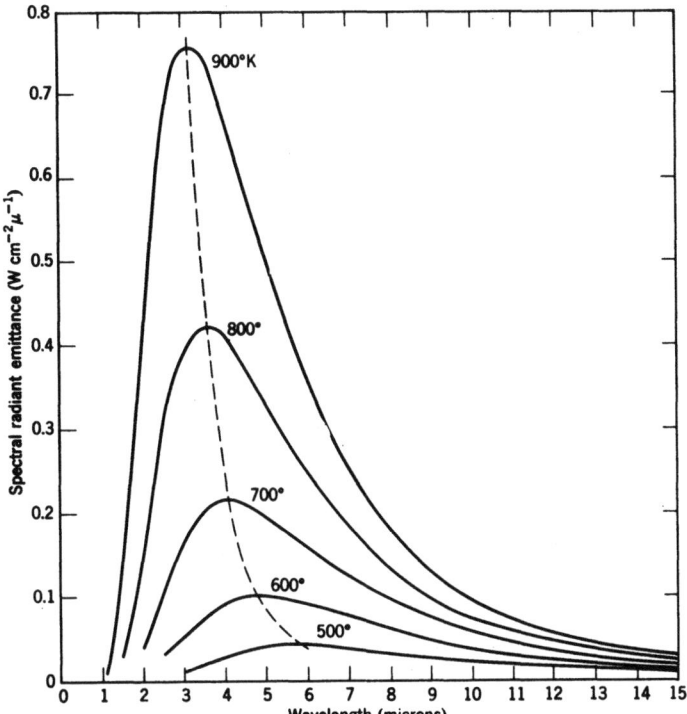

Fig. 2 Spectral radiant emittance versus wavelength with blackbody temperature as a parameter (after Hudson [1]).

In the infrared region, it is common to characterize objects with respect to the effective blackbody temperature. For example, the effective temperatures of the Earth and Sun are 290°K and 5900°K respectively. Wein's law gives $\lambda_{max} = 10\mu$ for the Earth and $\lambda_{max} = 0.49\mu$ for the Sun. Radiation from the Moon is a combination of blackbody radiation from the 400°K Moon surface and reflection of 5900°K Sunlight.

3. ATMOSPHERIC TRANSMISSION

Another important factor which must be considered in IR imaging is the transmission of IR radiation between the source and the image plane. For example, if the Earth were being viewed from within its atmosphere, the transmission characteristics of the atmosphere would greatly modify the spectrum reaching the sensor; for example, Fig. 3 shows the transmission characteristics of a 6000 ft. atmospheric path at sea level [1]. Note the two broad transmission "windows" between 3-5µ and 8-14µ. IR sensors which look through the Earth's atmosphere must be sensitive in one of these windows or perhaps a sub-window.

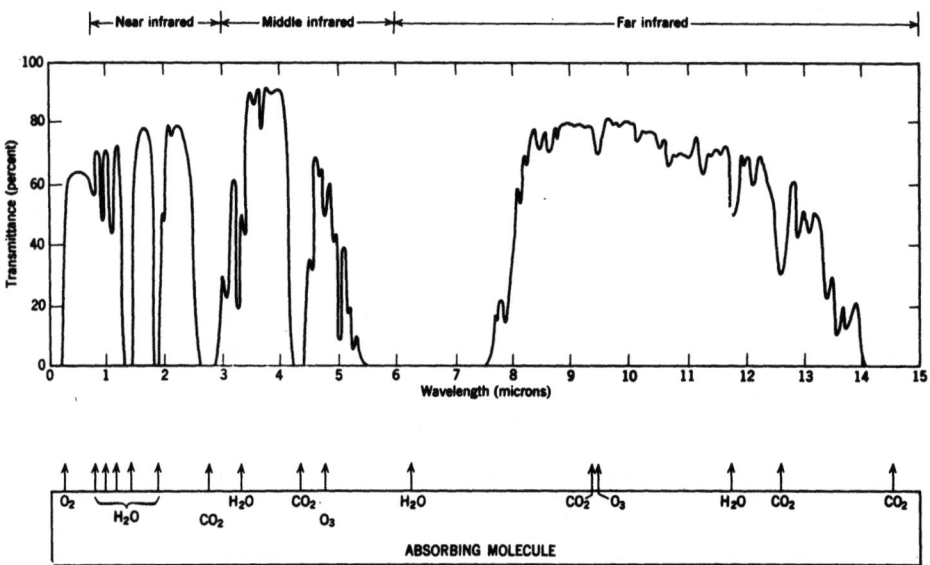

Fig. 3 Transmittance of a 6000 ft. atmospheric path at sea level as a function of wavelength (after Hudson [1]).

4. THERMAL IMAGING

A typical example of IR imaging is that of viewing the Earth in an attempt to resolve objects having temperatures 0.1 to 0.2°K higher than the average background temperature (290°K). If Eq. (2) for a 290°K blackbody is integrated from $\lambda = 0$ to $\lambda = \lambda_c$ and over a hemisphere, the resultant photon flux from $\lambda = 0$ to $\lambda = \lambda_c$ (Q_b) is given in Fig. 4. From Fig. 4, it can be seen that the background flux densities from the 290°K Earth are large for λ_c above a few microns [2]. For example, the flux density for $\lambda_c = 10\mu$ is approximately equivalent to the visible photon flux density in bright sunlight. For $\lambda = 3\mu$, the flux density is equivalent to the visible photon flux density of room light [3].

By integrating $Q_b\, dQ_b/dT$ from zero to λ_c, the contrast can be calculated. Fig. 5 shows the results of this calculation: Note that in for $\lambda_c = 5\mu$, the contrast is only 4% for every degree in temperature above 290°K, and for $\lambda_c = 14\mu$, the contrast is only 2% per °K. For a temperature difference of 0.2°K on a 290°K background, the contrast is 0.8% in 3-5μ band and 0.4% in the 8-14μ band. The main problems associated with IR imaging with the Earth as the background arise from the high background, low contrast situation.

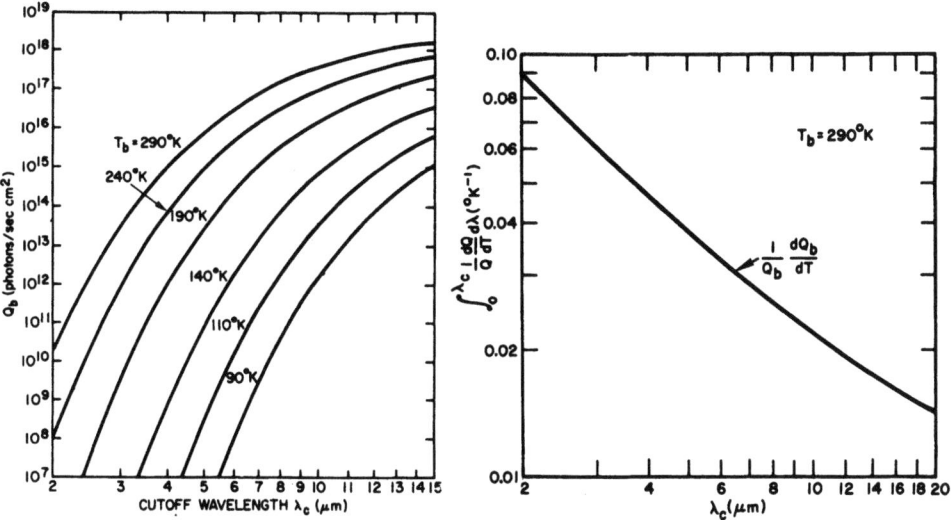

Fig. 4 Photon flux from $\lambda=0$ to $\lambda=\lambda_c$ versus λ_c (after Dimmock [2]).

Fig. 5 Contrast versus cutoff wavelength (after Dimmock [2]).

5. CONCEPT OF D^*

A concept which is often used in visible imaging to characterize the performance is the noise equivalent power (NEP) which is the power incident on the array to give a signal-to-noise ratio of unity. A concept commonly used in infrared imaging is the detectivity (D) of an array, where

$$D = (NEP)^{-1} \quad . \tag{4}$$

A more commonly used descriptor in infrared imaging is the specific detectivity (D^*) which normalizes D with respect to sensor area, A_d, and noise bandwidth, Δf; i.e.,

$$D^* = \frac{(A_d \, \Delta f)^{\frac{1}{2}}}{NEP} \qquad (5)$$

For example, a sensor having a NEP of 5×10^{-10} W, an area of 0.5 cm², and a noise bandwidth of 25 Hz would have $D = 2 \times 10^9$ W^{-1} and $D^* = 7.07 \times 10^9$ cm Hz$^{\frac{1}{2}}$ W^{-1}.

6. IR DEVICES WHICH USE THE CHARGE COUPLED CONCEPT

There are several devices which use the charge coupled concept which are potentially useful for IR imaging applications. These devices are: InSb CID, accumulation mode CCD, and the extrinsic silicon photoconductor CCD. The operation of these devices will be discussed below

6.1 InSb CID

The principle of operation of a InSb CID is the same as that of a silicon CID. The InSb CID uses chemically vapor deposited silicon oxynitride as the insulator. Interface state densities of less than 10^{11} cm^{-2} have been achieved. Storage times of 100 ms at 77°K have been reported (with no IR excitation). Measurements on single cell devices have resulted in values of D^* within a factor of 3 of background limited operation [4]. Fig. 6 shows the output voltage versus input irradiance for an InSb CID cell. Note that the input/output characteristic is linear.

6.2 Accumulation Mode CCD

If an MIS structure is cooled to temperatures below impurity freezout, and if the gate-to-substrate bias is of the polarity to attract majority carriers to the insulator-semiconductor interface, then this element can be used as the basic element of an accumulation mode CCD (AMCCD) or accumulation mode CID (AMCID) [5]. If the position of the doping level in the bandgap is ΔE with respect to the band edge, then photons of wavelength $\lambda = hc/\Delta E$ will be absorbed and will generate a majority carrier. If the doping density is large, the doping level will broaden into an impurity band in which case photon absorption will occur over a band of wavelengths with the absorption cutoff at $\lambda_c = hc/\Delta E$. Certain dopants in silicon have energy levels which correspond to IR wavelengths of interest; e.g., the cutoff wavelength for Ga and In doped Si occur at about 18µ and 8µ respectively. In this way the advantages of the well developed silicon technology can be used for IR imaging. The disadvantage is the low temperatures required.

Fig. 7 shows the energy bands for the basic MIS cell of an AMCCD below freezout. The energy bands have a linear dependence on distance into the silicon because of the absence of space charge. A 64-cell n-channel AMCCD has been fabricated on a $10^{15}\,cm^{-3}$ phosphorous doped substrate. The transfer inefficiency was 10^{-1} at 25 kHz and 4.2°K.

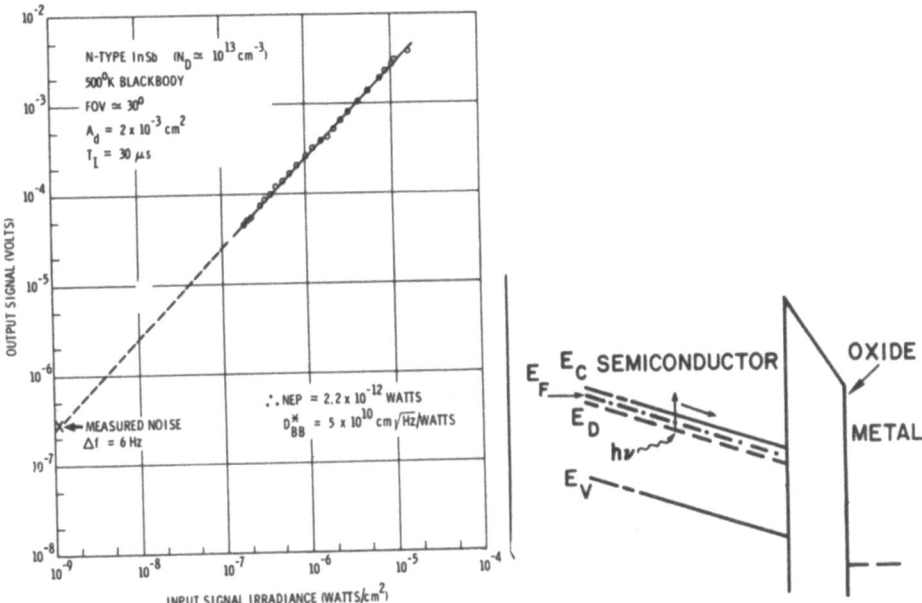

Fig. 6 Output voltage versus input irradiance for an InSb CID cell (after Kim [4]).

Fig. 7 Energy band diagram for an accumulation mode CCD (after Nelson [5]).

6.3 Extrinsic silicon photoconductor CCD

The extrinsic silicon IRCCD uses the extrinsic silicon substrate (In or Ga doped) for detection and a CCD fabricated in an n-type epitaxial layer for readout [6]. The basic cell is shown in Fig. 8. The temperature must be low enough so that the dopant sites are neutral. A positive bias is applied to the P+ layer on the backside of the substrate with respect to the frontside P+ regions. The screen gate and storage gate are biased so that the photoconductive currents in the detector regions are integrated in the CCD storage sites. Charge is read out of the storage sites by pulsing the transfer gates to allow the charge to flow into the CCD.

The major advantage of this approach is that it uses silicon. Disadvantages are crosstalk and the operating temperature. The crosstalk problem arises because the absorption coefficient in extrinsic silicon detector elements is low; thus, thick substrates (10-20 mils) are required for sufficient sensitivity. However, crosstalk increases with increasing substrate thickness. The low temperature is required because the extrinsic dopant sites in the detector region must be neutral. This requires, for example, $30^{\circ}K$ for Ga doped silicon and $60^{\circ}K$ for In doped silicon.

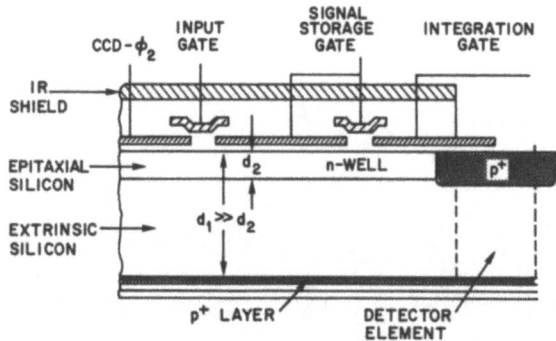

Fig. 8 The basic extrinsic silicon photoconductor cell (after Nummedal [6]).

7. IR IMAGING MODES

There are four IR imaging modes which can make use of CCD and/or CID technology: (1) staring mode, (2) parallel scan mode, (3) serial scan mode, and (4) serial parallel scan mode. Each of these will be treated separately.

7.1 Staring mode

In the staring mode the array integrates the full field of view for a period of time (1/60 second for two field TV compatible format) after which the field is read out while a second field is being integrated. The staring mode of operation is not suited for low contrast, high background thermal imaging because of the severe limits on the tolerable non-uniformity of response from element to element imposed by the combination of low contrast and high background. To illustrate this point, consider an array of IR elements operated in the staring mode. Assume that the mean responsivity of the elements is R and the deviation of the response of the ith element from the mean is ΔR_i. Let Q_b denote the photon background flux and C denote the contrast. After integrating for t_{integ} seconds the number of carriers in the ith CCD element is

$$N_i = (R + \Delta R_i) Q_b (1+C) t_{integ}$$

$$= \underset{(a)}{RQ_b t_{integ}} + \underset{(b)}{RQ_bC t_{integ}} + \underset{(c)}{\Delta R_i Q_b t_{integ}}$$

$$+ \text{negligible term} . \qquad (6)$$

Term (a) is a constant number of carriers in each cell and has no effect other than using dynamic range. Term (b) is the desired signal. Term (c) is the element-to-element variation in the number of carriers collected due to spatial non-uniformities of response. When the standard deviation of the distribution of terms (c) over the array is larger than (b), the signal will not be detectable. For example, in the 3-5μ range with a 290°K background, the contrast is 4%/°K. Thus, for a minimum resolvable temperature (MRT) of 0.2°K the maximum non-uniformity tolerable is 0.8%. This imposes severe constraints on material homogeniety and photolithographic tolerances [7]. Since the non-uniformity in response is a fixed distribution over the array, it's signature can be measured and stored in a memory. An element-by-element subtraction of the responsivity signature from the response to an actual IR scene can be used to reduce the MRT. The degree of improvement in the MRT depends on the accuracy of the signal processing (number of digital bits), and the signal processing may be expensive.

A second problem of staring mode operation in high-background IR imaging is that the high background flux densities may limit the integration to short times thereby limiting the signal-to-noise ratio to low values. For example, full-well CCD carrier densities are typically less than $10^{12} cm^{-2}$. From Fig. 4, the background flux density for $\lambda_c = 5\mu$ is approximately $10^{16} cm^{-2} sec^{-1}$. Therefore, the maximum integration time which could be used is 100 microseconds.

7.2 Parallel scanned mode

The parallel scanned mode uses a linear array of sensor elements with rotating optics providing parallel scanning of the scene across the array. The sensors are usually photovoltaic HgCdTe or PbSnTe. CCD's with multiple inputs can be used to multiplex these arrays.

The multiplexer implementation is shown in Fig. 9. Once per line time, the CCD, via the capacitively coupled input circuits, samples the detector output voltages, obtains charge in each CCD cell proportional to the corresponding detector output voltage, and shifts this charge configuration out. The object of this approach is to perform the multiplexing within the dewar with the least amount of power dissipation. In this way, the number of leads from the dewar to the outside world will be minimized and the heat load will be minimized. There are two problems associated with this approach - crosstalk between channels at the CCD output due to transfer inefficiency, and the low-noise injection of charge packets into the CCD which are proportional in charge to the voltage at the output of the detectors [8]. The crosstalk problem is largely solved by injecting into alternate CCD cells instead of injecting into each CCD cell.

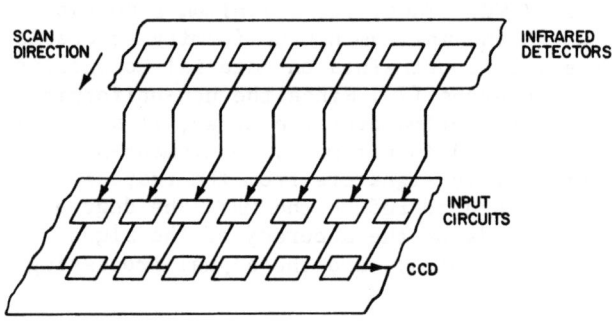

Fig. 9 Diagram showing the use of a parallel-in serial-out CCD for multiplexing a parallel scanned detector array.

If a single packet of charge containing charge Q is placed into a CCD well (i=0), then after N cell transfers, the distribution of charge in the cells i=0, 1, 2, ... is given by

$$D(i,N) = \frac{Q_i}{Q} = \frac{N!}{(N-i)!i!} (1-\alpha)^i \alpha^{N-i} \qquad (7)$$

When there is an input to every CCD cell, the crosstalk between adjacent channels (i=N and i=N-1) after N cell transfers is

$$D(N-1,N)/D(N,N) = N\alpha/(1-\alpha). \qquad (8)$$

If there is input to every other CCD cell, the crosstalk between adjacent channels (i=N, i=N-2) after N cell transfers is

$$D(N-2,N)/D(N,N) = [N(N-1)/2]\,[\alpha/(1-\alpha)]^2. \qquad (9)$$

Therefore, by injecting into alternate CCD cells, the crosstalk is reduced by the factor $2(1-\alpha)/(N-1)\alpha$ as compared with the crosstalk when there is no CCD isolating cell. For N=100 and $\alpha=10^{-4}$, the crosstalk reduction factor is 200.

The larger problem is to introduce charge into the CCD in response to low-level voltages. Surface channel CCD's would be surface-state noise limited; however, buried-channel CCD's with correlated double sampled output would be input noise limited. For an input capacitance of 0.2pf, the number of noise electrons due to input noise, $(kTC_{in})^{\frac{1}{2}}$, is 180 electrons. The corresponding noise voltage is $(kT/C_{in})^{\frac{1}{2}}$ which is 145μV. Since the input to the CCD is a voltage, it is the input noise voltage which must be minimized. This requires that C_{in} should be large.

7.3 Serial scanned mode

The serial scan mode utilizes one sensor element (or a few) to raster scan a field. The raster scan is achieved by optical scanning techniques [9]. The advantage as compared with parallel scan is better uniformity while the disadvantage is less sensitivity. A major advance was made when CCD delay-and-add techniques were applied to the serial scanned mode. This was called the discoid implementation. Today time delay and integration (TDI) techniques are implied in the term serial scan. The feasibility of this approach was proven using an array of InSb detectors and a buried-channel CCD as shown in Fig. 10. TDI was achieved by making a 1:1 interconnection between each detector element and the corresponding CCD element via the appropriate low-noise input circuits [9]. The charge injected into a CCD cell at a given point is proportional to the detector output voltage. The transfer of a charge packet along the CCD is synchronous with the velocity V_Q of the corresponding point in the image plane along the detector column. Thus, the effective signal is M times larger than the single-element signal, where M is the number of detectors in the column. Since the noise is proportional to $M^{\frac{1}{2}}$, the signal-to-noise improvement over the single detector case is $M^{\frac{1}{2}}$. For an N-column by M-element image plane, the number of interconnections between the detector array and the CCD array is M times N. Reliable fabrication of this large number of interconnections is a major concern for this hybrid approach. The implementation of this mode with the monolithic extrinsic silicon approach or the AMCCD would eliminate the interconnection problem.

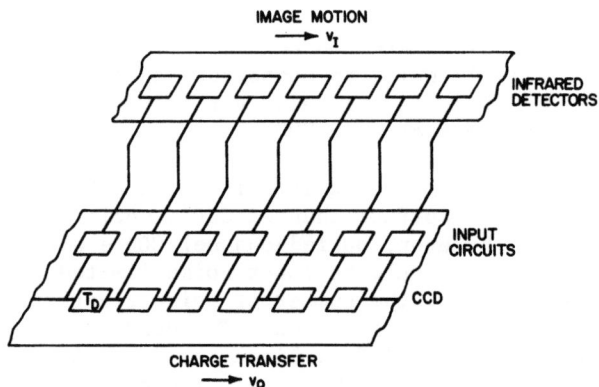

Fig. 10 Diagram showing the use of a parallel-in serial-out CCD to perform time delay and integration on a serial scanned detector array.

7.4 Serial-parallel scanning mode

The serial parallel scanning mode incorporates the advantages of the serial and parallel scanning; i.e., uniformity and sensitivity, Serial-parallel scanning incorporates parallel optical scanning and serial TDI. A serial-parallel scanned focal plane array requires a two dimensional array of detector elements. The number of detector elements perpendicular to the scan direction determines the number of lines which can be displayed. The number of elements parallel to the scan direction is the number of stages of TDI.

Monolithic focal plane arrays can be fabricated using the AMCCD or extrinsic silicon approaches; however, both of these arrays require very low temperatures ($< 77°K$). A serial-parallel scanned focal plane array operating in the 3-5μ range and operated at 77°K can be fabricated using small InSb CID arrays. As an example this focal plane array concept will be discussed in detail. The InSb CID module is a monolithic chip composed of 25 elements perpendicular to the parallel scan direction and 15 elements in the direction of parallel scan. A focal plane might, for example, be composed of seven of these chips positioned in such a way that they are optically contiguous. Each InSb CID chip would be read out by digital shift registers fabricated on silicon chips. Fabrication of the shift registers on the InSb CID chips would of course reduce the number of wire bonds. This has not yet been done in InSb although it now appears to be feasible. Each CID module would be read out through a single high performance amplifier into a silicon CCD chip which would perform the TDI operation. High pass CCD transversal filters could be used for background rejection, and a CCD multiplexer could be used to format the data for display. Fig. 11 shows the complete focal plane [10].

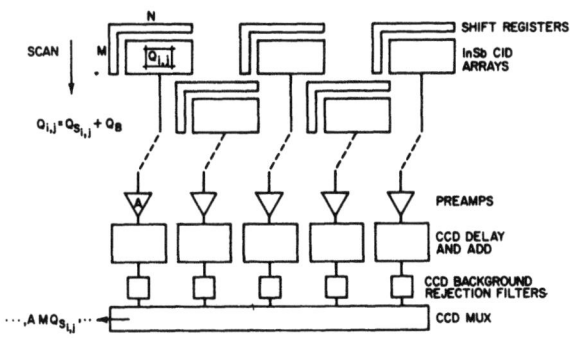

Fig. 11 Schematic diagram showing a focal plane array using InSb CID detector arrays and silicon CCD arrays for time delay and integration.

If such a focal plane were mounted at the focus of a telescope, the telescope could be rotated to provide the parallel scan for an IR search set. Fig. 12 illustrates the operation of the series-parallel mode. Only one channel having four delay-and-add elements is shown for simplicity. Note that for each CID column having M elements, two columns of M elements each are required in the silicon chip to provide the delay-and-add operation. As shown in Fig. 12, when the parallel scan moves one picture element, the data stored in the silicon shift registers is moved one element so that the charge from picture element A_i adds to that from picture element A_i from the previous CID readout. After an initial transient, the output from the silicon chip is MA_i (M=4 for the example of Fig. 12).

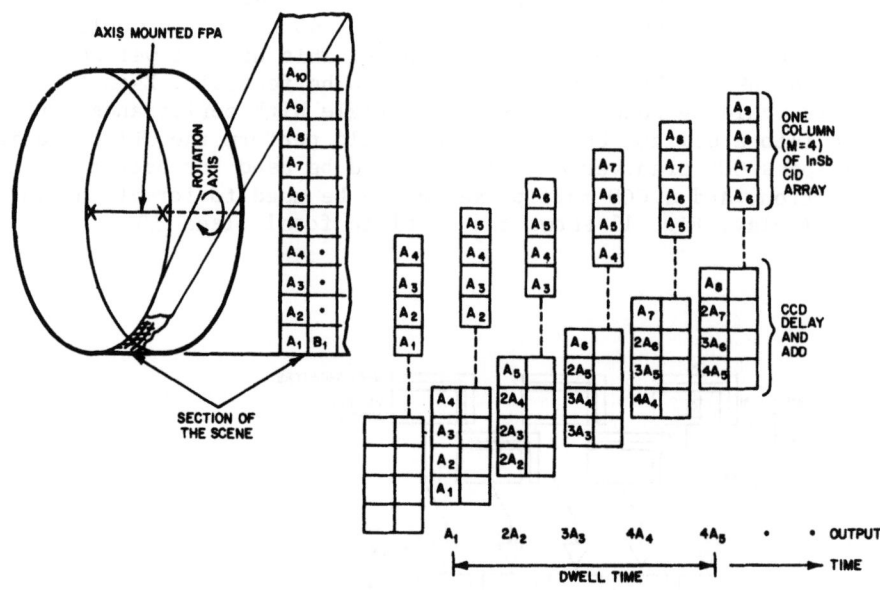

Fig. 12 Example of a serial parallel scanned InSb CID focal plane array operating in the search mode. Only four TDI elements and one channel are shown for simplicity.

In order to show the value of serial-parallel scanned focal plane arrays consider the detection range (R) of an IR search set. Assume that the target subtends one sensor element or less and that the sensor noise is dominant. Then the signal-to-noise ratio (S/N) is proportional to R^{-2}. If TDI is performed on M sensor elements, then S/N is also proportional to M^2; i.e.,

$$S/N \; \alpha \; R^{-2} \; \alpha \; M^{\frac{1}{2}}, \tag{10}$$

thus,

$$R \; \alpha \; M^{\frac{1}{4}}. \tag{11}$$

For example, using 16 TDI elements (M=16) would give an increase in range by a factor of two over the single sensor element case (M=1).

7. REFERENCES

1. R. D. Hudson, Jr., *Infrared System Engineering*, (Wiley, New York, 1969).
2. J. O. Dimmock, "Capabilities and Limitations of Infrared Imaging Systems," Proc. of the Society of Photo-Optic Instrumentation Engineers, 32, p. 9, 1972.
3. A. Rose, *Vision-Human and Electronic*, (Plenum, New York, 1973).
4. J. C. Kim, W. E. Davern, and T. Shepelavy, "InSb Surface-Charge Injection Imaging Devices," presented at the 22nd IRIS Meeting.
5. R. D. Nelson, "Accumulation-Mode Charge-Coupled Device," Appl. Phys. Lett., 25, pp. 568-570, 1974.
6. K. Nummedal, late news paper presented at the 1974 CCD Applications Conf., Edinburgh, Scotland.
7. D. F. Barbe, "Imaging Devices Using the Charge-Coupled Concept," Proc. IEEE, 63, pp. 38-67, 1975.
8. T. F. Cheek, et.al., "Design and Performance of CCD Multiplexers," in Proc. 1973 CCD Applications Conf., pp. 127-139.
9. D. M. Erb and K. Nummedal, "Buried-Channel Charge-Coupled Devices for Infrared Applications," in Proc. 1973 CCD Applications Conf., pp. 157-167.
10. A. F. Milton and M. Hess, "Series-Parallel Scan IR CID Focal Plane Concept," in Proc. 1975 CCD Applications Conf.

LOW LIGHT LEVEL PERFORMANCE OF CHARGE-COUPLED AREA IMAGING DEVICES

James A. Hall

Westinghouse Electric Corporation,
Advanced Technology Laboratory,
Baltimore, Maryland U.S.A.

ABSTRACT. The display signal to noise ratio concept permits both predicting the performance of proposed man viewed systems or designing them for specific performance goals and analyzing the performance of existing imaging devices in man-viewed systems. A specific Fairchild CCAID-244 is used for an example. Observed bar pattern visibility at 35 electrons per frame per element is analyzed to derive a noise equivalent signal charge of 46 electrons per element. A hypothetical camera using this device with a 44 mm f:1.5 lens should permit identifying an M-48 tank at about 390 meters by 0.3 x full moonlight. Maximum geometric identification range is 535 meters. Parameters considered include the scene and sensor spectral irradiances, the absolute spectral responsivity of the sensor, the square wave flux response of the sensor, and the contrast reducing effects of atmospheric scattering. Some of the CCD characteristics used here to make the example specific were deduced indirectly and should not be considered as typical values for other analyses unless checked with the device manufacturer.

1. INTRODUCTION

The chapter "Signal and Noise In The Display of Images" provides the tools to predict the performance of man-viewed imaging systems while these systems are being designed. Hence, using these tools one can design an imaging system to perform a specified visual task with confidence that the system will perform usefully. Conversely, one can analyze the observed performance of existing imaging systems to gain additional insights.

2. SYSTEMS ANALYSIS

To provide a concrete example, figure 1 shows images of a test chart taken at various sensor exposures with a Fairchild 190x244 element charge-coupled imaging device, CCAID-244, made in the laboratory by photographing the television monitor[1]. It must be emphasized that this figure, taken in the spring of early summer of 1975, is not intended to show the state of the CCD art, either then or now. As shown in figure 1, the finer patterns consisting of 3 black and 2 white bars on a white field are just visible in the photograph when the CCD exposure corresponds to a signal of 35 electrons per CCD element per 0.043 second frame time. From this information we will first attempt to calculate the noise equivalent charge signal, NES, the signal equal to the rms charge fluctuation per element per frame at this exposure. The bar pattern spatial frequency for the vertical bars of the finer groups was half the Nyquist frequency, so each bar width in the image equalled two element widths. As one can see, bar length is 5 times bar width in this pattern. The physical characteristics of the CCAID-244, listed in figure 2, show that the sensor cell is 30 μm wide by 18 μm high. Thus each bar image is 60 μm wide by 300 μm high, and includes 32 to 34 sensor cells.

According to Rosell and Willson[2], threshold visibility of a 5 bar pattern requires a display signal-to-noise ratio of approximately 3. The display signal to noise ratio equation for a conventional television camera was developed in the chapter "Signal And Noise In The Display Of Images" in the form

$$SNR_D = \left[2 t_e \Delta f_v \frac{a}{A} \right]^{1/2} \cdot SNR_V \qquad (1)$$

where t_e = integration time of the eye, about 0.1 seconds

f_v = analog video channel bandwidth

a = area of the image detail being viewed at the sensor input

A = scanned image area (raster area) at the sensor input

SNR_V = ratio of the peak to peak amplitude of the equivalent <u>square wave</u> video signal from the detail being viewed, to the rms video channel noise.

For this CCD sensor with its sampled data output, the expression becomes

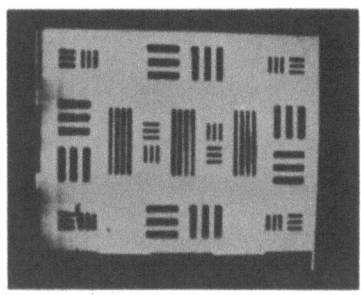
22K ELECTRONS

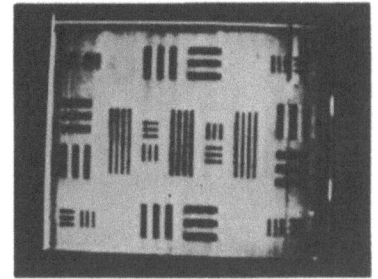
2200 ELECTRONS

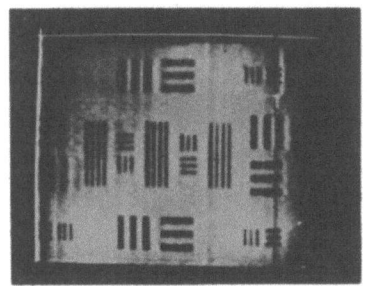
560 ELECTRONS

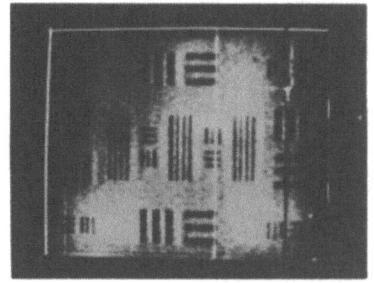
200 ELECTRONS

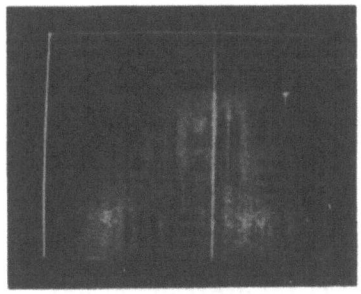
140 ELECTRONS

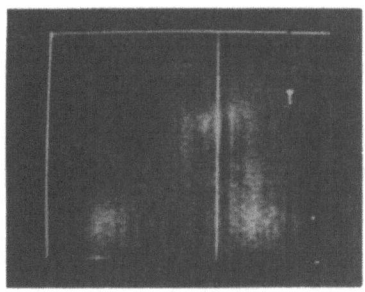
70 ELECTRONS

35 ELECTRONS

BAR PATTERN IMAGERY FROM THE DFGA OUTPUT OF THE 190 x 244 ELEMENT IMAGER

CONDITIONS:

0.50 and $0.25\ f_N$
$f_o = 2$ MHz
$t_{FRAME} = 43$ msec.
$T = +25°C$.

EXPOSURE TIME OF THE PHOTOGRAPHS=0.1sec.

75-0440-PA-114

Figure 1. Bar Pattern Image Taken with Fairchild CCAID-244. The Spatial Frequency of the Finer Groups is One-Half the Nyquist Frequency in the Horizontal Direction.

ELEMENT ARRAY	190 H x 244 V
CELL SIZE	30 μm H x 18 μm V
PHOTOSENSITIVE AREA	216 μm^2
PHOTOSENSITIVE FRACTION	0.5
CHARGE STORAGE AREA	76 μm^2
DARK CURRENT AREA	234 μm^2
SPECTRAL RESPONSIVITY	FIGURE 5
SATURATION SIGNAL	2×10^5 ELECTRONS
NOISE EQUIVALENT SIGNAL (PER S. CAMPANA)	~70 el
R_{SQ} AT F_N (EST)	0.24
AT $F_N/2$ (EST)	0.62

*VALUES CHOSEN ARE PRINCIPALLY FROM REFERENCE 1, AND ARE USED FOR THE EXAMPLE IN THE TEXT. THEY DO NOT REPRESENT THE STATE-OF-THE-ART, AND CURRENT CHARACTERISTICS FROM THE CCD MANUFACTURER SHOULD BE USED FOR ANY ACTUAL SYSTEM DESIGN CALCULATION.

Figure 2. Characteristics of Fairchild CCAID-244

$$SNR_D = \left[\frac{t_{exposure}}{t_{frame}} \cdot \frac{\text{area of bar}}{\text{area of element}}\right]^{1/2} \frac{Q_{sig} \; R_{SF}}{NES} \qquad (2)$$

where the coefficient of the bracket is the electrical signal to noise ratio in terms of electrons per element, and the bracket scales the sampled area up to the area of the detail (bar) being viewed and the sampling time up to the photographic exposure time. NES is the rms charge fluctuation in electrons per element and frame, and Q_{sig} is the measured signal for large area white and black bars in the high contrast test pattern. The square wave flux response factor, R_{SF}, is the relative square wave signal amplitude for a bar pattern input as a function of bar pattern spatial frequency, f_S, normalized to unity for a very coarse pattern.

One further refinement is in order. In a well designed charge-coupled device the system noise is small enough so that the fluctuation in the signal charge per element from frame to frame may add significantly to the measured noise. Thus

$$SNR_D = \left[\frac{t_e}{t_f} \cdot \frac{a}{a_e}\right]^{1/2} \frac{Q_{sig} \cdot R_{SF}}{\left[NES^2 + Q_{sig}\right]^{1/2}} \qquad (3)$$

where NES is now the system noise per element in the dark.

To calculate NES for the CCAID-244 in our example, we need a value for R_{SF}. Data on similar CCAID's indicates the peak to peak signal amplitude for a bar pattern input at the Nyquist frequency, f_N, is approximately $R_{SQ} = 0.24$ of the signal for a large white and black bar pair. Using linear interpolation for the decrement in R_{SQ}, $R_{SQ} = 1 - .76 \frac{f_S}{f_N}$, and the estimated peak to peak response at half Nyquist frequency is $R_{SQ} = 0.62$, when the optical bar pattern is optimally aligned with the sensor wells, and perhaps 0.45 for random alignment. This is also approximately R_{SF}, since the CCD elements average signal over their own widths to produce a square wave output signal. Substituting, at threshold

$$3 = \left[\frac{0.1}{.043} \cdot \frac{33}{1}\right]^{1/2} \cdot \frac{35 \cdot .45}{\left[NES^2 - 35\right]^{1/2}} \qquad (4)$$

and NES = 45.6, or about 46 electrons for this example. This compares with NES of about 70 electrons estimated by S.B. Campana[3] on similar devices a few months earlier.

It should be realized that this calculation does not say that bar pattern visibility is being limited by a random fluctuation of about 46 electrons per element per frame. Element to element dark current or response variations which can produce a pepper and salt or fine scale mottling appearance in the picture would have the same effect on visibility of the bar pattern. The calculation does say that the effect is the same as would have been produced by a 46 electron fluctuation on an otherwise perfect sensor.

3. PERFORMANCE PREDICTIONS FOR SOLID-STATE IMAGING SYSTEMS

A solid-state array imaging sensor forms part of an imaging system which includes the scene or object to be televised, the scene illumination, the atmospheric transmission path from the object to the camera, the objective lens, the imaging array itself, the scanning and signal amplifying circuits, the display device and finally the eye/brain combination of the human observer, all suggested in figure 3. The chapter "Signal and Noise

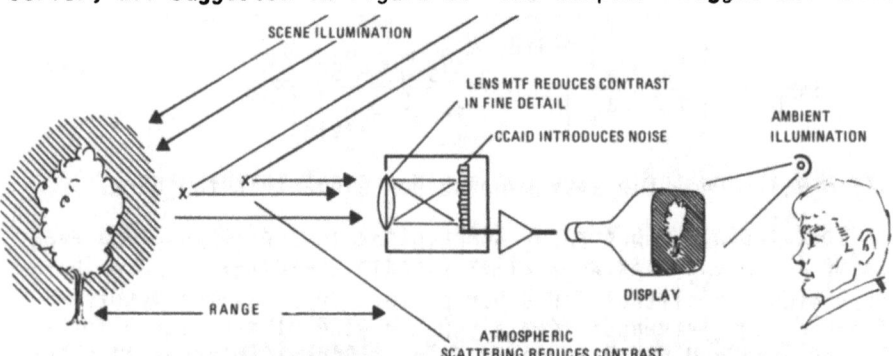

Figure 3. Even A Simple Imagery System Includes the Scene Illumination, the Scene Objective Being Televised, the Atmospheric Path, the Lens, the Sensor, the Circuitry, the Display, and Finally the Eye and Brain of the Human Observer Who Makes the Decisions.

In The Display Of Images" showed how observer performance on the simple visual tasks of detection, classification, and identification could be predicted from the display signal to noise ratio, and how the display signal to noise ratio could be calculated from the signal to noise ratio in the video signal produced by the imaging sensor. In this section, we consider how video signal to noise ratio can be predicted for a charge-coupled area imaging device under conditions like those found in military applications. Most reports and analyses have rated performance of charge-coupled imaging devices in terms of the noise equivalent signal charge within the device, the quantity calculated in the preceding example. Therefore one task is to relate the signal charge to the image irradiance on the sensor, or better to the characteristics of the scene being viewed, which will produce the required signal to noise ratio.

3.1 Statement Of The Problem

As a specific example, we shall calculate the range for television identification by 0.3 moonlight of a U.S. M-48 tank, with a TV camera using the specific Fairchild 190x244 element charge-coupled area imaging device described in the preceding example. Assume the camera uses an f:1.5 lens whose focal length provides a 6° horizontal field of view, like that of a starlight scope. The assumed characteristics of the CCAID-244 were given in figure 2 and in the preceding example. The table below gives scene data including visibility as defined by Middleton:[4]

Scene Information

0.3 moonlight	scene spectral irradiance as shown in figure 4
Object reflectance	5% ⎫ averaged over the spectral
Background reflectance	35% ⎭ region of CCD response
Object size	2.3 x 4.6 meters
Visibility	3 nmi

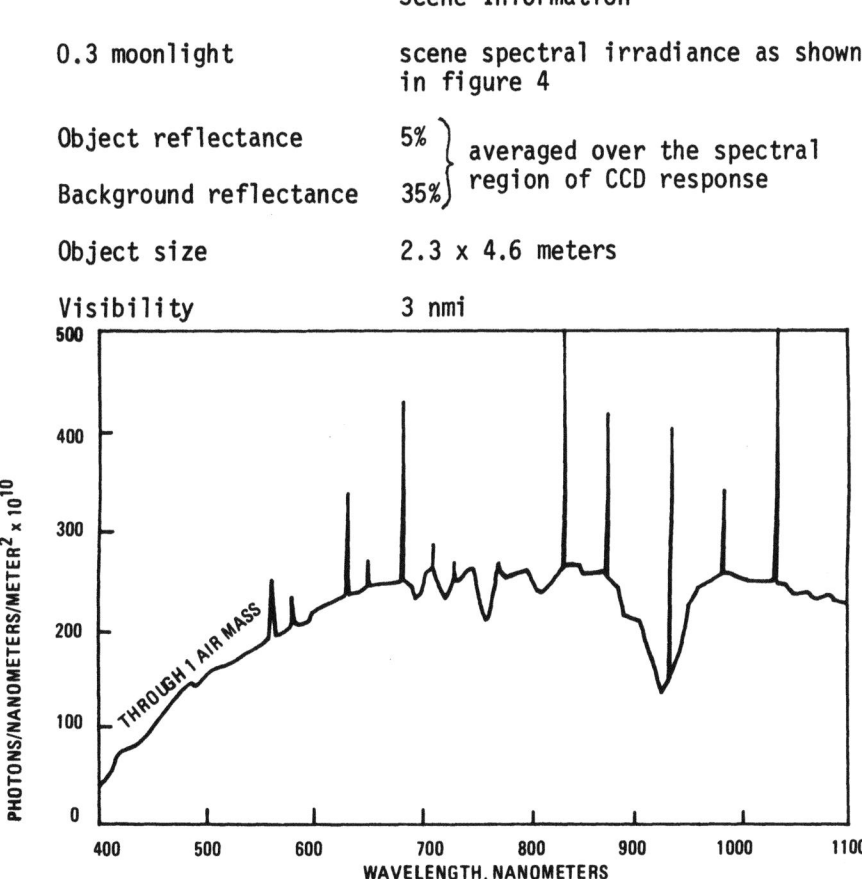

Figure 4. Spectral Irradiance at Earth's Surface From 0.3 Full Moon

Steps in the system design and performance prediction include:

3.1.1 Field Of View And Geometric Range Limitation

To provide a 6 degree horizontal field of view, with this sensor, the focal length of the lens is

$$f.\ell. = \frac{w/2}{\tan \theta/2} = \frac{5700 \times 10^{-3}/2 \text{ mm}}{\tan(6°/2)} = 54.4 \text{ mm}, \quad (5)$$

about 2.14 inches

since the object is assumed to be essentially at infinity. This calculation is diagramed in figure 5. Johnson's criteria[5] specifies that for identification one requires 13 resolvable elements in a distance equal to the smallest dimension of the vehicle, in this case the 2.3 meter height. Thirteen sensor elements occupy 0.234 mm as shown in figure 5. For a system with a 54.4 mm lens, the range for a 2.3 mm high object to produce a 0.234 mm high image is

$$R = \frac{54.4}{0.234} \cdot 2.3 = 535 \text{ meters} \quad (6)$$

Thus for this focal length lens, the maximum range for identification is limited by geometry to 535 meters for any scene irradiance level. The calculations in sections 3.1.2 and 3.1.3 are to learn whether the range is limited to smaller values by the available light.

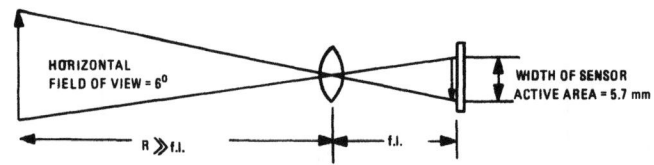

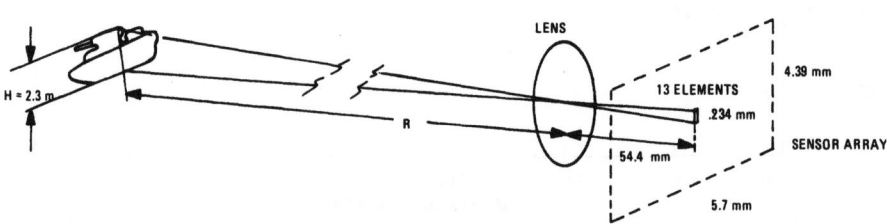

Figure 5. Sensor Field of View Determines Lens Focal Length and Geometric Range Limitations

3.1.2 Sensor Irradiance And Resulting Signal

The charge signal from a CCD sensor is approximately proportional to sensor exposure, the product of sensor irradiance or incident photon flux and integration time, with a constant of proportionality depending on the spectral distribution of the irradiance and the spectral responsivity of the sensor.

To calculate sensor irradiance or incident photon flux from scene irradiance, assume first no loss due to the atmosphere. The sensor spectral irradiance is:

$$H_D(\lambda) = \frac{R_S(\lambda) \, T_A(\lambda) \, T_L(\lambda)}{4f^2 \, (1+m)^2} H_S(\lambda) \tag{7}$$

where $H_D(\lambda)$ = spectral irradiance on detector, watts m^{-2} nm^{-1} or photons m^{-2} sec^{-1} nm^{-1}

$R_S(\lambda)$ = scene or object spectral reflectance, assumed to be diffusely reflecting

$T_A(\lambda)$ = transmission of atmosphere

$T_L(\lambda)$ = transmission of lens. Assume 80 percent, spectrally invariant

f = focal ratio or f number of lens

m = linear magnification of optical system, usually <<1

$H_S(\lambda)$ = spectral irradiance on scene, watts m^{-2} nm^{-1} or photons m^{-2} sec^{-1} nm^{-1}

The expression given is for spectral irradiance since all quantities in this equation except focal ratio and magnification are functions of wavelength. The expression must be evaluated at each wavelength to obtain the spectral distribution of irradiance on the sensor.

The sensor response per element may be calculated as

$$Q_{sig} = t_f \, a_e \int_0^\infty H(\lambda) \, R(\lambda) d\lambda = t_f \, a_e \int_0^\infty \frac{H(\lambda) \, \eta(\lambda)}{hc/\lambda} \, d\lambda \tag{8}$$

where $R(\lambda)$ is the sensor spectral response in amperes watt^{-1} or coul joule^{-1} referred to the sensor cell area

$\eta(\lambda)$ is the sensor external quantum efficiency referred to the sensor cell area.

a_e is the area of the sensor cell in m^2. (Not just the photosensitive fraction).

To limit the problem, we use values of scene and object reflectance averaged over the spectral response interval for a silicon sensor. This approach is risky, however, for a real system, since the signal from man-made objects with respect to background often reverses sign in the middle of the silicon response interval, reducing the effective contrast significantly.

The sensor spectral responsivities $R(\lambda)$ in amperes per watt on the sensor cell are shown in figure 6, which is a composite of 3 curves taken from reference 6, dated March 6-7, 1975. The upper curves show that for a thinned back illuminated frame transfer CCD sensor the larger effective photosensitive fraction, the use of an antireflection coating on the silicon, and accumulation of the first silicon surface to avoid carrier loss at surface states can produce a "best" device with significantly increased response compared to the front illuminated interline transfer device of unknown quality shown in the lowest curve. They also show that in early 1975 typical back illuminated devices fell significantly below the best device, whose curve was often quoted in papers written at that time. The lowest curve is actually that given for a Fairchild CCD-201 on a power per element sensitive area basis scaled to the percent photosensitive fraction of the CCAID-244 to refer the power to the entire sensor cell area. This extrapolation to provide information for this concrete example does not represent the state-of-the-art and may not be even approximately correct for a CCAID-244. Up to date information should be measured or obtained from the device manufacturer for any actual device comparison or systems analysis.

To apply the information to our example, we compare the lowest sensor response curve of figure 6 with the scene spectral irradiance curve of figure 5. Scene spectral irradiance is assumed proportional to sensor spectral irradiance since both object and scene have been assumed to have spectrally independent reflectance, and range is assumed short so atmospheric spectral effects can be neglected. Properly speaking, one should form the product of the response function, figure 6, and the sensor irradiance function, figure 5, at each point and integrate. For a first estimate, however, the curves are smooth enough so that average values can give useful information. Working in photons m^{-2} sec^{-1} nm^{-1}, in the spectral region from 500 to 1000 nm

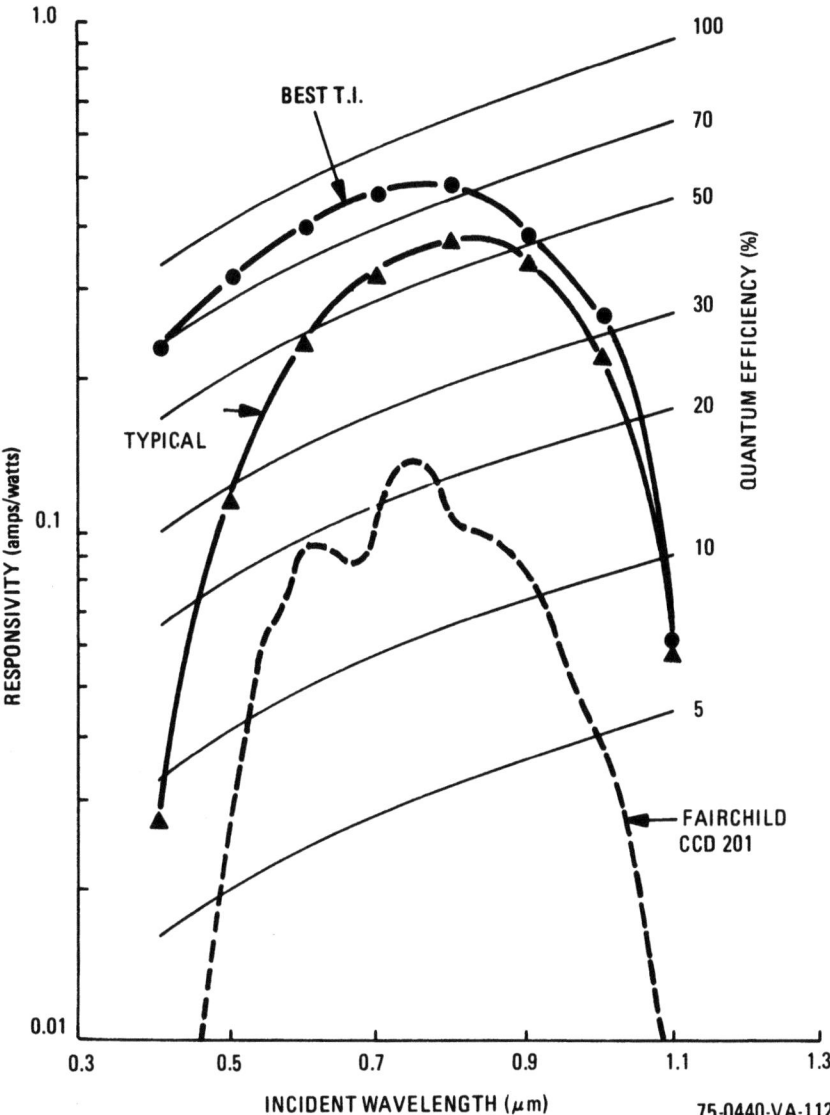

Figure 6. Spectral Responsivities for TI and Fairchild Charge-Coupled Area Imaging Devices. (Fairchild Curve is Extrapolated From CCD 201 and Should Not be Used for Systems Design.)

where the response is significant, the average quantum efficiency is about 0.13 and the average scene spectral irradiance is about 180×10^{10} photons m^{-2} sec^{-1} nm. Hence, for the more reflecting background,

$$Q_{sig} = t_f a_e \int_0^\infty \frac{H_D(\lambda)}{hc/\lambda} \eta(\lambda) d\lambda$$

$$= \frac{t_f a_e R_S T_A T_L}{4f^2 (1+m)^2} \int_0^\infty \frac{H_S(\lambda) \eta(\lambda) d\lambda}{hc/\lambda} \text{ electrons}$$

$$= \frac{.043 \text{sec frame}^{-1} \times 540 \times 10^{-12} \text{ m}^2 \times .35 \times 1 \times .8}{4 \times (1.5)^2 \times 1^2}$$

$$\times (180 \times 10^{10} \text{ photons m}^{-2} \text{ sec}^{-1} \text{ nm}^{-1} \times .13 \text{ elec photon}^{-1}$$

$$\times 500 \text{ nm})$$

= 84.5 electrons per element per frame, for the large area background. The signal from the object is 12.1 electrons per element per frame, and the difference signal, background over object, is about 72 electrons.

3.1.3 Display Signal To Noise Ratio

To determine at what range an observer could identify an M-48 tank using the CCD television camera described in the example, we apply Johnson's criterion that the image of a 13 bar test chart with horizontal bars filling the 4.6 x 2.3 meter tank outline must be resolved by the viewer, as described in the chapter "Signal And Noise In The Display Of Images". The equivalence is suggested in figure 7. The display signal to noise ratio expression is

$$SNR_D = \left[\frac{t_e}{t_f} \frac{a}{a_e} \right]^{1/2} \frac{Q_{sig} R_{SF}}{(NES^2 + Q_{sig})^{1/2}} \qquad (10)$$

where t_e = 0.1 sec, the eye integration time

t_f = 0.043 sec

a = area of one bar in the image at the sensor

$$= \frac{4.6 \times 2.3}{13} \frac{(54.4 \times 10^3)^2}{R^2} \mu m^2$$

$$= 2.41 \times 10^9 / R^2 \; \mu m^2 \text{ where R is in meters}$$

$a_e = 540 \; \mu m^2$

NES = 45.6 electrons

R_{SF} = 0.17 at f_N, 0.45 at $f_N/2$

Q_{sig} = 72 electrons

$$\text{and SNR}_D = \left[\frac{0.1}{0.043} \frac{2.41 \times 10^9}{540} \right]^{1/2} \frac{1}{R} \frac{72 \; R_{SF}}{(2079+72)^{1/2}} = 5 \times 10^3 \frac{R_{SF}}{R}$$

To provide an expression for R_{SF}, we assume the interpolating relationship used in section 2, $R_{SF} = 0.7 \left(1 - .76 \frac{f_S}{f_N}\right)$ and referring to figures 3 and 7, note for the 13 bar pattern that

$$f_S = f_N \frac{R}{535.} \tag{11}$$

Hence $R_{SF} = 0.7 - 0.53 \frac{R}{535} = 0.7 - 10^{-3} R$ \hfill (12)

when R is in meters. For recognition of the equivalent bar pattern, $SNR_D \approx 3$, and

$$SNR_D = \frac{5 \times 10^3}{R} (0.7 - 10^{-3} R) = \frac{3.5 \times 10^3}{R} - 5 = 3 \tag{13}$$

R = 440 meters

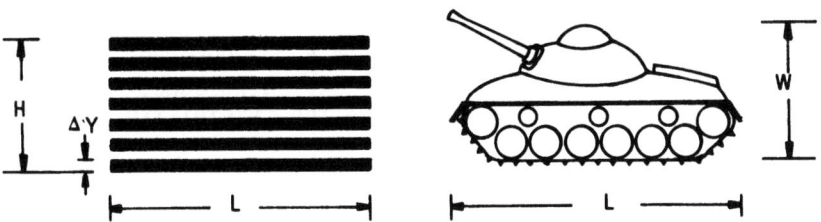

Figure 7. According to Johnson, Identification of a M-48 Tank is Equivalent to Recognition of a 13-Bar Test Chart.

Thus noise or nonuniformity in this CCAID would limit the range for identification of an M-48 tank by 0.3 full moonlight to 440 meters, just less than the geometric limit. This calculation also assumes the scene background is entirely uniform. The presence of background clutter would require a significantly higher signal to noise ratio and would approximately halve the range.

3.2 Atmospheric Effects

Neglected in this example are the effects of atmospheric light scattering on reducing image contrast at longer ranges. This effect has been observed by everyone who has lived in or journeyed through hilly forested country. The nearest hillsides appear dark green, the further hillsides appear a lighter gray-green because light from the sun and sky is scattered toward the observer's eye by dust, pollen, or moisture particles in the optical path, adding to the light coming from distant objects and making them appear less dark. It is this loss of contrast which limits the range at which the earth bound observer can see terrestrial objects except on very clear days where the earth's curvature may limit first.

By definition, visual range is defined as the range at which one can just distinguish a large black disc placed against the sky background above the horizon. The sky background near the horizon looks bright because light from the sun overhead is scattered toward the observer's eye by particles along the line of sight. It is not the distant background which is bright, but rather the whole cone of atmosphere along the line of sight whose brightness adds to produce the sensation in the observer's eye. There is a limit, however, since light is also scattered out of the line of sight by particles and very distant slices of atmosphere contribute no additional brightness.

A large black disc placed near the observer appears black. As the range increases the sky background remains unchanged but the disc appears lighter because of the luminosity of the intervening atmosphere, until at some range R the observer can no longer distinguish the disc from the sky background. This range is used by meteorologists to describe the "visibility" as e.g., 3 nmi in this example.

Middleton[4] shows in general for uniform illumination from sun and sky along the optical path that contrast, M, varies as

$$M = M_0 e^{-\sigma R} \tag{14}$$

where M_o is the contrast at the object

σ is a constant to be evaluated

R is the range to the object in e.g., nautical miles, nmi, or in meters

By agreement, the visual range or visibility is the distance at which a high contrast black object has an apparent contrast of 2 percent. Thus if meteorological visibility is V nmi, $0.02 = e^{-\sigma V}$ and $\sigma = 3.91/V$ nmi^{-1}. For three nautical mile visibility, the contrast of an object against background, viewed horizontally, varies as $M = M_o e^{-1.3R}$ if R is in nmi, and $M = M_o e^{-7.02 \times 10^{-4} R}$, if R is in meters. At 440 meters, contrast is reduced to 73 percent of its value at the object. Since the optical "signal" is proportional to constrast, such a contrast reduction can seriously limit the useful range for television reconnaissance in the visible even as it limits visual reconnaissance. In this case it would reduce the operating range to about 390 meters, as can be verified by solving the equation

$$SNR_D = \left(\frac{3.5 \times 10^3}{R} - 5\right) e^{-7.02 \times 10^{-4} R} = 3. \qquad (15)$$

REFERENCES

1. Figure 1 was provided by Dr. R. H. Dyck of the Fairchild Optoelectronics Division, Palo Alto, Ca.

2. Rosell, F.A. and Willson, R.H., "Recent Psychophysical Experiments and the Display Signal-to-Noise Ratio Concept", Chapter 5 in "Perception of Displayed Information", Ed L. M. Biberman, Plenum, New York, 1973. Also see the chapter "Signal And Noise In The Display Of Images".

3. Private communication. S. B. Campana of the Naval Air Development Center, Warminster, Pa., was responsible for CCD imager evaluation for the U.S. Navy who funded much of the Fairchild work.

4. Middleton, W.E.K., Vision Through The Atmosphere, University of Toronto Press, 1952.

5. Johnson, John, Image Intensifier Symposium, Fort Belvoir, Va., Oct. 6-7, 1958, AD22016D.

6. Proceedings, Symposium on Charge-Coupled Device Technology for Scientific Imaging Applications, March 6-7, 1975, Jet Propulsion Laboratory, California Institute of Technology, Pasadena, Ca.

COMPARISON OF SOLID-STATE IMAGERS AND ELECTRON BEAM SCANNING IMAGERS

James A. Hall
Westinghouse Electric Corporation
Advanced Technology Laboratory
Baltimore, Maryland U.S.A.

ABSTRACT. Are solid-state imagers now able to displace electron beam scanned imagers? In what applications? What is needed to compete for other applications? To answer these questions, the criteria for rating camera tube and solid-state array imagers were placed on a common basis as far as feasible. At this writing, for analog imaging, tubes lead over x-y array imagers for the greatest total number of resolvable elements per frame, for producing images with less element to element signal difference, for resistance to blooming, and for maximum output data rate on a single terminal. Solid-state imagers lead for greater signal to noise ratio at maximum signal, hence for larger effective dynamic range and greater theoretical ability to reproduce low contrast scenes, for freedom from lag, for resolving power independent of signal level (except when blooming), for geometric fidelity, for stability of characteristics and freedom from interactions, as well as for their obvious compactness, durability, and low operating voltages. A second major operating mode, photon counting, appears feasible with a thinned charge-coupled area imaging device operating in a vacuum tube and bombarded with photoelectrons. Here CCD's offer a significant advantage since lower amplifier noise permits photon counting imaging on arrays of modest size without use of auxiliary image intensifiers.

1. INTRODUCTION

While some companies are developing charge-coupled area imaging devices with entertainment television primarily in mind and a comparison could be made for that application, a comparison of solid-state and electron beam scanned imagers for quantitative scientific or military imaging applications yields more information. In the following sections, area imaging devices are compared primarily on the basis of their information handling capability.

Users of television techniques to obtain quantitative photometric or radiometric data or to detect faint objects against background soon find that their requirements are sufficiently different from those of entertainment television as to make the usual commercial performance criteria of little use. Where the commercial task is to present a subjectively satisfying picture with the constraints of 4.5 MHz channel bandwidth and need for flicker-free reproduction, the military man or the scientist, particularly the astronomer, needs to sense small constrast differences, small exposure differences in the presence of large backgrounds, and preferably to separate the effects of size and irradiance when performing radiometry on small details, be they spectral lines or star images. In turn, these requirements translate into need to integrate large signals from each sampled area, many electrons per picture element, so that the signal to photoelectron shot noise ratio will be high, and also so that the signal to sensor noise ratio will be high. Both the maximum number of stored electrons per element and the ratio of this number to the rms system noise per element in the dark, defined here as the dynamic range available in a single exposure, are criteria on which sensors will be compared. Another major criterion is the number of nearly independent picture elements in the sensor's field. Together these numbers measure the quantity of information which can be handled in each exposure. The third major criterion is the data rate which the sensor can accommodate.

To extend the total number of information quanta beyond that which can be stored in any analog imaging device one can perform photon counting, or better photoelectron counting, in which the image is scanned very rapidly and the individual photoelectron events are stored in digital form in a computer memory. Analog imaging devices will be discussed first, however, both to provide the motivation for, and to show the limitations of, the photoelectron counting technique.

2. ANALOG IMAGING DEVICES

This section compares Electron Bombarded Silicon (EBS)[1] camera tubes (figure 1), Secondary Electron Conduction (SEC)[2] camera tubes (figure 2), and charge-coupled area imaging devices, in that order.

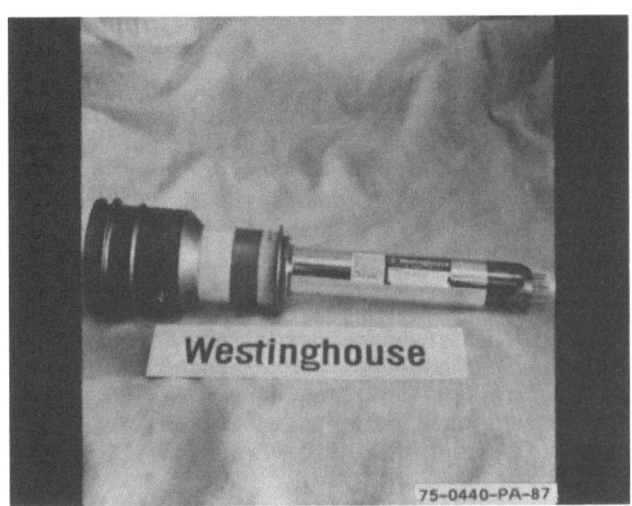

Figure 1. Westinghouse WX 32719 EBS Camera Tube Provides Prescanning of Up to 3,000 By Photo Electron Bombardment of a Silicon Diode Array Target

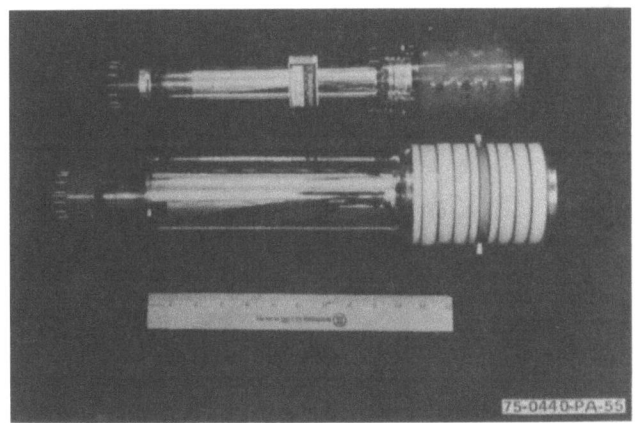

Figure 2. Westinghouse SEC Camera Tubes for Astronomical Applications Permit Long Integration Times and Provide Large Total Resolution

2.1 Information Storage Capacity - Number of Elements Per Frame

Television camera tubes are almost always evaluated under commercial broadcast conditions. Resolving power is traditionally given as limiting visually detected resolution on a monitor, an essentially useless figure for system design purposes. The most useful descriptor, the modulation transfer function, MTF, is almost never given by the camera tube manufacturer, but he does often give the response to a high contrast bar chart test pattern vs spatial frequency at a high exposure level, sometimes called the contrast transfer function CTF. Limansky[3] following Coltman[4] has demonstrated a straightforward conversion from CTF to MTF, and figure 3 shows MTF curves computed from CTF curves as measured for two typical Westinghouse EBS camera tubes, including the WX 32,719 whose 32 mm diagonal silicon target was the largest available when this was written, and figure 4 shows the same data for two large target SEC tubes designed for scientific applications.

The modulation transfer function, however, gives no single number descriptor to compare resolving power with self-scanned array sensors like CCD's or photodiode arrays. To provide a basis for such a comparison, consider an ideal array sensor shown schematically in figure 5. Each sensing element is assumed to have a square wave uniform response profile filling the entire

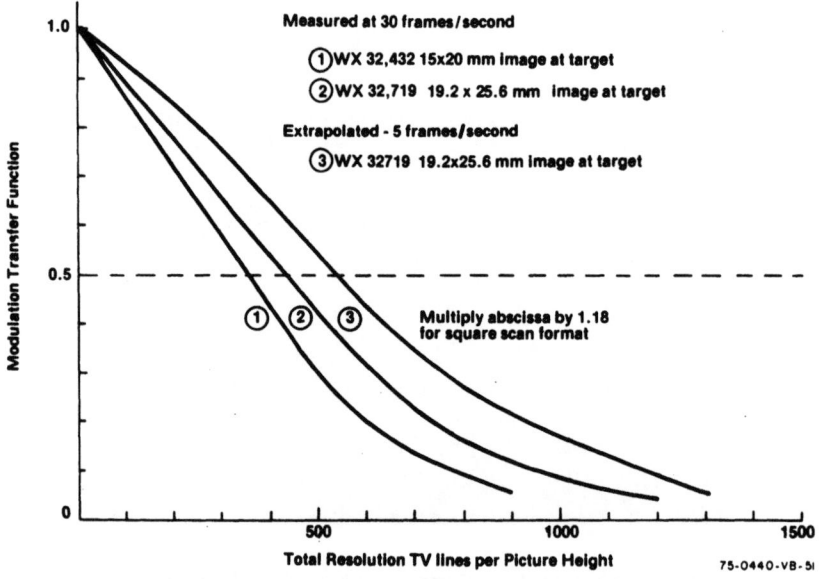

Figure 3. Modulation Transfer Function for Representative Westinghouse EBS Camera Tubes

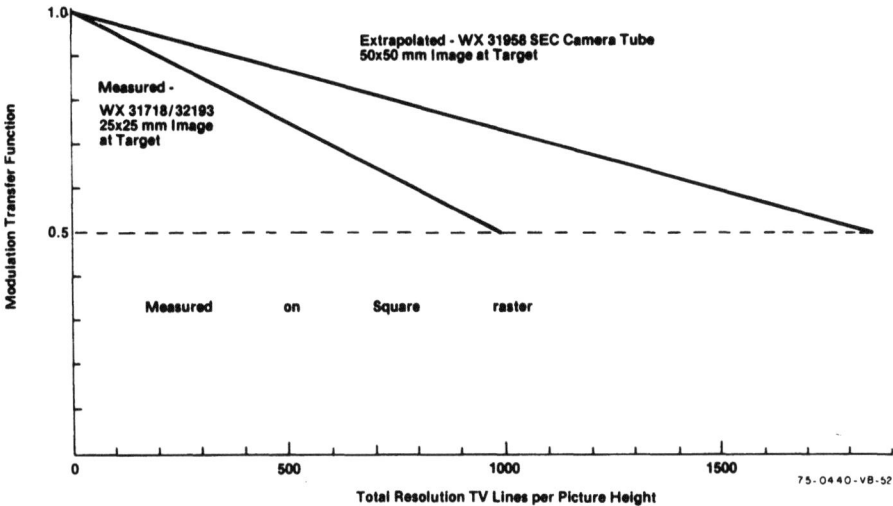

Figure 4. Modulation Transfer Function for Westinghouse SEC Camera Tubes

sensitive area allotted to that element. Nyquist's theorem tells us that the maximum image spatial frequency which can be meaningfully transmitted by an array is that which provides one sample per half/cycle. If a resolution bar chart image with a sinusoidal variation of irradiance vs position is presented to the array, the maximum meaningful spatial frequency is therefore $f_{s\ max} = 1/2p$, cycles/mm if p, the element pitch distance, is given in millimeters. The MTF for such an array is defined to be the maximum modulation, when the phase of the test pattern best matches the phase of the sensor elements:

$$\text{MTF} = \frac{\sin(\pi f_s \Delta X)}{\pi f_s \Delta X} \cdot \frac{\sin\left(\frac{\pi \Delta X}{2p} \cdot \frac{f_s}{f_{s\ max}}\right)}{\frac{\pi \Delta X}{2p} \cdot \frac{f_s}{f_{s\ max}}} \qquad (1)$$

where:

ΔX = width of the sensitive area of each element in millimeters

f_s = spatial frequency in cycles per mm

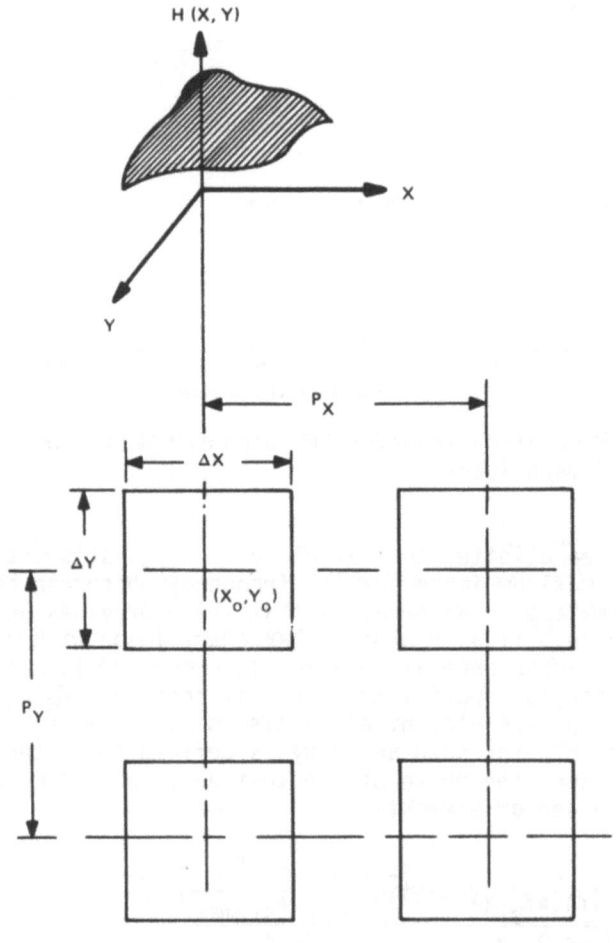

Figure 5. MTF for an Ideal Array Sensor is Calculated by Integrating the Input Waveform Over Each Sensing Element in Turn. At the Nyquist Frequency When Irradiance Peaks Fall on the Diodes $MTF_{f_s\ max} = \dfrac{\sin \pi \Delta X/2p}{\pi \Delta X/2p}$

p = array pitch in mm

$f_{s\ max} = \frac{1}{2p}$ = spatial Nyquist frequency for the array

For the ideal case of figure 5, at the Nyquist frequency, if the sensitive element width ΔX equals the pitch p, the sensitive elements are as large as possible,

$$MTF_{f_{s\ max}} = \frac{\sin \pi/2}{\pi/2} = 0.637 \qquad (2)$$

Note that by definition this is not a statistical average over all array positions.

Actual measurements on diode array sensors with close to 100 percent area efficiency show a white light MTF ≈ 0.5 at the Nyquist frequency for front surface illumination, lower than the theoretical value because the incident radiation is absorbed over a significant distance into the silicon and the carriers diffuse laterally before they are collected. CCD sensors give worse results, partly because the depletion regions extend less far into the silicon, and partly because of charge transfer inefficiency which acts to reduce contrast in fine detail. Fairchild[5] has reported MTF of only 0.19* at the Nyquist frequency for their 1000 element linear array, CCLID-1000B, while at the JPL conference in March TI[6] reported MTF = 0.31* at the Nyquist frequency for their 160x100 element array. However, because MTF of about 50 percent has been achieved in diode arrays and seems possible with further development for CCD's, the 50 percent MTF point is chosen in this chapter to define the resolution element size for camera tubes which is used in the comparison chart of figure 6.

A word of caution. The assumed camera tube resolution element size is based on the MTF, hence on the contrast transfer function, measured in the center of the image. In all television camera tubes designed to date the focus quality is less good toward the edges of the scanned area. Thus the element size chosen may be too small at the corners of the image, although it is probably conservatively large in the center.

---

*Both reported response of their arrays to bar chart test patterns. Coltman's conversion from CTF to MTF was used to calculate the stated values.

Camera Tubes	Image Size mm	Elements		Saturation Charge Level Q_s el cm^2	Q_t Electrons	Q_e Electrons	E1
EBS WX 32,432	15 x 20	360 x 480	30 fps	1.5×10^{11}	4.5×10^{11}	2.6×10^{6}	53 μm
WX 32,719	19.2 x 25.6	430 x 573	30 fps	1.5×10^{11}	7.37×10^{11}	3.0×10^{6}	56 μm
	19.2 x 25.6	530 x 707	5 fps	1.5×10^{11}	7.37×10^{11}	1.97×10^{6}	45 m
SEC WX 31718/ 32193	25 x 25	1000 x 1000	1/15 fps	1.6×10^{10}	1×10^{11}	1×10^{5}	25 μm
WX 31958	50 x 50	2000 x 2000	1/15 fps	1.6×10^{10}	4×10^{11}	1×10^{5}	25 μm
CCD Fairchild	4.4 x 5.7	190 x 244	60 fps	1.29 to 3.1×10^{10}	3.25 to 7.9×10^{9}	0.7 to 1.7×10^{5}	18 x 30
	3.1 x 4.2	100 x 100	30 fps	3.1×10^{10}	4×10^{9}	4×10^{5}	31 x 42 (20 x 30 Sens.)
TI	2.29 x 3.66	100 x 160		3.82×10^{11}	3.2×10^{10}	2×10^{6}	22.5
Theoretical				3×10^{12}		1.9×10^{7}	25

75-0440-VA-62

Figure 6. Information Capacity - Elements at 50 Percent MTF for Camera Tubes

2.2 Information Storage Capacity - Electrons Per Element

The charge stored per element for each camera tube is calculated from the maximum signal which can be read from the tube under slow scan conditions where the scanning electron beam does not limit the signal. Stored charge is then simply $i_{sig}t_f$. This charge for the whole target, Q_T, is divided by the area to express the results as charge/cm^2, Q_S, and by the number of elements to yield Q_e, the saturation charge per element. All saturation charge levels in figure 6 are expressed as a number of electrons, since the basic information quantum is the electronic charge.

The corresponding figures for CCD sensors are taken from recent reports and papers published by Fairchild[5] and Texas Instruments[6]; principally the JPL conference. The last line of figure 6 relates to earlier calculations by Barbe[7] and others who predicted that the absolute charge density limit for CCD's would be set by the dielectric strength of the silicon dioxide layer under the gates to about 10^{13} electrons/cm^2. However, the active well area of each sensor cell is only 25-35 percent of the total cell area so the theoretical figure becomes 3×10^{12} cm^{-2}. If one assumes a 25 μm square sensor cell, this becomes 1.9×10^7 electrons per element at saturation for an "ideal practical" charge-coupled area imaging device. As shown, TI reported saturation at about 2×10^6 electrons per element, an order of magnitude less, for a back illuminated 100x160 element CCD. Fairchild's buried channel devices saturated at 0.7 to 4×10^5 electrons per sample, two orders of magnitude below first order theory. This lower value is due partly to the shallower wells inherent in buried channel devices, although the Fairchild report suggested other causes for lower saturated signals in these particular arrays which may be correctible later.

As shown in figure 6, based on the criterion of total charge per picture element alone, the Westinghouse EBS camera tube offers substantial advantages over the buried channel charge-coupled device for information capacity. Based on the total information stored per frame, the charge per element times the number of elements, both the EBS and large SEC tubes handle substantially more information quanta than the listed charge-coupled area imaging devices. The larger TI 400x400 element CCD, when available, should be within a factor of two of the Westinghouse EBS tube for total information storage.

A word of qualification is required, however. When making measurements of low contrast images in the presence of background, one exposes to a major fraction of the saturation signal to maximize the signal to noise ratio. But at this condition any variation in responsivity across the tube photocathode or in gain across the target can cause a nonuniform signal from a uniform irradiance pattern. Slightly darker shading toward the edge of a television picture due to such response variations is very common and of no concern in commercial television because the human eye is a very poor photometer as long as there are no abrupt discontinuities. Such variations may be due to non-normal landing of the deflected electron beam which results in a nonuniform potential distribution on the scanned side of the camera tube target in the dark, so that diodes near the edge are back biased less strongly than those near the center. The diode depletion regions are then less extensive and the collection efficiency, hence the gain, may be slightly lower. This effect should be small if the focus and deflection coils are optimally designed, but a special design effort is required, since commercial television imposes no such strict requirement for uniformity of response across the field. Camera tubes with electrostatically-focused image sections and fiber-optic face plates which are thin in the center and thicker near the edge can show an added shading effect due to greater light losses in the longer fibers. Although both these effects can be made small, they represent causes of nonuniform response which are not present in charge-coupled area imaging devices or in electron bombarded CCD's which employ either magnetic or proximity focusing. On the other hand, element to element response nonuniformities in diode array and SEC targets are largely ignored since the electron scanning beam size sets the responsive area in a camera tube, while in a CCD or diode array imager, element to element response variations produce "fixed pattern noise" which acts to mask picture details. See the chapters on Display Signal to Noise Ratio and Low Light Level Performance for quantitative treatment.

2.3 Read Out Noise and Dynamic Range

Another area of comparison concerns the noise sources inherent in the readout process which limit performance of analog devices at low exposure levels. For television camera tubes, as shown in figure 7(b), a large camera tube load resistor is followed by a low noise amplifier. The camera tube acts like a current generator, hence the signal voltage developed at the amplifier input falls as the frequency increases because of the falling reactance of the shunt capacitance to ground. This capacitance, in turn, includes the camera tube target electrode capacitance, the wiring capacitance, and the input capacitance of the gate of the first transistor. To provide a flat signal response over a passband Δf up to a selected maximum frequency, the amplifier is high peaked for a rising gain vs frequency characteristic

$G = G_0 \sqrt{1+\omega^2 R_L^2 C_S^2}$, rising at 6 dB per octave above some corner frequency.

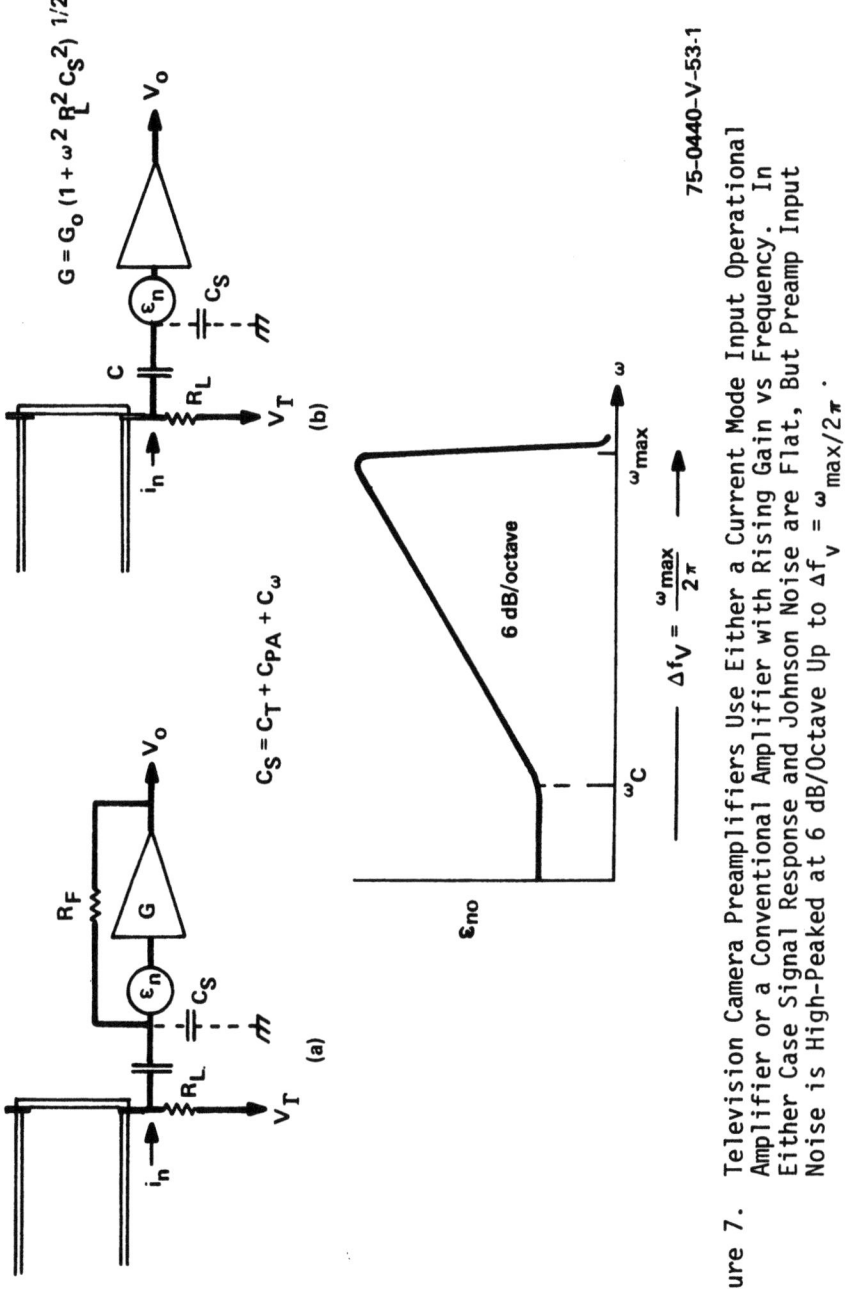

Figure 7. Television Camera Preamplifiers Use Either a Current Mode Input Operational Amplifier or a Conventional Amplifier with Rising Gain vs Frequency. In Either Case Signal Response and Johnson Noise are Flat, But Preamp Input Noise is High-Peaked at 6 dB/Octave Up to $\Delta f_v = \omega_{max}/2\pi$.

There are three variants on this basic scheme. A cascode input stage is sometimes used to eliminate Miller capacitance to reduce C_S. A current mode operational amplifier may be used to provide a low effective input impedance to make high peaking unnecessary, as shown in figure 7(a). It can be shown that neither of these alternatives changes the noise analysis indicated in figure 8. The only approach suggested thus far which does reduce the noise is the series peaking system proposed by Percival[8] and used in Great Britain for some EMI cameras using the CPS Emitron. Details on these load circuit and preamplifier designs are given in the chapter on Preamplifier Noise.

The noise sources in the basic circuit are the Johnson noise in the load resistor and the noise in the first transistor. As shown in figure 7, for a field effect transistor input stage the transistor noise can be represented by a voltage generator internal to the transistor in series with the gate, whose value varies as $g_m^{-1/2}$. For best signal to noise ratio R_L should be made large enough so that the Johnson noise equals the transistor noise. The latter is high peaked by the rising gain vs frequency characteristic of the amplifier, and its total contribution varies as $\Delta f_v^{3/2}$ rather than the $\Delta f_v^{1/2}$ variation found for white noise sources. For an optimum design the shunt capacitance, the transistor noise, and the bandwidth determine the rms noise current. The tube output electrode capacitance C_T and the wiring capacitance C_W should be reduced as far as possible. The best transistor design will then be one whose active gate capacitance matches the tube plus wiring capacitance, and which has very little stray capacitance. Assuming that wiring capacitance and transistor stray capacitance can be made negligibly small, $C_G = C_T$ and $C_S = C_T + C_G = 2C_T$. Then

$$i_n = \sqrt{2 \cdot \frac{4}{3} \pi^2 (2C_T)^2 \Delta f_v^3 \epsilon_n^2} \qquad (3)$$

and the video frequency passband is the variable available for the experimenter to follow the rule of thumb that slower is better. As shown, if system requirements permit scan rates of several seconds per frame, as is often the case for astronomy, for a tube electrode capacitance of 10 pF and a corresponding $\epsilon_n = 1.3 \times 10^{-9}$ V/Hz$^{1/2}$, use of a 10 kHz bandwidth can reduce the load circuit noise to about 40 electrons per Nyquist sample, a very favorable figure. For fast scan like that required for photon counting or simply for high definition television, a 10 MHz bandwidth with the same tube would produce noise of 1280 electrons per sample. Substantial prescanning gain is then required so that photoelectron pulses will be distinguishable from noise pulses.

$$i_n = \sqrt{\frac{4KT}{R_L} \Delta f_V + \frac{\varepsilon_n^2}{R_L^2} \Delta f_V + \frac{4}{3}\pi^2 C_S^2 \varepsilon_n^2 \Delta f_V^3}$$

Load Resistor Johnson Noise | "White" component of preamp noise - often negligible | High frequency peaked preamp noise

Optimum preamplifier design uses large R_L so Johnson noise equals preamp noise. For given C_T, say 10 pf, choose input transistor for high g_m, low noise, so that $C_{PA} \approx C_T + C_W$, and minimize C_W.

Then $i_n \approx \sqrt{2\left(\frac{4}{3}\pi^2(2C_T)^2 \Delta f_V^3 \varepsilon_n^2\right)}$, and $\varepsilon_n \approx 1.3 \times 10^{-9}$ V/Hz$^{1/2}$
$R_{eq} \approx 106\,\Omega$

Then noise current depends on scan rate, Δf_V

Δf_V	10^4 Hz	10^5 Hz	10^6 Hz	4.5×10^6 Hz	10^7 Hz
i_n	1.3×10^{-13} amp	4.1×10^{-12}	1.3×10^{-10}	1.23×10^{-9}	4.1×10^{-9}
q_n	40.3 e	128	403	854	1280

Figure 8. Optimum Preamplifier Design Gives $q_n \propto \Delta f_V^{1/2}$, But $i_n \propto \Delta f_V^{3/2}$. Noise is Determined by $C_S \approx 2C_T$.

The comparable amplifier situation for CCD's is shown in figure 9. Each information element appears as a separate charge packet and is transferred through a gated diode to the gate node of an on-chip MOSFET first amplifier stage which Westinghouse calls an electrometer amplifier because of its high input impedance. The electrometer operates in the common drain configuration and drives an off-chip current mode input operational amplifier. After the voltage change produced at the electrometer gate by each signal charge is read, the gate node is reset before transferring the next charge packet.

The resetting operation introduces a major noise contribution directly parallel to the Johnson noise in the camera tube load resistor. When the reset switch, actually an MOS transistor, is closed, its channel Johnson noise appears across C_S with a spectrum band limited by the impedance of the channel resistance and C_S in parallel. Each time the reset switch is opened a sample of this noise is stored on the gate node, and added to the following signal packet. It can be shown that $Q_n = \sqrt{kTC_S}$, and even for a minimal C_S = 0.17 pF, Q_n = 2.61x10^{-7} coul = 163 electrons. To eliminate this contribution, Westinghouse engineers have devised correlated double sampling. As shown in figure 9, after the gate node is reset the clamp switch is closed to store the negative of a voltage equal to the noise charge times circuit gain on the clamp capacitor. When the next signal charge packet is transferred to the gate it is added to the pre-existing noise charge, but the signal at the output of the clamp capacitor is due to signal charge alone, and this signal is then sampled and held for subsequent processing.

This scheme has several important advantages. Beside removing the reset noise, it also removes much of the excess 1/f noise of the on-chip amplifier since the signal is the difference between two samples taken close together in time. Lastly, the sample transmitted to following amplifiers can be taken when all switching transients have damped out so that switching or clocking noise has no effect on following stages or on the display.

There remains the noise contribution of the first amplifier stage. For a MOSFET this is again represented as a voltage generator within the transistor in series with the gate electrode. Optimum noise performance is again obtained when the active gate capacitance equals the driving circuit capacitance, since $\varepsilon_n = K/\sqrt{C_{GA}}$. For the assumed C_S = 0.17 pF, ε_n = 20x10^{-9}V/Hz$^{1/2}$ and for a 1 MHz operational amplifier effective bandwidth $q_n = \varepsilon_n \sqrt{\Delta f C_S}$ = 21 electrons, with a typical 19 electrons from off-chip amplifiers. Thus amplifier noise in an optimized system using coherent double sampling and an "effective bandwidth" of 1 MHz should be only about 30 electrons per sample.

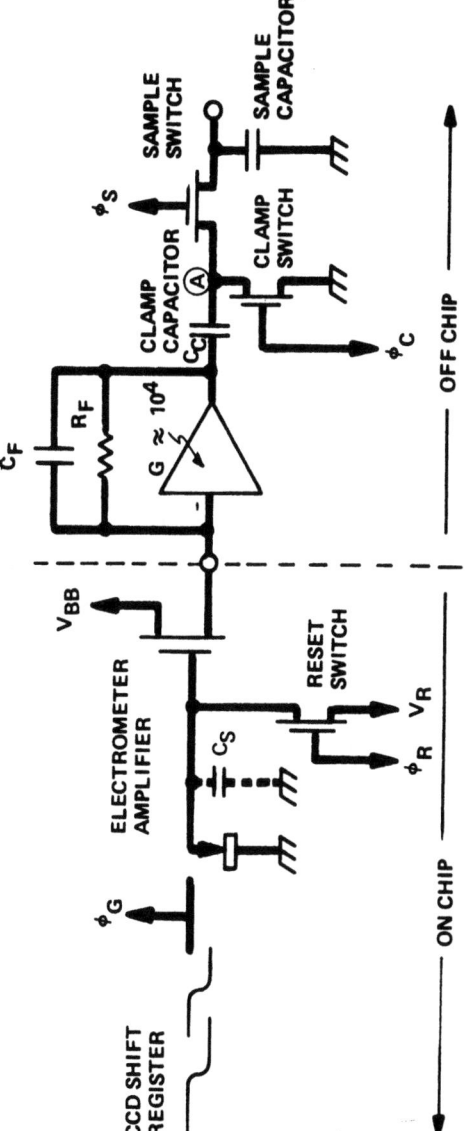

Figure 9. Correlated Double Sampling Removes Reset Noise at CCD Output

Double sampling, however, samples the noise of the on-chip electrometer and the off-chip operational amplifier twice for each signal sample, doubling the noise power added to the signal, or equivalently making the noise bandwidth double the required signal bandwidth. Since $Q_n \alpha \varepsilon_n \Delta f^{1/2} C_s$, use of 10 MHz signal bandwidth to scan, say, a 190x244 array at 100 fps for photon counting or a larger array at 60 fps for interlaced TV would raise amplifier noise to about 130 electrons per sample for a CDS system. This is the noise component which the distributed floating gate amplifier is intended to reduce. By in effect paralleling many input amplifier stages without increasing the input capacitance, the DFGA, at least in theory, can provide the signal gain to make off-chip noise components less significant and the effective transconductance high so that its own noise is very small. Fairchild has reported amplifier noises of 10 to 20 electrons even at frequency bandwidths of 7 MHz.

For any amplifier the shunt capacitance at the device output, C_s, sets the noise equivalent signal for an otherwise ideal device, and here again the chip designer must strive to eliminate stray capacitance to realize good performance.

In fact, effective CCD noise today is set by many other factors: leakage current shot noise, fat zero shot noise in surface channel devices, trapping and release noise which smears out the transfer process, and of course signal shot noise. In addition, for an imaging device variations in dark current or in response from element to element constitute a disturbance pattern which reduces the intelligibility of the reproduced image. When all these factors are included, the effective noise equivalent signal today is about 70 electrons, as shown. Reports that bar patterns are visible in reproduced images at lower signal levels do not contradict this statement. As shown by Rosell[9] and others, the eye integrates in space and time so that a bar pattern can be seen when the video signal to noise ratio is significantly smaller than unity. The relationship between video signal to noise ratio and display signal to noise ratio is treated in a chapter on that subject.

Based on all the foregoing, figure 10 shows the noise equivalent signal, the dynamic range, and the high exposure signal to noise ratio obtainable with analog devices for a set of conditions which seem typical for a scientific application of each device, 2 frames per second and a 1 MHz data rate for an EBS tube operating at about 237°K, 1/15 fps and 1.5×10^5 Hz for the 50x50 mm SEC tube as used in astronomy, and a 1 MHz data rate for the 100x160 TI CCD, operating at 230°K. Dynamic range under these conditions is greatest, 3×10^4, for the CCD. The EBS tube can equal this figure by cooling further and scanning slower. The 210:1 dynamic

Camera Tubes			max. q_e	Dyn. Range	Max. SNR		
EBS	Δf_r	q_n			G = 1	G = 100	G = 1000
WX 32,719 - 2 fps	10^6	400	2×10^6	5×10^3	1360	140	45
SEC					G = 1	G = 10	G = 100
WX 31958 1/15 fps	1.5×10^5	150	1×10^5	667	350	100	31.7
CCD - T.I.					G = 1	G = 100	G = 1000
		400/70	2×10^6	3×10^4	1410	141	45

Figure 10. NES and Dynamic Range

range for the SEC tube looks small by comparison, but figure 6 showed the reason for the significant usefulness of this device for applications like astronomy, the very large number of elements which provides a large total information storage capacity for this device.

So far solid-state and electron beam scanned imaging devices have been compared on the capacity of the analog storage member and the noise associated with reading out the signal because these appear fundamental characteristics. The analysis so far applies to CCD's or ICCD's, to diode array target vidicons or to EBS tubes. To increase responsivity on EBS or SEC tubes or on CCD's by operating them in the electron bombarded mode as ICCD's so that a few photons can result in a signal above noise, a vacuum tube is used which includes a photoemissive cathode and a high voltage ($\sim$10 kV) image section to make each photoelectron energetic enough to create 10, 100, or 1000 carrier pairs in the target. This provides a signal over readout noise from a low exposure, but at the cost of a reduction in effective information storage capacity, since each single primary photoelectron event now results in a 10, 100, or 1000 electron information quantum to be stored in the target. In the last columns of figure 10 are shown the signal to noise ratio capability of the three analog devices set by a combination of primary photoelectron shot noise and by readout noise. As shown, analog signal to noise ratio cannot exceed 50:1 for a saturation exposure at a gain of 1000 for either the EBS tube or the ICCD. Gains over 100 would make no sense for the SEC tube.

Here is the crunch. If data rate problems rule out use of long time integration to build signal above readout noise, one can provide prescanning gain to increase responsivity of analog image sensors, but at the cost of a significant reduction in the number of independent photoelectron events which can be stored in a single exposure. Thus, even for devices which have large high capacitance storage targets, providing high gain with an image section severely limits their capability to measure low contrast images.

3. PHOTON COUNTING SENSORS

To provide an image integration capability governed by one's pocketbook rather than by the laws of physics, one can image by "photon counting" or better "photoelectron counting." Here the requirements as of today are straightforward. One must provide enough prereadout gain so that each photoelectron event results in a signal pulse which is larger than essentially all the noise pulses so that amplitude discrimination can eliminate the noise and pass the signal. Further, the imaging device selected must be capable of being scanned rapidly enough so that at the maximum sensor irradiance which must be measured no more than 1 photon event will normally be registered on any resolution element in a frame time. In practice, this means that the frame frequency must be 5 or more times faster than the average rate per element of arrival of photons or generation of primary photoelectrons. Lastly, of course, one must interface the device with a computer type memory to record photoelectron events by location in the image for long enough to accumulate sufficient information to perform meaningful relative radiometric measurements. Photon counting over an extended image is obviously restricted to very faint images. Some of the problems in photon counting are apparent in the treatment of the previous section. To accommodate as large a photon flux as possible, the device must be read rapidly, and there must be many small picture elements. But this in turn dictates a large video signal bandwidth, which means more readout noise per sample, which may require higher prereadout gain so that the signal pulses will be clearly above the noise pulses.

Specifically, at 100 frames per second, a WX32,719 EBS camera tube with C_T = 22 pF would require a video bandwidth of about 16 MHz, and have an rms readout noise of about 3400 electrons per sample, or 174 nA. Thus a target gain of 4000:1 would not be sufficient to make photoelectron events comfortably larger than system noise and an added intensifier stage would be required to boost prescanning gain by 10 to 20 fold. But an added intensifier stage causes a loss in MTF which would reduce the effective number of resolution elements and a loss in time discrimination due to lag in the output phosphor screen. To keep noise below, say 500 electrons per sample to allow noise discrimination without using an added image intensifier stage, one can restrict the scan rate or use a smaller tube like the WX31960 with a 16 mm target and C_T = 12 pF and use the series peaking and video notch filter technique of Percival and James[8] to reduce readout noise by $1/\sqrt{2}$ to 1/2. But in fact the ICCD solution chosen for photon counting in the large space telescope project is somewhat more promising from a noise viewpoint. For a 160x100 element unit being read at 100 frames per second, the element rate is 1.6×10^6 and an effective bandwidth of 2.5 MHz should provide adequate settling time. Amplifier noise calculates to 48 electrons for a conventional amplifier with correlated double sampling, confirmed by Fairchild's reported 70 electrons per sample at 7 MHz and TI's 70 electrons at 1 MHz so far, to provide more than adequate margin below a 1000 to 4000 electron signal pulse in an ICCD for adequate signal to noise discrimination.

The task certainly becomes more difficult due both to the charge transfer inefficiency and increased bandwidth as the array size is increased, but Fairchild has reported their 190x244 element CCD can be scanned at a 7.9 MHz rate, equivalent to over 240 frames per second, while still maintaining a 70 electron per element rms noise figure. The device has also been scanned at up to 20 MHz with a subjectively pleasing picture and no gross evidence of charge transfer inefficiency. Granted that the Fairchild device uses both buried channel CCD shift registers and a distributed floating gate amplifier, hence represents the latest state of the art, the result certainly indicates feasibility for future ICCD photon counting devices up to the order of a 244x380 array.

4. PHOTO RESPONSE

The response to light of a silicon diode array sensor or a back illuminated charge-coupled area imaging device is compared with that of a good tri-alkali photoemissive cathode in figures 11 and 12. Figure 11 shows absolute spectral response, S_λ, in amperes per watt at each wavelength, and figure 12 shows the quantum efficiency at each wavelength. The silicon diode array or back illuminated CCD sensor has a quarter wave anti-reflection coating peaked in the visible. Front illuminated CCD's or CID's would have 10 percent to 30 percent of this response. The tri-alkali photocathode plotted has a response to CIE illuminant "A" of about 300 μA/lumen.

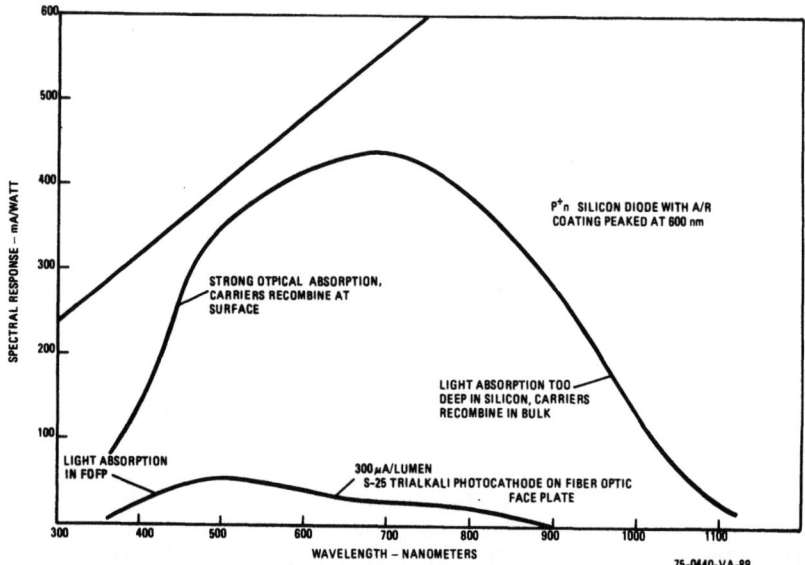

Figure 11. Spectral Response in mA/watt for a Silicon Diode with Quarter Wave A/R Coating is an Order of Magnitude Larger than for an S-25 Photocathode

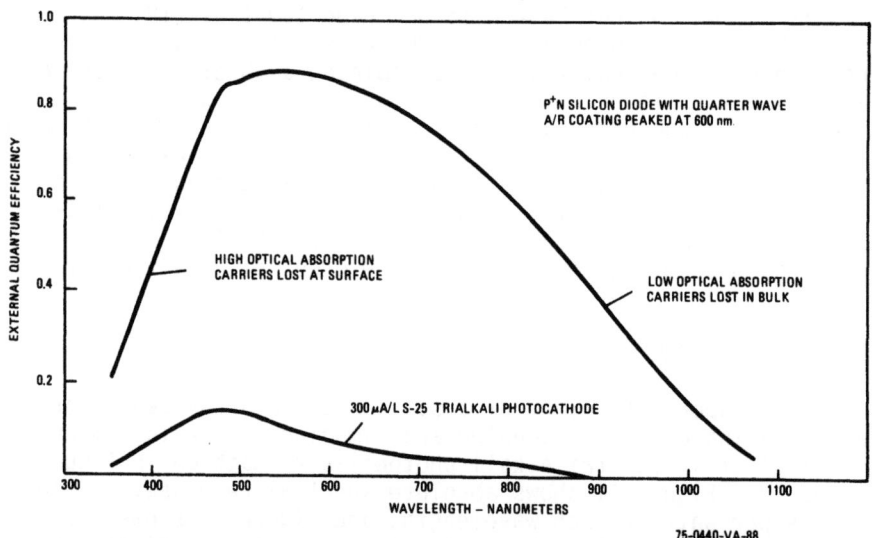

Figure 12. Quantum Efficiency of a Silicon Diode with Quarter Wave Anti-Reflection Coating Averages Nearly 80 Percent Through the Visible Spectrum Compared to About 9 Percent for a Good S-25 Photoemissive Cathode

To compare the conversion efficiencies of either sensor to polychromatic radiation from a real scene, one must perform the integration

$$R = A \int_0^\infty H_\lambda S_\lambda d\lambda = A \int_{\lambda_1}^{\lambda_2} H_\lambda S_\lambda d\lambda \qquad (4)$$

where:

- R = response in amperes
- A = effective sensor area, in e.g. m^2
- H_λ = spectral irradiance in watts m^{-2} μm^{-1}
- S_λ = spectral response in amperes per watt

λ_1 and λ_2 in μm bound the spectral interval over which the integrand is nonzero. Response is often reported to a specific spectral distribution of irradiance, like that to CIE illuminant "A", a tungsten filament source operated at a color temperature of 2854°K, but this is of little value for most military or scientific purposes, and the integral must be evaluated for each scene when more than an order of magnitude accuracy is required.

For monochromatic radiation through most of the visible, however, the quantum efficiency of a good anti-reflection coated silicon diode or similar sensor varies from 70 percent to 90 percent, while the semi-transparent photoemissive cathode varies from 4 percent to 14 percent. Hence for a typical sun illuminated outdoor scene, the average signal from the silicon sensor will be of the order of 8 times larger, using "eyeball integration."

Since the basic information content of the frame is measured by the number of independent charge carriers collected during an exposure, the higher quantum efficiency of the silicon sensor permits obtaining the same information content for a sun illuminated scene with an exposure only 1/8 as long. However, the reason for using a photocathode at all is either a requirement for response in the ultraviolet where the silicon response is low because the radiation is absorbed essentially at the surface and carriers are lost to surface states, or in those cases where the image must be amplified to compete with dark current or readout noise. It is only in certain marginal cases that the relative quantum efficiency decides the choice of placing a simple silicon array at the image plane or adding the complexity of a vacuum envelope, a photoemissive cathode, electron optics, and perhaps an electron gun.

5. DATA RATE

Television camera tubes like diode array vidicons or EBS camera tubes are rarely operated at maximum signal levels above 1500 nA, and the manufacturer usually rates performance at signal levels of only 600 to 800 nA. This is because the scanning electron beam current must be somewhat larger than the maximum highlight signal current, and demanding larger beam current usually means producing a more diffuse beam with a larger effective diameter, hence reducing resolving power. Thus it is the electron gun and the associated electron optics in the scanning section which actually limit the maximum signal current which can feasibly be read from a television camera tube, at least at commercial television scanning rates and above. This fact surprises many workers, since the scanning beam current is rarely more than 2 or 3 μA. However the typical camera tube electron gun uses near the thermonic cathode a beam defining aperture operating at about 300 volts whose diameter is 0.0015" to 0.0018", cross sectional area is 1.14 to 1.64×10^{-5} cm^2, and the average beam current density there is therefore 0.122 to 0.175 amp/cm^2, a respectable current density. And in order to obtain this current one must draw electrons from a significant area of the thermonic cathode and converge them to a small waist or crossover near the defining aperture, and spherical aberration limits the current density there. Lastly, the higher radial velocity components for beam electrons emitted further from the axis tend to broaden the axial velocity spread, raising the effective beam temperature and causing a type of chromatic aberration which increases the effective diameter of the electron beam at the target. This subject is treated in detail by Vine[10] and Moss[11].

As shown above and in the chapter on Preamplifier Noise, for an optimized conventional preamplifier,

$$i_n = \sqrt{\frac{8}{3} \pi^2 (2C_T)^2 \Delta f_v^3 \epsilon_n^2} = 2.7 \times 10^{-19} \Delta f_v^{3/2}$$

and (5)

$$q_n = \frac{1}{e} \sqrt{\frac{2}{3} \pi^2 (2C_T)^2 \Delta f_v \epsilon_n^2} = 0.846 \Delta f_v^{1/2}$$

and if the $\epsilon_n = 1.2 \times 10^{-9}$ V/Hz$^{1/2}$ and $C_T = 22$ pF, as for the WX32719 EBS tube, $i_n = 2.71 \times 10^{-9} \Delta f_v^{3/2}$ and

Δf_v =	1 MHz	4.5 MHz	10 MHz	20 MHz	
i_n =	2.71×10^{-10}	2.59×10^{-9}	8.57×10^{-9}	2.42×10^{-8}	amp
q_n =	847	1800	2680	3780	electrons

Thus the combination of a peak highlight signal current in the order of 1500 nA and preamplifier noise of 24 nA would alone limit the maximum dynamic range in a single exposure for such a camera tube in a 20 MHz system to about 60:1, even for the maximum nonsaturated exposure with an optimized conventional preamplifier equal to the state of the art. Use of a Percival coil[7] and notch filter might double this figure. For some high performance television applications, for example, a 1029 line TV system needed to reproduce a page of typescript, or for photon counting, such data rates are realistic. Hence, while present camera tube designs are well optimized for entertainment television applications, especially those of the United States and the United Kingdom, the data rate capabilities of direct beam reading tubes like vidicons and EBS tubes do set limits on achieving substantially higher performance, i.e., more than 4 or 5 times the data rate of broadcast television.

The situation on charge-coupled area imagers is not so clear. As suggested in the chapter on Preamplifier Noise, the far lower capacitance of CCD output circuits combined with correlated double sampling results in a noise charge per sample from the amplifiers in the order of

$$q_n = \frac{1}{e} \epsilon_n C_S \sqrt{\Delta f_v} = 28.4 \sqrt{\Delta f_v} \qquad (6)$$

electrons where Δf_v is in megahertz and is approximately twice the signal bandwidth. For a video sample rate of 10^7 sec^{-1}, the analog amplifier bandwidth would have to be at least 20 MHz, the effective noise bandwidth 40 MHz, and the amplifier noise charge would be at least 180 electrons. Actually it is not at all certain that correlated double sampling will work at this rate, or that CCD transfer inefficiency will permit obtaining quantitative imaging data at such data rates, although Fairchild has reported a subjectively satisfying picture at 2×10^7 samples/second with no obvious evidence of charge transfer inefficiency. This point of comparison is therefore left open. Camera tubes with direct beam readout and target electrode outputs should operate up to 20 MHz video bandwidths, and with electron multiplier outputs and perhaps with isocon scan at least a factor of two higher. Charge-coupled area imaging devices at this writing are still not proven. The comparison should be reviewed again in two or three years as more data becomes available.

6. BLOOMING

Silicon diode array target camera tubes exhibit blooming when sensor irradiance in parts of the image exceeds saturation irradiance by several orders of magnitude. The effect is thought to be made more severe by the Webster effect in which one back-biased diode at the brightest image point is completely discharged early in a frame time, then acts as a source of noise to help discharge the next diode, and so forth, so that the lateral spread of holes is in a sense propagated instantaneously to the far side of each diode in turn and diffusion need take place only across the relatively narrow n regions between the p+ diffusions. Westinghouse has reduced blooming substantially by building mesa-type diode array targets, shown in figure 13, where the p+ diffusions and the depletion regions are separated by physical valleys to substantially lengthen the diffusion paths for excess holes. Other manufacturers have reduced blooming by use of back surface treatment for an intermediate surface recombination velocity to reduce lateral hole transport at a slight cost in reduced short wavelength response, or by use of epitaxially formed diodes. At this time an overload of two orders of magnitude above saturation will produce an apparent spot diameter growth of only 3:1 on the Westinghouse tube.

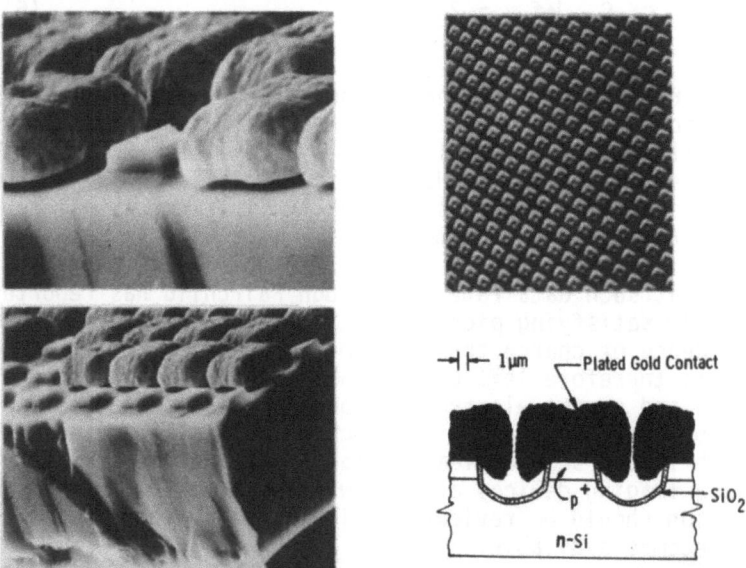

Figure 13. Westinghouse Deep Etched Metal Cap EBS Target Uses Mesa Type Diodes for Reduced Blooming and Gold Caps for Improved Beam Acceptance and for Shielding Against X-Ray Damage

Comparative blooming performance data on charge-coupled area imaging devices is still very limited, and the designs and design philosophies for CCAID's are still evolving, but the following data is indicative of performance demonstrated when this was written. RCA's Rodgers and Giovachino[12] report testing their SID 51232 512 x 320 element charge-coupled silicon imaging device at saturation, 10X and 100X saturation. They used a light spot diameter of 0.25 mm on a 12.2 mm diagonal image, 2 percent of the image diagonal, and figure 1 in their paper shows:

Exposure	Sat	10X Sat	100X Sat
Image Diameter	1	1.6	1.7

This is for a surface channel device where phase gates adjoining each sensor well have been placed in the accumulation mode to positively limit blooming to the neighborhood of the sensor element. Noise equivalent charge was about 500 electrons per sample for this device.

Fairchild, on the other hand, has chosen to use a channel stop structure which limits blooming to a single column, together with a sink for excess carriers at the end of each column. Element saturation, according to their report ED-AX-61[5], occurs at about 1 to 2×10^5 electrons per well, while noise equivalent signal is only about 50 electrons per well. This useful dynamic range is 2 to 4×10^3, and high quality images are reported at 2×10^4 electrons per well, an order of magnitude below saturation. Thus this device should normally be used far below saturation and the incidence of highlights orders of magnitude above saturation should be rare.

For comparison, a Westinghouse WX32,719 EBS tube with a low blooming target measured

Exposure	Sat	10X	100X	1000X
Image Diameter	1	1.9	3.3	5.2

This test was performed with an optical image diameter only 1 percent of raster diagonal, a somewhat more rigorous test. In comparing results, one must also remember that the area efficiency for conversion of incoming photons is essentially 100 percent for this device. The linear dynamic range of this device depends on the video bandwidth, hence on the scanning rates. At normal TV rates dynamic range is about 300:1.

7. CONCLUSIONS

This chapter has attempted to compare available TV camera tubes and charge-coupled area imagers as both analog and photon counting electronic output image sensors and to set forth a reasonable basis for such a comparison. CCD's, ICCD's, SEC camera tubes, and EBS camera tubes, all offer viable analog options. For the largest number of elements in a field, the choice certainly goes to the 50x50 mm SEC tube, and for most stored information the EBS WX32719 leads at the time this was written, provided both can be operated at fairly slow scan rates. For photon counting, amplifier noise in tube type cameras becomes inconveniently large at the required large bandwidths, and usually requires at least one added image intensifier stage. Here the ICCD clearly appears the best long range choice, although further ICCD development is required to reach this potential.

The paper also suggests that as other sensors are considered, similar yardsticks be applied. For analog work, the information storage capacity must be large. And for photon counting, the physics of the device should permit high speed readout without excessive readout noise so that compound image sections can be avoided.

REFERENCES

1. L. Biberman and S. Nudelman, ed., Photo Electronic Imaging Devices, "Early Stages in the Development of Camera Tubes Employing the Silicon-Diode Array as an Electron-Imaging Charge-Storage Target," Chapter 12, Plenum Press, New York, Volume II, pp. 253 ff (1971).

2. L. Biberman and S. Nudelman, ed., Photo Electronic Imaging Devices, "Camera Tubes Employing High-Gain Electron-Imaging Charge-Storage Targets," Chapter 11, Plenum Press, New York, Volume II, pp. 217 ff (1971).

3. Igor Limansky, "A New Resolution Chart for Imaging Systems," The Electronic Engineer, (June, 1968).

4. J.W. Coltman, "The Specification of Imaging Properties by Response to Sine-Wave Input," JOSA44(6), 468-471 (June 1954).

5. CCD Photosensor Array Development Program (Phase II), Final Report, April 1975 on contract N00039-73-0015 to Naval Electronic Systems Command by Fairchild Space and Defense Systems. The work reported was completed by 30 August 1974.

6. Antcliffe et al, "Large Area CCD Imagers for Spacecraft Applications," in *Proceedings of Symposium on Charge-Coupled Device Technology for Scientific Imaging Applications*, Mar. 6-7, 1975, Jet Propulsion Laboratory, Pasadena,

7. D.F. Barbe and W.A. Schmidt, "Infrared Charge-Coupled Imagers," IRIS Imaging Speciality Group, 28-29 November, 1972.

8. Percival, W.S., Brit. Patent #528,179 (1939). For a more complete treatment see I.J.P. James, "Fluctuation Noise in Television Camera Head Amplifiers," (Brit) Journal IEE, No. 20, Part III-A, pp. 796-803, 1952.

9. L. Biberman, ed., *Perception of Displayed Information,* "Recent Psychophysical Experiments and the Display Signal to Noise Ratio Concept," Chapter 5, Plenum Press (1973).

10. L. Biberman and S. Nudelman, ed., *Photoelectronic Imaging Devices*, "Electron Optics," Chapter 10, Plenum Press, New York, Volume I, p. 193 ff (1971).

11. H. Moss, "Narrow Angle Electron Guns and Cathode Ray Tubes," in *Advances in Electronics and Electron Physics,* Supp. No. 3, Academic Press, New York (1968).

12. R.L. Rodgers, III, and D.H. Giovachino, "Cooled Slow Scan Performance of a 512 x 320 Element Charge-Coupled Imager," in *Proceedings of Symposium on Charge-Coupled Device Technology for Scientific Imaging Applications*, Mar. 6-7, 1975, Jet Propulsion Laboratory, Pasadena, California.

LIST OF PARTICIPANTS

Ahlbom, S.H., AB VOLVO, Göteborg, Sweden
Akselrod, B., Tel Aviv University, Ramat-Gan, Israel
Al Khatib, A., ABC Engineering Company, Beirut, Lebanon
Anagnostopoulos, C., Eastman Kodak Company, Rochester, U.S.A.
Aslund, N., Royal Institute of Technology, Stockholm, Sweden
Austvoll, I., Norwegian Institute of Technology, Trondheim, Norway
Bartelink, D.J., Xerox, Palo Alto, U.S.A.
Bates Jr., C.W., Stanford University, Stanford, U.S.A.
Bejar, J., British Columbia Research Council, Vancouver, Canada
Benard-Dende, J.C., SECRE, Paris, France
Bereczki, J., CNES, Toulouse, France
Berglund, B., AB L.M. Ericsson, Stockholm, Sweden
Bertemes, P., CNES TPE/EP, Toulouse, France
Bettens, J.P., Usines Balteau, Beyne-Heusay, Belgium
Blanchard, Y., O.N.E.R.A., Châtillon, France
Bordoloi, K.C., University of Louisville, Louisville, U.S.A.
Bos, A., Observatory for Radioastronomy, Dwingeloo, Netherlands
Brewer, R.J., Mullard Research Laboratories, Surrey, England
Brown, G., University of Surrey, Surrey, England
Burrow, N.G., Manchester Polytechnic, Manchester, England
Buser, A., Reticon Corp., Regensdorf, Switzerland
Castagne, M., Université des Sciences et Techniques, Montpellier, France
Copeland, M.A., Carleton University, Ottawa, Canada

Corsi, C., Consiglio Nazionale Ricerche, Rome, Italy
Croman, J., University of Cincinnati, Clemson, U.S.A.
Crosta, G., Gilardoni Raggi X., Como, Italy
Cutler, S., Riverside Research Institute, New York, U.S.A.
D'Amico, Consiglio Nazionale Ricerche, Rome, Italy
Dey, S.K., University of Saskatchewan, Saskatoon, Canada
Drevet, J., Thomson CSF, Issy-les-Moulineaux, France
Elmasry, M.I., University of Waterloo, Ontario, Canada
El Said, M.A., Cairo University, Orman, Egypt
Farah, R., ABC Engineering, Beirut, Lebanon
Forgacs, G., Hungarian Academy of Sciences, Budapest, Hungary
Frick, W., Universität Stuttgart, Stuttgart, BRD
Friedman, W., University of Rochester, Rochester, U.S.A.
Frohmader, K.P., Universität Erlangen, Erlangen, BRD
Goss, A., English Electric Valve Co., Chelmsford, England
Green, R., University of Bradford, Bradford, England
Gunshor, R., Purdue University, West Lafayette, U.S.A.
Gustin, P., MBLE, Bruxelles, Belgium
Herbst, Siemens AG, München, BRD
Hazendonk, T.J., Philips Research Laboratories, Eindhoven, Netherlands
Hoeberechts, A., Philips Research Laboratories, Eindhoven, Netherlands
Hornung, J., Institute of Applied Solid State Physics, Freiburg, BRD
Housri, O., ABC Engineering, Beirut, Lebanon
Kolenko, S., ISKRA, Ljubljana, Jugoslavia
Kress, K., Bedford Engineering Corporation, Bedford, U.S.A.
Leclercq, J., Belgian Army Technical Service, Bruxelles, Belgium
Lecrosnier, D., C.N.E.T., Lannion, France
Leduc, Y., Université Catholique de Louvain, Couillet, Belgium
Leparquier, Thomson-CSF, Paris, France
Letzring, S., University of Rochester, Rochester, U.S.A.
Levitt, R.S., AMPEREX Electronic Corp., Slatersville, U.S.A.

Levy, J., Israel Armament Development Authority, Haifa, Israel
Lohstroh, J., Philips Research Laboratories, Eindhoven, Netherlands
Losee, D.L., Eastman Kodak Co., Rochester, U.S.A.
Losehand, R., Siemens AG, München, BRD
Lumbroso, L., SECRE, Paris, France
Lux, P., Philips Research Laboratories, Hamburg, BRD
McCaughan, D.V., Royal Radar Establ., Great Malvern, England
Mazzone, A., C.N.R.-LAMEL Laboratory, Bologna, Italy
Meusemann, B., RWTH Inst. für Halbleitertechnik, Aachen, BRD
Michel, J.P., Limeil-Brevannes, France
Michon, G., General Electric Co., Schenectady, U.S.A.
Moreno., J., LETI/MEA CENG, Grenoble, France
Nadler, M., Consulting Engineer, La-Celle-Saint-Cloud, France
Nussbaum, A., University of Minnesota, Minneapolis, U.S.A.
Ordung, P., University of California, Santa Barbara, U.S.A.
Oren, R., Israel Electronic Industries, Ltd., Tel Aviv, Israel
Peaker, A.R., University of Manchester, Manchester, England
Rabethge, H., Inst. für Halbleitertechnik, Stuttgart, BRD
Raina, J.P., Catholic University of Louvain, Louvain-la-Neuve, Belgium
Richou, F., C.N.E.T., Lannion, France
Rigaux, G., LETI/ME CENG, Grenoble, France
Roberts, P., University of Southampton, Southampton, England
Robin, L., Thomson-CSF, Issy-les-Moulineaux, France
Ryan, W.D., The Queen's University, Belfast, N.Ireland
Seifert, F.J., Institute of Physical Electronics, Wien, Austria
Sclar, N., Rockwell International, Anaheim, U.S.A.
Shenk, L.H., Astro Consultants, Palms Springs, U.S.A.
Shibata, A., Sony Corp., Yokohama, Japan
Smithies, S.A., National Electrical Engineering Research Institute, Pretoria, South Africa
Strasilla, U., Swiss Federal Institute of Technology, Zürich, Switzerland
Talley, H., University of Kansas, Kansas, U.S.A.

Taylor, W., University College London, London, England

Thompson, L., NASA/Goddard Space Flight Center, Greenbelt, U.S.A.

Touboul, A., Université des Sciences et Techniques, Montpellier, France

Urbach, I., Israel Armament Development Authority, Haifa, Israel

van der Spiegel, J., Université Catholique de Louvain, Heverlee, Belgium

Verbrugghe, W., Université Catholique de Louvain, Louvain-la-Neuve, Belgium

Wang, K., University of Houston, Houston, U.S.A.

Weckler, G., Reticon Corporation, Sunnyvale, U.S.A.

Wedam, A., Universa v Ljubljani, Ljubljana, Yugoslavia

Werner, W., Inst. für Theoretische Elektrotechnik, Aachen, BRD

Wittman, H., U.S. Army Research Office, Durham, U.S.A.

- - - - - - - -

LECTURERS

Barbe, D.F., Naval Research Laboratory, Code 5260, Washington D.C. 20390, U.S.A.

Dyck, R.H., Fairchild Research & Development Laboratories, 4001 Miranda Avenue, Palo Alto, Ca. 94304, U.S.A.

Esser, L.J.M., Philips Research Laboratories, Building W.O., Eindhoven, Netherlands

Goetzberger, A., Inst. für Angew. Festkörperphysik, Eckerstr. 4, D-7800 Freiburg 1, West Germany

Hall, J.A., Westinghouse Defense and Electronics Systems Center, Advanced Technology Lab., Elkridge Site, Box 1521, Mail Stop 3531, Baltimore, Md. 21203, U.S.A.

Jespers, P., Université Catholique de Louvain, Laboratoire de Microélectronique, Bâtiment Maxwell, 1348 Louvain-la-Neuve, Belgium

Sequin, C.H., Bell Laboratories, 600 Mountain Avenue, Room No. 2A-243, Murray Hill, N.J. 07974, U.S.A.

van de Wiele, F., Université Catholique de Louvain, Laboratoire de Microélectronique, Bâtiment Maxwell, 1348 Louvain-la-Neuve, Belgium

White, M.J., Westinghouse Defense and Electronics Systems Center, Advanced Technology Laboratory, Eldridge Site, Box 1521, Mail Stop 3531, Baltimore, Md.21203, U.S.A.

MIX
Papier aus verantwortungsvollen Quellen
Paper from responsible sources
FSC® C105338

If you have any concerns about our products,
you can contact us on
ProductSafety@springernature.com

In case Publisher is established outside the EU,
the EU authorized representative is:
**Springer Nature Customer Service Center GmbH
Europaplatz 3, 69115 Heidelberg, Germany**

Printed by Libri Plureos GmbH
in Hamburg, Germany